Springer Water

Series Editor

Andrey Kostianoy, Russian Academy of Sciences, P. P. Shirshov
Institute of Oceanology, Moscow, Russia

The book series Springer Water comprises a broad portfolio of multi- and interdisciplinary scientific books, aiming at researchers, students, and everyone interested in water-related science. The series includes peer-reviewed monographs, edited volumes, textbooks, and conference proceedings. Its volumes combine all kinds of water-related research areas, such as: the movement, distribution and quality of freshwater; water resources; the quality and pollution of water and its influence on health; the water industry including drinking water, wastewater, and desalination services and technologies; water history; as well as water management and the governmental, political, developmental, and ethical aspects of water.

More information about this series at http://www.springer.com/series/13419

Martina Zeleňáková · Katarzyna Kubiak-Wójcicka ·
Abdelazim M. Negm

Editors

Quality of Water Resources
in Poland

 Springer

Editors
Martina Zeleňáková
Institute of Environmental Engineering
Technical University of Košice
Košice, Slovakia

Katarzyna Kubiak-Wójcicka
Nicolaus Copernicus University
Toruń, Poland

Abdelazim M. Negm
Faculty of Engineering
Zagazig University
Zagazig, Egypt

ISSN 2364-6934   ISSN 2364-8198   (electronic)
Springer Water
ISBN 978-3-030-64894-7   ISBN 978-3-030-64892-3   (eBook)
https://doi.org/10.1007/978-3-030-64892-3

This Springer imprint is published by the registered company Springer Nature Switzerland AG
The registered company address is: Gewerbestrasse 11, 6330 Cham, Switzerland

Preface

This volume is the second part of the book "Water Resources in Poland" which focused on "Quality." As the name implies, this volume is devoted to the quality and protection of water resources.

This book consists of 19 chapters, which are presented in 5 parts. This is the result of a teamwork of scientists from various institutions and research centers who, in their scientific research, address the issue of water quality in surface and underground waters and their changes under the influence of natural and anthropogenic factors.

Water quality research is a basic responsibility of the state, which includes monitoring the quality of water resources and assessing water quality. These studies in Poland are carried out under the State Environmental Monitoring and are publicly available data. The State Environmental Monitoring includes a system of research and assessment of the state of the environment ensuring registration of changes that occur in it. Reliable information on the degree of degradation of individual elements of the natural environment is necessary for making optimal decisions by state administration at all levels. Tests are carried out at designated facilities or points in accordance with the adopted schedule and testing procedure. On rivers and lakes not covered by the State Environmental Monitoring, research is conducted by various water users and scientists. The scientific material collected in this book is unique. It presents the results of studies developed by the authors of the chapters based on publicly available data obtained from the State Environmental Monitoring or the Central Statistical Office. These studies are supplemented by the results obtained during fieldwork carried out for selected rivers or lakes by researchers. Often these are the only such studies conducted for these facilities. They can provide a basis for including a given object in systematic monitoring tests.

The scientific material presented in this book is the most recent information on the topic that can be useful to both practitioners and experts in quality of water resource in Poland. We hope that designers, employees not only dealing with water management, as well as society will find useful information on the quality of water resources, necessity and possibilities of their protection. Systematic examination and assessment of the quality of surface and underground waters in Poland will allow for a reliable diagnosis of the existing condition, forecasting these changes in the future and taking protective measures. This is a necessary requirement to ensure the conditions necessary for sustainable development.

This volume consists of five parts. The book certainly does not exhaust the entirety of the issues contained in the title, which is why each chapter contains a bibliography extending the problems discussed.

Part I consists of "**Introduction**," which was written by the volume's editors. The goal is to familiarize the reader with the research issues that have been discussed in this volume.

Part II discusses "**Key Issues of Water Resources Protection in Poland**." The first chapter in this part was presented by Katarzyna Kubiak-Wójcicka from Faculty of Earth Sciences and Spatial Management, Nicolaus Copernicus University in Toruń. Chapter 2 is entitled "**Assessment of Water Resources in Poland**." This chapter discusses the general issue of the amount of water resources in Poland and their use. Various ways of reducing water consumption, including wastewater treatment activities, have been identified.

Chapter 3 entitled "**Development and Protection of Water Resources in Protected Areas in Poland in Pursuit of Sustainable Development**" was presented by Roman Cieśliński from the Faculty of Oceanography and Geography, University of Gdańsk. This chapter discusses the issue of protecting water resources located in protected areas such as national parks. The author points to the short-comings of conservation measures due to the lack of an individual approach to individual protected areas resulting from the varying availability of water resources and biodiversity.

In Chap. 4, Mariusz Rzętała from the Faculty of Earth Sciences, University of Silesia in Katowice presented "**Anthropogenic Water Reservoirs in Poland**." It discusses the origin and water resources accumulated in anthropogenic reservoirs in Poland. The social and economic importance of standing water was assessed against the background of limnic processes, e.g., thermal processes, oxygen conditions, ice phenomena, changes in water fertility, salinity.

The last chapter in this part is entitled "**Irrigation and Drainage in Polish Agriculture: State, Problems and Needs**." This chapter was prepared by Leszek Łabędzki from Institute of Technology and Life Sciences in Falenty, Edmund Kaca and Andrzej Brandyk from Warsaw University of Life Sciences. The paper discusses the conditions as well as the current status, techniques, trends and perspectives of irrigation and drainage in agriculture sector of Poland. Irrigation and drainage infras-tructure was found to be in poor condition over the whole country. The study indicates what various actions should be taken to stimulate the development of irrigation and drainage, to adapt existing systems to extreme meteorological phenomena resulting from climate change, to improve their functioning and management.

Ten chapters are presented in **Part III** entitled "**Water Quality Evaluation**."

Chapter 6 entitled "**Quality Assessment of Water Resources of River in Poland**" was prepared by Adam Solarczyk from Faculty of Earth Sciences and Spatial Management, Nicolaus Copernicus University in Toruń. This chapter assesses the quality of river water resources in Poland based on selected physicochemical indi-cators studied in 2010–2015. The spatial diversity of the degree of river pollution in

102 catchments was presented, and its condition was assessed in accordance with the WFD.

Chapter 7 **"Water Quality in Main Dam Reservoirs in Poland"** was written by a team composed of Damian Absalon, Magdalena Matysik from the Faculty of Natural Sciences, University of Silesia in Katowice and Michał Habel from Department of Inland Waterways Revitalization, Institute of Geography, Kazimierz Wielki University in Bydgoszcz. The chapter assesses the state of water quality in 22 selected dam reservoirs in Poland based on physical, chemical and biological parameters. The possibilities and selected results of modern continuous monitoring of one of the dam reservoirs were also presented.

Moreover, Chapter 8 with the title **"Water Quality and Ecosystem Modeling: Practical Application on Lakes and Reservoirs"** was written in cooperation with the Institute of Environmental Protection—National Research Institute represented by Rafał Ulańczyk, Agnieszka Kolada and Silesian Water Centre, Faculty of Earth Sciences, University of Silesia in Katowice, represented by Bartosz Łozowski, Damian Absalon and Andrzej Woźnica. The chapter presents mathematical models as a tool supporting the management of lakes and reservoirs, especially in the case of complex environmental problems or lack of data required for pressure analyzes and forecasts.

Chapter 9 **"Assessment of Pollution of Water Resources and Process of Pollution Spreading"** was written by Jacek Kubiak and Sylwia Machula from West Pomeranian University of Technology in Szczecin. The study discusses the problem of eutrophication and protection of lake waters, which is largely associated with the development and use of the catchment. The paper presents the results of long-term research enabling tracking of changes occurring in the lakes of West Pomerania. For selected lakes, the intensity of eutrophication, trophy level and role of the catchment were determined, which allowed determining the susceptibility of the studied lakes to degradation.

Chapter 10 **"Ingression and Ascension of Saline Waters on the Polish Coast of the Baltic Sea"** was presented by Arkadiusz Krawiec from Faculty of Earth Sciences and Spatial Management, Nicolaus Copernicus University in Toruń. The paper presents the extent of the impact of sea (salt) waters on fresh groundwater resources in the coastal zone of the Polish Baltic coast. This area is located in various geological units and different hydrogeological regions: islands, sandspits, coastal lowlands, cliff coasts and the Vistula delta (Żuławy). Therefore, the problem of groundwater salinity in these areas is varied.

Chapter 11 **"Monitoring of Groundwater Quality in Poland"** was described by Izabela Jamorska from the Faculty of Earth Sciences and Spatial Management, Nicolaus Copernicus University in Toruń. As part of this publication, the chemical properties of the Quaternary, Paleogene and Neogene groundwater, Cretaceous and Jurassic aquifers in regional and point perspective are presented. In addition, an assessment of the chemical properties of groundwater in Poland over the past five years was carried out for selected observation wells and sources.

Chapter 12 **"Sediment Management in River Basins: An Essential Element of the River Basin Management Plans"** was written by the team of Michał Habel,

Dawid Szatten, Zygmunt Babiński and Grzegorz Nadolny from Department of Inland Waterways Revitalization, Institute of Geography, Kazimierz Wielki University in Bydgoszcz. This chapter describes the type of data, modeling and analysis tools that are required to qualitatively and quantitatively describe sludge dynamics. A brief overview of the types of sediments transported by the Vistula and Oder rivers as well as their tributaries and anthropogenic impact on sediment transport were presented.

Chapter 13 entitled "**Environmental and Anthropogenic Determinants of Water Chemistry in the Carpathians**" was written by Joanna P. Siwek from Department of Hydrology, Institute of Geography and Spatial Management, Jagiellonian University in Kraków. The chapter discusses the quality of water in streams and groundwater in small Carpathian groundwater catchments. The impact of the main factors determining the diversity of water chemistry in streams and groundwater is presented, among others geological structure, surface runoff from rainfall and snow melting. The impact of tourist traffic and related infrastructure on water quality was determined in order to effectively protect against degradation.

Chapter 14 "**Surface Water Eutrophication in Poland: Assessment and Prevention**" was written by Elena Neverova-Dziopak from Department of Environmental Management and Protection, Faculty of Surveying and Environmental Engineering, AGH University of Science and Technology in Kraków. The paper presents the essence of the eutrophication process, analysis of methodological problems of trophic state assessment and assessment methodology based on simple and inexpensive indicators ensuring continuous monitoring and control of eutrophication. The development of eutrophication of surface waters in Poland as well as the principles of its prevention and management were also presented.

The last chapter in this part of the volume entitled "**Monitoring of Small Catchments in Poland Under the Integrated Environmental Monitoring Programme: The Functioning of the Struga Toruńska River Agricultural Catchment**" was written by a team of authors Marek Kejna, Ireneusz Sobota, Joanna Uscka-Kowalkowska from Faculty of Earth Sciences and Spatial Management, Nicolaus University in Toruń and by Henryk Wojtczak from Voivodship Inspectorate of Environmental Protection in Bydgoszcz. The chapter presents the results of research conducted as part of the Integrated Environmental Monitoring Program within the Struga Toruńska drainage basin, located in central Poland. The catchment area is an example of a small agricultural catchment area that operates under strong anthropopressure, mainly under the influence of agriculture.

Part IV of the book is entitled "**Water and Wastewater in Urban Areas**" and consists of three chapters.

Chapter 16 "**The State of Water and Sewage Management in Poland**" was written by Katarzyna Kubiak-Wójcicka and Monika Kielik from Faculty of Earth Sciences and Spatial Management, Nicolaus Copernicus University in Toruń. This chapter discusses issues related to water and sewage infrastructure in Poland. The time and spatial differentiation of the density of the water supply, sewage and sewage treatment plants are presented in detail. The study includes issues related to the possibilities of financing activities leading to the improvement of wastewater treatment.

Chapter 17 "**Water Problems in Urban Areas**" was written by Tadeusz Ciupa and Roman Suligowski from Division of Environmental Research and Geo-Information, Institute of Geography and Environmental Sciences, Jan Kochanowski University in Kielce. The chapter presents selected issues of the water environment of Polish cities. Attention was paid to the causes of periodic excess rainwater and surface runoff, with particular emphasis on enclosed areas. The impact of these areas on the dynamics and size of flood waves was demonstrated, and consequently on flood risk.

The last chapter in this part "**The Multidimensional Aspect of Water Resources Management in Metropolitan Areas (a Case Study of the Poznań Metropolis)**" was written by Renata Graf from Department of Hydrology and Water Management, Institute of Physical Geography and Environmental Planning, Adam Mickiewicz University in Poznań, Institute of Physical Geography and Environmental Planning and Krzysztof Pyszny from EnviMap in Poznań. This chapter presents and discusses the model of integrated water resources management, including multi-level e-management and management of the "catchment area." The authors discuss the need to develop a new water resource management model in Poland, taking into account the principle of a common area and water policy in metropolitan areas.

The last part of the book entitled "**Conclusions**" was written by the editors. Chapter 19 contains "**Updates, Conclusions, and Recommendations for "Quality of Water Resources in Poland"**" and closes the volume of the book with the main conclusions and recommendations of the volume, as well as an update of some arrangements.

Special thanks to all who contributed in to making this high-quality volume a real source of knowledge and the latest findings in the field of Quality of Water Resources in Poland. We would love to thank all the authors for their invaluable contributions. Much appreciation and great thanks are also owed to the editors of the Earth and Environmental Sciences series at Springer for the constructive comments, advice and the critical reviews. Acknowledgments are extended to include all members of the Springer team who have worked long and hard to produce this volume.

The volume editor would be happy to receive any comments to improve future editions. Comments, feedback, suggestions for improvement or new chapters for next editions are welcome and should be sent directly to the volume editors.

Košice, Slovakia   Martina Zeleňáková
Toruń, Poland   Katarzyna Kubiak-Wójcicka
Zagazig, Egypt   Abdelazim M. Negm
September 2020

Contents

Part I
Introduction

Chapter 1
Introduction to the "Quality of Water Resources in Poland"

Katarzyna Kubiak-Wójcicka, Martina Zelenakova, and Abdelazim M. Negm

Abstract This chapter presents the main features of the book "**Quality of Water Resources in Poland**" and related current problems and research topics implemented by scientists dealing with water resources in Poland. The discussed research issues were divided into 5 thematic blocks. These are: Introduction (part I), Key Issues of Water Resources Protection in Poland (part II), Water Quality Evaluation (part III), Water and Wastewater in Urban Areas (part IV) and Conclusions (part V). The main technical elements of each chapter are presented under the appropriate topic.

Keywords Quality · Surface and underground water resources · Eutrophication · Pollution · Water and sewage infrastructure · Wastewater treatment · Urban area · Poland

1.1 Poland: A Brief Background

In addition to providing the right amount of water to satisfy life processes and the development of the national economy, the quality of water accumulated in rivers, reservoirs and underground waters is an extremely important element. This is extremely important from the point of view of health security, especially in the context of the consumption of water by the population [1–3]. Ensuring health security in Poland is the responsibility of the state, which is why these issues have been

K. Kubiak-Wójcicka
Department of Hydrology and Water Management, Faculty of Earth Sciences and Spatial Management, Nicolaus Copernicus University, Lwowska 1, 87-100, Toruń, Poland
e-mail: kubiak@umk.pl

M. Zelenakova
Department of Environmental Engineering, Faculty of Civil Engineering, Technical University in Košice, Vysokoškolská 4, 042 00, Košice, Slovakia
e-mail: martina.zelenakova@tuke.sk

A. M. Negm (✉)
Water and Water Structures Engineering Department, Faculty of Engineering, Zagazig University, University Street, Zagazig 44519, Egypt
e-mail: Amnegm@zu.edu.eg

© Springer Nature Switzerland AG 2021
M. Zeleňáková et al. (eds.), *Quality of Water Resources in Poland*, Springer Water,
https://doi.org/10.1007/978-3-030-64892-3_1

regulated by various legal acts with a long tradition. One of the most important events in this area was the establishment of the State Environmental Monitoring. The State Environmental Monitoring was created by the Act of 20 July 1991 on the Inspection for Environmental Protection to provide reliable information on the state of the environment [4]. As part of the monitoring, tests were established and carried out in accordance with the previously adopted methodology and indicators for surface and groundwater [5, 6]. Changes in this area began in 2003, when Poland signed the accession treaty on accession to the European Union. Poland, being a candidate country for full European Union membership, had to adapt its legislation to the EU standards. The provisions of Directive [7] have been implemented into national law and have been reflected in a number of laws and regulations related to water and sewage management. The basic instrument for implementing the Directive is the National Programme for Municipal Waste Water Treatment [8]. In the Polish legal system, all issues related to sewage management, rational shaping and protection of water resources are regulated by the Water Law Act and executive regulations to this Act [4]. The legal, organizational and economic solutions contained in the Act are addressed to both water owners and users as well as public administration bodies, they are intended to achieve good ecological status of waters, i.e. preservation of a rich and balanced ecosystem.

Currently, the State Environmental Monitoring is implementing tasks related to water quality in reservoirs, rivers and underground waters based on the 2001 Environmental Protection Law Act [9] and the Water Law Act of 20 July 2017 [10]. In addition, the basic legal acts in Poland relating to health safety issues are the Act of 7 June 2001 on collective water supply and sewage disposal and the Regulation of the Minister of Health of 7 December 2017 [2] on the quality of drinking water. These legal acts are in accordance with EU directives that regulated the issues of river water quality, drinking water, its quantity and purity standards.

These studies are supplemented by the results obtained during field work carried out for selected rivers or lakes by researchers. Often these are the only such studies conducted for these facilities. They can provide a basis for including a given object in systematic monitoring tests. This is particularly true of small waterbodies, including ponds and small lakes, low-order streams, ditches and springs that are critical for freshwater diversity. They play an important role in providing ecosystem services and require better protection [11–14].

The purpose of monitoring is to determine the quality status of waters, and thus the grounds for taking action to improve it if necessary. It is the basis for water protection, especially protection against pollution, both pollution leading to eutrophication, mainly from the domestic and municipal sector and agriculture (biogenic pollution), as well as industrial pollution. As the assessment of surface water status is used for integrated water management, water monitoring applies to surface and groundwater.

1.2   Main Themes of the Book

The book includes the following main parts: Key Issues of Water Resources Protection in Poland (part II), Water Quality Evaluation (part III), and Water and Wastewater in Urban Areas (part IV). In addition to the introduction and conclusions parts. The second part consists of 4 chapters. Part III contains 10 chapters, while Part IV consists of 3 chapters. In the sections below, the main features of each of them will be presented under the appropriate topic.

1.3   Key Issues of Water Resources Protection in Poland

The Section 1.3 presents the general state of water resources in Poland and their economic use in Poland. The current state and quality of water resources is the result of water management over many years in accordance with the applicable water resources management system and the organizational structure of water management in Poland.

Chapter 2 "**Assessment of Water Resources in Poland**" discusses the size of water resources in Poland, which are among the lowest among European countries. Water resources are very variable, both temporally and spatially, which results from the temporal and spatially variable distribution of precipitation, which is random. The average annual total outflow from Poland in the years 1951–2015 was about 61 km^3, from 37.5 km^3 in 1954 to 89.9 km^3 in 1981. The average unit outflow in the years 1951–2015 was 5.4 dm^3/skm^2 in the Vistula basin area and 4.7 dm^3/s km^2 in the Oder basin. The study indicates ways to reduce water surpluses and deficiencies, including in the form of small reservoir retention. A separate problem related to the use of water resources and its quality is the problem of water intake and consumption by various sectors of the national economy. Practical solutions used in the world and to a small extent in Poland for recovery and reuse of water in industrial plants, agriculture and newly built residential buildings were indicated.

Chapter 3 entitled "**Development and Protection of Water Resources in Protected Areas in Poland in Pursuit of Sustainable Development**" introduces the reader to the general assumptions of protecting and shaping water resources. The author pays special attention to legally protected areas, which are characterized by high landscape value. The current study focuses on lakes occurring in national parks because they are the least stable and most vulnerable places in terms of environmental changes in the geographical area. The best way to preserve areas that are valuable for the environment is to give them legal status based on environmental goals. However, not all environmentally valuable areas are legally protected or may be legally protected. The author indicates that in order to protect unique water resources a local spatial development plan is needed.

Chapter 4 entitled "**Anthropogenic Water Reservoirs in Poland**" discusses the capacity and quality of water in artificial water reservoirs found in Poland. Among artificial bodies of water, the following types are most often distinguished: reservoirs impounded by dams, flooded mineral workings, levee ponds, water bodies formed in subsidence basins and hollows, artificial pools and industrial ponds. Water bodies in Poland function in quasi-natural areas as well as under conditions of varied agricultural or urban-industrial human pressure. Their environment determines the course of limnic processes which are characteristic of lentic waters, e.g. water circulation, water level fluctuations, thermal processes, oxygen conditions, ice phenomena, changes in water fertility, salinity levels and others. Artificial water bodies in Poland mostly serve numerous functions despite the many environmental protection problems related to the lentic water environment, e.g. eutrophication, salinisation. Together with their immediate surroundings, they fulfil important natural and landscape roles. Storage reservoirs are used primarily for economic purposes such as water supply, flood protection, recreation and tourism, the breeding of fish and other aquatic organisms, energy production, inland transport, the extraction of mineral resources, etc.

Chapter 5 entitled "**Irrigation and Drainage in Polish Agriculture: State, problems and needs**" discusses the need for irrigation and drainage of agricultural areas in Poland. Irrigation and drainage systems and devices are in poor condition in Poland. Melioration equipment and systems on reclaimed agricultural land are usually in poor technical condition. Detailed melioration equipment is maintained on less than half of the meliorated area of agricultural land. Economic situation in agriculture and lack of sufficient funds for farmers, farmer associations and local government units responsible for amelioration in the catchment, are the main reasons for the negligence of the proper use of the melioration systems and facilities. There is a reduced interest in the use of water devices. The authors believe that various measures should be taken to adapt existing drainage systems to extreme meteorological events arising from climate change, to improve operation, maintenance and management.

Summing up this part of the volume, it should be pointed out that Poland's water resources are small and therefore require proper management. Proposals to increase the amount of water resources within the catchment area indicate specific technical actions.

1.4   Water Quality Evaluation

Part II of the volume is entitled "**Water Quality Evaluation**". This section consists of 10 chapters. The chapter "**Quality Assessment of Water Resources of River in Poland**". The chapter presents the assessment of water quality in Polish rivers examined in the years 2010–2015 as part of the State Environmental Monitoring Program. The chapter presents an assessment of water quality in Polish rivers as examined in 2010–2015 under the State Environmental Monitoring Programme. Using selected averaged physicochemical indicator values, an analysis was made of the spatial variability of mineral pollution (SEC, Total hardness), organic pollution (BOD, TOC) and biogenic pollution (TN and TP) in the rivers of the catchments and subcatchments of the Vistula, the Oder, and the Baltic Sea. Graded indexation of six quality characteristics (SEC, Total hardness, BOD5, TOC, TN, TP) characterising particular types of pollution has shown that there are rivers with high levels of pollution in catchments in Wielkopolska, Kujawy and Górny Śląsk. There are very low and low levels of pollution in mountain and foothill rivers of the Carpathians and Sudetes, and in the Pomorze region of northern Poland. The longest rivers in Poland (the Vistula and the Oder) and their main tributaries have different water qualities that vary along their courses.

The chapter **"Water Quality in Main Dam Water Reservoirs in Poland"** presents the assessment of water quality in the main Polish water reservoirs based on data available from the State Environmental Monitoring. The chapter also contains selected aspects of water quality determinants in water reservoirs, indicating hydrological factors and factors resulting from the method of reservoir management. The water quality features presented in the chapter are based on the analysis of the following physical, chemical and biological parameters: dissolved oxygen, BZT5, TOC, conductivity, total nitrogen, total phosphorus, nitrate nitrogen, phosphates, phytoplankton (IFPL) and phytobenthos (IO). On this basis, the status of dam reservoirs was classified and assessed. The general classification of the rated reservoirs indicates a good general condition of the reservoirs located in the Carpathian Mountains. Other tested reservoirs have poor water condition. The last part of the chapter presents a continuous monitoring system carried out on the "Goczałkowice" reservoir, which is one of the main sources of water supply for the Silesian Metropolis. The principles of construction of this system were discussed and selected results and recommendations were presented indicating the purposefulness of using such a method of assessing the quality of water in tanks used for drinking water supply.

The chapter in this part of the work concerns **"Water Quality and Ecosystem Modeling: Practical Application on Lakes and Reservoirs"**. The chapter presents mathematical models as a tool supporting the management of lakes and reservoirs, especially in the case of complex environmental problems or lack of data required for pressure analyzes and forecasts. Several practical applications of 3D dynamic models were presented and the need and usefulness of such applications were confirmed. The most attention was paid to three-dimensional, dynamic models of aquatic ecosystems, in particular to the AEM3D model and the combination of ELCOM and CAEDYM models. Examples of using these models in Poland include one natural lake (Łękuk Wielki), which is part of the Integrated Monitoring of the Natural Environment in Poland. The remaining four examples are dam reservoirs from 0.1 to 32 km^2, i.e. Rogoźnik I, Paprocany, Kozłowa Góra and Goczałkowice. In the case of Lake Łękuk, the Great Model proved to be a useful tool supplementing the monitoring system with detailed information on the dynamics of temperature and water flow, recognized as key factors affecting water quality and the water ecosystem. In the case of water reservoirs, the presented applications of the models include an online water quality information system; analysis of the impact of tank dredging on the transport of pollutants to the water intake; simulation of the impact of damming levels, winds, inflow speed and nutrient load reduction on retention time and water quality; and finally, an assessment of the effects of restoring the natural catchment area and an assessment of the operation of the solid barrier partially isolating the bathing area.

The chapter titled "**Assessment of Pollution of Water Resources and Process of Pollution Spreading**" discusses the issues of lake water pollution. Excessive supply of nutrients, organic matter and pollution to lake waters associated with the development of industrialization, tourism and intensive agricultural production into lake waters causes a continuous deterioration of their quality. The ineffective methods of lake protection also contribute to this. The progressing eutrophication process has led to disturbance of the biocenotic balance of waters, and even embraced especially protected waters—in national parks. Currently, lake water quality assessment for monitoring purposes includes only selected hydrochemical indicators whose measurements are made during spring circulation and summer stagnation. Water quality in lakes should be carried out annually, in all limnological seasons. In assessing trophies and the quality of lake waters, it is also important to know their natural susceptibility to degradation.

The chapter "**Ingression and Ascension of Saline Waters on the Polish Coast of the Baltic Sea**" presents the problem of salinity of groundwater in the Polish Baltic coast. In order to determine the extent of penetration and accumulation of salt and saline waters, geophysical surveys were carried out using electrical resistance tomography. Water samples were taken in selected wells and exploration wells for detailed physicochemical analysis, determination of isotopes and inert gas concentrations. Lowering the water level or piezometric pressure of groundwater caused by the exploitation of water intakes in the coastal zone causes an increased risk of salt water intrusion from the lower rock ground or the risk of seawater penetration. The penetration of salt water from the Baltic Sea occurs locally only in shallow aquifers and has a limited range. Until now, they occurred mainly in coastal regions, in holiday locations supplied with water from local groundwater intakes. On the Baltic coast, there is a clear seasonal diversity in terms of groundwater extraction. In summer, extraction is locally 3 times higher than the annual average.

The next chapter is entitled "**Monitoring of Groundwater Quality in Poland**". As a result of the research conducted within the past five years, it can be concluded that the groundwater of the Jurassic, Cretaceous as well as Palaeogene and Neogene aquifers has had a stable chemical composition. The greatest variability is revealed by the groundwater of the Quaternary aquifer. It is a consequence of high concentrations of chloride and sulphate ions, which are of anthropogenic origin and frequently permeate into water-bearing levels. Taking into consideration the groundwater quality classes, it can be claimed that the quality of groundwater on all water-bearing levels either has increased or has remained stable.

The chapter "**Sediment Management in River Basins: An Essential Element of River Basin Management Plans**" discusses the problem of understanding hydrodynamic and morphodynamic effects in sediment transport, which are therefore the necessary scientific basis for optimal sediment management. Sustainable management in the catchment is not only about the rational management of water resources (in terms of quantity and quality). It should take into account the functioning of the entire river system, in which sediment management plays a very important role. At the macro-scale, it plays an important role in delivering debris to the seas and the World's Ocean. Activities for sediment management are primarily necessary to meet environmental requirements that directly result from the provisions of European law (including WFD) but also from the national level (including RBMPs). The influence of anthropopressure on the dynamics of river sediment transport in the longitudinal profile is presented on the example of the two largest Polish rivers (Vistula and Oder). The authors own research results presented here. In addition, the investments listed in RBMPs that have a potential impact on the transport of sludge have been characterized.

The next chapter presents "**Environmental and Anthropogenic Determinants of Water Chemistry in the Carpathians**". Factors affecting stream water and groundwater chemistry in the Carpathians were identified mainly on the basis of hydrochemical case studies from numerous small headwater catchments. The most important environmental factor affecting water chemistry in the Carpathians is noted to be geology. Complex geologic structure of Carpathians determines significant variances in stream water and groundwater chemistry across the region. Hydrologic factor (changes in discharge) controls stream water and spring water chemistry changes in time. Elevation and geographic location of streams and springs in given climate zone and vegetation zone are natural factors affecting water chemistry in Carpathians in some extend. Anthropogenic factors affecting water chemistry in the Carpathians include acid rain, deforestation, agriculture, and tourist-generated wastewater.

The chapter "**Surface Water Eutrophication in Poland: Assessment, Impacts and Mitigation**". Anthropogenic eutrophication belongs to global processes which intensity has increased significantly in recent decades, affecting the surface waters in many countries and leading to deterioration of their ecological status and functional properties. Now anthropogenic eutrophication has become the research focus for the specialists in the field of water protection, ecological engineering and water resources management. Diagnostics of eutrophication causes and factors, the control and prognosis of its development are of great theoretical and practical importance. Regardless of the broad-scale projects implemented in Poland and other Baltic Sea basin countries, aimed at reduction of biogenic loads discharged to the sea and related huge capital expenditures, the expected effects in eutrophication abatement were not achieved and cost-benefit ratio is rather high. The chapter presents an overview concerning the essence of the eutrophication process, its causes and effects, provides a review of methodological approaches to trophic status assessing, discusses the methods of protection against eutrophication and a comprehensive analysis of this problem in Poland.

The next chapter presents "**Monitoring of Small Catchments in Poland as Part of the Integrated Environmental Monitoring Program: Functioning of the Agricultural Catchment of Struga Toruńska**". The Struga Toruńska catchment represents the dominant form of land use in Poland—arable land. The catchment is under strong pressure from agricultural activity. The influence of the nearby city—Toruń is significant. Here, the processes of the typical suburban area take place: land use changes, residential houses are being built, the industry develops, service facilities are created, the network of roads and highways is expanding. The functioning of the agricultural catchment of Struga Toruńska is influenced by climate change, air temperature increases, there is often no thermal winter, and the growing season has extended. Atmospheric precipitation is characterized by considerable variability from year to year. Droughts or months with high rainfall are more and more frequent. Lack of permanent snow cover changes the water cycle and affects the functioning of the biosphere, including the cultivation of plants.

1.5 Water and Wastewater in Urban Areas

The next part consists of three chapters. The first of them is "**Water And Wastewater Management Condition in Poland**". The authors believe that progress in improving the state of water and wastewater management has been visible in recent years. The main reason for this is the implementation of the WFD regulations, which resulted from Poland's accession to the European Union on May 1, 2004. This gave the opportunity to benefit from the EU funds, which in the first place constituted support for the construction of water and sewerage infrastructure. The analysis shows that the availability of water supply and sewerage infrastructure in Poland is not sufficient. The disproportion between the length of the water supply system and the sewerage system is particularly evident. Despite a significant improvement in the condition of the infrastructure, which consisted in the increase in the length of the sewerage network and the population served by sewage treatment plants, the process is slow and does not completely solve the problems associated with sewage disposal.

Chapter "**Water Problems in Urban Areas**." The chapter presents selected issues of the water environment of Polish cities. Attention is drawn to the causes of periodic excess of rainwater and surface runoff, with particular emphasis on sealed areas. The influence of these areas on the shaping of dynamics and the size of flood waves, and on flood risks as a consequence is demonstrated. The occurrence of floods of various genesis is discussed. The importance of small streams in the geoecosystem of a city with regard to their functioning and management is presented. The importance of the human-water environment interaction with the maintenance of the sustainable development of the city in terms of ensuring the safety of its inhabitants is referred to. On the example of selected cities, a series of activities of an administrative, technical and pro-ecological nature are presented.

The last chapter is "**Multidimensional Aspect of the Water Resources Management in Metropolitan Areas**." This chapter discusses new challenges in the management of water resources in urban agglomerations. As urban agglomerations grow in Poland, they face water resources management challenges. The increase in demand for good quality water in highly urbanised areas, increasing threats connected with the pollution of water and droughts or floods, has heightened the need for elaborating a new model of integrated water management. In Poland, the multifacetedness and interdisciplinarity of water resources management in a metropolitan area follows from the priority of the water policy of the member states of the European Union indicated in the Water Framework Directive. As a priority are considered actions, which concern limiting the degradation of waters by attaining and maintaining the proper state and potential of waters and the ecosystems connected therewith. The sustainable utilisation of water resources in metropolises is made possible by comprehensive surface water management systems which enable the pro-environmental and economic reutilisation of water, and thereby water security management for both man and the natural environment. The minimisation of flood risk in the metropolitan area can be attained through the implementation of a cohesive programme targeted at changing the method of usage of the river valleys: the biotechnical method of development of precipitation water, afforestation, or limiting the sealing of surfaces of urbanised areas.

The book ends with conclusions and recommendations, which are contained in Chapter 19.

Acknowledgements The authors would like to acknowledge the authors of the chapters for their efforts during the different phases of the book including their inputs in this chapter.

References

1. Parafińska K (2015) Hazards for water supply systems and their impact on public health. Problemy Higieny i Epidemiologii 96(1):92–100 (In Polish)
2. Regulation of the Minister for Health on the Quality of Water Intended for Human Consumption (2017) Official Journal No 2294. http://prawo.sejm.gov.pl/isap.nsf/download.xsp/WDU 20170002294/O/D20172294.pdf (In Polish)
3. Kubiak-Wójcicka K, Machula S (2020) Influence of climate changes on the state of water resources in Poland and their usage. Geosci 10(8):312. https://doi.org/10.3390/geosciences1 0080312
4. National Environmental Monitoring Program For 2013–2015 (2012) Główny Inspektor Ochorny Środowiska, Warszawa. http://www.gios.gov.pl/images/dokumenty/pms/pms/PPM S2013-2015_str.int.pdf (In Polish)
5. Soszka H (2009) Methodical problems associated with the assessment of lake eutrophication for designation of zones vulnerable to nitrates. Water-Environ Rural Areas 9(25):151–159 (In Polish)
6. Bielak SR, Borek K, Pijar K, Staszkiewicz K (2012) Application of macrophytes in assessment and classification of ecological state of upland rivers and streams in south Poland, in accordance with requirements of the Water Framework Directive. Tech Trans 109(23):17–32 (In Polish)
7. EU Directive 2000/60/EC of the European Parliament and of the Council of 23 October 2000 establishing a framework for Community action in the field of water policy. OJEU, L327,1; 2000
8. Minister of Environment. National Programme for Municipal Waste Water Treatment (2017) http://prawo.sejm.gov.pl/isap.nsf/download.xsp/WMP20170001183/O/M20171 183-02.pdf (In Polish)
9. Regulation of the Minister of Environment on Water Law (2017) Official Journal No 1556. http://prawo.sejm.gov.pl/isap.nsf/download.xsp/WDU20170001566/U/D20171566Lj.pdf (In Polish)
10. Regulation of the Minister of Environment on Environmental Protection Law (2001). Official Journal No 62 Item 627. http://prawo.sejm.gov.pl/isap.nsf/download.xsp/WDU20010620627/ U/D20010627Lj.pdf (In Polish)
11. Pęczula W, Banach B (2013) Small water bodies and lakes protected under EU Habitats Directive—results of the pilot wildlife monitoring in the Lubuskie Region. Teka Komisji Ochrony i Kształtowania Środowiska Przyrodniczego 10:306–317

12. Biggs J, von Fumetti S, Kelly-Quin M (2017) The importance of small waterbodies for biodiversity and ecosystem services: implications for policy makers. Hydrobiologia 793:3–39. https://doi.org/10.1007/s10750-016-3007-0
13. Korzeniewska E, Harnisz M (eds) (2020) Polish River Basins and Lakes—Part I. hydrology and hydrochemistry. Springer, Cham
14. Korzeniewska E, Harnisz M (eds) (2020) Polish River Basins and Lakes—Part II. biological status and water management. Springer, Cham

Part II
Key Issues of Water Resources Protection in Poland

Chapter 2
Assessment of Water Resources in Poland

Katarzyna Kubiak-Wójcicka

Abstract The paper analyzes the size of water resources in Poland, which are among the lowest in Europe. Water resources are very variable, both temporally and spatially, which results from the temporally and spatially variable distribution of atmospheric precipitation, whose course is random. The average annual total outflow from Poland in the years 1951–2015 was about 61 km^3, ranging from 37.5 km^3 in 1954 to 89.9 km^3 in 1981. The average unit outflow from the years 1951–2015 was 5.4 dm^3/s·km^2 in the Vistula basin, and 4.7 dm^3/s·km^2 in the Oder basin. One way to reduce excesses and deficits of water may be to undertake small-retention reservoir projects within agricultural catchments and in urban areas. Such works have been implemented in Poland as part of a small-retention programme by Lasy Państwowe (State Forest Holding) in mountainous and lowland areas. A separate problem related to the use of water resources is the level of water intake and consumption by various sectors of the national economy. The reduction of water consumption can be achieved in various ways. One of the simplest, but also the most severe for all users, is to increase the price of water. However, the most practical solution is to invest in the most modern water recovery and reuse technology in industrial plants, agriculture, and newly built residential buildings. Nowadays, technical solutions using rainwater in single-family houses are becoming increasingly popular in Poland. This water is most often used for watering gardens. However, there is little interest in the construction of purification and reuse systems for previously used water ("grey water"), mainly due to high installation costs. Grey water recycling or rainwater management not only significantly reduces operating costs, but also has tangible environmental benefits.

Keywords Water resources in Poland · Outflow · Flow variability · The Vistula river · The Oder river

K. Kubiak-Wójcicka (✉)
Department of Hydrology and Water Management, Faculty of Earth Sciences and Spatial Management, Nicolaus Copernicus University, Lwowska 1, 87-100, Toruń, Poland
e-mail: kubiak@umk.pl

© Springer Nature Switzerland AG 2021
M. Zeleňáková et al. (eds.), *Quality of Water Resources in Poland*, Springer Water,
https://doi.org/10.1007/978-3-030-64892-3_2

2.1 Introduction

In this chapter, attention will mainly be devoted to the conditions that determine water resources in Poland, because Poland is one of the nine countries in the Baltic Sea basin area. The total area of the Baltic Sea basin is 1,380,900 km^2, including 311,900 km^2 in Poland. The area of Poland from which water flows to the Baltic Sea constitutes 22.6% of the total Baltic Sea basin [1]. Hydrographically, Poland has exceptionally favourable conditions, because the country's administrative borders match its hydrographic boundaries. The Baltic Sea basin accounts for 99.7% of the total area of Poland. Such a hydrographic system facilitates water management in terms of catchments and basins, and in terms of regulating water regimes within a given area.

Water management in Poland, and in particular the formation and protection of water resources, the use of water and the management of water resources, is regulated by the Polish Water Law. Water management tasks are implemented by various state institutions.

Systematic measurements and hydrological and meteorological observations are carried out in Poland by the Institute of Meteorology and Water Management— National Research Institute (*in polish* IMGW-PIB) in Warszawa [2], using basic metering systems and networks. Its activity dates back to 1919, with the establishment of the National Hydrological Service and the Institute of Water Management. IMGW-PIB continuously provides current information on hydrospheric and atmospheric conditions, as well as forecasts and warnings in normal and potentially hazardous conditions. Up-to-date hydrological and meteorological data are available at http://www.pogodynka.pl, while selected data on water resources can be found on the website of Poland's Central Statistical Office [1].

Surface water quality tests cover physicochemical, chemical and biological parameters and are conducted in accordance with the Water Law under the aegis of the Voivodeship Inspector for Environmental Protection (*in polish* Wojewódzki Inspektor Ochrony Środowiska). Water quality monitoring is carried out according to a six-year water management cycle, as provided for by national law, which reflects the requirements of Directive 2000/60/EC of the European Parliament and of the Council of 23 October 2000, which in turn establishes the Community framework for action on water policy know as the Water Framework Directive 2000. Monitoring is conducted based on designated "water bodies" which are understood as separate and significant surface waters and which constitute the basic unit of water management.

This chapter focuses on the temporal variability of water resources and their geographical distribution and extent of use. Poland's water resources are compared in size to those of other countries in the Baltic Sea basin and other European countries.

2.2   The Water Resources of Poland

The size of global water resources depends primarily on climatic conditions. According to the International Glossary of Hydrology Dictionary [3], water resources are defined as the totality of currently and potentially available waters of the quantities and qualities required to meet specific needs. However, water resources are usually only taken to be those waters involved in the circulation of precipitation falling on land and its transformation into outflow (the land phase of the hydrological cycle). Many studies, therefore, equate water resources with the river outflow of a given area.

Water resources are very variable, both temporally and spatially, which results from the temporally and spatially variable distribution of atmospheric precipitation, whose course is random (Fig. 2.1). Changes in the distribution of atmospheric precipitation may cause extreme phenomena, including floods and droughts. The occurrence of sudden and frequent precipitation can lead to frequent floods. In turn, a lack of precipitation may lead to drought and, consequently, to desertification. Water resources are, therefore, extremely sensitive to climate change. Climate change in Poland is contributing to the increased frequency of droughts and floods resulting from deficits or high levels of precipitation [4].

The amount of water resources is best determined by total outflow. It is calculated based on long-term observation of flow in entire basins—including outside Poland—from a total area of 351,028 km^2, which is 12.3% larger than the country itself.

The average annual total outflow from Poland in the years 1951–2015 was about 61 km^3, ranging from 37.5 km^3 in 1954 to 89.9 km^3 in 1981 [5]. For the shorter period 1975–2015, the average annual total outflow was 63.2 km^3. Figure 2.1 shows

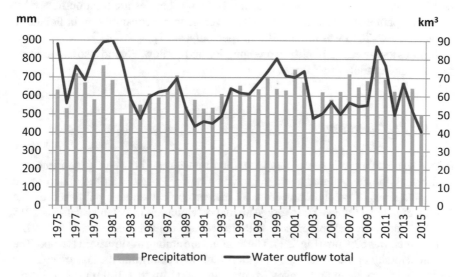

Fig. 2.1 Precipitation and outflow, 1971–2015 (own study based on the data from [1])

a marked irregularity in the total outflow, which manifests as alternating cycles of years of high outflows and then low. There were clear periods of high outflows in the years 1977–1982, 1997–2002 and 2010–2011, while low outflows occurred in the years 1988–1993 and 2003–2009. Since 2001, there has been an increased frequency of alternating high and low outflows. In 2010, the outflow was exceptionally high, totalling 86.9 km^3, while in 2015 it was exceptionally low, at 40.8 km^3.

Poland's water resources include the country's own indigenous resources from precipitation within the country, and resources inflowing from outside Poland.

Indigenous resources constitute 87.5% of total resources. Poland's resources from outside the country constitute the remaining 12.5%. In terms of water management and ability to control water resources, a high proportion of indigenous resources is more favourable. The majority of total resources (95.5%) flows directly to the Baltic Sea and the remaining part (4.5%) to neighbouring countries [6].

It is possible to determine whether available water resources can meet the needs of the population and the economy by using a population-related measure. The *per capita* total resources in the analysed multi-year period of 1951–2015 averaged 1,600 m^3/person per year, while in the period 1975–2015 this value was slightly higher, at approximately 1,700 m^3/person. In this respect, Poland ranks 4th from last among European countries. According to data from [1], only Malta, Cyprus, and Czech Republic have lower *per capita* resources. The average *per capita* resources in Europe are three times higher than in Poland. In the Baltic Sea countries such as Latvia, Sweden, and Finland, *per capita* resources are even 10 times higher than in Poland, totalling over 17,000 m^3/person. In dry years, total resources for Poland are even lower, totalling 1,100 m^3/person (e.g., in 1990 and 2015), while in wet years the values are higher, at 2,600 m^3/person (1975) and 2,257 m^3/person (2010). It should be noted that in 1990–2015 there was a systematic increase in population in Poland, which resulted in *per capita* water resources decreasing despite total outflow rates being comparable to those of 1975–1989. Average water consumption in Poland in the years 1999–2015 was around 300 m^3/person/year [7].

Besides long-term variability in precipitation and outflow, Poland also experiences seasonal and spatial variability.

2.2.1 Precipitation

Poland's prevailing climate is temperate and transitional between a maritime and continental climate. Annual average precipitation for Poland was calculated based on annual sums of precipitation from 32 meteorological stations published by the Central Statistical Office. The average annual precipitation in Poland for 1951–2015 was 625 mm, which is comparable with the average annual precipitation for 1975–2015, which was 629 mm (Fig. 2.1). These values are among the lowest in Europe. The spatial diversity of the annual precipitation sums that determine the size of Poland's resources is very large and ranges from around 500 mm to 1,100 mm. The lowest precipitation is recorded in the Greater Poland region, where annual precipitation is below 500 mm. In central Poland, in the lowlands, annual precipitation ranges from 500 to 550 mm. Higher precipitation is recorded in the north and south of Poland. In the north-east (the Masurian Lake District) annual precipitation is from 600 to 650 mm, while in the north-west (the Pomeranian Lake District) annual precipitation is higher than in the Masurian Lake District, ranging from 600 to 750 mm. In upland areas, precipitation totals about 600 mm, while in mountainous and foothill regions it is above 800 mm, and in high mountain areas reaches 1,100 mm.

The spatial distribution of annual precipitation totals in Poland for 1951–2015 is shown in Fig. 2.2.

The highest annual precipitation totals in Poland for 1975–2015 were recorded in 2010, at 803 mm, while the lowest annual precipitation totals were for 2003, at 489 mm.

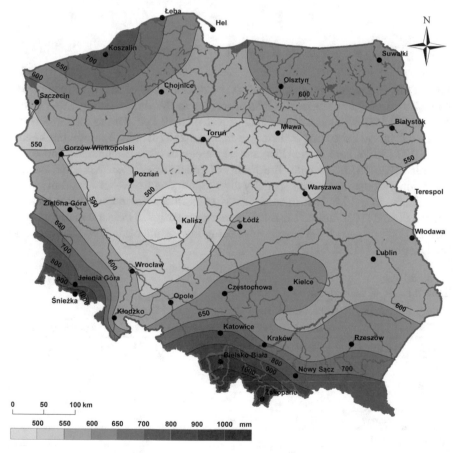

Fig. 2.2 Spatial variation of precipitation in Poland, 1971–2015 (own study based on the data from [2])

The maximum precipitation is in the summer months (June–August). The July maximum is greatest, but in some stations (mainly in the north and west of the country) the maximum happens in August. During this period, precipitation is about 2–3 times greater than in winter periods (over 4 times greater in the Tatra mountains). The minimum precipitation occurs from January to March [8].

In addition to sums of annual precipitation, average air temperature is an important factor in the size of water resources. The spatial and temporal variability of thermal conditions in Poland in the period 1951–2015 was analysed by [9]. The annual average air temperature ranged from 6.4 °C in north-eastern Poland to 8.9 °C in western Poland. In meteorological stations on the Baltic Sea and in central Poland, the annual average air temperature is not lower than 8 °C. In the most elevated region of the country (the Tatra Mountains, southern Poland), the average annual air temperature was-0.5 °C at Kasprowy Wierch (1,986 m above sea level).

2.2.2 The Hydrographic Network of Poland

Poland's hydrographic network comprises rivers, canals, lakes, and man-made reservoirs. There are 4,656 rivers in Poland according to IMGW, including 2,913 watercourses (62.6%) with a catchment area of less than 50 km^2. Of the total 488 rivers with a catchment exceeding 200 km^2, as many as 250 (over 51%) are in the Vistula basin, 177 (36.3%) are in the Oder, and 61 belong to other catchments. Those 488 rivers total 33,456 km in length. The density of Poland's river system is about average for Europe, although it is very spatially diverse, ranging from over 3 km/km^2 in the Carpathians to 0.2 km/km^2 in sandy areas of northern Poland and karst areas of eastern Poland [10]. Table 2.1 lists the largest river basins in Poland—those of more than 10,000 km^2—with the largest areas being the Vistula basin (194,000 km^2) and the Oder basin (119,000 km^2).

Table 2.1 List of the largest rivers in Poland [1]

River	Parent	Basin area (km^2)		Length (km)	
		Total	of which in Poland	Total	of which in Poland
Vistula	Baltic Sea	193,960	168,868	1,022	1,022
Oder	Baltic Sea	119,074	106,043	840	726
Narew	Vistula	74,527	53,846	499	443
Warta	Oder	54,520	54,520	795	795
Bug	Narew	38,712	19,239	774	590
Noteć	Warta	17,302	17,302	391	391
San	Vistula	16,877	14,426	458	457
Wieprz	Vistula	10,497	10,497	349	349

A system of man-made connections and reservoirs supplements Poland's natural hydrographic network. The longest canals in Poland include the 140-km-long Wieprz-Krzna canal, which is used for drainage. The second-longest—the Augustowski canal (80 km)—is a navigation canal, as are others: the Elbląski canal (62.5 km), the Gliwicki canal (41.2 km), the Ślesiński canal (32.0 km), the Notecki canal (25.0 km) and the Bydgoski canal (24.5 km). In recent years, rivers and channels have been little used for water transport, especially for long-distance transportation [1, 11, 12] (Fig. 2.3).

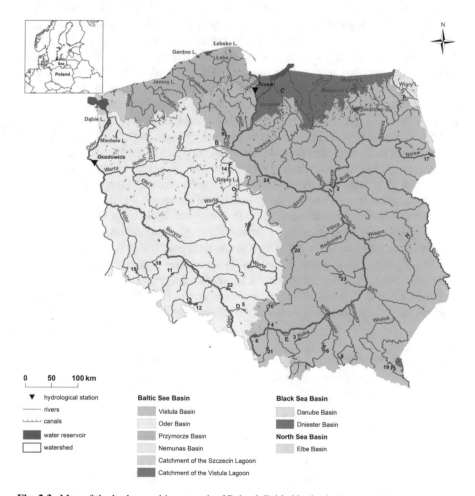

Fig. 2.3 Map of the hydrographic network of Poland divided by basin (own study)

2.2.3 Lakes and Man-Made Reservoirs

The lakes in Poland are not the largest in the Baltic Sea basin. The largest lake in Poland—Śniardwy—is 157 times smaller than Europe's largest lake—Lake Ladoga in Russia.

The largest accumulation of lakes in Poland is in the north, which is connected with the extent of the last glaciation. According to [13], compared to Europe as a whole, northern Poland is classified as an average area for lakes, both in terms of percentage coverage of lakes and density of occurrence. Polish lakes have two characteristic features. Firstly, there is an extremely high proportion of post-glacial lakes and, secondly, they are relatively small in area. Despite their small areas, Poland's post-glacial lakes are classified as bodies of considerable depth. The deepest lake in Poland—Hańcza—has a maximum depth of 108.5 m. Poland has 7,081 lakes with an area of over 1 ha [14]. The largest accumulations of lakes are in the Masurian Lake District, the Pomeranian Lake District and the Kuyavian–Greater Poland Lake District. Table 2.2 lists Poland's 11 largest lakes, each of an area above 20 km^2.

Man-made reservoirs supplement the network of lakes in Poland. Table 2.3 lists the 24 largest man-made reservoirs, each with a total capacity of more than 30 hm^3. These are mainly dam reservoirs that were created by a river or river valley being intersected by a dam. The largest retention capacity is the Solina reservoir (472.4 hm^3) in south-eastern Poland, although in terms of the area it is only 8th. The largest reservoir is Włocławek with an area of 75.0 km^2 and a capacity of 453.6 km^3. A second group consists of reservoirs created by barraging natural lakes. The largest of the reservoirs listed in Table 2.3 is the Pakość reservoir, which is located in central Poland [16]. A separate group of man-made water reservoirs is that of post-mining reservoirs. The largest was created in the Upper Silesian region, e.g., Dzierżno Duże, Dziećkowice, and Kuźnica Warężyńska. According to [17], Upper Silesia has more than ten such reservoirs with a capacity of more than 1 hm^3, but the total number

Table 2.2 The largest lakes in Poland [15]

Lake	Drainage basin	Area (km^2)	Max. depth (m)
Śniardwy	Pisa	113.4	23.4
Mamry	Węgorapa	102.8	43.8
Łebsko	Łeba	71.4	6.3
Dąbie	at the estuary of the Oder	56.0	4.2
Miedwie	Płonia	35.3	43.8
Jeziorak	Drwęca	32.2	12.0
Niegocin	Pisa	26.0	39.7
Gardno	Łupawa	24.7	2.6
Jamno	Baltic Sea	22.4	3.9
Gopło	Noteć	21.5	16.6
Wigry	Czarna Hańcza	21.2	73.0

Table 2.3 Man-made reservoirs in Poland [15]

	Reservoir	River	Year of entry into service	Total capacity at maximum accumulation (hm³)	Area at maximum in accumulation (km²)	Height of accumulation (m)
1	Solina	San	1968	472.4	22.0	60.0
2	Włocławek	Vistula	1970	453.6	75.0	12.7
3	Czorsztyn-Niedzica	Dunajec	1997	231.9	12.3	54.5
4	Jeziorsko	Warta	1986	202.0	42.3	11.5
5	Goczałkowice	Mała Wisła	1956	161.3	32.0	13.0
6	Rożnów	Dunajec	1942	159.3	16.0	31.5
7	Dobczyce	Raba	1986	141.7	10.7	27.9
8	Otmuchów	Nysa Kłodzka	1933	130.5	20.6	18.4
9	Nysa	Nysa Kłodzka	1971	124.7	20.7	13.3
10	Turawa	Mała Panew	1948	106.2	20.8	13.6
11	Tresna	Soła	1967	96.1	9.6	23.8
12	Dębe	Narew	1963	96.0	33.0	7.0
13	Dzierżno Duże	Kłodnica	1964	94.0	6.2	11.2
14	Sulejów	Pilica	1973	84.3	23.8	11.3
15	Koronowo	Brda	1960	80.6	15.6	22.0
16	Siemianówka	Narew	1991	79.5	32.5	9.2
17	Mietków	Bystrzyca	1986	71.9	9.1	15.3
18	Dziećkowice	Woda z Soły	1976	52.5	7.1	14.5
19	Pilchowice	Bóbr	1912	50.0	2.4	46.7
20	Kuźnica Warężyńska	Przemsza	2005	46.3	4.8	2.3
21	Pakość	Noteć Zachodnia	1974	42.6	13.0	4.8
22	Klimkówka	Ropa	1994	42.6	3.1	33.3
23	Słup	Nysa szalona	1978	38.7	4.9	19.1
24	Wióry	Świślina	2007	35.0	4.1	23.4

1 hm³ = 1 mln m³

of anthropogenic lakes reaches several thousand, and their surface area exceeds 100 km^2.

A significant proportion of the man-made reservoirs was created in the 1960s and 1970s. Of the 69 largest reservoirs with a capacity of more than 5 hm^3, only 6 were created after 2000. Most man-made reservoirs are used in energy production. The hydroelectric power plant in Włocławek (162 MW) has the greatest power of the hydropower plants. The highest installed power is in pumped storage power plants such as Żarnowiec (716 MW) and Porąbka-Żar (500 MW), which use both natural and man-made reservoirs.

The total amount of water accumulated in the 69 man-made reservoirs of capacity exceeding 5 hm^3 is 3,678 hm^3, which is about 6% of the average total annual outflow from the years 1951–2015. According to many authors, reservoir retention in Poland is too small to limit the effects of a lack or excess of water [18–20]. The largest man-made reservoir in central Poland—the Włocławek Reservoir—has very limited capacity to retain flood waters. The turnover of water in the reservoir is very quick, at every 5 days, and with high flood flows above 4,000 m$^3 \cdot$s^{-1}, the turnover period drops below one day [9].

2.3   The Legal Classification of Inland Surface Waters

Water management in Poland is regulated by the Water Law Act of 20 July 2017 (Journal of Laws of 2017, item 1566, as amended). The Act regulates matters relating to the ownership of waters and lands covered with waters, as well as the principles for managing these components as the property of the State Treasury. According to this law, inland surface waters are divided into flowing inland waters and standing inland waters. Flowing inland waters are waters in natural watercourses and the springs from which these watercourses originate; lakes and other natural water reservoirs with continuous or periodic natural inflow or outflow of surface waters; and man-made water reservoirs located on flowing waters and canals. Standing inland waters are inland waters in lakes and other natural water reservoirs not directly connected with flowing inland surface waters. Waters are owned by the State Treasury, and other natural or legal persons. Inland flowing waters classed as public waters are not subject to civil trade, except in cases specified in the Act. The issues of ownership of surface waters in Poland and its legal consequences are addressed in the study by [21, 22].

2.4   Unit Outflow

As mentioned earlier, the size of water resources depends mainly on climatic conditions, which are variable. The unit outflow from the river basins of the main Polish rivers is a measure of the diversity of water resources. The average unit outflow from

Table 2.4 Flows of the largest rivers in Poland in the years 1951–2015 (own study based on the data from [1, 2])

River	Gauging station	Kilometre on course of river	Basin area up to gauging point (km^2)	Average flow (m^3/s)	Unit outflow (dm^3/s·km^2)
Vistula	Tczew	908.6	193,806.5	1,044	5.4
Oder	Gozdowice	645.3	109,802.3	522	4.7
Narew	Ostrołęka	146.8	21,921.0	108	4.9
Warta	Gorzów Wielkopolski	56.4	52,364.7	207	4.0
Bug	Wyszków	33.8	38,384.0	154	4.0
Noteć	Nowe Drezdenko	38.0	15,917.0	72	4.5
San	Radomyśl	10.3	16,837.6	133	7.9
Wieprz	Kośmin	17.9	10,328.6	38	3.6

the years 1951–2015 was 5.4 dm^3/s·km^2 in the Vistula basin, and 4.7 dm^3/s·km^2 in the Oder basin. In comparison with other rivers that flow to the Baltic Sea, the Vistula and the Oder have relatively small water resources [23]. The average outflow from the basins of lowland rivers such as the Warta, Bug, or Wieprz is even lower than that of the Vistula or Oder, at less than 4 dm^3/s·km^2 (Table 2.4).

The size of the outflow is different in different regions of Poland. The largest unit outflow (12 dm^3·s^{-1}·km^{-2}) is in the catchments of the mountain tributaries of the Vistula and Oder rivers. The smallest outflow (4 dm^3·s^{-1}·km^{-2}) occurs in the catchment areas of the central lowlands, in the Kuyavia and Greater Poland regions, where it falls locally to even 1 dm^3·s^{-1}·km^{-2}) [24]. The highest average annual unit outflows were recorded in the Tatras (over 50 dm^3·s^{-1}·km^{-2}) [25], while at the foot of the Tatras, they ranged from 20 to 35 dm^3·s^{-1}·km^{-2}) [26]. The distribution of unit outflows in Poland depends on the sums of atmospheric precipitation, as well as on supply from groundwaters and the retention capacity of the catchment. The share of supply from groundwaters in Poland is on average 55%, i.e., it is more significant to outflow than surface water supply. Groundwater outflow is greater in northern Poland. Lake catchments, as with the rivers of the coastal basins of Pomeranian rivers, have a specific outflow structure in which the supply of underground water predominates. In turn, in mountainous areas, supply from precipitation prevails: retention and groundwater supply have little effect on the outflow structure [27–31].

2.5   Flow Variability in the Years 1951–2015

In determining the abundance of water, it is extremely important to know both the long-term and the annual variability of the hydrological regime. The flows at the mouths of the Vistula and the Oder showed different rhythms of changes in the 65-year study period. To compare the flows of the Vistula and Oder, the k coefficient is used, which is the ratio of the average annual flow to the average flow from 1951–2015. This coefficient has been used in other comparative studies of Baltic rivers [23, 32]. For $k > 1.0$, it is a wet year, in which the flow is above the long-term average, while $k < 1.0$ denotes a dry year. Wet years are characterised by high flows during the year, while dry years correspond to periods of low flows. During the 65-year study period, dry years dominate for the Vistula and the Oder. The driest was 1954, both on the Vistula ($k = 0.63$) and on the Oder ($k = 0.57$). The wettest year was 2010 on the Vistula ($k = 1.62$), and 1977 on the Oder ($k = 1.57$). The long-term period of variability consists of wet phases ($k > 1$) and dry phases ($k < 1$). One cycle has one wet phase and one dry. The Vistula and the Oder have a similar cyclical variability of flows. Fourteen-year cycles prevail, but in recent years, these cycles have shortened significantly (Figs. 2.4 and 2.5).

Figures 2.5 and 2.6 show the flow regimes of the Vistula and the Oder for 1951–2015. The Vistula's maximum flows at Tczew ranged from 1,600 m^3·s^{-1} to 6,430 m^3·s^{-1}. Average flows ranged from 646 m^3·s^{-1} to 1,663 m^3·s^{-1}. The narrowest range of variability was for low flows, which ranged from 264 m^3·s^{-1} to 754 m^3·s^{-1}. The trend of flows was level over the multi-annual period [33].

The largest floods on the Vistula occurred in 1962 (6,430 m^3·s^{-1}) and in 2010, at 6,360 m^3·s^{-1}. These were pluvial floods. The 2010 flood wave in the Vistula estuary section was not large, as a result of flood embankments near Płock bursting, which flooded large areas of the region [34].

Oder flows at Gozdowice in the 1951–2015 hydrological years have a considerable spread. The maximum Oder flows range from 471 m^3·s^{-1} to 3,180 m^3·s^{-1}. The

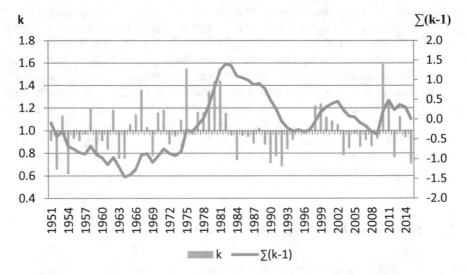

Fig. 2.4 K coefficient of average annual flows and the sum (k−1) for the Vistula at Tczew, 1951–2015 (own study based on the data from [2])

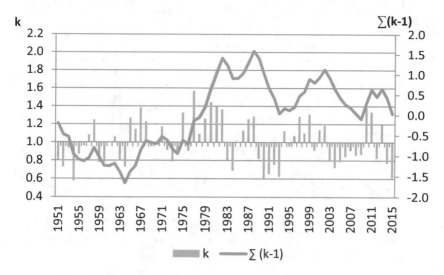

Fig. 2.5 K coefficient of average annual flows and the sum (k−1) for the Oder at Gozdowice, 1951–2015 (own study based on the data from [2])

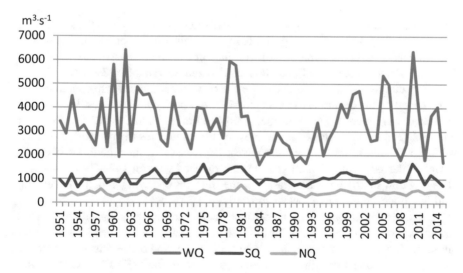

Fig. 2.6 Maximum (WQ), average (SQ) and minimum (NQ) annual flow (in $m^3 \cdot s^{-1}$) of the Vistula at the Tczew station, 1951–2015 (own study based on the data from [2])

average flows range from 300 $m^3 \cdot s^{-1}$ to 819 $m^3 \cdot s^{-1}$, and the minimum flows range from 124 $m^3 \cdot s^{-1}$ to 390 $m^3 \cdot s^{-1}$. The biggest floods on the Oder in the last 65 years are pluvial floods, which occurred, among others, in 1977, 1997 and 2010. A record flood occurred in July 1997, which in addition to significant losses to the economy also resulted in fatalities [35–37]. Due to its drastic course, this flood was referred

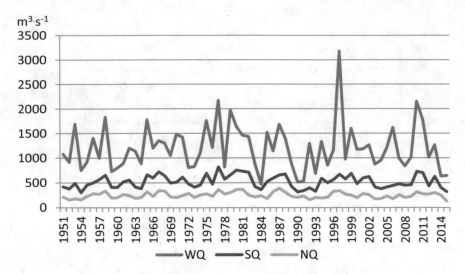

Fig. 2.7 Maximum (WQ), average (SQ) and minimum (NQ) annual flow (in $m^3 \cdot s^{-1}$) of the Oder at the Gozdowice station, 1951–2015 (own study based on the data from [2])

to as "the flood of the millennium." It was caused by heavy rains in the south of the country on July 3–8, 1997. A special feature of this atmospheric situation was the high intensity and long duration of precipitation over an extensive territorial range covering Poland, the Czech Republic, Austria, and Slovakia [38, 39].

Important information relevant to the use of water resources for economic needs is provided by the occurrence of exceptionally dry periods. The lowest Vistula flows at Tczew were recorded in December 1960 (264 $m^3 \cdot s^{-1}$), in 1962 (269 $m^3 \cdot s^{-1}$) and in 1992 (266 $m^3 \cdot s^{-1}$). However, the most troublesome was the dry period of 2015, which, although it did not have the lowest flow, lasted from June to October. The lowest Oder flows in the multi-year research period were in 1952, 1954 and 2015. The low flow of 2015 was the most extreme in terms of having both the lowest flow and the longest duration of low water levels.

The hydrological year has one distinct peak marking the surge of melt waters, which reaches its maximum in April. Flow deficits are mainly seen in summer and autumn. The steadiest flows in the whole analysed period on the Vistula occurred in July, and the highest volatility was observed in April flows (Fig. 2.7). On the Oder, the steadiest flows occurred in May, while the highest volatility was recorded in August (Figs. 2.8 and 2.9).

2.6 Changes in Water Resources in the Immediate Future

In light of the latest studies on global climate change, it is noted that an increase in the occurrence of extreme phenomena should be expected in the coming years [40].

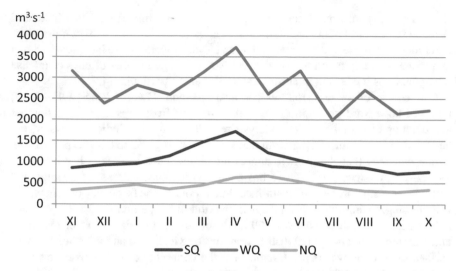

Fig. 2.8 Maximum (WQ), average (SQ), and minimum (NQ) monthly Vistula flows (in $m^3 \cdot s^{-1}$) at Tczew, 1951–2015 (own study based on the data from [2])

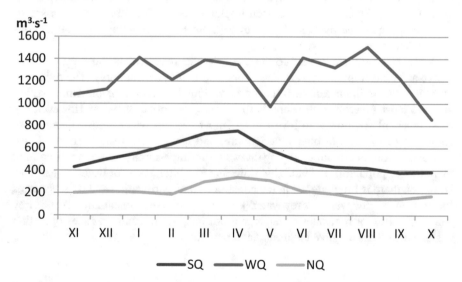

Fig. 2.9 Maximum (WQ), average (SQ) and minimal (NQ) monthly Oder flows (in $m^3 \cdot s^{-1}$) at Gozdowice, 1951–2015 (own study based on the data from [2])

Kundzewicz [41] believes that compared to other countries around the world, Poland is not particularly vulnerable to global warming, but the problem of floods (mainly pluvial) and droughts may increase.

Recent research indicates that extreme weather conditions should be expected. Comparative analysis of colder years (1961–1990) and warmer years (1991–2015) showed that the maximum daily precipitation for the summer half-year increased at many stations in Poland and that the increases in the summer half-year are more numerous than those in the winter half-year. In addition, there are increases in 5-day and monthly precipitation totals at several stations. The number of days of intense precipitation is increasing in north-western Poland in particular [42].

The course of air temperatures has also changed. Owczarek and Filipiak [9] recorded downward trends in the annual number of frost days, and in the number and duration of cold waves. The analysis revealed symptoms indicating systematic and persistent warming. The growth rate varies from 0.18 °C to 0.34 °C per decade, which indicates that the average annual air temperature in Poland has increased by 1.1–2.2 °C since 1951. There has been a significant increase in average and extreme air temperatures. The confirmed that in Poland, the frequency of heat waves is increasing, mainly in July and August (maximum temperature ≥ 25 °C) [43, 44]. It is expected that these changes will adversely affect the size of water resources and their temporal and spatial distribution, limiting access to and the availability of surface and ground waters [6]. Examples of the variability of Polish river flows can be found in detailed studies [24, 45–47] and reviews [48, 49].

One way to reduce excesses and deficits of water may be to undertake small-retention reservoir projects within agricultural catchments [18, 50–53] and in urban areas [54, 55]. Such works have been implemented in Poland as part of a small-retention programme by Lasy Państwowe (State Forest Holding) in mountainous and lowland areas.

A separate problem related to the use of water resources is the level of water intake and consumption by various sectors of the national economy. The reduction of water consumption can be achieved in various ways. One of the simplest, but also the most severe for all users, is to increase the price of water. However, the most practical solution is to invest in the most modern water recovery and reuse technology in industrial plants, agriculture, and newly built residential buildings. Nowadays, technical solutions using rainwater in single-family houses are becoming increasingly popular in Poland. This water is most often used for watering gardens. However, there is little interest in the construction of purification and reuse systems for previously used water ("grey water"), mainly due to high installation costs. Grey water recycling or rainwater management not only significantly reduces operating costs, but also has tangible environmental benefits.

2.7 Summary and Conclusions

Compared to other countries in the Baltic Sea basin and other European countries, Poland's water resources are small. Surface water resources depend on climate and anthropopressure. In periods of low outflow, there is a lack of water, which may inhibit economic development in some parts of Poland. Therefore, the proper management of excess water and its use during shortages is very important.

This is possible through appropriate retention of rainwater within the catchment, which is supported by the favourable layout of the hydrographic network and by building small retention reservoirs. Small retention projects have already been implemented, but there is a need for further investments in this area, not only in forest areas but also in agricultural and, above all, urban areas.

On the other hand, there is the issue of the economical management of water and the reuse of previously used water. Although grey-water recovery technology is not new, water-recycling installations for residential buildings are not cheap in Poland. This situation is expected to change with the introduction of the new Water Law. In the light of new regulations that have been in force since January 1, 2018, significant fees have been imposed on industrial and commercial recipients for water intake and sewage disposal. At the same time, additional fees were introduced for draining rainwater from hard surfaces around buildings. The introduced changes may induce potential water consumers to use water more economically and to use new technological solutions for the proper management of rainwater and the recovery of used water. Monitoring the amount of water used and wastewater collected will be helpful in this respect. It is hoped that the proposed activities will contribute to the protection of existing water resources and their rational use in the near future.

2.8   Recommendations

The chapter is addressed to water users, water managers and decision makers. Small water resources in Poland and their high variability in time require economical use of water and good management of water resources. The improvement of the situation may be caused by the construction of small retention reservoirs in areas with limited resources, but also by economical water management and reuse of previously used water. The introduced changes may encourage potential water consumers to use water more economically and to apply new technological solutions for proper rainwater management and recovery of used water.

References

1. Główny Urząd Statystyczny (GUS). Central Statistical Office. http://stat.gov.pl/
2. Institute of Meteorology and Water Management—National Research Institute in Warszawa. https://www.imgw.pl/
3. International Glossary of Hydrology. World Meteorological Organization (2012)
4. Kundzewicz ZW, Matczak P (2012) Climate change regional review: Poland. Wiley Interdisc Rev Clim Change 3(4):297–311. https://doi.org/10.1002/wcc.175
5. Michalczyk Z (2004) Rola obszarów wiejskich w tworzeniu i wykorzystaniu zasobów wodnych w Polsce. Woda-środowisko-obszary wiejskie 4(2a) (11):13–24
6. Gutry-Korycka M, Sadurski A, Kundzewicz ZW, Pociask-Karteczka J, Skrzypczyk L (2014) Zasoby wodne a ich wykorzystanie. Nauka 1/2014:77–98. http://www.pan.poznan.pl/nauki/N_114_07_Gutry.pdf

7. Kubiak-Wójcicka K, Piątkowski K (2015) Analiza zmian poboru wody w woj. kujawsko-pomorskim na tle kraju. Ekologia i Technika, XXIII, 5:397–304

8. Kożuchowski KM (2015) Obfitość opadów w Polsce w przebiegu rocznym. Przegląd Geofizyczny z1-2:27–38

9. Owczarek M, Filipiak J (2016) Contemporary changes of thermal conditions in Poland, 1951–2015. Bull Geogr Phys Geogr Series 10:31–50. https://doi.org/10.1515/bgeo-2016-0003

10. Jokiel P (2004) Zasoby wodne środkowej Polski na progu XXI wieku. Wydawnictwo Uniwersytetu Łódzkiego, ISBN83-7171-825-X

11. Kubiak-Wójcicka K (2014) Stopień wykorzystania infrastruktury liniowej w żegludze śródlądowej na przykładzie bydgoskiego odcinka międzynarodowej drogi wodnej E-70 (Usage degree of the line infrastructure in inland navigation based on the example of Bydgoszcz's part of the international waterway E-70). Logistyka 6/2014:12820–12832. https://www.czasopism ologistyka.pl/artykuly-naukowe/send/316-artykuly-na-plycie-cd-5/5810-artykul (in Poland)

12. Kubiak-Wójcicka K, Pokropski T (2019) Natężenie ruchu żeglugowego na dolnej Wiśle na podstawie analizy śluzowań we Włocławku w latach 1997–2016 (Intensity of inland navigation on the lower Vistula river based on the analysis of lockage in Włocławek in the years 1997–2016). Pr Geogr 156:79–99. https://doi.org/10.4467/20833113PG.19.004.10308

13. Choiński A (2007) Limnologia fizyczna Polski, Wydawnictwo Naukowe UAM, Poznań

14. Choiński A (2006) Katalog jezior Polski. Wydawnictwo Naukowe UAM, Poznań

15. Ochrona Środowiska (2017) Informacje i Opracowania Statystyczne (Environment. Statistical Information and Elaborations). Główny Urząd Statystyczny, Warszawa

16. Grobelska H (2006) Ewolucja strefy brzegowej zbiornika pakoskiego (Pojezierze Gnieźnieńskie). Prace Geograficzne 205, IG i PZ PAN, p 122

17. Rzętała M (2008) Funkcjonowanie zbiorników wodnych oraz przebieg procesów limnicznych w warunkach zróżnicowanej antropopresji na przykładzie regionu górnośląskiego (Functioning of water reservoirs and the course of limnic processes under conditions of varied anthropopres-sion a case study of Upper Silesian Region). Wydawnictwo Uniwersytetu Śląskiego, Katowice, p 171

18. Mikulski Z (2001) Wzrost retencji zbiornikowej w Polsce (The increase in reservoir retention in Poland). Gospodarka Wodna 3:110–113

19. Babiński Z (2005) Renaturyzacja dna doliny dolnej Wisły metodami hydrotechnicznymi (Renaturisation of the lower Vistula valley using the hydrotechnical method). Przegl Geogra 77(1):21–36

20. Glazik R (1978) Wpływ zbiornika wodnego na Wiśle we Włocławku na zmiany stosunków wodnych w dolinie. Dokumentacja Geograficzna IGiPZ PAN, pp 2–3

21. Kubiak-Wójcicka K, Marszelewski M (2012) Definitions and evolutions of the terms "flowing and stagnant waters" in the context of the proprietorship of the lakes in Poland. Limnol Rev 12(4):191–197

22. Kubiak-Wójcicka K, Brózda S, Sznajder A (2017) Hydrographic changes in a river system and their influence on the legal classification of watercourses, exemplified by selected tributaries of the San river. Bull Geogr Phys Geogr Series 12:5–17. https://doi.org/10.1515/bgeo-2017-0001

23. Gailiušis B, Kriaučiūnienė J, Jakimavičius D, Šarauskienė D (2011) The variability of long–term runoff series in the Baltic Sea drainage basin. Baltica 24(1):45–54

24. Jokiel P, Stanisławczyk L (2016) Zmiany i wieloletnia zmienność sezonowości przepływu wybranych rzek Polki (Long-term changes and variability in the seasonality of river discharge for selected Polish rivers). Pr Geogr 144:9–33. https://doi.org/10.4467/20833113PG.16.001.5126

25. Pociask-Karteczka J, Baścik M, Nieckarz Z (2010) Zróżnicowanie przestrzenne i zmienność odpływu ze zlewni tatrzańskich. Nauka a zarządzanie obszarem Tatr i ich otoczeniem, tom I. Zakopane, pp 123–130. https://tpn.pl/filebrowser/files/T1_17.pdf

26. Łajczak A (1988) Opady i odpływ w polskich Tatrach w świetle pomiarów wieloletnich. Czasopismo Geogr 59(2):137–170

27. Dynowska I (1971) Typy reżimów rzecznych w Polsce. Zesz. Nauk. UJ, 268, Pr Geogr 28:50

28. Tomaszewski E (2001) Sezonowe zmiany odpływu podziemnego w Polsce w latach 1971–1990. Acta Geographica Lodziensia, p 79
29. Wrzesiński D, Perz A (2016) Cechy reżimu odpływu rzek w zlewni Warty. Badania Fizjograficzne. R.VII-Seria A. Geografia fizyczna, 289–304. https://doi.org/10.14746/bfg.2016.7.21
30. Jokiel P, Tomalski P (2017) Sezonowość odpływu z wybranych zlewni karpackich (Seasonality of outflow in selected Carpathian catchments). Przegl Geogr 9(1):29–44. https://doi.org/10.7163/PrzG.2017.1.2
31. Kubiak-Wójcicka K, Bąk B (2018) Monitoring of meteorological and hydrological droughts in the Vistula basin (Poland). Environ Monit Assess 190:691. https://doi.org/10.1007/s10661-018-7058-8
32. Kļaviņš M, Rodinov V, Timukhin A, Kokorīte I (2008) Patterns of river discharge: long-term changes in Latvia and the Baltic region. Baltica 21(1–2):41–49
33. Kubiak-Wójcicka K (2019) Long-term variability of runoff of Vistula River in 1951–2015. Air and water—components of the environment: conference proceedings, Cluj-Napoca, Romania, pp 109–120. https://doi.org/10.24193/awc2019_11
34. Dorzecze Wisły monografia powodzi maj czerwiec 2010 (2011) Redakcja P, Maciejewski M, Ostojski MS, Walczykiewicz T (eds) IMGW-PIB, Warszawa
35. Odry Dorzecze (1999) Powódź 1997. Międzynarodowa Komisja Ochrony Odry Przed Zanieczyszczeniem, Wrocław
36. Dorzecze Odry: monografia powodzi 2010 (2011) In Maciejewski M, Ostojski MS, Tokarczyk T (eds) Warszawa
37. Cyberski J (1982) Charakterystyka hydrologiczna (Hydrological characteristics, in Polish). In Dolina Dolnej Wisły (Lower Vistula Valley) Augustowski B (ed) Ossolineum, Gdańsk, pp 103–153
38. Kundzewicz ZW, Szamałek K, Kowalczak P (1999) The great flood of 1997 in Poland. Hydrol Sci 44(6):855–870
39. Marsalek J, Watt WE, Zeman E, Sieker F (2000) Flood issues in contemporary water management, vol 71, Springer-Science + Business Media B.V., 432 p
40. IPCC (2013). Climate change 2013: the physical science basis. contribution of working group I to the Fifth Assessment Report of the Intergovernmental Panel on Climate Change. In: Stocker TF, Qin D, Plattner GK, Tignor M, Allen SK, Boschung J, Nauels A, Xia Y, Bex V, Midgley PM (eds) Cambridge University Press, Cambridge, UK and New York, NY, USA
41. Kundzewicz ZW (2011) Zmiany klimatu, ich przyczyny I skutki – obserwacje i projekcje. Land Anal 15:39–49
42. Pińskwar I, Choryński A, Graczyk D, Kundzewicz ZW (2018) Observed changes in extreme precipitation in Poland: 1991–2015 versus 1961–1990. Theoret Appl Climatol. https://doi.org/10.1007/s00704-018-23-72-1
43. Wibig J (2017) Heat waves in Poland in the period 1951–2015: Trends, patterns and driving factors. Meteorol Hydrol Water Manage 6(1):1–9
44. Wypych A, Sulikowska A, Ustrnul Z, Czekierda D (2017) Temporal variability of summer temperature extremes in Poland. Atmosphere 8:51. https://doi.org/10.3390/atmos8030051
45. Wrzesiński D, Sobkowiak L (2018) Detection of changes in flow regime of rivers in Poland. J Hydrol Hydromech 66(1):55–64. https://doi.org/10.1515/johh-2017-0045
46. Piniewski M, Marcinkowski P, Kundzewicz ZW (2018) Trend detection in river flow indices in Poland. Acta Geophys. 1–14. https://doi.org/10.1007/s11600-018-0116-3. https://link.springer.com/article/10.1007%2Fs11600-018-0116-3
47. Kubiak-Wójcicka K (2020) Variability of air temperature, precipitation and outflows in the Vistula basin (Poland). Resources 9(9):103. https://doi.org/10.3390/resources9090103
48. Jokiel P, Marszelewski W, Pociask-Karteczka J (eds) (2017) Hydrologia Polski. PWN, Warszawa
49. Kundzewicz ZW, How Ø, Okruszko T (eds) (2017) Zmiany klimatu i ich wpływ na wybrane sektory w Polsce. Poznań. http://scrwer1557491.home.pl/autoinstalator/wordpress/wp-content/uploads/2017/06/Zmiany-klimatu-i-ich-wp%C5%82yw-na-wybrane-sektory-w-Polsce.pdf

50. Mioduszewski W (ed) (2013) Odbudowa melioracji i rozwój retencji wodnej w świetle potrzeb rolnictwa i środowiska. Falenty. Wydaw. ITP. ISBN 978-83-62416-46-2, p 107
51. Mioduszewski W (2014) Small (natural) water retention in rural areas. J Water Land Dev 20(I–III):19–29. https://doi.org/10.2478/jwld-2014-0005
52. Kowalewski Z (ed) (2014) Metody retencjonowania wody na obszarach rolniczych i warunki ich stosowania. Falenty. Wydaw. ITP. ISBN 978-83-62416-82-0, p 162
53. Przybyła Cz, Sojka M, Mrozik K, Wróżyński R, Pyszny K (2015) Metodyczne i praktyczne aspekty planowania małej retencji (Methodical and practical aspects of small water retention planning). Poznań. Bogucki Wydaw. Nauk. ISBN 978-83-7986-057-9, p 204
54. Ciupa T, Suligowski R (2020) Impact of the city on the rapid increase in the runoff and transport of suspended and dissolved solids during rainfall—the example of the Silnica River (Kielce, Poland). Water 12(10):2693. https://doi.org/10.3390/w12102693
55. Hejduk L, Kaznowska E (eds) (2016) Hydrologia obszarów zurbanizowanych. Monografia Komitetu Gospodarki Wodnej Polskiej Akademii Nauk, z. 39, p 291

Chapter 3
Development and Protection of Water Resources in Protected Areas in Poland in Pursuit of Sustainable Development

Roman Cieśliński

Abstract In order to protect and shape the geographic environment, and this includes water environments, it is necessary to maintain as many as possible areas characterized by harmonious evolution and high landscape value. This is especially true of environmentally valuable areas viewed as equivalent to areas protected by law. When planning and designing environmental protection efforts, it is necessary to identify the relevant water balance relationships in the target area. Water plays a leading role, as it strongly affects other living and non-living elements of the natural environment and remains dependent on some of them. This prompts the following question: Is it possible to both develop and protect water resources in the context of sustainable growth in geographic areas protected by law? The study area, in this case, consists of national parks—areas which provide the highest level of legal protection to the natural environment in Poland. The current study focuses on lakes found in national parks, as these are the least stable and most susceptible sites in terms of environmental change in the geographic domain. The best way to maintain environmentally valuable areas is to grant them legal status based on environmental protection goals. However, not all environmentally valuable areas are legally protected or can be legally protected. In addition, the low effectiveness of many forms of environmental protection, especially lower level protected status, shows that only local forms of protection are highly effective. A local spatial management plan is needed in order to protect unique areas. Most commune strategic documents list main tasks that include the protection of the natural environment, utilization of tourist attractions and natural resources, maintenance of biodiversity and its the impact on it, and protection of environmentally valuable areas and sites. Hence, it is important to motivate local government officials to take responsibility for locally valuable areas in the form of local laws or local spatial management plans. In summary, national parks in Poland are characterized by different water availability. Some are dominated by water balance issues, while others are not. This is why each park needs to have a unique protection strategy. This is due to differences in water availability and threats that may affect each given

R. Cieśliński (✉)
Department of Hydrology, Faculty of Oceanography and Geography, University of Gdańsk, Bażyńskiego 4, 80-309, Gdańsk, Poland
e-mail: georc@univ.gda.pl

© Springer Nature Switzerland AG 2021
M. Zeleňáková et al. (eds.), *Quality of Water Resources in Poland*, Springer Water,
https://doi.org/10.1007/978-3-030-64892-3_3

national park. Despite these differences, it is necessary to pursue sustainable development whose overall goal is to protect the natural environment including its water areas in a way that allows both social and economic targets to be achieved.

Keywords Poland · Protection · Development · Water resources · National park · Development · Sustainable

3.1 Introduction

Currently the most important problem in the relationship between man and the natural environment, in addition to the actual state of the environment, is the formulation of a method of resolving social and economic problems without generating conflict while maintaining both living and non-living environments in the best state possible, and this includes the hydrosphere [1]. The state of the natural environment is first and foremost the result of human impact. With proper environmental management practices, man can produce a harmonious landscape. In the opposite case, a damaged natural landscape is the outcome. Hence, the pursuit of an ecologically-inspired spatial order based on sustainable "proportions" of natural, cultural, and technological elements appears to be both a reasonable and attainable goal [2]. In order to protect and shape the geographic environment, and this includes water environments, it is necessary to maintain as many as possible areas characterized by harmonious evolution and high landscape value [3].

This is especially true of environmentally valuable areas viewed as equivalent to areas protected by law [4] such as nature reserves and national park. When planning and designing environmental protection efforts, it is necessary to identify the relevant water balance relationships in the target area. Water plays a leading role, as it strongly affects other living and non-living elements of the natural environment and remains dependent on some of them [2]. Water is a constituent of virtually all ecosystems and is responsible for the production, transport, exchange, and processing of organic matter into inorganic matter. It also impacts all sectors of the national economy including industry, agriculture, and forest management. In addition, water is a key component of tourism and recreation as well as science and education. This high level of water usage leads to changes in water balance relationships and water circulation patterns in many geographic areas. Water circulation may become altered by changes in land use, building construction, and water use in various technological processes that utilize clean water and yield polluted water [5]. This prompts the following question: Is it possible to both develop and protect water resources in the context of sustainable growth in geographic areas protected by law? The study area, in this case, consists of national parks—areas which provide the highest level of legal protection to the natural environment in Poland. The current study focuses on lakes found in national parks, as these are the least stable and most susceptible sites in terms of environmental change in the geographic domain.

3.2 Development and Protection of Water Resources

Poland is one of the most water-poor countries in Europe. It is ranked 26th in Europe in terms of water resources according to EUROSTAT and 72nd in the world. The average amount of water per year per capita in Poland is about 1,600 m^3, while in the European Union overall it is more than 4,500 m^3 (global average: about 7,300 m^3). Water resources vary significantly from country to country across Europe. For example, the amount of water per one resident of Sweden is more than 20,000 m^3, while the amount per resident of Austria is more than 10,000 m^3. The amount per German resident is about 1,800 m^3, while the amount per resident in the Czech Republic is roughly the same as in Poland. Access to water at less than 1,500 m^3 per one person per year is commonly believed to be very low and triggers serious problems in water resource management.

Water shortages are described using the concept of the water exploitation index (WEI). This is an index that relates annual water usage to available water volumes [6]. When water usage exceeds 20% of reserve capacity, water management needs to be thought of as a critical element of the national economy. In recent years Poland has been ranked 11th in Europe based on WEI (Fig. 3.1), which remains close to the dangerous boundary value of 20%, above which water shortages may be expected. The WEI value for Poland is 18.2% at present [7].

Poland is a country with "hydrographic boundaries." Only 13% of the country's water resources emerge outside of Polish territory. The amount of water "exported" by Poland is about half that. The average surface water volume available across Poland is about 63 km^3, and may decline to less than 40 km^3 in a very dry year while increasing to 90 km^3 in a very wet year [8]. Hence, Poland's water supply is highly variable over time and is unevenly distributed geographically. At a 95% water security level less untouchable supplies, the real water supply level in Poland in dry periods hovers at only about 250 m^3 per person per year. This low supply level calls for water retention practices and continuous monitoring of the quantity and quality of the country's water supply. Water usage per resident of Poland in 2012 stood at about 300 m^3, while in other European countries it ranged from less than 100 m^3 (Luxembourg and Malta) to more than 1,200 m^3 in Estonia. According to GUS (Central Statistical Office), Poland's main statistical agency, the industry is the leading water customer in the country (67%), with municipal services at 21%, and agriculture and forestry at 12%.

Poland's water supply consists mainly of natural lakes (33.0 km^3), reservoirs (3.07 km^3), rivers (1.30 km^3), and fish ponds (0.60 km^3). Poland's water storage needs are estimated to be at about 7 km^3 including flood management capacity ranging from 1.5 to 2.0 km^3. Despite good natural conditions in the country, water retention capacity in Poland is very limited. The total volume of retention reservoirs is about 4 billion m^3, which is sufficient to store only 6.5% of mean annual runoff in the country. This value may potentially increase to 15%. Water storage must increase in Poland in order to secure the country's water supply. This may be accomplished

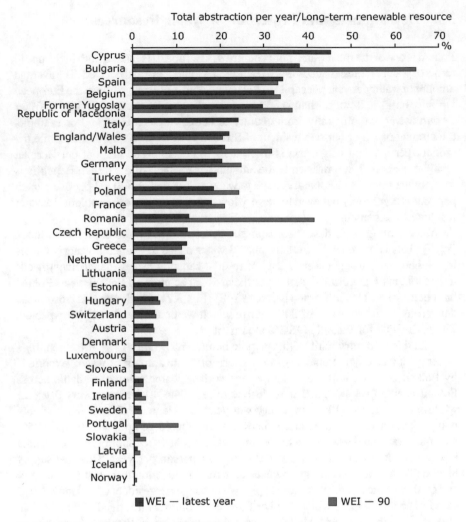

Fig. 3.1 Water exploitation index (WEI) [9]

in a number of ways including the utilization of existing outflow-free depressions and peat bogs [10].

According to a UN report on development issues in 2007 called GEO-4, water usage in developing countries will increase as much as 50% by 2025, while the increase for developed countries will be close to 18%. Another UN report from 2015 states that global water demand will increase by 55% by 2050, while the global water supply will continue to decline. Hence, the need for effective water management is of crucial importance, especially in the era of climate change, which may manifest itself in the form of a growing water deficit in Poland as well as changes in water quality. This will also apply to areas protected by law where water is the main component. In this context, the implementation of the principles of sustainable development is the most important in protected areas. The highest priority needs to be the deceleration of damage to natural landscapes and the biosphere generated by human impact as well as the reduction of the number of disruptions to natural water circulation including major changes in surface runoff patterns, infiltration, and evaporation. Decisions made at the local level are the key to reaching strategic goals described in EU directives and national laws and programs.

3.3 Sustainable Development

The last few decades have seen rapid economic development including development of water areas on the World. Frequently irresponsible human actions have led to the emergence of major threats to the natural environment and man himself. The current global trend is to introduce sustainable development wherever possible [11]. This is an economic doctrine that permits quality of life based on current development levels. It also assumes it is possible to have the type of development that allows current generations to meet their needs, but without making it difficult for future generations to meet their needs. In order for this proper form of development to become a reality, the international community has been struggling for the last five decades to create solutions that would help humanity develop while at the same time maintaining a key supply of natural capital, without which future generations may not survive. The very concept of sustainable development became widely recognized following the publication of a report by the World Commission on the Environment and Development in 1987. This report focused on the need to link economic goals with social goals (Fig. 3.2) and made a note of the fact that human society has reached a development level that is sustainable given proper management practices.

This type of development model assumes a certain level of awareness of the connection between economic growth, concern both for the natural and manufactured environment, and the quality of life including human health. It also assumes a certain type of relationship between the aforesaid "components." In Poland the most commonly accepted definition of sustainable development is that of a socioeconomic system that integrates political, economic, and social actions with a focus on ecological balance and permanence of fundamental environmental processes in order to assure that both current and future generations will be able to meet their basic needs.

3.4 Protected Areas in Poland

The Nature 2000 program covers 21% of Poland's territory. This includes 817 habitat areas and 142 bird protection areas (Fig. 3.3). Nature 2000 areas are not closed areas. In contrast to popular opinion, Nature 2000 areas are not characterized solely by

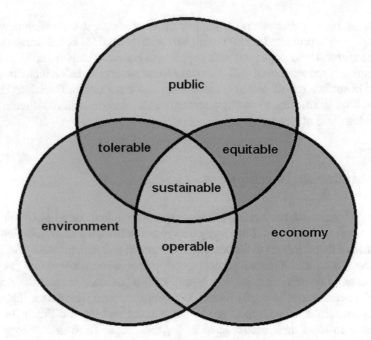

Fig. 3.2 Sustainable development and the link between economics, society, and the environment [12, changed]

bans and restrictions. In fact, most of these areas are governed by standard economic processes, with limitations only on some forms of economic activity—forms that are deemed to be highly harmful to the given area protected by law. Article 36 of the Environmental Protection Resolution dated April 16, 2004 (with changes) states the following: "Nature 2000 areas do not restrict activity associated with the maintenance of flood control facilities and equipment as well as economic activity in the form of agriculture, forestry, hunting, and commercial and amateur fishing unless it yields a significant negative impact on stated protection goals associated with Nature 2000 areas." It is important to note that some actions planned for Nature 2000 areas need to be approved based on environmental impact assessments. Examples of actions include drainage works and relevant maintenance of drainage ditches, construction of water impoundment structures, afforestation, construction of bed and breakfast inns and agritourism facilities, farm expansion or remodeling, and tree and bush removal. Economic activity in Nature 2000 areas part of national parks and nature reserves is permitted to the extent permitted by legal restrictions placed upon these areas. In cases where some forms of economic activity need to be limited to some extent in protected areas, it is possible to receive compensation for the related loss of income.

Poland is home to 23 national parks. Other forms of legal protection of the environment include nature reserves, landscape parks, landscape protection areas, and also natural landscape zones.

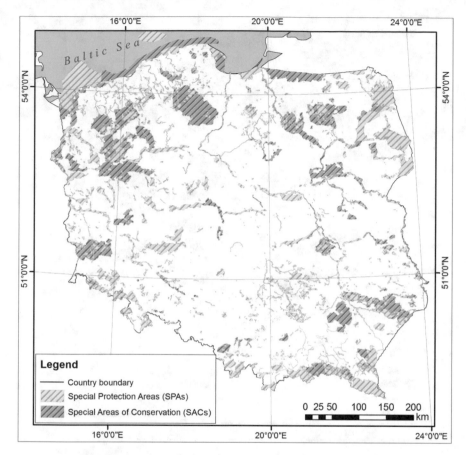

Fig. 3.3 Nature 2000 areas in Poland [own study]

National parks (Fig. 3.4) are defined as protected areas with a surface area no smaller than 1,000 hectares and characterized by a significant environmental, scientific, social, cultural, and educational value. The importance of national parks and nature reserves is underscored by the legal framework created to manage these natural areas. When a national park or nature reserve is created or expanded, this is done on the basis of the Resolution dated August 21, 1997, which articulates the public interest in the real estate market [13, with changes]. The reason for creating national parks is to protect the whole of nature as well as specific landscape characteristics in a given area. It is also feared that areas surrounding a national park will yield a strong negative impact upon it. This is why laws are passed to help national parks by designating a near-park zone around each national park in order to isolate it from negative impacts in adjacent areas. National parks in Poland cover a total area of 3,167 km^2.

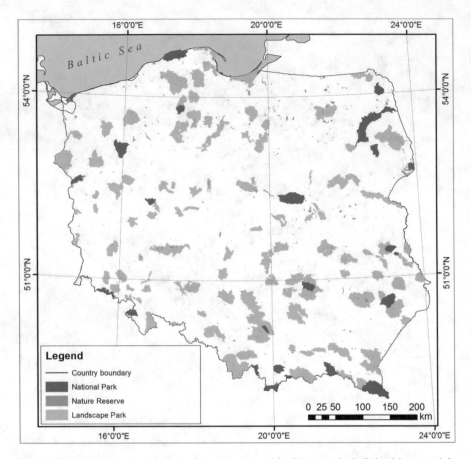

Fig. 3.4 Location of national parks, nature reserves, and landscape parks in Poland [own study]

Nature reserves (Fig. 3.4) are defined as areas maintained in the natural state or slightly altered, ecosystems, natural habitats, plant habitats, animal habitats, mushroom habitats, as well as inorganic formations and elements characterized by unique environmental, scientific, and cultural value or landscape value. Restrictions placed upon nature reserves are exactly the same as those for national parks. A nature reserve differs from a national park in the purpose of the near-reserve area and the absence of a minimal surface area requirement. Poland is home to a total of 1,481 nature reserves with a total surface area of 165,733 hectares.

Landscape parks (Fig. 3.4) are established due to their environmental, historical, cultural and landscape value. Farmland and forests, as well as other types of real estate, found within the limits of a landscape park may be used for economic purposes. What is characteristic of parks such as these is the degree of influence exerted over them by local governments. Environmental protection laws in Poland only provide a list of suggested restrictions for landscape parks. The number of landscape parks in Poland is 122, with a total surface area of 26,100 km². The first landscape park established in Poland was Suwalski Landscape Park (1976). The youngest park is Nadgoplański Millennium Park (2009). Barycz Valley Landscape Park is the largest, while Stawki Landscape Park is the smallest landscape park in the country.

Protection plans established for each form of environmental protection consist of four stages. Each plan serves as the basis for the functioning of each protected area. The first step is to catalogue the resources of each target area making it possible to assess the state of the natural environment. The second step consists of the identification of threats and potential threats that may occur in the future. Given that protected areas are not closed areas, the third step involves the formulation of a social and economic assessment and the issuance of an opinion of the state of spatial management in the target area. The last step of the plan formulation process consists of an analysis of the effectiveness of currently used environmental protection methods. Plans for the three types of protected areas discussed herein are created for periods of 20 years. This time perspective takes into account the fact that natural processes occur on a different timetable than economic processes.

Environmental protection laws in Poland identify **Protected Landscape Areas** as those featuring unique landscapes with different types of ecosystems, which are valuable due to their ability to meet the tourism and recreational needs of society or and ability to serve as ecological corridors. The law provides a list of potential restrictions that may be imposed upon protected landscape areas—similarly as with landscape parks. The functioning of "protected landscape areas" is significant first and foremost from a spatial planning perspective, whereby economic activity is not restricted unless it is deemed to be highly burdensome. Poland is home to a total of 409 protected landscape areas covering 22.7% of the territory of Poland.

Dig sites, ecological sites, and environmental-landscape complexes are forms of natural environment protection formulated by commune councils. The law provides for a common for all three forms of protection list of restrictions, which can be used as guidance by lawmakers reviewing appropriate protections for target areas.

Dig sites include scientifically or educationally important places of occurrence of some geological formations, plant fossils, animal fossils, collections of fossils or mineral formations, caves, surface mines, and underground mines. This is the only form of environmental protection that does not cover surface areas. Poland is home to 165 dig sites with an area of 905 hectares.

Ecological sites are established across remainders of ecosystems strongly relevant to the preservation of biodiversity and may include natural bodies of water, ponds, wetlands, peat bogs, ancient river channels, and environmental habitats. Poland is home to 7,029 ecological sites with a total area of 51,843 hectares.

Environmental-landscape complexes include parts of natural and cultural landscapes that merit protection due to their aesthetic or vantage point value. Poland is home to 184 such complexes with a total area of 95,869 hectares.

Water management in protected areas requires first and foremost a certain water level determined by predefined environmental goals as well as the determination of priorities related to water usage in a given area. In addition, predetermined environmental goals need to be based on limits on water usage. It is important to put limits on the use of dams, use of groundwater and surface water, the release of wastewater into surface water and groundwater, employment of drainage works in construction, and the establishment of hydroelectric facilities. In addition, some forms of environmental protection provide a guarantee that no construction will take place and no homes will be built. This applies to national parks. Other forms of environmental protection provide for limited construction opportunities. The greatest problem for protected areas is new construction in so-called near-park or near-reserve areas.

3.5   Lakes and Threats to Lakes in National Parks in Poland

This section of the paper provides information on lakes, current threats, and potential threats to lakes found in national parks with the largest water capacity in Poland.

Babiogórski National Park—characterized by the presence of numerous small landslide lakes recharged by groundwater. Of the 18 lakes in the area, 13 are permanent. At high water stages, water covers an area of about 0.4 hectares. The largest lake, Mokry Stawek, has an average area of 450 m^2 with a maximum of 1,500 m^2. Its average depth is 2.5 m with a maximum of 5 m. It does dry out or spill out occasionally [14]. The lake used to be a swimming hole and boating area before World War II. Today the lake is still popular with tourists, which has not yet produced any deterioration in the lake basin [15]. Lakes in Babiogórski National Park were also used as watering holes for sheep and cattle until the mid-twentieth century. This practice ended when the park was officially established [16]. The mineral content of the park's lake water is low. The lake water pH is neutral. The lakes feature high oxygen content and low biogenic substance content [17], which also means that water quality is very high. The highest mineral content is observed in winter and the lowest during the spring snowmelt season [15]. The main threats to the lakes today are air pollution arriving from nearby areas, wastewater treatment facilities needed to serve a growing number of tourists, and increasing vehicular traffic [14, 18].

Drawieński National Park features 20 lakes characterized by variable morphology and trophic state. The largest lake is called Ostrowiec and has an area of 369.9 hectares and a maximum depth of 28.5 m [19]. Another large lake is called Płociczno, which is characterized by high susceptibility to degradation due to its large catchment area to lake area ratio [20] and contaminated sediments [21]. The park also includes hydrologically unique Lake Czarne—a charophyte lake—whose waters undergo exchange once every 20 years. The park features more lakes of this type but also features dystrophic lakes [20]. Fishing is possible on a seasonal basis in some lakes in the park (Ostrowite, Sitno). Tourists may also kayak on some of the lakes or camp lakeside. Water quality in the park's lakes varies. Charophyte lakes are characterized by very good water quality.

On the other hand, some of the park's lakes are strongly eutrophic and feature poor quality water, gauged via high concentrations of biogenic substances, low water transparency, oxygen deficits, and high trophic state. The main threats to surface waters in Drawieński National Park include kayaking, vehicular traffic, parking spaces, and fishing, which remains a major burden at 2,583 man-days per year. In order to improve the quality of water in these lakes, it is necessary to modernize wastewater systems, limit pollution flows from agricultural areas, and manage tourism and recreational development [20].

Karkonoski National Park features postglacial lakes in the form of ponds and small peat bog lakes. The latter reflect the pH of their parent bogs and tend to be acidic at 3.0–3.5 due to the presence of humus acids [22]. PTTK (Polish Tourist and Sightseeing Society) lodges in the park and private lodges use water from the park's lakes. Surface water quality is low due to significant acidity and elevated concentration of selected ions [23]. Given the low mineral content in the park's waters and their strong acidity, they are sensitive to external factors. The main factor is polluted precipitation water, although the acidity of precipitation has been decreasing since the 1990s [24].

Bory Tucholskie National Park features 23 natural lakes and one artificial reservoir for a total capacity of 41 mln m^3. Lake water quality is good and very good. The park's lakes feature low mineral content and low nitrogen and phosphorus content— concentrations 30–40% lower than in other regions in Poland [25, 26]. The only problem in some of the lakes is low oxygen content in summer and winter [27]. The lakes' hydrochemical and biological state varies significantly. Some of the lakes are humus lakes, while others are transitional lakes at various stages of trophic evolution. Humus lakes are also outflow-free lakes characterized by cascading patterns relative to one another and strong water acidity due to the influx of humus substances and acid rain. Surface waters in the park are affected by groundwater whose chemistry is driven by the local climate water balance [26]. Lake oligotrophication or dystrophication may be an outcome of changes in hydro-geologic conditions and land cover [28].

Lake dystrophic processes are also accelerated by human impact in the form of forest clearing and plowing of arable land before afforestation as well as the construction of drainage ditches in the park [29]. On the other hand, transitional lakes are gutter lakes connected with each other via the Struga Siedmiu Jezior. It is important to note that the trophic state of lakes increases rapidly with the course of the Struga [26]. The only water usage in the park is that for fish ponds and fish management catches. Some water is also used by a tree farm in Klosnów. This is the only surface water collection site in the park—it is found on Lake Ostrowite. Discharge ranges from 4,000 m^3 to 6,000 m^3 per month. Finally, the park's hydrologic system is sensitive to external changes of all types.

Ujście Warty National Park—its current surface water system is the outcome of a strong human impact in the region [30]. The outcome is a weak hydrographic network, as manifested by the current surface area of Lake Kostrzyńskie (37.2 ha), which used to cover the area between the Odra and Warta rivers [31]. The existence of Lake Kostrzyńskie and four other lakes formed from ancient river channels depends on water levels in the Odra and Warta rivers, flood stages, and groundwater levels [30]. Also, there is one lake in the park located on the Northern Polder—Lake Sierżant— with a surface area of 2.9 ha. The park first and foremost functions as a water retention area. Water tourism and fishing are also available in the park. Water in the park is strongly eutrophic, and the trophic state of the park's lakes depends on the quality of river water, which is rated less than good. Hence, the main threats to water quality in the park are river water quality and any changes in the hydrographic network [30].

Poleski National Park features four łęczyńsko-włodawskie lakes District and numerous ponds. An additional eight lakes are found in the near-park zone, with one lake transformed into a retention reservoir. Three of the four lakes found in the park are dystrophic lakes, while one is a eutrophic lake [32]. All the lakes are shallow and their surface area ranges from 3.52 ha to 131.63 ha. Fluctuations in water levels slightly exceed 40 cm [33]. Water usage in the park is low and includes mostly local water collection for economic purposes. A limited amount of commercial fishing is allowed on several ponds. Amateur fishing is allowed on one pond. Lakes in the park are not available for recreational purposes and only serve research and teaching purposes. Commercial fishing is limited to selection and control catches [33]. The water balance in the park today is the product of human impact leading to lowered water levels [34]. One of the main problems in the park is unregulated wastewater management in the form of wastewater releases into some lakes as well as the use of fertilizer and pesticides in local agriculture [33]. Other problems include drainage ditches dug in the 1950s and two subsequent decades as well as corresponding changes in the land use structure [35].

Roztoczański National Park is characterized by the presence of numerous ponds. Kościelny Pond is one of the more interesting of the ponds with a historic water mill at its outlet and also a baroque church on an island. Other interesting ponds include Echo, Czarny Staw, and Staw Florianiecki. The park also includes Rudka recreation pond whose water temperature in summer is only 15 °C, which limits its usefulness as a swimming hole. A retention reservoir is used by a paper mill in Zwierzyniec. Water usage in the park is limited to household use by park ranger stations and surface water use by the Echo, Czarny, and Florianiecki ponds. In the summer, the park is visited by large numbers of tourists driving cars [36]. The main threat to surface water quality in the park is the influx of agricultural and household-type pollutants along with low emissions polluting the local air. Other threats include a decreasing supply of groundwater, drainage works performed in the park in the 1960s and 1970s, and an increase in overall human impact [37].

Słowiński National Park is unique among national parks in Poland in that it is characterized by the highest share of water areas at more than 60% [38]. This is the largest water-based national park in Poland due to the presence of two large lakes—Lake Łebsko and Lake Gardno—and additionally three smaller lakes— Dołgie Wielkie, Dołgie Małe, Smołdzińskiego. The two larger lakes were formed as a result of sand migration from the Łeba sandbar, eventually covering parts of a lagoon in the northeastern corner of present-day Lake Gardno. The lagoon existed until the seventeenth century. The third lake became separated from Lake Gardno in the mid-twentieth century due to the land formation [39]. The park's hydrographic network was subject to strong human impact beginning in the mid-eighteenth century when a local water balance program was set into motion in the Pomerania region (Fig. 3.5).

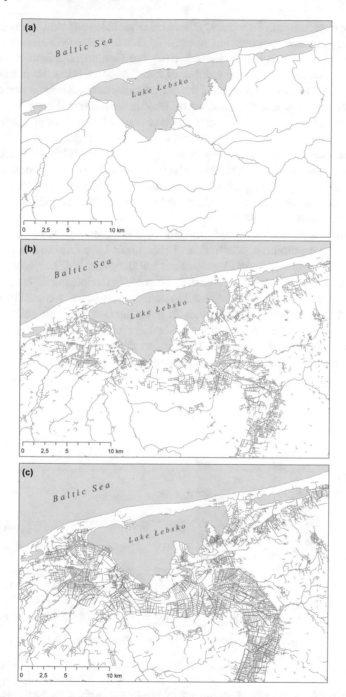

Fig. 3.5 Changes in the drainage network surrounding Lake Łebsko in (a) the nineteenth century, (b) early twentieth century, and (c) late twentieth century [own study]

Surface water usage in Słowiński National Park includes different forms of activity in some ways related to protection work [40]. This includes irrigation of meadows and refilling ponds with fish in lakes Łebsko and Gardno. The goal is to balance the species composition and the number of fish in the two lakes. Swimming is not allowed in the park, as is sailing and the use of motorboats. Also not permitted are water sports, fishing, and the catching of sea creatures along the sea coast. Water sports are permitted at selected sites along lakes Łebsko and Gardno. Fishing along the sea coast is also permitted along tourist trails. Lake fishing is also permitted [41]. Research has shown that the lakes are strongly eutrophicated and overall water quality is poor. This is largely due to the influx of biogenic substances. One other major factor in lake water quality in the park is periodic seawater intrusions manifested by high concentrations of chloride, magnesium, sodium, potassium, and sulfate as well as high specific conductivity [42].

Chloride is a particularly good indicator of Baltic Sea effects on the two lakes (Fig. 3.6). The maximum concentration of chloride in Lake Łebsko can reach about 1,900 mg•dm^3, while in Lake Gardno about 1,500 mg•dm^3 [43].

Given the unique location of the park and especially the location of lakes across its lower elevations (Fig. 3.7), main threats to surface water quality come from outside the park. The Łeba and Łupawa rivers flowing through the park collect most of the pollutants present in the catchment. Seawater intrusions may also pollute the two lakes in some cases. However, the greatest threat to water quality is wastewater releases and unregulated wastewater management near the park. Areal pollution is another source of problems for the park, especially agricultural pollution [44].

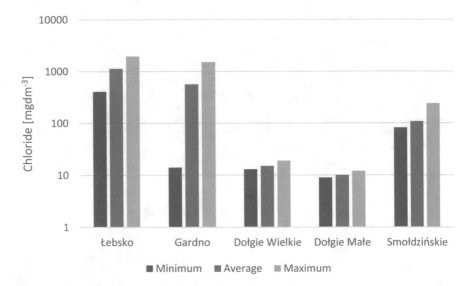

Fig. 3.6 Minimum, mean, and maximum concentration of chloride in lake water in Słowiński National Park in the years 2002–2008

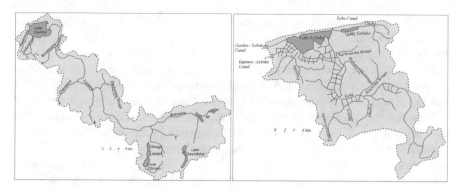

Fig. 3.7 Location of lakes Gardno and Łebsko along the lower reaches of the Łupawa and Łeba rivers [39]

Other negative factors affecting the park's waters include mass tourism and recreation, especially in the summer, aeolian processes, lowering of the groundwater level, industrial-scale peat extraction, and lack of treatment plants for wastewater [41].

Tatra National Park—surface water retention in the form of lakes is quite important in the park [45], which features postglacial lakes, mostly old glacial cirques and cauldrons filled with water. Some of the lakes exist in deep valleys enclosed by rock walls. In some cases, karst depressions become filled with water. In other cases, former glacier sites can become filled with water [46]. The largest lake district in the park is called the Valley of Five Polish Ponds with the deepest lake in the Tatra Mountains—Great Polish Pond [47]. Surface waters may be classified as super freshwater or ultra freshwater, with some classified as simply freshwater [48]. Water quality in the park's lakes is good. One threat to water quality in the park is that of acid rain in the area, although this problem is much smaller today than in the first half of the twentieth century [49]. Another problem in the park is human waste found near tourist trails. The park is visited by more than 2.5 million tourists per year. The park is reducing the scale of this problem by installing portable sealed toilets at some locations. The park also includes a number of water collection sites on lakes Morskie Oko and Przedni Staw Polski. The water is used by tourist lodges present at selected locations in the park [46]. Other key threats to water quality in the park include releases from wastewater plants in the area. This is leading to local changes in the natural environment, lake water pollution due to mass tourism near Lake Morskie Oko, and changes in thermal stratification and lake water circulation due to the collection of water from Lake Wielki Staw Polski by a local power generation plant [50].

Wielkopolski National Park features 11 lakes characterized by different trophic states. Most of the lakes in the park are gutter lakes. Lake Chomęcickie features a swimming area and the only campground in the park. Lake Jarosławieckie also features a swimming area, although the lake is experiencing rapid overgrowth [51]. The largest lake in the park is called Łódzko-Dymaczewskie and features a swimming area along with two recreational centers. The park allows tourists to use three lakes—Witobelskie, Łódzko-Dymaczewskie, Jarosławieckie. The three lakes experience strong human impact from both swimmers and people fishing. The only lake in the park to allow sailing and water sports is Lake Łódzko-Dymaczewskie. A place to build a campfire is also available along lake shores. Most lakes in the park are polluted to a substantial degree [52]. The location of the park close to the Poznań metropolitan area yields the risk of overuse by urban visitors. Years of human impact due to a lack of systematic wastewater treatment and lakeside tourism have contributed to significant degradation of the park's lakes. Another growing problem in the park area is increasing traffic. Other issues include increasing beaver activity in the lakes' direct catchment areas, which may contribute to the washing out of biogenic substances from exposed areas. The catchments of many of the lakes are agricultural, which further increases the influx of biogenic substances. However, the greatest threat to surface water quality in the park is unregulated wastewater that enters lakes from summer lodges located along lakes Łódzko-Dymaczewskie and Witobelskie [52].

Wigierski National Park—the most valuable features of the park are its lakes and dystrophic postglacial lakes. The park includes 42 lakes with a total surface area of 28 km^2. The largest lake in the park is Lake Wigry. One group of lakes in the park is unique on the Central European Plain—dystrophic postglacial lakes known as suchary. These are small woodland lakes without an outlet. Their waters are acidic and poor in calcium and rich in humus substances. The park includes 20 suchary lakes. Also, there are woodland lakes often surrounded by low bogs. The park permits various forms of tourism including hiking, biking, skiing, and water sports. Lakes in the park play a multitude of roles including environmental, educational, research, and recreational roles. Eight lakes may be used for water sports. Sailing is also allowed, as is fishing. The park is home to 15 tent areas and a beach. There are also a number of businesses that rent water sports equipment as well as agritourism farms [53]. Lakes in the park may be placed in two categories: (1) transitional lakes characterized by a rise in fertility, (2) non-transitional lakes or suchary. Lake water in the park is subject to a variety of factors including human impact and global climate change. Illegal fishing is also a problem, as is the increasing number of tourists. The park attempts to counter these challenges via action designed to maintain ecosystem diversity and water organism diversity as well as by returning the park to its natural state of functioning [53].

Woliński National Park includes 5 lakes in the Woliński Lake District and two artificial lakes called Turkusowe Lake and Stara Kredownia Lake. Water in the park is systematically used for recreational purposes. Tourism remains an important part of the park's functioning. The largest tourist traffic is encountered along lakes Turkusowe, Czajcze, and Grodno [54]. The surface waters of the park are used for swimming on a regular basis as well as for limnology and biology research [55]. Water quality in the park's lakes is medium or low, even though some improvements in water chemistry have been observed. The greatest threats to the park's waters are polder fishing, groundwater use, influx of household wastewater, poor state of local wastewater treatment plants, decreasing water levels in lakes, and road infrastructure [56].

3.6   Main Threats to Water Quality in Protected Areas

The Preamble of the Framework Water Directive issued by the European Union states: "…water is not a commercial product as are other products, but an inherited good that must be protected and treated as such…." Hence, the most important goals for European countries are to meet the water needs of their populations, farm sector, and industrial sector, and to promote sustainable water use and water protection along with the protection of ecosystems that remain in good ecological shape. Sustainable water management takes into account entities competing for resources in a way that water is allocated in a fair manner in order to fully meet the needs of the various entities utilizing water for a variety of purposes. Water management in protected areas in an exceptionally important part of sustainable water management.

The Framework Water Directive [57] makes reference to a variety of issues, but its environmental goals section includes a direct reference to protected areas among other goals such as not worsening water quality, reaching a good chemical and ecological state of waters, and reducing the effects of floods and droughts. The reference to protected areas states that "it is important to meet the requirements of protected areas in terms of the threat of pollution by nitrogen compounds arriving from agricultural areas, designation for recreational use, drinking water collection, threat of eutrophication due to household wastewater influx, designation for protection of marine animals with economic significance, and protection of habitats and species whose survival depends on good water quality. FWD Article 6 underscores the requirement to create a separate registry of protected areas for tributary catchment areas".

These areas are defined in Attachment IV and include areas marked for drinking water collection, areas designated for the protection of marine animals with economic significance, parts of waters designated for recreational purposes including areas designated as swimming areas based on Directive 76/160/ECM [58], areas sensitive to biogenic substances including areas designated as sensitive areas based on Directives 91/676/ECM [59] and 91/271/ECM [60], areas marked for habitat and species protection, where the maintenance of water quality or its improvement is a key factor in their protection, including selected sites part of the Nature 2000 program that were identified based on Directive 92/43/ECM (Habitat Directive) [61] and Directive 79/409/ECM (Bird Directive) [62]. Despite the protections afforded by these directives—and direct and indirect protections provided by national parks and other forms of environmental protection—there are still many threats to the natural environment of these areas. These threats are particularly strong with respect to surface waters including standing waters. Two leading threats are tourism and water sports. In the opinion of more than 60% of tourists, the natural environment represents the "main attraction" [63].

This is particularly true of water areas, which are very popular recreational sites across Poland. However, increased tourist traffic, especially around lakes, means more human impact on lakes and their ecotones [64]. According to Krzymowska–Kostrowicka [63], there are three primary types of human impact on the natural environment:

1. Physical impact associated with the movement of machines and trampling,
2. Chemical impact in the form of the introduction of foreign substances into an environment, and
3. Biological impact in the form of the introduction of foreign animal and plant species into an environment as well as the removal of native animal and plant species from an environment.

Other key threats can include a lowering of groundwater levels which leads to the lowering of lake water levels, atmospheric pollution, poor state of equipment designed to prevent reductions in water resources, improper wastewater treatment systems, drainage works, fishing, lack of restrooms at tourist facilities, agriculture, surface water collection, water sports, excessive numbers of tourists, roads, sewage disposal, and climate change.

In the aspect of development and protection of water resources against the background of sustainable development, different needs of particular public groups are observed, which are not always convergent. Also their role is not the same. Man strives for development that would satisfy his existential, cultural and financial needs. Unfortunately, this does not always go hand in hand with environmental protection, although in recent years there has been a desire to live in an unpolluted environment with high landscape and natural values. At the moment, public groups have little capacity to co-decide in which direction the development of their inhabited areas is to go, especially those related to the aspect of water. Also the creation of local space by the community is quite limited, although in recent years there has been a tendency to become more and more involved in local matters. The voice of people is often noticed by the government, who use good ideas from the community. They are also often overlooked. The role of the community is to propose own solutions that create geographic space and enforce their implementation by local and regional authorities. In turn, the role of the government is to conduct a sustainable development policy, i.e. one that allows the community to develop without harming the natural environment. This policy should be based on the introduction of modern technologies to human life, which are at the same time pro-ecological, ordering the law related to the protection of valuable natural areas, as well as listening to the needs of local communities and, if possible, implementation. The government must have a far-reaching vision that would take into account the possible changes in the natural environment resulting from global climate change, as well as potential shortages of water suitable for use. Therefore, it should safeguard the rational distribution of existing water resources. The last group that is important in running a sustainable development policy are investors. They should remember that economic development is important, but it should not cause significant changes in the environment. Nowadays, the living should leave the current state of the geographical environment for future generations that could use it as much as they do today. They should therefore cooperate with the government and public groups. In Fig. 3.8 added a flowchart to summarize the needed actions from each party to help them to achieve this objective.

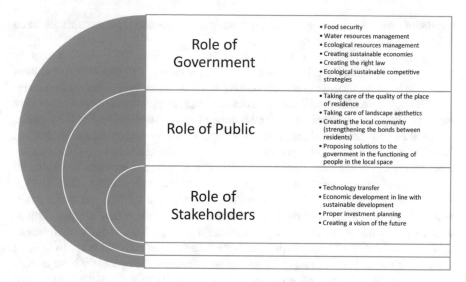

Fig. 3.8 The needed actions from public, government and stakeholders to achieve sustainable development [own study]

3.7 Conclusions

The best way to maintain environmentally valuable areas is to grant them legal status based on environmental protection goals. However, not all environmentally valuable areas are legally protected or can be legally protected. In addition, the low effectiveness of many forms of environmental protection, especially lower level protected status, shows that only local forms of protection are highly effective. A local spatial management plan is needed in order to protect unique areas. Most commune strategic documents list main tasks that include the protection of the natural environment, utilization of tourist attractions and natural resources, maintenance of and impact on biodiversity, and protection of environmentally valuable areas and sites. Hence, it is important to motivate local government officials to take responsibility for locally valuable areas in the form of local laws or local spatial management plans [65]. Currently, environmentally valuable areas are not defined in local laws, although this concept is mentioned in the Resolution on Environmental Protection [66, 67] and Environmental Protection Law [67].

Environmentally valuable areas protected by law are components of a broader system of environmental protection in Poland. The Resolution on Environmental Protection [66] requires that for some areas a separate planning document is created that regulates how the most environmentally valuable areas are used—national parks, landscape parks, nature reserves, and Nature 2000 areas. For all other forms of environmental protection, a fairly general list of bans and requirements serves as the leading set of guidelines. These other areas include protected landscape areas, ecological areas, and environmental and landscape complexes. It is noteworthy that regulations for these other areas do not specify how bans and requirements are to be included in local spatial management plans.

The current version of the Resolution on Spatial Planning [68] does not directly address environmentally valuable areas. Instead, it requires all planning documents including local documents to outline rules of environmental protection and limits on land development. The Resolution implies that local spatial management plans may serve as a key instrument in the maintenance of environmentally valuable areas at the local level in the case of areas not formally covered by environmental protection laws. On the other hand, rules and regulations that apply to inland surface waters are linked with the Water Law Resolution [69]. The Resolution includes a number of regulations of varying precision that apply to inland waters including lakes.

In summary, national parks in Poland are characterized by different water availability. Some are dominated by water balance issues, while others are not. This is why each park needs to have a unique protection strategy. This is due to differences in water availability and threats that may affect each given national park. In spite of these differences, it is necessary to pursue sustainable development whose overall goal is to protect the natural environment including its water areas in a way that allows both social and economic targets to be achieved.

3.8   Recommendation

At the end of the work, I would like to present some recommendations for development and protect water resources in the context of sustainable growth in geographic areas protected by law. Below are a few of them:

– Water management in protected areas requires first and foremost a certain water level determined by predefined environmental goals as well as the determination of priorities related to water usage in a given area.
– Sustainable development requires an integrated approach and a holistic perspective, in which a structure of inter-linked components is taken into account. This structure contains not only hydrological or water resources components but also a number of other components, such as environmental, economic, demographic, socio-cultural and institutional subsystems [11].
– It is necessary to create regulatory mechanisms and ensure security against the growing civilization threats.
– One central water authority should be created at the national level, which would deal with all aspects of water management, e.g. Ministry of Water Resources.
– It should be increased coordination between national bodies currently dealing with water resources.
– Increase the effectiveness of many forms of nature protection (change of legal provisions).
– Increase the role of local communities that decide on the development of water resources (subsidiarity principle).

- Mobilize local authorities to take responsibility for valuable natural areas with the introduction of appropriate provisions to local law documents, i.e. local spatial development plan.
- Change the philosophy associated with meeting the demand for water (water is not a free good) (water is not a free good any more but rather an economic good).
- It is necessary to change the provisions indicating the principles of protection and shaping of valuable natural areas, and not covered by protection in accordance with the Nature Conservation Act, which are dispersed in many documents—statutes and regulations.
- It is necessary to clearly define in the strategic documents of communes the manner of implementing provisions regarding the protection and shaping of valuable natural sites.
- Adequate funding must be guaranteed to ensure the implementation of sustainable development objectives.
- New concepts of international cooperation and assistance need to be developed.
- The principle of treating the duty of environmental protection as an element of the proper functioning of the economy should be introduced, and the breach of this obligation should be regarded as a betrayal of the rules of good management.
- An effective way to implement the principles of sustainable development seems to be the process of greening the economy
- Large risks must be avoided, the consequences of which are irreversible damage to ecosystems and social systems that cannot be estimated,

Acknowledgements I would like to thank Mrs. Alicja Olszewska for help in the performance of some of the figures.

References

1. Sun S, Wang Y, Liu J, Cai H, Wu P, Geng Q, Xu L (2016) Sustainability assessment of regional water resources under the DPSIR framework. J Hydrol 532:140–148
2. Jokiel P, Pociask-Karteczka, J (2012) Dlaczego wody? In: Water in national parks of Poland. Institute of Geography and Spatial Management of the Jagiellonian University Cracow, pp 9–14 (in Polish)
3. Kates RW, Parris TM, Leiserowitz AA (2005) What is sustainable development? Environ Sci Policy Sustain Dev 47(3): 8–21
4. Steiner F (1999) The living landscape 1999. An ecological approach to landscape planning. ISLANDPRESS, Washington, Covelo, London
5. Song M, Cen L, Zheng Z, Fisher R, Liang X, Wang Y, Huising D (2017) How would big data support societal development and environmental sustainability? insights and practices. J Cleaner Prod 142(2):489–500
6. Pedro-Monzonís M, Solera A, Ferrer J, Estrela T, Paredes-Arquiol J (2015) A review of water scarcity and drought indexes in water resources planning and management. J Hydrol 527:482–493
7. Wyszkowska D, Artemiuk D, Godlewska A (2017) Including the quality of life in the study of the state of the green economy. Wroclaw Econ Rev 23(4):11–26 (in Polish)

8. Gutry-Korycka M, Sadurski A, Kundzewicz ZW, Pociask-Karteczka J, Skrzypczak L (2014) Water resources and their use. NAUKA 1:77–98 (in Polish)

9. EEA Report—EEA signals: key environmental issues facing Europe (2009)

10. Major M, Cieśliński R (2015) Retentivity as an indicator of the Capacity of Basins without an outlet to accumulate water surpluses. Polish J Environ Stud 24(6):2503–2514

11. Kundzewicz ZW (1997) Water resources for sustainable development. Hydrol Sci J 42(4):467–480

12. http://hlmssustainability.com/sustainable-development-important/

13. Journal of Laws: No. 261, Item 2603 (2004)

14. Łajczak A (2012a) Babiogórski National Park. Water in national parks of Poland, Institute of Geography and Spatial Management of the Jagiellonian University Cracow, pp 14–30 (in Polish)

15. Łajczak A (2012b) Hydrography of Babia Góra. resources, circulation and water features, explanations for the hydrographic map. Publisher of the Institute of Botany named after W. Szafer Polish Academy of Sciences, Cracow (in Polish)

16. Łajczak A (1995) Mother of Improperness. Babia Góra's paths. Colgraf-Press, Poznań (in Polish)

17. Pasternak K (1983) Surface waters. Studia Naturae B series 29:367–378 (in Polish)

18. Pasierbek T, Lamorski T, Omylak J (2006) Characteristics and scope of threats to high-altitude nature in the Babiogórski National Park. Roczniki Bieszczadzkie 14:247–265 (in Polish)

19. Piotrowicz R, Klimaszyk P (2009) Problems with the classification of ecological status of lakes with different hydrographical and morphometric-catchment parameters (Lake Ostrowieckie, Drawieński National Park). In: Diagnosing the state of the environment: research methods, forecasts. BTN Bydgoszcz, pp 121–130 (in Polish)

20. Ławniczak AE, Strzelczak A, Choiński A (2012) Drawieński National Park. In: Water in national parks of Poland. Institute of Geography and Spatial Management of the Jagiellonian University Cracow, pp 74–92 (in Polish)

21. Mielnik L, Piotrowicz R, Klimaszyk P (2009) Chemical properties of bottom sediments in through flow lakes located in Drawieński National Park. Oceanol Hydrobiol Stud 38(3):69–76 (2009)

22. Czerwiński J, Mazurski KR (1992) Giant mountains. Sport and Tourism Publisher, Warsaw (in Polish)

23. Mazurski KR (2012) Karkonosze National Park. In: Water in national parks of Poland. Institute of Geography and Spatial Management of the Jagiellonian University Cracow, pp 120–136 (in Polish)

24. Mazurski KR (1998) Ecological aspects of the presence of shelters in the Karkonosze National Park. In Geoecological problem of the Giant Mountains. Arcus Poznań, pp 279–284 (in Polish)

25. Zdanowski B, Stawecki K, Prusik S, Hutorowicz J (2004) Physicochemical properties of water. In: Water ecosystems of the "Bory Tucholskie" National Park, Publisher of the Inland Fisheries Institute Olsztyn, pp 17–32 (in Polish)

26. Marszelewski W, Nowicka B (2012) National Park "Bory Tucholskie". In: Water in national parks of Poland. Institute of Geography and Spatial Management of the Jagiellonian University Cracow, pp 178–200 (in Polish)

27. Marszelewski W (2006) Physical and chemical parameters of the Seven Lakes Streams. In: Lakes and Mires of Bory Tuchoskie National Park. Oficyna Wydawnicza FOREST Józefów, pp 127–136

28. Milecka K (2005) History of lobelia lakes in the western part of the Tuchola Forest against the background of post-gradual development of forestry. UAM Scientific Publisher, Poznań (in Polish)

29. Bociąg K (2003) The impact of acidic organic matter on the diversity of underwater vegetation in soft water lakes. Acta Soc Botanicorum Pol 72(3):221–229

30. Choiński A, Ławniczak AE, Ptak M (2012) Ujście Warty National Park. In: Water in national parks of Poland. Institute of Geography and Spatial Management of the Jagiellonian University Cracow, pp 224–238 (in Polish)

31. Piasecka J (1974) Hydrographic changes of the Warta valley during the last 200 years. Czasopismo Geograficzne 45(4):229–238 (in Polish)
32. Chmiel S (2009) Hydrochemical evaluation of dystrophy of the water bodies in the Łęczna and Włodawa area in the years 2000–2008. Limnol Rev 9(4):153–158
33. Michalczyk Z, Mięsiak-Wójcik K, Turczyński M (2012) Poleski National Park. In: Water in national parks of Poland. Institute of Geography and Spatial Management of the Jagiellonian University Cracow, pp 250–264 (in Polish)
34. Wilgat T, Michalczyk Z, Turczyński M (1992) Łęczyńsko-Włodawskie Lakes. Studia Ośrodka Dokumentacji Fizjograficznej Polish Academy of Sciences 19:23–140 (in Polish)
35. Chmielewski TJ, Chmielewski S (2010) The process of lake and bog ecosystems disappearance in the region of Polesie National Park since the second half of XX ct. and the prospects of their protection. Proble Ekologii Krajobrazu 26:121–134 (in Polish)
36. Bartoszewski S, Michalczyk Z (2012) Roztoczański National Park. In: Water in national parks of Poland. Institute of Geography and Spatial Management of the Jagiellonian University Cracow, pp 264–280 (in Polish)
37. Bartoszewski S (2004) The condition and threats to the water relations of the Roztoczański National Park. In: Geographical research in understanding the environment. UMCS Lublin, pp 273–277 (in Polish)
38. Partyka J, Pociask-Karteczka JL (2008) Water resources of national parks in Poland-size, structure and threats. In: Water in protected areas, Institute of Geography and Spatial Management of the Jagiellonian University, Hydrological Commission of the Polish Geographical Society Cracow, pp 15–29 (in Polish)
39. Cieśliński R (2011) Geographical conditions of hydrochemical variability of lakes of the southern Baltic coast. Publisher of the University of Gdańsk, Gdańsk (in Polish)
40. Cieśliński R, Olszewska A (2012) Exploitation and protection of the Polish southern Baltic coastal zone lakes and their potential for recreation. Pol J Nat Sci 27(4):377–392
41. Bogdanowicz R, Izydorek I (2012) Slowiński National Park. In: Water in national parks of Poland. Institute of Geography and Spatial Management of the Jagiellonian University Cracow, pp 280–296 (in Polish)
42. Cieśliński R, Olszewska A (2018) New insight into defining the lakes of the southern Baltic coastal zone. Environ Monit Assess 190(2):102
43. Cieśliński R (2009) Hydrological assessment of the Baltic Sea impact on the Polish coastline. GEOLOGIJA 51(3–4):146–152
44. Cieśliński R (2007) Natural and anthropogenical threats of lakes of polish coastal zone. Arch Environ Prot 33(1):15–27
45. Choiński A, Macias A (2008) Lakes in national parks, nature reserves and landscape parks. In: Water in protected areas. Institute of Geography and Spatial Management of the Jagiellonian University, Hydrological Commission of the Polish Geographical Society Cracow, pp 31–42 (in Polish)
46. Kot M, Pociask-Karteczka J (2012) Tatra National Park. In: Water in national parks of Poland. Institute of Geography and Spatial Management of the Jagiellonian University Cracow, pp 310–328 (in Polish)
47. Lange W, Maślanka W, Nowiński K (2000) Separation and physical and limnological diversity of the Polish Tatras. In: Geographical research in the Polish Tatras. University of Gdańsk Publisher Gdańsk, pp 99–126 (in Polish)
48. Oleksynowa K, Komornicki T (1996) Water chemism. In: Nature of the Tatra National Park. TPN Cracow—Zakopane, pp 197–214 (in Polish)
49. Żelazny M, Kasina M, Siwek J, Siwek JP, Kot M (2008) Factors shaping seasonal changes in the chemical composition of Tatra stream water. In: Chemism of atmospheric precipitation, surface and underground water. Publisher of University of Lodz Łódź, pp 35–45 (in Polish)
50. Balon J, Jodlowski M, Pociask-Karteczka J (2008) Ways of preventing the degradation of waters of Morskie Oko Lake. In: Water in protected areas. Institute of Geography and Spatial Management of the Jagiellonian University, Hydrological Commission of the Polish Geographical Society Cracow, pp 317–320 (in Polish)

51. Pełechaty M (2006) Do physical-chemical water properties reflect the macrophyte and open water included spatial heterogeneity of a shallow lake? Oceanol Hydrobiol Stud 34(4):369–386
52. Ławniczak AE, Choiński A (2012) Wielkopolski National Park. In: Water in national parks of Poland. Institute of Geography and Spatial Management of the Jagiellonian University Cracow, pp 328–342 (in Polish)
53. Bajkiewicz-Grabowska E, Górniak A (2012) Wigierski National Park. In: Water in national parks of Poland. Institute of Geography and Spatial Management of the Jagiellonian University Cracow, pp 342–360 (in Polish)
54. Lewicki I (2000) Activity of the Wolin National Park in 1996–1997. Klify 4:257–263 (in Polish)
55. Kubiak J (2001) Hydrochemistry of the Wolin Island Lakes. Folia Agric Stettin 218(28):63–76
56. Machula S, Choiński A, Ławniczak AE (2012) Woliński National Park. In: Water in national parks of Poland. Institute of Geography and Spatial Management of the Jagiellonian University Cracow, pp 360–375 (in Polish)
57. EU: Water Framework Directive No. 2000 60 EC. Parliament and Council of the European Union (2000)
58. Directive 76/160/ECM
59. Directive 91/676/ECM
60. Directive 91/271/ECM
61. Directive 92/43/ECM
62. Directive 79/409/ECM
63. Krzymowska – Kostrowicka A (1999) Geoecology of tourism and leisure. PWN, Warsaw (in Polish)
64. Skłodowski J, Sater J, Strzyżewski T (2006) Impact of water tourism activity in forest/waterside ecotones on the example of the Bełdany Lake. SYLWAN 10:65–71
65. Cieszewska A (2008) Environmentally valuable sites protection as a base of landscape structure planning on master plan level. Problemy Ekologii Krajobrazu 21:239–250 (in Polish)
66. Journal of Laws, 2004, No. 92, Item 880
67. Journal of Laws, 2006, No. 129, Item 902
68. Journal of Laws, 2003, No. 80, Item 717
69. Journal of Laws, 2017, Item 1566

Chapter 4
Anthropogenic Water Reservoirs in Poland

Mariusz Rzętała

Abstract Anthropogenic water bodies play an increasingly important role in the development of water resources. In 2018, there were 50 water bodies in Poland with capacities greater than $10\,hm^3$ ($mln\,m^3$) each. Their total retention capacity is slightly more than $3.7\,km^3$, and their total area is $530\,km^2$. Among artificial bodies of water, the following types are most often distinguished: reservoirs impounded by dams, flooded mineral workings, levee ponds, water bodies formed in subsidence basins and hollows, artificial pools and industrial ponds (for some bodies of water, multiple origins can be indicated). Water bodies in Poland function in quasi-natural areas as well as under conditions of varied agricultural or urban-industrial human pressure. Their environment determines the course of limnic processes which are characteristic of lentic waters, e.g. water circulation, water level fluctuations, thermal processes, oxygen conditions, ice phenomena, changes in water fertility, salinity levels and others. Artificial water bodies in Poland mostly serve numerous functions despite the many environmental protection problems related to the lentic water environment, e.g. eutrophication, salinisation. Together with their immediate surroundings, they fulfil important natural and landscape roles. Storage reservoirs are used primarily for economic purposes such as water supply, flood protection, recreation and tourism, the breeding of fish and other aquatic organisms, energy production, inland transport, the extraction of mineral resources, etc.

Keywords Water body · Reservoir · Water balance · Limnic processes · Thermal conditions · Ice phenomena · Oxygen conditions · Eutrophication · Salinisation · Flood control · Water supply · Leisure · Recreation · Poland

4.1 Introduction

Among the hydronyms which are often used when classifying the origins of lentic waters, there are also "ponds" (without their origins being clearly distinguished; this

M. Rzętała (✉)
Institute of Earth Sciences, Faculty of Natural Sciences, University of Silesia in Katowice, Będzinska 60, 41-200 Sosnowiec, Poland
e-mail: mariusz.rzetala@us.edu.pl

© Springer Nature Switzerland AG 2021
M. Zeleňáková et al. (eds.), *Quality of Water Resources in Poland*, Springer Water,
https://doi.org/10.1007/978-3-030-64892-3_4

classification is mostly made on the basis of their small area) and "storage reservoirs", which denote a water body type [1–3]. The term "pond" is useful in hydrobiological classifications, where it is used to denote a natural or artificial "inland, shallow and usually freshwater body in which, unlike in a lake, it is not possible to distinguish the deep water (profundal) zone or the open water (pelagic) zone, and the entire bottom of the water body is usually overgrown with macrophytes, and thus the only ecological zone present is the littoral one" [4]. The meaning of the term "storage reservoir" is also ambiguous in terms of the criterion of the basin's origin. Storage, i.e. temporary retention of water, maybe an attribute of any limnic object, and thus the term "storage reservoir" can be considered to refer more to the hydrological characteristics of a body of water, its natural significance or economic function rather than its origin [3, 5].

The most popular criteria for classifying water bodies concern their morphogenesis (basin origin), surface area, capacity and practical significance (tasks, functions), etc. Other classification criteria are used as well by analogy to the rules of classifying lakes in terms of their thermal, dynamic and trophic characteristics.

4.2 Origin of Water Bodies

The formation of artificial water bodies is directly or indirectly related to human activity (Fig. 4.1), and their typology is usually based on the origin of the basin in which the water accumulates (morphogenesis). Among artificial bodies of water, the following types are most often distinguished: reservoirs impounded by dams, flooded mineral workings, levee ponds, water bodies formed in subsidence basins and hollows, artificial pools and industrial ponds (e.g. with concrete, earth or metal basins). For some bodies of water, multiple origins can be indicated, e.g. in the case of those created as a result of river regulation, lakes impounded by dams, etc.

Reservoirs impounded by dams (often referred to as lakes impounded by dams) are formed as a result of damming up river waters (Fig. 4.2). Dams are constructed in locations where geological and morphological conditions of the river valley are favourable for the erection of such structures. The morphometric characteristics of reservoirs impounded by dams depend on the surface areas and retention capacities of the valleys flooded with river water. Depth distributions of reservoirs of this type are determined by the shapes of river valleys and by their longitudinal and transverse gradients [6–8]. The deepest point of the basin is situated near the bottom outlet of the dam, and the reservoir becomes shallower towards its edges which are adjacent to valley slopes and also towards the backwater zone. Reservoirs impounded by dams have precisely defined tasks already at the stage when their construction is planned, and a precondition for their proper operation is that a sufficient amount of water is fed into the reservoir [9]. The largest reservoirs impounded by dams in Poland are found within the valleys of major rivers—the Vistula River and its tributaries, and also within the tributaries of the Oder River [10]. Most of these can be found in the mountains in the south of Poland and also in the lake districts in the northern part of

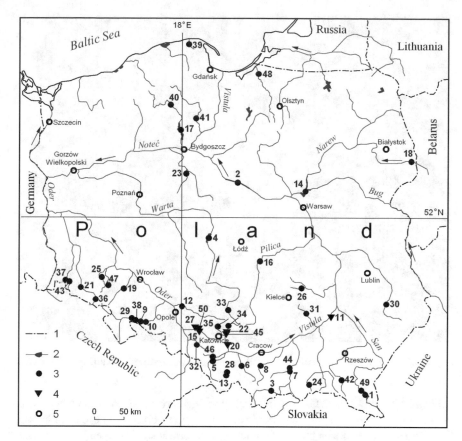

Fig. 4.1 Important artificial water reservoirs in Poland (water body designations—see Table 4.1): 1—watercourses and bodies of water; 2—state border; 3—important localities

the country. There are around a dozen reservoirs impounded by dams in Poland with maximum areas of more than 10 km^2 (Table 4.1). Twelve of those have capacities greater than 100 hm^3, and in the case of the largest one (the Solina Reservoir on the San River), this figure reaches 472.4 hm^3.

In Poland, flooded mineral workings are commonly found whose formation was determined by the locations and dimensions of depleted mineral deposits and also by the scope of works carried out as part of land reclamation and development in order to remedy the environmental damage caused by opencast mining operations (Fig. 4.3). These workings have common names which derive from the type of the material that was extracted at the location in question, e.g. flooded sand/gravel/clay/peat pits. There are also water bodies which formed in the hollows left after the opencast mining of granite, limestone, dolomites, sulphur and hard coal. The peculiar characteristics of flooded mineral workings result from the fact that their morphometry and bathymetry (shape, size, depth distribution) reflect the dimensions of the depleted

Fig. 4.2 Czorsztyn Reservoir on the Dunajec River—the view from the front dam Photo: M. Rzętała

deposit. Although small flooded mineral workings are most common, there is an increasing number of large water bodies of this type in Poland. These have capacities of up to several dozen hm^3 and are located in former sand pits (e.g. the Dzierżno Duże, Dziećkowice, Kuźnica Warężyńska Reservoirs) and at locations where sulphur (e.g. the Machowski Reservoir) or lignite (e.g. water bodies near Konin) used to be mined, etc. The end of lignite mining in Bełchatów will involve the construction of the largest flooded mineral working in Poland with a target capacity of around 3 km^3 and an area of nearly 39 km^2 whose average and maximum depths will exceed 78 metres and 200 metres, respectively [3].

Levee ponds are anthropogenic water bodies formed by constructing low levees (earth embankments) which enable water to be retained. These occur mainly in river valley bottoms and are rarely constructed on valley slopes or plateaux. In terms of their morphometric features, levee ponds are distinguished by the flat bottoms of their basins (which are the result of small height differences on floodplains) and their limited depth (usually less than one metre); the latter characteristic usually results in mean depth to maximum depth ratios which are close to one. Despite the limited retention capacities of individual levee ponds, they play a very significant role in shaping the landscape. They are also a very important part of small retention.

Water bodies in subsidence basins and hollows are an unintended consequence of economic activity and land subsidence or collapse (Fig. 4.4). Where mineral resources are extracted from shallow depths, discontinuous deformations usually

Table 4.1 The largest artificial water bodies in Poland in 2018

Reservoir[a]	Catchment	Water level	Commissio-ning date	Key roles[g]	Maximum area [km^2]	Total capacity [hm^3]
1	Solina[b]	San	60.0	1968	R, E, F	22.0	472.4
2	Włocławek[c]	Wisła	12.7	1970	F, E, R	75.0	453.6
3	Czorsztyn[b]	Dunajec	54.5	1997	F, R, E	12.3	231.9
4	Jeziorsko[b]	Warta	11.5	1986	F, S, E, R	42.3	202.0
5	Goczałkowice[b]	Mała Wisła	13.0	1956	S, F, T	32.0	161.3
6	Świnna-Poręba[b]	Skawa	50.0	2017	S, F, R, B, E	10.3	161.0
7	Rożnów[b]	Dunajec	31.5	1942	F, E, S, R, T	16.0	159.3
8	Dobczyce[b]	Raba	27.9	1986	F,S,E	10.7	141.7
9	Otmuchów[b]	Nysa Kłodzka	18.4	1933	F, E, R	20.6	130.5
10	Nysa[b]	Nysa Kłodzka	13.3	1971	F, R, E	20.7	124.7
11	Machowski[e]	Wisła	0.0	2009	R	4.5	111.2
12	Turawa[b]	Mała Panew	13.6	1948[d]	F, R, E	20.8	106.2
13	Tresna[b]	Soła	23.8	1967	E, R, F,	9.6	96.1
14	Dębe[c]	Narew	7.0	1963	T, E, S, R, F	33.0	96.0
15	Dzierżno Duże[e]	Kłodnica	11.2	1964	T, F	6.2	94.0
16	Sulejów[b]	Pilica	11.3	1973	F, E, R	23.8	84.3
17	Koronowo[b]	Brda	22.0	1960	R	15.6	80.6
18	Siemianówka[b]	Narew	7.0	1991	S, E, B, R	32.5	79.1
19	Mietków[b]	Bystrzyca	15.3	1986	F, E, S, R	9.1	71.9
20	Dziećkowice[e]	Przemsza/Soła	14.5	1976	S, R	7.1	52.5
21	Pilchowice[b]	Bóbr	46.7	1912	F, E, R	2.4	50.0

(continued)

Table 4.1 (continued)

Reservoir[a]		Catchment	Water level	Commissio-ning date	Key roles[g]	Maximum area [km^2]	Total capacity [hm^3]
22	Kuźnica Warężyńska[e]	Przemsza	2.3	2005	F, R	4.8	46.3
23	Pakość[b]	Noteć Zachodnia	4.8	1974	F, S, R	13.0	42.6
24	Klimkówka[b]	Ropa	33.3	1994	E, F, R, S	3.1	42.6
25	Słup[b]	Nysa Szalona	19.1	1978	S, F	4.9	38.7
26	Wióry[b]	Świślina	23.4	2007	F, E	4.1	35.0
27	Pławniowice[e]	Potok Toszecki	2.2	1975	R, T	2.4	29.2
28	Porąbka[b]	Soła	21.2	1936	E, R, F	3.3	27.2
29	Topola[b]	Nysa Kłodzka	7.8	2003	F, E, R	3.4	26.5
30	Nielisz[b]	Wieprz	9.6	1997[d]	E, R	8.3	25.6
31	Chańcza[b]	Czarna Staszowska	12.8	1985	R, F, S	4.7	24.2
32	Rybnik[b]	Ruda	11.8	1972	E, R, F,	4.6	23.5
33	Poraj[b]	Warta	12.0	1978	S, R, F	5.1	20.8
34	Przeczyce[b]	Przemsza	12.5	1963	R, F, S	4.7	20.4
35	Kozłowa Góra[b]	Brynica	6.5	1939	S, F	5.8	17.6
36	Bukówka[b]	Bóbr	22.4	1987	S, F, E	2.0	16.8
37	Leśna[b]	Kwisa	35.8	1907	R, E	1.4	16.8
38	Kozielno[b]	Nysa Kłodzka	8.0	2003	F, E	3.5	16.4
39	Żarnowiec[f]	Piaśnica	16.0	1983	E	0.9	16.4
40	Mylof[b]	Brda	10.4	1972[d]	E, R	26.0	16.2
41	Żur[b]	Wda	15.5	1929	R, E	3.0	16.0
42	Besko[b]	Wisłok	30.0	1978	F, S, R	2.1	14.2
43	Złotnicki[b]	Kwisa	22.0	1924	E, R	1.2	12.4
44	Czchów[b]	Dunajec	10.0	1948	F, E, R, S	2.4	12.0
45	Pogoria III[e]	Pogoria	0.0	1976	R, S	2.1	12.0
46	Łąka[b]	Pszczynka	6.0	1986	F, S	3.4	12.0
47	Dobromierz[b]	Strzegomka	24.0	1986	F, S	1.0	11.4

(continued)

Table 4.1 (continued)

Reservoir[a]	Catchment	Water level	Commissio-ning date	Key roles[g]	Maximum area [km²]	Total capacity [hm³]
48 Pierzchały[b]	Pasłęka	(–)	1936	E, R	2.4	11.5
49 Myczkowce[f]	San	15.0	1961	E	2.0	11.0
50 Dzierżno Małe[e]	Drama	(–)	1938	T, R, F	1.2	10.0

Explanatory notes (–) no data available, [a]ranked in descending order of total capacity (at maximum water level); [b]reservoir impounded by dam, [c]barrage; [d]reservoir commissioning year after modifications; [e]flooded mineral working; [f]upper reservoir of a pumped storage power plant; [g]economic role (R—tourism and recreation, E—energy generation, F—flood protection, S—water supply, T—transport or the improvement of navigation conditions, B—fisheries management). (1 hm³ = 1 mln m³)
Source [11]; amended and supplemented

Fig. 4.3 Pogoria III Reservoir—an example of a flooded mineral working Photo: M. Rzętała

occur, and depressions, ditches, sinkholes, crevices or thresholds are formed. On the other hand, so-called continuous deformations are formed where the overlying rocks are plastic, or the mining takes place at greater depths; these result in subsidence troughs being formed whose shape is roughly circular or elliptical and their morphometric characteristics depend on factors related to the mining methods used

Fig. 4.4 Żabie Doły on the Silesian Upland—an example of water bodies in subsidence basins
Photo: M. Rzętała

and geology. At the bottom of depressions created in this manner, wetlands and water bodies form. Their emergence is usually contributed to by the continuous subsidence processes, the inflow of groundwater as well as the sheet flow of rainwater and melt-water. Numerous water bodies of this type are present in mining areas, e.g. in the Upper Silesian Coal Basin [12]. Water bodies in subsidence basins and hollows are highly dynamic in terms of their numbers and surface areas [13]. In hydrological terms, these water bodies are either endorheic or rheolimnic (with a limited water retention period). They are usually up to a few metres deep and have surface areas of up to several hectares.

Another type of anthropogenic bodies of water is artificial pools. These are water-filled cavities dug in the ground, which are constructed as a result of social or economic demand for water retention. The ground excavated during the construction of an artificial pool is treated as (reusable) waste, which is a fundamental distinction between such pools and flooded mineral workings whose origins are inseparably connected with the extraction of mineral deposits. Despite the fact that they are progressively liquidated, artificial pools continue to prevail in rural areas.

Typically anthropogenic water bodies are various industrial ponds—these are reservoirs with man-made basins made of concrete or other materials (Fig. 4.5). The morphometric characteristics of such industrial ponds include basins with bottoms which are flat or have uniform gradients, and also the presence of smooth (vertical or

Fig. 4.5 Łosień Reservoir in Dąbrowa Górnicza—a link in the chain of water transfer for industrial purposes Photo: M. Rzętała

steep) walls. Their shores often have regular shapes. This type of ponds can usually be found in urban and industrial areas, near roads, etc. These are usually facilities which have been specially constructed to fulfil specific roles.

Basins of many water bodies were formed as a result of multiple, complex conditions and factors. A case in point are bodies of water created as a result of river regulation—curved former sections of river channels which were cut off as a result of the regulation works conducted in the valley floor. These are classified as anthropogenic lakes because the change from the river environment to the limnic one was caused by technical (regulation) measures rather than by fluvial processes that occurred naturally, despite the fact that the basins of such water bodies were shaped by natural meandering processes. Water bodies created as a result of river regulation have retained many morphometric, biotope and biocoenotic characteristics in common with ox-bow lakes, which are natural bodies of water. They are quite common in valleys of major regulated rivers, e.g. in the Vistula River valley. Lakes impounded by dams are found mainly in northern Poland. These are most often lakes of glacial origins whose basins were adapted in order to retain more water by constructing embankments or water discharge and damming structures. One of the largest lakes of this type in Poland is the Pakość Reservoir.

4.3 Number of Water Bodies and Morphometric Parameters

The number of artificial water bodies is difficult to determine, mainly due to the fact that we do not know how many small water storage facilities there are. Inventories of water bodies carried out within some catchments or within individual administrative units point to rapid changes in the number and capacity of such reservoirs. The number, water area and retention capacity of the largest anthropogenic water bodies have also been increasing steadily. In 1998, there were 40 water bodies in Poland with capacities greater than 10 hm^3 each, which had a total capacity of 3.2 km^3 and an area of 463 km^2, and there were another 61 water bodies with capacities ranging from 1 to 10 hm^3 each [14]. Two decades later, there are already 50 water bodies with a capacity greater than 10 hm^3 each. Their total retention capacity is slightly more than 3.7 km^3, and their total area is 530 km^2 (Table 4.1). Choiński and Skowron [14] state that the total capacity of water bodies (with capacities greater than 1 hm^3) in Poland accounts for around 18% of the water resources retained by Polish lakes and around 6% of the volume of water flowing out of the territory of Poland annually, and the total area of these water bodies accounts for nearly 18% of the total area of lakes in Poland. The total area of all artificial water bodies in Poland is estimated at around 1,000 km^2, and the corresponding water resources amount to several km^3 [14].

Some areas in Poland deserve to be called anthropogenic lake districts owing to the number of water bodies present there and their total area [15, 16]. This label is used to denote the Muskau Bend—an area situated close to the border in south-western Poland where there are many such water bodies in the former lignite mining area [17]. A similar name is also given to the Silesian Upland and the Oświęcim Basin together with the outskirts of the neighbouring regions in southern Poland [18]. The Upper Silesian Anthropogenic Lake District is the largest such district in Poland (Fig. 4.6). It covers an area of 6,800 km^2 and includes approximately 4,700 water bodies. There are 70.5 bodies of water per 100 km^2 in the area, its lake density amounts to 2.74%, and the total water surface area is 185.4 km^2 [18]. In the 1960s, only 3,100 water bodies with a total area of 113.4 km^2 were present within the Upper Silesian Anthropogenic Lake District. Lake density amounted to 1.68%, and there were on average 45.77 bodies of water per 100 km^2 [19]. Peculiar features of the Upper Silesian Anthropogenic Lake District include the diverse origins of the water bodies found there and the high urbanisation level of the area (the Lake District coincides, among others, with the Upper Silesia-Dąbrowa Basin Metropolis).

4.4 Selected Issues of Water Balance

The water balance of anthropogenic water bodies is highly complex, which results from the complexity of both natural (e.g. climate) and anthropogenic (e.g. water transfers, dynamic changes in water body surface) conditions that determine this

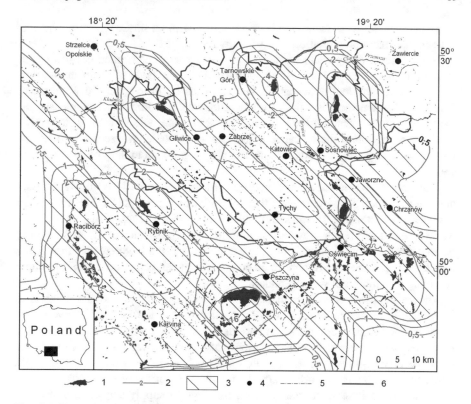

Fig. 4.6 Upper Silesian Anthropogenic Lake District ([18]; amended and supplemented): 1—watercourses and bodies of water, 2—contour lines reflecting the percentage share of water body surface per 100 km², 3—the extent of the anthropogenic lake district, 4—important localities, 5—state border, 6—boundary of the Upper Silesia-Dąbrowa Basin Metropolis

balance [20]. In the case of anthropogenic water bodies, usually unlike in lakes, the natural water cycle is subject to changes resulting from the performance of current water and sewage management tasks, which are directly related to land use in the area in question. The water cycle is commonly modified in both quantitative and qualitative terms, although some bodies of water have already been assimilated into the surrounding environment and have hydrological regimes which are similar to lakes.

The annual supply of water to these water bodies from rainfall is estimated at about 640 mm. Absolute rainwater supply values differ since areas of the water bodies in question are different as well. The percentage share of precipitation in the total supply of water to lake basins ranges from a few to several dozen percent, and this variation is mainly due to the size of the water body in question and the amount of surface water inflow, although it also depends on the amount of underground supply.

The annual inflow of surface water from the catchment area to the water bodies varies to a similar extent. As a rule, the share of surface water inflow in total supply is greater than 80%. An exception is water bodies without surface inflows where all water comes from precipitation and groundwater supply.

In the case of many water bodies, surface outflow is determined not just by the amount of water released or flowing naturally out of the body of water in question, but also by the amount of water sourced from the basin to meet the water needs related to economic activity. Total surface outflows (including artificial releases and the water sourced) account for more than 90% of water loss in most water bodies (with the exception of the so-called endorheic water bodies where water losses are mostly caused by evaporation and sometimes by significant underground outflow as well).

The loss of water caused by evaporation is on average around 440 mm. Water losses due to evaporation are higher where thermal pollution is present [18, 21] or the water body exhibits intensive overgrowth resulting in increased evapotranspiration [22]. Nevertheless, the amount of water evaporating from the surface of the vast majority of flow-through water bodies is usually estimated at just a few percent of the water lost, although in the case of endorheic bodies of water it is the most important item depleting the amount of water retained.

The amount of water retained in the basin varies widely for individual water bodies in Poland; the maximum value is 472.4 million m^3 for the Solina Reservoir. The water retained is also replenished in various ways. In this respect, the Dziećkowice Reservoir is exceptional. The most important source of water supply to this lake since the very beginning of its operation has not been its own river catchment but rather water transfers from the Carpathian tributaries of the Vistula River, i.e. the Soła and Skawa Rivers. In annual terms, this type of supply accounts on average for 86% of the total water supplied to the reservoir [23]. Precipitation, inflow from the river catchment and from the direct catchment of the reservoir and underground supply account for around 14% on average. In the 1970s, around 20 hm^3 of water were supplied annually by way of such transfers. In the next few years, this amount varied from 30 to more than 50 hm^3 per year. In the late 1980s and in the early 1990s, an increase in water supply to nearly 70 hm^3 followed. The maximum level of water supply from the Soła River to the reservoir was reached in the mid-1990s (in 1995, it was 92 hm^3). From the very beginning of its operation, the main user of the Dziećkowice Reservoir was the Katowice Steelworks, which also transferred water from the reservoir to other external users such as the coking plant and water treatment plants using its own water transfer system.

4.5 Selected Limnic Processes

The functioning of water bodies involves limnic processes. These are processes characteristic of stagnant waters which result in changes in the physical, chemical and biological conditions within water bodies. These include variations in water cycle conditions, water mass dynamics, water temperature and oxygenation and ice phenomena as well as changes in water fertility, shore processes, bottom sediment formation and water body longevity. Among all limnic processes which take place in water bodies in Poland, of particular importance from the point of view of the economic utility of these water bodies are variations in thermal conditions, ice phenomena, water oxygenation, eutrophication and water salinity.

4.5.1 Thermal Conditions

The thermal conditions prevailing in anthropogenic water bodies are influenced by both natural and man-made factors; natural thermal systems have only been significantly transformed in some bodies of water. Thermal systems typical of the temperate climate zone predominate. Water bodies with depths of at least a few (or more than ten) metres are dimictic with a clear and typically anothermal stratification in summer, catothermal stratification in winter and two homothermal periods. Shallow water bodies with a medium depth not exceeding 2 to 4 metres are polymictic with more frequent occurrences of homothermal stratification, which is caused mainly by wind mixing.

A clear example of human impact on lentic waters is water bodies burdened with so-called thermal pollution, which exhibit clearly increased or lowered temperature (in relation to the natural temperature of standing or flowing waters under given conditions) caused by human activity, e.g. heating caused by the inflow of water, sewage or sediments. A common phenomenon is the burdening of water bodies with municipal, industrial or household sewage, which raises water temperature in zones where the sewage flows into the water. Just a few water bodies in Poland are subject to extreme anthropogenic thermal loads. Their thermal regimes are primarily shaped by discharges of heated water, and the thermal characteristics of their water have been completely transformed [21]. Examples of water bodies which retain heated water are the reservoirs near Konin or the Rybnik Reservoir. Water is sourced from the Rybnik Reservoir, it is heated in the process of cooling the power units of a conventional power plant and discharged back into the reservoir basin, which causes water temperature to rise to approximately 6 °C in winter and to 20 °C–35 °C in summer depending on the exact sector of the reservoir. The temperature of water in the reservoir is even 10 °C higher than the average temperature of other bodies of water (Fig. 4.7). As a result of its water being heated, the reservoir permanently exhibits a stratification similar to the anothermal one with varying stability with respect to its vertical range and spatial distribution (in the absence of discharges of heated water, spring homothermia, anothermia, autumn homothermia and catothermia would reflect thermal transformations within the reservoir). Therefore, the reservoir may be described as an anthropomictic and anthropothermal one [18].

Local human pressure caused by storing waste or sewage within the water body basin may result in meromixia; this leads to a peculiar vertical temperature distribution, which is present, e.g. in the Górka Reservoir in Trzebinia. The reason for the

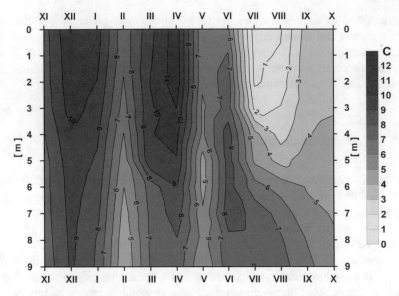

Fig. 4.7 Anthropothermy in Rybnik water reservoir, i.e. caused by anthropopression increase in water temperature against a background of normal thermal situation, i.e. thermal limnological background—thermoisopleths expressed in °C ([18]; amended)

occurrence of the few meromictic stratification systems in water bodies in Poland is the high water density caused exclusively by man-made factors (waste storage, supply of high-density pollutants), which, when combined with the small area and usually considerable depth of the water body in question and the presence of land features which shield it result in the permanent stagnation of its waters [24].

4.5.2 Ice Phenomena

Ice phenomena are construed as the occurrence of ice in water irrespective of its structure, form and duration [25]. The issue of icing of lakes and other water bodies has been relatively rarely discussed in the Polish limnological literature. The most frequently cited research results concerning limnic ice phenomena include data from studies conducted in individual water bodies, less frequently regional studies or analyses of ice phenomena in lakes carried out within a broader context. Of all the ice phenomena observed in water bodies, the presence of a complete ice cover is the one most readily noticed [26].

In water body, the vertical position of ice cover changes as a consequence of the freezing of water resulting in ice thickness increasing from below and from above and also as a result of ice melting, and ice cover evolution during the winter season depends mainly on so-called external factors (mainly air temperature changes and precipitation). The ice cover in water bodies changes its location during the winter season depending on air temperature, the amount of precipitation and the retained volume of limnic waters. The main factor determining the vertical variation in the structure of the ice cover is precipitation (the snowier the winter, the more complicated the structure of the ice cover) (Fig. 4.8).

The spatial variation in ice cover thickness in a water body is a consequence of the influence of natural factors (e.g. spatial differences in the pressure of snow on ice which translates into superimposed ice accretion, the circulation of water under ice, the amount of heat released from sediments in particular areas of the water body) as

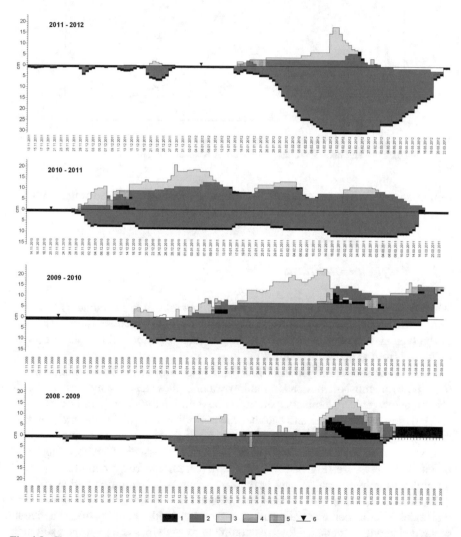

Fig. 4.8 Changes in the ice cover of the reservoir in Czeladź in the winter season 2008–2012 ([27]; suplemented): 1—water, 2—ice, 3—snow, 4—snow saturated with water, 5—snow ice or remaining ice supersaturated with water, 6—height of the upper part of the ice cover at the beginning stage

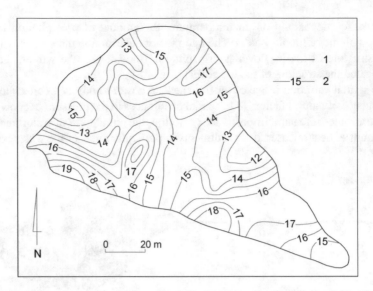

Fig. 4.9 Variation in ice cover thickness in Czeladź water reservoir in January 15, 2010 ([28]; suplemented): 1—water body shoreline, 2—ice cover isopachytes [cm]

well as anthropogenic factors (e.g. discharges of thermally polluted waters and the operation of installations which oxygenate the water body) (Fig. 4.9).

4.5.3 Oxygen Conditions

Water bodies in Poland represent different patterns related to gas contents of water. These are directly related to the scale of human influence, which is expressed mainly as the type and amount of sewage discharged into water or soil and the inflow of pollutants from agricultural sources.

Full oxygen saturation (around 100%) or at least normal oxygen saturation levels (60–100%) are most often found in shallow water bodies that are polymictic and thus also homooxygenic. A similar situation can be found in dimictic water bodies, especially during homothermic and catothermic periods when good water oxygenation is favoured by higher solubility of oxygen in water from autumn to spring.

Oxygen shortfall or total disappearance is clearly associated with agricultural or industrial pressure (Fig. 4.10). In extreme cases, anaerobic conditions prevail throughout the hypolimnion, and at the same time supersaturation and significant daily fluctuations of oxygen content are present in the epilimnion, with an oxycline that exhibits considerable gradients at a shallow depth. Equally often, deficient oxygenation with a tendency towards complete oxygen loss and hydrogen sulphide secretion is present during the catothermic period as a result of oxygen consumption by oxidation processes and in the absence of oxygen supply (where a complete

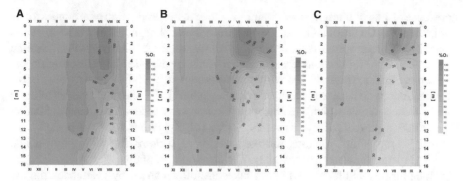

Fig. 4.10 Mean saturation of limnic waters by oxygen in the hydrological years 1998–2007 ([18]; amended): A—Pogoria III water reservoir (reservoir effectively insulated from the inflow of pollutants), B—Pławniowice water reservoir (agricultural catchment), C—Dzierżno Duże water reservoir (urban-industrial catchment)

ice cover is present). In turn, water supersaturation with oxygen results from its mechanical oxygenation (e.g. through cascade flow, wind mixing and locally as a result of motor boats passing), but in the near-surface layer it is most commonly a consequence of the development of eutrophication processes, which lead to a transformation in the plant and animal species present and also the growth of phytoplankton on a massive scale, i.e. so-called algal blooms. It is to the presence of phytoplankton organisms that the summer supersaturation of the near-surface water layer with oxygen is attributed (through intensive photosynthesis), which is accompanied by a decrease in oxygen content to values even below 10% in the bottom zone (reduced light, oxygen consumption through biochemical and chemical processes); the latter endangers the life of aerobic organisms and activates undesirable processes of anaerobic decomposition of organic matter under anoxic conditions [25].

4.5.4 Eutrophication

Eutrophication plays an important role in the evolution of water bodies in Poland [29, 30]. Undesirable growth in lentic water fertility results in water body basins is becoming shallower and finally in their disappearance [25]. Eutrophication processes are particularly intense in anthropogenically transformed areas, which is contributed to by sewage discharges, the intensification of agriculture, deforestation, air pollution, etc. Where lentic water becomes eutrophicated on a large scale, many natural processes (thermal, aerobic, oxidation and reduction, sedimentation, sedentation processes, etc.) are modified, and eutrophication also makes the operation of reservoirs and the use of their shores difficult [18, 31].

The trophic status of lentic water is primarily determined by phosphorus and, to a small extent, nitrogen content [29, 32]. Kajak [29] cites the statement that the presence of 1 kg of phosphorus in water may lead to the production of around 1 tonne of fresh algae. On the other hand, the concentration of phosphorus at which so-called algal blooms may appear (as a result of excessive phytoplankton growth) is around $20–30\ \mu g\ P/dm^3$ for lowland reservoirs impounded by dams [33]. In oligotrophic and oligomezotrophic water bodies without anthropogenic pollution loads, phosphorus concentrations do not exceed around a dozen $\mu g/dm^3$ [33]. Threshold values of basic water indices above which the eutrophication process occurs have been defined at the following levels for lentic waters: above $0.1\ mg\ P/dm^3$ for total phosphorus, above $1.5\ mg\ N/dm^3$ for total nitrogen, above $25\ \mu g/dm^3$ for chlorophyll α and below 2 metres for water transparency. The development of this process is evidenced by long-term algal blooms, which in lakes are often caused by cyanobacteria, the growth of periphyton algae on a massive scale, the disappearance of oxygen from the hypolimnion (with possible hydrogen sulphide secretion), and finally the reduction in the diversity and abundance of macrophytes, invertebrate fauna and fish. Most of these indices refer directly or indirectly to the water trophic classifications used worldwide, among which Organization for Economic Cooperation and Development guidelines are a common standard [34].

The content of chlorophyll α is the parameter that best expresses the volume of primary production [35]; in the case of most water bodies, this figure indicates that the water is being enriched with nutrients and intense phytoplankton growth is taking place. In some water bodies burdened with agricultural pollutant runoffs, chlorophyll α reaches average annual concentrations which are only slightly higher than the threshold values, but maximum concentrations have exceeded these values many times. Therefore, it is difficult to find water bodies in Poland which would be similar in terms of their chlorophyll α concentrations to reservoirs impounded by dams that are located in quasi-natural environments, e.g. the Wisła Czarne (1–$1.5\ \mu g/dm^3$) and Wapienica (ca. $1\ \mu g/dm^3$) Reservoirs in the Beskidy Mountains [18].

Concentrations of phosphorus and nitrogen compounds in water bodies in Poland vary greatly—from levels considered natural to those which indicate comprehensive degradation of the aquatic environment; there are also a number of water bodies which represent intermediate ranges of nutrient concentrations [18]. In general, water bodies in urban-industrial catchments exhibit processes typical of polytrophy and hyper-trophy (e.g. the Dzierżno Duże Reservoir) [18, 36, 37], while in water bodies with catchment areas used for agricultural purposes, the eutrophication problem stems mainly from their pollution with nitrogen compounds from agricultural sources, e.g. the Kozłowa Góra Reservoir [38], Sulejów Reservoir [39].

An additional criterion reflecting the eutrophication of water bodies in Poland is water transparency. The most transparent water bodies were found in quasi-natural catchments which were insulated from the inflow of pollutants, and the lowest transparency was found in strongly eutrophicated waters in catchments used for agricultural purposes or in urban-industrial catchments.

Compared to the general trophic status of lentic waters, the hypertrophy and polytrophy of the Dzierżno Duże Reservoir stand out, which is the effect of supply of pollutants from the Kłodnica River [37]. Thus, the environment in which fish and other aquatic organisms in this lake live is rich in nutrients. In the eutrophic environment, there is a small population of predators and quite large specimens of tench, crucian carp and carp are present as well. The fish in the reservoir are unsuitable for human consumption owing to the high content of toxic metals in the meat and liver of some fish species present there, as found by Kostecki [40].

4.5.5 Salinisation

The degrading effect of salinity on the aquatic environment is particularly common in industrialised and urban areas [41]. The presence of large amounts of salt in water has an adverse effect on the development of most plants, but on the other hand, it stimulates the growth of saltwater plant species on the shores of saline anthropogenic lakes. This is the most vivid manifestation of salinity caused by the anthropogenically controlled salt cycle. Complex natural and man-made conditions determine the level of salinity of limnic waters as exemplified by a number of water bodies in Poland, but on the Silesian Upland and its outskirts, the problem of increase in water salinity is exceptionally pronounced [18].

Surface water salinity is typically found in urban and industrialised areas, but the values found in the anthropogenic water bodies of the Silesian Upland and on its outskirts are many times higher due to the supply of saline groundwater, which is caused by the draining of underground mines. While the salinity of surface waters considered natural in the region reflects concentrations of chlorides ranging from around ten to several dozen mg/dm^3, the impact of urbanisation, industrialisation and transport results in an increase in Cl-concentration levels to even several hundred mg/dm^3. As a result of the draining of mines, water reaches the surface with mineralisation levels ranging from around 0.5 g/dm^3 through several or several dozen g/dm^3 to even 372 g/dm^3 in subsequent hydrogeochemical zones of the Upper Silesian Coal Basin sedimentary basin [42]. Saline water is retained in reservoirs near mines, from which it is dosed into surface watercourses within the Vistula and Oder River basins, where this water is diluted, but still causes an increase in mineralisation and thus also salinity levels. Similarly, high water salinity parameters are found in small reservoirs used as settling tanks, waste dumps, decanting tanks, etc., where concentrations of chlorides, sulphates and total dissolved solids reach levels of several thousand mg/dm^3, several hundred mg/dm^3 and more than ten thousand mg/dm^3, respectively.

The heightened salinity of surface waters caused by the inflow of water originating from the mining drainage of salt-bearing Miocene or Carboniferous formations results in an increase in the salinity of the water retained in water bodies, also those located outside mining areas. The largest flooded mineral working in Poland in terms of water storage capacity (from 42 to 94 hm^3), i.e. the Dzierżno Duże Reservoir, exhibits a chloride content of 1.7 g/dm^3, while the hydrogeochemical background for water in the region is several dozen mg Cl/dm^3. On the other hand, the increase in the salinity of water in the Pławniowice Reservoir indicates several possible sources of salt supply, the most probable of which is salt leaching into water from saline

Tertiary sediments present in the bottom of the basin and also from the waste used to backfill the flooded mineral working in the initial period of land reclamation. Such an increase in salinity between reservoir inflow (c—513.0–602.4 μS/cm, Na^+—8.3–19.1 mg/dm^3, Cl^-—17.0–43.0 mg/dm^3) and outflow (c—522.0–658.0 μS/cm, Na^+—21.5–39.2 mg/dm^3, Cl^-—63.2–121.0 mg/dm^3) has not been documented in limnological studies to date and points to the effects of man-made transformations of the environment which create new directions for water migration via old mines which are now filled with water [18]. The process of transfer of saline water outside Upper Silesia via surface watercourses should be considered a highly undesirable phenomenon.

4.6 Role of Water Bodies

Artificial water bodies in Poland mostly serve numerous functions despite the many environmental protection problems related to the lentic water environment [7, 10, 18, 24, 25, 43]. Together with their immediate surroundings, they fulfil important natural and landscape roles, e.g. serving as locations where birds breed and nest, shaping the local climate and microclimate, influencing the water self-purification process and favouring the formation of mineral deposits [10, 44, 45]. Storage reservoirs are used primarily for economic purposes such as water supply (industrial, municipal, agricultural), flood protection, recreation and tourism, the breeding of fish and other aquatic organisms, energy production, inland transport, the extraction of mineral resources, etc. [46].

4.6.1 Flood Control Importance

Flood protection is among the most important functions of anthropogenic reservoirs. In practice, many water bodies are of little or no flood protection relevance owing to their limited or absent water damming potential (such water bodies are very similar to natural lakes). Most water bodies can be made to serve flood protection purposes, but usually, only the largest reservoirs enable the effective reduction of flood waves. It is especially those reservoirs which offer flood control storage, and a large part of their usable storage often remains available during high water stages. The amount of flood control storage is determined to take into account environmental conditions in the water body catchment area. Therefore, it varies and ranges from a few hm^3 (e.g. the Dzierżno Duże, Przeczyce, Porąbka, Poraj, Sulejów, Dębe, Topola, Kozielno, Bukówka Reservoirs) through around a dozen hm^3 (e.g. the Wióry or Turawa Reservoirs) to several dozen hm^3 (e.g. the Goczałkowice, Tresna, Dobczyce, Czorsztyn, Rożnów, Solina, Włocławek, Nysa, Otmuchów Reservoirs).

So-called dry reservoirs and flood polders, in which water is retained only during high-water stages, are also used to control flooding. The largest polders are Buków,

Fig. 4.11 Stronie Śląskie—an example of a dry reservoir in the Sudety Mountains Photo: M. Rzętała

Oława-Lipki, Połupin-Szczawno and Krzesin-Bytomiec on the Oder River. Dry reservoirs exist, for instance, in Stronie Śląskie (Fig. 4.11), Sobieszów, Cieplice, Mysłakowice, Jarnołtówek.

4.6.2 Water Supply

The primary function of reservoirs is water supply and protection against drought. Water from the reservoirs is used for municipal, industrial and (to a lesser extent) agricultural purposes.

In Poland, it was as early as the nineteenth century that the water sourced from rivers and existing water bodies proved insufficient to meet the demand for water, especially in industrialised and urban areas, e.g. in the Bielsko-Biała Industrial District or in the Upper Silesian Industrial District [47]. Already in the first half of the twentieth century, the small Wapienica Reservoir in the Beskidy Mountains was commissioned and started to be used as a water reservoir for the city of Bielsko-Biała. The Międzybrodzie Reservoir with a dam at Porąbka was constructed on the Soła River. In the mid-twentieth century, the water shortfall became so significant that a decision was made to build the Goczałkowice Reservoir on the Vistula River. The

Kozłowa Góra Reservoir within the Przemsza River basin was also adapted for water supply purposes. In the second half of the twentieth century, the Przeczyce, Pogoria III, Dziećkowice, Pławniowice, Łąka Reservoirs and several others were integrated into the water supply system (Fig. 4.12). The largest reservoirs impounded by dams and flooded mineral workings on the Silesian Upland and on its outskirts are major components of the largest water management system in Poland [23].

Similar functions are performed by the Dobczyce, Zegrze, Sulejów Reservoirs and many others. First of all, these reservoirs store the water which is subsequently transferred using pipelines for industrial and municipal needs, or they provide the flow required for the operation of infiltration (surface or groundwater) intakes situated further downstream. Water from other reservoirs (e.g. the Siemianówka and Jeziorsko Reservoirs) is used for irrigation of agricultural land. In agriculture, water is also used on farms and in livestock rearing facilities as well as for irrigation of crops in greenhouses and tunnels, vegetable plantations, mushroom farms, for cleaning agricultural machinery, etc. Agricultural purposes also include filling fish ponds with water. Irrigation in forestry involves the creation of small standing water reservoirs, so-called waterholes with an area of up to 1 ha and a capacity of up to several thousand m^3 (mid-field reservoirs, rural ponds). These are extremely important in terms of

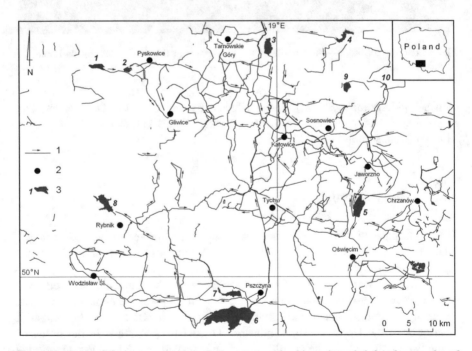

Fig. 4.12 Upper Silesian water-economical system with objects intended for the transfer of pure waters ([18]; amended and suplemented): 1—transport water mains of different standing, 2—more important localities, 3—more important water reservoirs in the sysyem of pure waters transfer (1—Pławniowice, 2—Dzierżno I and Dzierżno Małe, 3—Kozłowa Góra, 4—Przeczyce, 5—Dziećkowice, 6—Goczałkowice, 7—Łąka, 8—Rybnicki, 9—Pogoria III, 10—Łosień)

improving water supply conditions, ensuring environmental protection and enabling small water retention as well as in terms of flood control measures in agricultural and forest areas [3].

4.6.3 Leisure, Recreation, Nature Protection

The recreational significance of water bodies in Poland can be considered in terms of possibilities for organising leisure activities and engaging in sports: sailing, canoeing, fishing, powerboating, etc. Inland water bodies are intermediate or final destinations for tourist migrations, and their significance is frequently considered in terms of sight-seeing, landscape architecture, environmental protection, etc. The development of tourism and recreation around water bodies stimulates the establishment of catering and accommodation, tourist facilities and those related to tourist traffic as well as transport infrastructure [48].

Sailing is one of the most common pastimes associated with artificial water bodies in Poland. Various types of equipment with sails can be used for this purpose, and thus there is e.g. yachting, windsurfing and iceboating. Such water bodies can be used for kayaking, including longer kayaking trips.

Amateur fishing in accordance with legal regulations and the rules determined by fishing associations is also a common pastime which can be pursued on the shores of water bodies at any time of the year.

Water bodies are often used for other recreational purposes: they provide bathing beaches, venues for sports events (e.g. the Malta Reservoir in Poznań), diving facilities (e.g. the Koparki Reservoir in Jaworzno), venues for powerboat racing or amateur powerboating (e.g. the Jeziorsko Reservoir) and venues for lakeside recreation with the use of facilities such as cable skiing systems (Fig. 4.13) as well as providing so-called silent zones.

4.6.4 Breeding of Fish and Other Aquatic Organisms

The breeding of fish, and occasionally also of other aquatic organisms (e.g. molluscs, crustaceans, edible algae) takes place in breeding ponds (this is referred to as aquaculture). These facilities may operate separately or in clusters (Fig. 4.14). Breeding ponds are generally characterised by high water fertility levels and peculiar water management methods, which consist in filling their basins in spring and emptying them in autumn. In aquaculture, a number of hydraulic structures operate which are typical of hydrological facilities of this type.

Aquaculture most often involves the inland breeding and farming of salmonid fish (e.g. trout, salmon, grayling); ponds for cyprinid and similar fish, in which e.g. carp, pike, zander, crucian carp, tench and roach are bred and farmed; so-called

Fig. 4.13 Stawiki Reservoir in Sosnowiec—an example of recreational use of the reservoir Photo: M. Rzętała

recirculating aquaculture systems for breeding and farming e.g. sturgeon or African sharptooth catfish.

The largest complex of fish ponds in Europe with an area of about 8,000 hectares is the Milicz Ponds in the Barycz River valley. There are also many such ponds in the Mała Wisła River valley (from the vicinity of Skoczów to the vicinity of Oświęcim).

4.6.5 Other

Another use of lentic waters involves the production of electricity in a technological process (i.e. as a renewable energy source). Water from the reservoirs is used to drive electric generators at pumped storage power plants situated next to them (at Żarnowiec with a capacity of 716 MW, Porąbka-Żar—500 MW, Solina—200 MW, Żydowo—150 MW, Niedzica—93 MW, Dychów—90 MW) as well as at flow-through hydropower plants (e.g. Włocławek—162 MW, Żur—9 MW, Czchów—8 MW). More efficient use of flowing water as an energy resource can be made where a group of several neighbouring water reservoirs is used, such as in the case of the Soła River cascade with three hydropower plants, including two flow-through hydropower plants (in the Żywiec Reservoir dam and the Porąbka Reservoir dam)

Fig. 4.14 Stawy Podgórzyńskie in the Jelenia Góra Basin—a breeding pond complex Photo: M. Rzętała

and one pumped storage power plant with its upper reservoir on the Żar Mountain in the Beskid Mały range and the Porąbka Reservoir as the lower reservoir [49]. Of indirect importance for energy generation is the use of reservoir waters for cooling power units at conventional power plants (Fig. 4.15). Water sourced from the reservoir as the cooling medium is discharged back into it after being used in the power plant installation, which results in thermal pollution. A model example of such use of water bodies is the Rybnik Reservoir whose waters are used to cool the nearby power plant, and the resulting flow sometimes exceeds 30 m³/s.

Reservoirs also play a role from the point of view of transport. This is, however, limited both in terms of transporting people and goods. This function usually consists of arranged tourist and recreational cruises on individual reservoirs, e.g. the Solina, Żywiec or Czorsztyn Reservoir. Anthropogenic lakes play an indirect transport role, supplying inland waterways with water. For example, the Dzierżno Duże, Dzierżno Małe and Pławniowice Reservoirs within the Kłodnica River basin supply the Gliwice Canal with water through a system of hydraulic structures and improve navigation conditions on the Oder River.

The extraction of mineral resources from water body basins is usually of minor economic importance. The period for which these reservoirs have existed is too short for significant deposits of mineral resources to have formed within them. Nowadays, vegetation growing on the banks is sometimes used for economic purposes (e.g. as

Fig. 4.15 Rybnik Reservoir—water source for CHP plant cooling purposes Photo: M. Rzętała

wicker). Debris deposits in backwater zones of some reservoirs, e.g. at the place where the Dunajec River flows into the Czorsztyn Reservoir, are most commonly mined. The minerals extracted are used to produce various types of aggregate, which is most often employed as a raw material in road construction or for constructing single-family residential buildings. Of lesser economic importance are bottom sediments removed as part of reclamation measures or during the dredging or cleaning of water body basins. Such activities are periodically carried out in the backwater zone of the Dzierżno Duże Reservoir in the south of Poland [50]. Within that reservoir, a delta has formed, which mainly consists of sewage sludge and coal dust. The area of the delta has been estimated at around 1 km^2 during low water stages, and it may be more than ten metres thick. The volume of the delta alone is estimated at slightly more than 2 million m^3, and this is compounded by more than an additional 2 million m^3 of bottom sediments accumulated in its immediate vicinity. The entire material deposited meets the criterion of an anthropogenic deposit, and this deposit is mined as a source of high-energy fuel for the local CHP plant.

Other economic functions of water reservoirs most commonly include the following: fire protection, compensation—water level stabilisation (Fig. 4.16), anti-debris barriers, industrial (e.g. settling tanks for various types of water), municipal (e.g. ponds at sewage treatment plants), military or defensive (e.g. the Kozłowa Góra Reservoir in the Przemsza River basin, which was constructed in the 1930s as part of fortifications of the Fortified Area of Silesia).

Fig. 4.16 Sromowce Wyżne Compensation Reservoir—the view from the Czorsztyn Reservoir dam Photo: M. Rzętała

4.7   Conclusion

Water bodies in Poland function in quasi-natural areas as well as under conditions of varied agricultural or urban-industrial human pressure. Their environment determines the course of limnic processes which are characteristic of lentic waters, e.g. water circulation, water level fluctuations, thermal processes, oxygen conditions, ice phenomena, changes in water fertility and salinity levels. The course of these processes as well as the manner in which these water bodies are used and their natural and socio-economic significance have been modified by changes in natural conditions and also by human activity in the last decades.

Basically, two types of water bodies can be distinguished on the basis of their origins and the limnic processes taking place. Water bodies of the first type retain the attributes of lakes where such processes occur under the influence of natural conditions, e.g. thermal processes, phenomena related to the presence of oxygen and ice in quasi-natural reservoirs. The second type, which is largely influenced by human activity, only retains some of those attributes. In water bodies of the second type, the following phenomena may occur: disturbances in the natural variability of water inflow and outflow, changes in the volumes of water retained as a result of human activity, an increase in water temperatures and limited occurrence of ice phenomena, advanced eutrophication, deterioration of oxygen conditions, prevalence of high salinity and a number of other consequences as well.

Such large spectrum of quantitative-qualitative changes in reservoirs of lentic waters at their large numerical force identified with the multiplicity of scenarios of their functioning and use, creates the possibility to predict the course of degradation processes of limnic waters quality in the objects, which contemporarily function under incomparably smaller environment transformation and to undertake the protection and recultivation activities.

4.8   Recommendations

Knowledge of the issues related to the functioning of water bodies in Poland (e.g. their occurrence and number, morphological and hydrological origins, limnic processes and socio-economic importance) is indispensable at various stages of education (e.g. in the field of earth and environmental sciences) as well as can be leveraged for economic purposes where the measures undertaken are informed by scientific evidence. This is because artificial reservoirs are an integral part of Poland's water resources and their quantitative and qualitative condition affects spatial management within the region in which they are present. Therefore, thorough quantitative and qualitative identification of limnic processes and the socio-economic importance of inland water bodies provides the basis for:

- the assessment of the role of water bodies in the structure and functioning of the overall landscape;
- optimising the use of water bodies and managing the areas in their vicinity (in accordance with the principles of sustainable development, environmental protection and landscape architecture);
- the reclamation, revitalisation and protection of water bodies and the proper use of their catchment areas;
- the liquidation of water bodies by draining and backfilling them with soil and the subsequent land reclamation and development efforts;
- the construction of new reservoirs.

References

1. Dynowska I, Tlałka A (1976) Hydrografia [Hydrography], I. University scripts, 237. Uniwersytet Jagielloński, Kraków [in Polish]
2. Bajkiewicz-Grabowska E, Mikulski Z (2007) Hydrologia ogólna [General Hydrology]. Wydawnictwo Naukowe PWN, Warszawa [in Polish]
3. Rzętała M (2017) Sztuczne zbiorniki wodne i ich funkcje [Artificial water bodies and their functions]. In: Jokiel P, Marszelewski W, Pociask-Karteczka J (eds) Hydrologia Polski [Hydrology of Poland]. PWN, Warszawa, pp 240–246

4. Żmudziński L, Komijów R, Bolałek J, Górniak A, Olańczuk-Neyman K, Pęczalska A, Korze-niewski K (2002) Słownik hydrobiologiczny. Terminy, pojęcia, interpretacje [Hydrobiological dictionary. Terms, concepts, interpretations]. PWN, Warszawa [in Polish]
5. Jaguś A, Rzętała M (2010) Zbiorniki Czorsztyński i Sromowiecki – położenie, charakterystyka, nazwy [Czorsztyn and Sromowce Reservoirs—location, characteristics and nomenclature]. In: Soja R, Knutelski S, Bodziarczyk J (ed) Pieniny – Zapora – Zmiany. Monografie Pienińskie, 2. Pieniński Park Narodowy, Krościenko nad Dunajcem, pp 9–22 [in Polish with English summary]
6. Grobelska H (2006) Ewolucja strefy brzegowej Zbiornika Pakoskiego (Pojezierze Gnieźnieńskie) [The evolution of the Pakość reservoir shore zone (Gniezno Lakeland)]. IGiPZ PAN, Warszawa [in Polish with English summary]
7. Wiejaczka Ł (2011) Wpływ zbiornika wodnego Klimkówka na abiotyczne elementy środowiska przyrodniczego w dolinie Ropy [The influence of the Klimkówka water reservoir on the abiotic elements of the natural environment in the Ropa valley]. IGiPZ PAN, Warszawa
8. Kaczmarek, H (2018) Ewolucja strefy brzegowej nizinnych zbiorników zaporowych w warunkach dużych wahań poziomu wody na przykładzie Zbiornika Jeziorsko na Warcie [Evolu-tion of the coastal zone of lowland water reservoirsin the conditions of Significant water level fluctuations on the example of the Jeziorsko reservoir on the river Warta (central Poland)]. IGiPZ PAN, Warszawa [in Polish with English summary]
9. Głodek J (1985) Jeziora zaporowe świata [Reservoirs impounded by dams worldwide]. PWN, Warszawa [in Polish]
10. Kasza H (2009) Zbiorniki zaporowe. Znaczenie – Eutrofizacja – Ochrona [Dam reservoirs. Role—Eutrophication—Protection]. Akademia Techniczno-Humanistyczna, Bielsko-Biała [in Polish]
11. Rocznik statystyczny Rzeczypospolitej Polskiej, 2014 [Statistical Yearbook of the Republic of Poland 2014] (2014) Główny Urząd Statystyczny [Statistics Poland] Warszawa
12. Machowski R (2010) Przemiany geosystemów zbiorników wodnych powstałych w nieckach osiadania na Wyżynie Katowickiej [Transformations of geosystems of water reservoirs orig-inated in subsidence depressions (a case study of the Katowice Upland)]. Wydawnictwo Uniwersytetu Śląskiego, Katowice [in Polish with English summary]
13. Jankowski AT, Molenda T, Rzętała M (2001) Reservoirs in subsidence basins and depression hollows in the Silesian Upland—selected hydrological matters. Limnological Rev 1:143–150
14. Choiński A, Skowron R (2020) Zasoby wodne wód stojących w Polsce [Lentic water resources in Poland]. In: Zelenakova M, Kubiak-Wójcicka K, Negm A (eds) Management of Water Resources in Poland. Springer, Berlin
15. Flis J (1998) Słownik szkolny. Terminy geograficzne [School dictionary. Geographical terms]. WSiP, Warszawa [in Polish]
16. Czaja S (2003) Zbiorniki i pojezierza antropogeniczne [Water body and antropogenic lake district]. In: Szczypek T, Rzętała M, (eds) Człowiek i woda [The Man and water]. Polskie Towarzystwo Geograficzne – Oddział Katowicki, Sosnowiec 22–30
17. Jędrczak A (1992) Skład chemiczny wód pojezierza antropogenicznego w Łuku Mużakowskim [Chemical composition of waters in the anthropogenic lake district situated in the Muskau Bend] Wydawnictwo Wyższej Szkoły Inżynierskiej w Zielonej Górze, Zielona Góra [in Polish]
18. Rzętała M (2008) Funkcjonowanie zbiorników wodnych oraz przebieg procesów limnicznych w warunkach zróżnicowanej antropopresji na przykładzie regionu górnośląskiego [Functioning of water reservoirs and the course of limnic rocesses under conditions of varied anthropopres-sion a case study of Upper Silesian Region]. Wydawnictwo Uniwersytetu Śląskiego, Katowice [in Polish with English summary]
19. Rzętała M, Jaguś A (2012) New lake district in Europe: origin and hydrochemical character-istics. Water Environ J 26:108–117
20. Korzeniewska E, Harnisz M, (eds) (2020) Polish River Basins and Lakes—Part I. In: Hydrology and Hydrochemistry. Springer, Cham
21. Jankowski AT, Kuczera A (1992) Wpływ zrzutu wód podgrzanych na warunki termiczne, tlenowe i przezroczystość wody w Zbiorniku Rybnickim [The effect of dischargingheated

waters to the Rybnik Reservoir on the thermal and oxygen conditions and on water transparence]. Wydawnictwo UŚ, Katowice [in Polish with English summary]
22. Rösler A (2002) Comparison of evaporation conditions from a sunken and floating pans on Lake Sława. Limnological Rev 2:333–341
23. Bok M, Jankowski AT, Michalski G, Rzętała M (2004) Zbiornik Dziećkowice – charakterystyka fizycznogeograficzna i rola w górnośląskim systemie wodno-gospodarczym [Dziećkowice Reservoir—physico-geographical characteristics and role in the Upper Silesian water management system]. Polskie Towarzystwo Geograficzne – Komisja Hydrologiczna, Warszawa [in Polish]
24. Molenda T (2011) Naturalne i antropogeniczne uwarunkowania zmian właściwości fizyczno-chemicznych wód w pogórniczych środowiskach akwatycznych (na przykładzie regionu górnośląskiego i obszarów ościennych) [Natural and anthropogenic factors conditioning changes in physical and chemical properties of waters in post-mining aquatic environments—the case of the Upper Silesian region and neighbouring areas]. Uniwersytet Śląski, Wydawnictwo Gnome, Katowice [in Polish]
25. Choiński A (2007) Limnologia fizyczna Polski [Phisical limnology of Poland]. Wydawnictwo Naukowe UAM, Poznań [in Polish]
26. Solarski M, Rzętała M (2020) Ice regime of the Kozłowa Góra Reservoir (Southern Poland) as an indicator of changes of the thermal conditions of ambient air. Water 12(2435):1–16
27. Rzętała M (2014) Ice cover development in a small water body in an undrained depression. In: International Multidisciplinary Scientific Geoconferences, 14th GeoConference on Water Resources. Forest, Marine and Ocean Ecosystems. Albena, Bułgaria, pp 397–404
28. Rzętała M, Solarski M (2011) Codzienne obserwacje terenowe źródłem identyfikacji nowych form i procesów zlodzenia zbiornika wodnego [Daily field observations as a identification source of the water reservoir ice cover new forms and processes]. Z badań nad wpływem antropopresji na środowisko [Res Eff Anthropogenic Impact Environ] 12:155–161 [in Polish with English summary]
29. Kajak Z (1979) Eutrofizacja jezior [Eutrophication of lakes]. PWN, Warszawa [in Polish]
30. Korzeniewska E, Harnisz M (eds) (2020) Polish River Basins and Lakes—Part II. Biological Status and Water Management. Springer, Cham
31. Rzętała MA, Solarski M, Pradela A, Rzętała M (2014) Eutrophication of water reservoirs under anthropogenic, agricultural and industrial impacts (example of Southern Poland). Res J Chem Environ 18(4):49–55
32. Wiśniewski RJ (1994) Fosfor w zbiornikach zaporowych – zasilanie, kumulacja, wymiana między osadami dennymi i wodą [Phosphorus in reservoirs impounded by dams—supply, accumulation, exchange between bottom sediments and water]. In: Zalewski M (ed) Zintegrowana strategia ochrony i zagospodarowania ekosystemów wodnych [Integrated strategy for the protection and management of aquatic ecosystems]. WIOŚ, ZE S UŁ, Łódź, pp 49–60 [in Polish]
33. Kajak Z (1995) Eutrofizacja nizinnych zbiorników zaporowych [Eutrophication of lowland reservoirs]. In: Zalewski M (ed) Procesy biologiczne w ochronie i rekultywacji nizinnych zbiorników zaporowych [Biological processes in the conservation and restoration of lowland reservoirs]. Biblioteka Monitoringu Środowiska, PIOŚ, WIOŚ, ZES UŁ, Łódź, pp 33–41 [in Polish with English summary]
34. Vollenweider RA, Kerekes JJ (1982) OECD: Eutrophication of waters—monitoring, assessment and control. OECD, Paris
35. Kajak Z (1998) Hydrobiologia - Limnologia. Ekosystemy wód śródlądowych [Hydrobiology—Limnology. Inland water ecosystems]. PWN, Warszawa [in Polish]
36. Rzętała M (2000) Bilans wodny oraz dynamika zmian wybranych zanieczyszczeń zbiornika Dzierżno Duże w warunkach silnej antropopresji [Functioning of water reserviors and the course of limnic rocesses under conditions of varied anthropopression a case study of Upper Silesian Region]. Wydawnictwo Uniwersytetu Śląskiego, Katowice [in Polish with English summary]

37. Kostecki M (2003) Alokacja i przemiany wybranych zanieczyszczeń w zbiornikach zaporowych hydrowęzła rzeki Kłodnicy i Kanale Gliwickim [Allocation and transformations of selected pollutants in dam reservoirs of the hydro-junction of the Kłodnica and Gliwice Canal]. Prace i Studia IPIŚ PAN 57 Zabrze [in Polish with English summary]

38. Jaguś A, Rzętała M (2003) Zbiornik Kozłowa Góra - funkcjonowanie i ochrona na tle charakterystyki geograficznej i limnologicznej [Kozłowa Góra water reservoir—functioning and protection against a background of geographical and limnological characteristic]. Komisja Hydrologiczna Polskiego Towarzystwa Geograficznego, Warszawa [in Polish with English summary]

39. Galicka W (1996) Limnologiczna charakterystyka nizinnego zbiornika zaporowego na Pilicy w latach 1981–1993 [Limnological characteristics of the lowland reservoir impounded by dam on the Pilica River in the years 1981–1993]. Wydawnictwo Uniwersytetu Łódzkiego, Łódź [in Polish]

40. Kostecki M (2000) Zawartość metali ciężkich w mięsie i wątrobie niektórych gatunków ryb z antropogenicznego zbiornika Dzierżno Duże (woj. śląskie) [Heavy metals in flesh and liver of some fish species in Dzierżno Duże dam-reservoir (Upper Silesia)]. Archiwum Ochrony Środowiska [Arch Environ Prot] 26(4):109–125 [in Polish with English summary]

41. Jankowski AT (1986) Antropogeniczne zmiany stosunków wodnych na obszarze uprzemysłowionym i urbanizowanym (na przykładzie Rybnickiego Okręgu Węglowego) [Anthropogenic changes in the water relations in the industrialized and urbanized area for example Rybnik Coal]. Wydawnictwo UŚ, Katowice [in Polish]

42. Różkowski A (ed) (2004) Środowisko hydrogeochemiczne karbonu produktywnego Górnośląskiego Zagłębia Węglowego [Hydrogeochemical environment of coal measures in the Upper Silesian Coal Basin]. Wydawnictwo UŚ, Katowice [in Polish with English summary]

43. Banach M (1994) Morfodynamika strefy brzegowej zbiornika Włocławek [Morphodynamics of the Włocławek reservoir coastal zone]. Prace Geograficzne IGiPZ PAN, vol. 161. Zakład Narodowy im. Ossolińskich Wydawnictwo Polskiej Akademii Nauk, Wrocław – Warszawa – Kraków [in Polish]

44. Jaguś A, Rahmonov O, Rzętała M, Rzętała MA (2004) The essence of cultural landscape transformation in the neighbourhood of selected artificial water reservoirs in Southern Poland. In: Kirchner K, Wojtanowicz J (eds) Cultural landscapes. Brno, Regiograph, pp 37–55

45. Jankowski AT, Rzętała M (2009) Warunki wodne [Water conditions]. In: Polska, jej zasoby i środowisko [Poland, its resources and environment]. Stowarzyszenie Rozwoju Społeczno-Gospodarczego „Wiedza", Warszawa, pp 39–50 [in Polish]

46. Mikulski Z (1998) Gospodarka wodna [Water management]. Wydawnictwo Naukowe PWN, Warszawa [in Polish]

47. Czaja S (1999) Zmiany stosunków wodnych w warunkach silnej antropopresji (na przykładzie konurbacji katowickiej) [Changes in water relations under the conditions of strong antropopression (a case study of Katowice conurbation)]. Wydawnictwo Uniwersytetu Śląskiego, Katowice [in Polish with English summary]

48. Rzętała M (2016) The new evaluation proposal of tourist-recreational attractiveness of water reservoirs. In: 3rd International Multidisciplinary Scientific Conferences on Social Sciences & Arts SGEM 2016. Book 2. Political Sciences, Law, Finance, Economics & Tourism. Conference Proceedings, vol. IV. Economics & Tourism. Sofia, Bulgaria, pp 773–780

49. Machowski R, Rzętała MA, Rzętała M, Wistuba B (2005) Zbiornik Żywiecki – charakterystyka fizycznogeograficzna i znaczenie społeczno-gospodarcze [Żywiec Reservoir—physico-geographical characteristics and socio-economic significance]. Polskie Towarzystwo Geograficzne – Oddział Katowicki, Sosnowiec [in Polish]

50. Rzętała MA, Jaguś A, Machowski R, Rzętała M (2015) The development of freshwater deltas and their environmental and economic significance. Ecol Chem Eng S 22(1):107–123

Chapter 5
Irrigation and Drainage in Polish Agriculture: State, Problems and Needs

Leszek Łabędzki, Edmund Kaca, and Andrzej Brandyk

Abstract The paper discusses the conditions as well as the current status, techniques, trends and perspectives of irrigation and drainage in agriculture sector of Poland. Irrigation and drainage infrastructure was found to be in poor condition over the whole country. Moreover, the secondary melioration equipment is maintained on less than half of the meliorated area of agricultural land. It is also stressed that economic situation and lack of sufficient funds by farmers, farmer associations and responsible, local government units became main reasons for the negligence of proper use of water management systems. In particular, sufficient maintenance of irrigation/drainage systems has been abandoned, restricting possibilities of controlled water table management. It is a fact, that the interest in the utilization of water facilities has gradually decreased. For these reasons, it was discussed in this paper, that irrigation and drainage systems should be rebuilt and improved for better control of the outflow of water. Another goal, that was deeply analysed, is to create favourable conditions for water retention and finally more effective water management in agricultural lands. Different actions should be taken to stimulate the development of irrigation and drainage, to adapt the existing systems to extreme meteorological events resulting from climate change, to enhance their operation and management.

Keywords Irrigation · Drainage · Agricultural water management · Polish agriculture

L. Łabędzki (✉)
Institute of Technology and Life Sciences, Kuyavian-Pomeranian Research Centre, Glinki 60, 85-174 Bydgoszcz, Poland
e-mail: l.labedzki@itp.edu.pl

E. Kaca
Warsaw University of Life Sciences—SGGW, Institute of Environmental Engineering, Nowoursynowska 159, 02-776 Warsaw, Poland
e-mail: edmund_kaca@sggw.edu.pl

A. Brandyk
Warsaw University of Life Sciences, Water Centre, Ciszewskiego 6, 02-776 Warsaw, Poland
e-mail: andrzej_brandyk@sggw.edu.pl

© Springer Nature Switzerland AG 2021
M. Zeleňáková et al. (eds.), *Quality of Water Resources in Poland*, Springer Water,
https://doi.org/10.1007/978-3-030-64892-3_5

5.1 General Background

5.1.1 Agricultural Land Cover and Use

Located within the Central European Plain, Poland constitutes a lowland country (75% of the area lies below 200 m above mean sea level). It should be stressed, however, that its' landscape is diversified as a result of geographical setting and glaciation-related processes. Poland covers the area of 312684 km^2 (ca 31 mln ha), from which 60% is farmland, and 29% constitutes a woodland. Arable land is present on the area of 14046000 ha, including: meadows—2598000 ha, pastures—1480000 ha, orchards—268000 ha. Over the last years (from 1991) the agricultural land area has decreased by about 250000 ha [1].

The major agricultural region is situated on the Central Plains which stretch across the entire width of Poland. There are such provinces where agriculture became the main source of income, while the productive agricultural land of ca 14.6 mln ha supports a population of 38.4 mln people (about 0.389 ha per person). Private farms have still been the most significant land ownership form, occupying 92% of the agricultural areas. Although 15% of the total employed persons belong currently to the agriculture sector, the share in GDP is estimated to be only 3% [1–3].

The main areas of agriculture activity in Poland are crop and dairy production, and animal husbandry, cattle and livestock directed at meat production. The share in total sowing area of basic cereals is considered to be 57.1%, the potatoes occupy 9.6%, oil-bearing—2.7%, sugar beets—2.7%, field vegetables—2.0%, and orchards pertain to 1.0% respectively [3, 4].

5.1.2 Soils

Poland area is characterized by high diversity of the parent rocks, geological as well as pedological processes, leading to a number of soil types and sub-types existence over polish territory [5]. Among the most recognized soil types there are brown soils, acid brown soils, grey brown podsolic soils, rusty soils and podsolic soils. A much smaller area is occupied by chernozem soils, rendzina soils, black soils as well as alluvial soils. With respect to organic soils, that occupy about 4% of Poland's territory and are usually covered by diverse grassland areas, the main types embrace: alluvial muck soils, muck soils and peat soils. It is a fact, that the majority of soils originate from loose-structured, post-glacial rocks with relatively small area occupied by massive bedrock. Taking into consideration the cultivation potential, all soils considered to be proper either excellent for agricultural use amount to 23% of the whole arable lands. The soil deposits that fall into moderate category, occupy about 47%. Marginal and unsuitable ones are present on 30% of the land. In some extent, the unsuitable Polish soil types have still been cultivated, however, the abandoned land has been increasing ever since.

It was stressed, that on average, the soil quality in Poland can be assessed as respectively low. There is 27% of arable soils belonging to good or very good (classes I–III), while poorest soils (classes V–VI) occupy more than 33% of the arable land. Soil quality classification for grasslands receives even lowest merit: soils of classes I–III account for only about 14%, soils of class IV—for 39%, while soils of class V–VI are most common, covering as much as 46% of total grassland area [4].

5.1.3 Agro-Climatic Conditions

Poland lies in the moderate transitional climate zone. There are noticed impacts of a dry continental climate from the east and a mild oceanic climate from the western side. They bring about considerable variability in long-term weather trends but also short term ones (days, weeks) [6]. The average annual air temperature value is +7.5 °C and it ranges from slightly above 5 °C in the north-eastern part of the country and mountain areas to about 9°C in the south-western part [7]. Average, long-term annual precipitation (1961–2015) is equal to 605 mm and in the vegetation season (April–September) reaches 388 mm. The average of 210 days is noticed for the growing period, during which potential evapotranspiration over most of the country exceeds precipitation, taking effect in continuous water deficit on light soils, in particular, that are attributable to rather low water-holding capacity. The driest regions of Poland, however, involve the whole central region, but also north-western and mid-eastern parts. It is a fact, they have yet been most threatened by frequent and severe droughts, associated by the annual rainfall amount not reaching 300 mm. Low precipitation and high potential evapotranspiration influence water quantity and generate water management problems in these regions [2]. Only in the mountain zone and the sea coast average precipitation is high enough to satisfy the requirements of agricultural plants. Moreover, there have been such occurrences in Poland, locally and periodically, when excessive rainfall water contributed to extensive waterlogging and flooding of soils.

5.1.4 Agricultural Water Resources

The average annual precipitation of 605 mm gives the volume of 192 km^3 of water in one year. At the same time, estimated annual runoff becomes about 62 km^3/yr with a variability ranging from 38 to 90 km^3/yr, and in the end, the available water resources (assumed to be equal to 40% of mean runoff) amount to 26 km^3/yr. It should also be stressed, that there are over 8000 lakes with an area of over 1 ha and the total capacity of 17 km^3, while the capacity of artificial reservoirs amounts to 4.5 km^3. There exist 4700 fish ponds, covering the total area of 45000 ha, and providing the capacity of 0.5 km^3. When it comes to groundwater, the renewable resources in usable layers have gone to the level of 11 km^3. To sum up, natural water resources in Poland are found to be very small. Surface water resources in Poland, that are renewable, rate 1600 m^3/cap/yr, and this is three times less than the average value in Europe (4560 m^3/cap/yr).

The fact should not be omitted, that some surface waters are distinguished for their significance to agriculture. They include small rivers, irrigation canals and ditches, canals and ditches draining water from fields, small natural lakes and artificial reservoirs in rural areas. Alteration as well as management of those water resources became a domain of melioration. Under polish conditions, two types of land reclamation facilities are distinguished: primary watercourses and hydraulic structures and secondary facilities. The first ones embrace in general small rivers being a source of water for irrigation or a water receiver from drained fields and hydraulic structures associated with them; the other—canals and ditches with different facilities and hydraulic structures located directly in the fields. It should be mentioned that this division existed until recent times and is no longer involved in the new Polish Water Act [8]. The length and number of these facilities, providing for water resources significant to agriculture, are given in Table 5.1. The analyses point at total length of canals and rivers of local importance for agriculture to be 75300 km, of which 43400 km became regulated. Furthermore, the length of other small watercourses and ditches is estimated as 279000 km. Available surveys show that in 2016 there were 4176 small agricultural water reservoirs with the usable capacity about 200000 dam^3 [9]. Not all reservoirs are or have become operational. In some extent, the designed capacity could not be achieved. According to the estimates of provincial wetlands and water facilities management, in 2013 the requirements of improvement or reconstruction referred to water reservoirs with a capacity of 9% of the existing one.

Table 5.1 Primary melioration facilities

Facility	Parameter	Unit	Amount/number
Streams	Length	thous. km	75.3
Regulated streams	Length	thous. km	43.4
Canals	Length	thous. km	6.8
Melioration ditches	Length	thous. km	279
Agricultural reservoirs	Useful capacity	mln m^3	199
	Number	pieces	4176
Melioration pump stations	Efficiency	m^3·s^{-1}	920
	Number	pieces	579

Source [9]

5.1.5 Water Use in Agriculture

With respect to current water resources status in Poland, the total consumption in Poland is equal to about 11 km³/yr, which is the equivalent of 300 m³ per 1 inhabitant per year. The structure of present water consumption is as follows: industry—70%, domestic consumption by water-line supply systems—20% and agriculture, forestry and fish ponds—10%. Most of the water (83%), required to supply the above mentioned sectors, is withdrawn from surface resources [1, 2]. However, the decrease in water use has been observed already from the nineties of the XX century. The causes lie in the implementation of new water-saving technologies in the branches of industry, reduction of industrial production, the decrease in irrigated area as well as more effective use and savings of water by population once the measuring system of domestic water consumption and water prices have been introduced. It is expected that this trend will become stronger due to stabilization of national economy.

It is thought, indeed, that agriculture is characterized by relatively large water requirements. As it can be estimated from the water balance of a country or a region/catchment, a significant amount of water is consumed by agricultural areas in the process of evapotranspiration. It equals about 65 km³/yr [10], which stands for 34% of the water resources coming from rainfall. On the other hand, the amount of water used for agricultural irrigation seems to be minimum and amounts to about 90 hm³/yr, which is about 1% of the total water intake in the national economy. This describes the role of irrigation in agriculture as marginal, since only a very small part of agricultural land is irrigated. At present, on average, 0.5% of the total agricultural land is actually subject to irrigation [10]. If we add to this the irrigation amount in forestry, and also water consumption of the fish ponds, we obtain about 1000 hm³/yr of water used, which accounts for 10% of the whole consumption in the national economy.

The current issue refers to more water required per unit area and probably for unit crop productivity, but still with no guarantee of stable and high yield and likely to restrict crop water productivity. The occurring water scarcity in agriculture may stimulate the contemporary tendencies to increase local water resources and their availability, sustainable irrigation and advance in the efficiency of water consumption.

5.2 Water Deficits in Agriculture: Irrigation Water Requirements

The observed variability of meteorological conditions (the periods of severe droughts either floods and water excess), but also soil types, contribute to crop water insecurity through restricted water availability for irrigation. It is worth to be stressed, that irrigation water requirements have been variable on a time scale of a few days, and there is no chance to create proper, fixed guidelines to control irrigation systems. An assumption can be made, such that irrigation water demands for most irrigated crops in Central Poland are approximately equal to 100–300 mm in an average year. The irrigation water requirement gained exceptionally high rate during very dry periods since rapid moisture content decline is to be observed, and water resources become limited at the same time.

Summing up, the values of mean water demand of highly productive crop species under climatic conditions of Central Poland are as follows:

– 300–400 mm for crops of short growing period (by the end of July) and small water demand, e.g. cereals, early potatoes
– 400–500 mm for crops of longer growth (by the end of September) and moderate water demands, e.g. root crops
– 500–600 mm for crops of long growing season (until the end of September) and high water demands, e.g. fodder crops, vegetables.

With regard to soil moisture retention in the form of readily available soil water equal to 50–100 mm in light and medium textured soils as well as mean precipitation equal to 300 mm, the crop water deficits, pointing at net irrigation water requirements, can be found (Table 5.2).

Table 5.2 Net irrigation water requirements N in a growing period in Central Poland

Crop	Growing period	N (mm)
Rye	April–July	0–50
Winter wheat	April–July	50–100
Oat	April–July	0–50
Early potatoes	April–July	50–100
Late potatoes	April–September	100–150
Grass	April–September	160–210
Yellow lupine	April–July	100–150
Vetch	April–July	200–250
Sugar beet	April–September	150–200
Fodder beet	April–September	200–250
Carrot	May–September	190–240
Maize	April–September	100–150
Alfalfa	April–September	80–130
Early vegetables	May–July	50–200
Late vegetables	May–September	200–300
Berry crops	Varied	170–250
Orchards	Varied	200–400
Permanent grassland	April–September	150–200

5.3 Present Status of Irrigation and Drainage

5.3.1 Role of Irrigation and Drainage in Agriculture

In general, the role of irrigation in Poland is considered to be supporting and inter-ventional, put into operation in short periods during the growing season. Proper systems of irrigation mainly are located in regions of severe and frequent droughts. The driest regions of Poland, most threatened with droughts, cover almost the entire central region as well as north-western and mid-eastern ones. Statistically, irrigation is needed once in three years. It should be stressed that very seldom some of the crops cultivated in Poland do not require irrigation in the growing season. Nevertheless, in extremely dry years (e.g. 1992 and 2000) around 40% of the country area becomes constrained by drought. Obviously, it is related with the status of agricultural produc-tion, pointing at a mean decrease in crop yield to range 10–40% in those years in reference to the average year. The economic cost of drought to agriculture can be roughly estimated as 0.1% of GDP. Irrigation management in Poland is supposed to counteract the development of drought locally, in areas of light soils with valu-able crops, where water scarcity periods most probably occur, leading to substantial yields' decline. But still irrigated agriculture has a small role in food production [11].

In respect of spatial distribution of yield reduction of all crops, the central, central-east and central-west part of Poland are remarkable, where agriculture drought risk is evaluated as the greatest one. These regions of Poland are most threatened by water deficits causing the most significant crop yield losses. In two very dry years: 1982–1983, an average decrease in grain crops reached 5–30% in various regions of Poland and that of potatoes even 10–40% as compared to the average crops amounts in 1985–1987 [11]. On account of significant drought in 1992 the overall value of crops decreased by 25%. Besides, the obtainable crops of grain, potatoes and bulk fodder as expressed in cereal units dropped in 1992 by 31% than those gathered in 1991. Meadow hay crops (mean for the country) were reduced by 27% in comparison to the average rates of 1986–1990. Droughts in years 2005–2006 hindered crop production significantly, particularly in the central part of Poland, from the west to the east. In 2006, however, due to few week hot weather strip and shortage of precipitation in June and July, the most significant drop in yields of spring cereals, maize, potatoes, sugar beets and grasslands was noted. According to the evaluations made by the Ministry of Agriculture and Rural Development, the reduction in meadows and pastures yield ranged from 40–100%, spring cereals—20–60%, winter cereals—15–50%, rape—15–45%, potatoes and sugar beets—20–60%, and vegetables—30–60% respectively. It was also noticeable that the drought in 2015 caused 12% yield loss of cereals, 20% of vegetables, 25% of potatoes and 50–70% of silage maize.

However, the functioning of irrigation systems on agricultural lands has still been minor on account of a very small occupied surface (0.5% of the total agricultural lands). In addition to a vital role in eliminating the effects of drought and stabilizing crop production, in recent years the importance of sub-irrigation systems for the protection of organic soils and natural resources has been stressed.

On the other hand, taking into account the climatic and geomorphological variability of the Poland's territory, there occurs local and periodical excess of water that creates the need for drainage. Areas inundated by excessive rainfall and high groundwater levels have negative impacts on agricultural production and call for drainage infrastructure and management to mitigate those effects [12]. As it was already stressed by the authors, the positive effects of agricultural drainage on the land resources as well as its impacts on agricultural productivity and farm income are well known. Drainage also plays a significant role in mitigating the adverse effects of droughts on crop yields. The role of drainage systems involves lowering of the groundwater table in the spring, which allows field works to be started earlier and thus earlier plant growth and root development, triggering the root systems to take up water from the deeper layers of the soil profile during prevailing drought conditions. However, an awareness needs to be maintained that if water is drained from the soil, less water remains in its' profile and drought could appear earlier.

5.3.2 Irrigated and Drained Area

In Poland, about 6.4 million hectares (about 4.6 million hectares of arable land and 1.8 million hectares of grassland) are meliorated which is 44% of total area of agricultural land (Table 5.3). All areas that have been subject to land reclamation include drained and irrigated areas. A substantial area of agricultural land in Poland is equipped with hydraulic structures, mainly for soil drainage. Irrigated land (equipped

Table 5.3 Area of reclaimed agricultural land in Poland (2016)

Land reclamation and water control facilities	Registered area (thous. ha)	Percentage share of the registered area (%)*	
		Requiring reconstruction or modernization	Maintained
Reclaimed agricultural land	6370	22.7	46.6
Reclaimed arable land	4593	18.7	51.0
Reclaimed grassland	1777	33.3	34.9
Drained arable land	3973	–	49.7
Drained grassland	411	–	29.6
Agricultural land adapted for irrigation	422	–	26.9
Arable land adapted for irrigation	47	–	27.4
Grassland adapted for irrigation	375	–	26.8

Source [7], *acc. to the Polish Ministry of Agriculture and Rural Development (2017)

Table 5.4 Irrigated area in Poland and water withdrawal for irrigation

Year	Total	Irrigation method				
		Subirrigation	Sprinkling	Microirrigation	Flood irrigation	Other
	Irrigated area, ha					
1990	301500	284950	10300	–	2550	3700
2016	73202	64077	8737	155200	168	220
	Water withdrawal, hm^3					
1990	519	–	–	–	–	–
2016	89.9	83.8	5.4	–	0.03	0.04

Source [7]

with irrigation systems and facilities) is about 422000 ha of which—375000 ha with gravity systems on grasslands and 47000 ha with pumped systems on arable lands. It is noticed, that most of the agricultural land is drained without possibility to be irrigated.

The area irrigated in 2016 (Table 5.4) was not significant and totaled only 73200 ha which is 0.5% of the total agricultural land. Since 1989, due to transformation in national economy and the conversion of economic conditions in agriculture, the systematic decrease in actually irrigated area and the use of irrigation systems and facilities has been observed. Irrigated area decreased by 76% in comparison with 1990 (Table 5.4).

Referring also to sprinkling irrigation of arable lands under field production and forests, it is found to cover almost 9000 ha, while other forms, such as subirrigation of permanent grasslands in wetlands and inland valley bottoms, pertain to 64000 ha. Flood irrigation, along with other irrigation methods covers about 390 ha and receives much less importance. Official government publications provide no data on micro-irrigation (mainly drip irrigation) in particular. These techniques are used to irrigate orchards (60000 ha), ground vegetables (40000 ha) and vegetables under covers (16200 ha in greenhouses and 39000 ha in foil tunnels) [13]. The total area of irrigated agricultural land can be now estimated at about 230000 ha.

As it was mentioned in the above sections, there are two sources of irrigation water: surface water and groundwater. Surface sources include lakes, reservoirs, streams, water-user association distribution facilities and wastewaters. It is exactly 90% of irrigation systems' area that becomes supplied from surface water sources and 10% from groundwater sources respectively. Total water withdrawal for agricultural and forest irrigation as well as filling fish ponds is equal to about 1 km^3/year, which was already considered. In 2016, for example, water withdrawal only for agricultural irrigation totalled about 90 hm^3, of which 84 hm^3 dedicated for subirrigation and 5 hm^3 devoted to sprinkler irrigation. Since 1990 water withdrawal for irrigation has decreased by 83%.

The possibility of irrigation and drainage is provided by various types of channels that have a direct connection with watercourses. Their length is 6.8 thousand km (Table 5.1), while they are maintained at 65% of their length, and at 23% of the length they are eligible for reconstruction.

The combined systems—drainage-irrigation ones (i.e. double-acting) that are usually put into force on permanent grasslands, clearly belong to a minority. The statistical data show that in Poland about 4 million hectares are covered with permanent grasslands (meadows, pastures). Only 9% of this area can be irrigated, and as much as 36% of this area is equipped with only drainage systems, without the possibility of water retention, e.g. rain or thawing water on-site.

5.3.3 Irrigation and Drainage Techniques

The core of irrigation techniques in Poland from sixties of the XX century onwards involved permanent valley grasslands. Large drainage-subirrigation networks were put into operation there, covering the approximate surface of 0.5 mln ha. Not before than in the middle of the sixties, the first attempts to implement the pressure irrigating devices in the field crop cultivation were completed. Initially, in the sixties, the portable sprinkler systems, semi-portable and stationary systems were launched. With the passage of time, the equipment became more automatic; mainly side-roll wheel systems and reel hose travelling rain guns were introduced. On the state-owned, large-area farms, the center pivot pipelines of a continuous rotation around the central axis were employed, on condition the source of water was finally found: usually wells or hydrants of underground feeding pipeline. On account of transformation in the national economy after 1989, big state-owned farms were held back followed by sprinkling irrigation systems abandonment, and finally their exploitation came to a standstill.

At the end of the seventies and in the eighties several forms of micro-irrigation were introduced. The first system of that type of irrigation was a set of equipment for irrigations under the crown of trees in orchards, with the application of micro-jets. It was found to be widely applied in orchards, greenhouses and plastic tunnels as well as tree and shrub nurseries. By the end of the 1980s, it used to be the main system, applied to irrigate orchards in Poland. At the same time, the development of sub-crown micro-jets was followed by drip irrigation rapidly taking place in orchards and "under the roof". Simultaneously with the introduction of sub-crown micro-jet irrigation, the attempts were undertaken in Poland to make drip irrigation systems more popular which would take advantage of inspirations and solutions, being practised in other countries.

At present, the irrigation takes place in considerably smaller areas, that is in orchards and within vegetable fields in particular. Drip irrigation systems are taking steps ahead much faster than any other type of irrigation in Poland. At the same time action is being taken to restore part of the drained river valleys and introduce solutions dedicated to increase retention capacity. There is a number of introduced solutions, aimed at water conservation. Nonetheless, sprinkler irrigation is situated commonly in the western and central part of the country, covering regions of relatively low precipitation and fertile soils. On the other hand, referring to subirrigation systems typically used in river valleys on permanent grasslands, occupy largely the lowland regions (North and Central Poland).

Systems of irrigation in agriculture in Poland, serving for mitigation of water deficits in the soil are treated as a part of land reclamation (amelioration or land-water improvement). Under prevailing climatic, soil and economic conditions in Poland, and especially economical and structural sector in agriculture, two main tendencies in the field of irrigation systems can be distinguished [14]:

1. Gravity systems of (driven by gravity—non pressure—schemes) irrigation on the areas of permanent grasslands within the borders of river valleys, located in the neighborhood of lowland watercourses on vast post-swampy ecosystems, dependent on subsoil irrigation from the network of ditches and drains if necessary. Maintenance of appropriate groundwater table in those schemes is feasible by: controlled drainage, sub-irrigation with a constant water level, sub-irrigation with a regulated water table depth (groundwater hydrologic alterations).
2. Pressurized irrigation systems (equipped with puming devices). They involve sprinkler and all types of on-site irrigation, i.e. drip, micro jets, "sweating" hoses etc. of intensive and semi-intensive root crops, industrial crops, horticulture crops in open air, greenhouse crops and orchards.
3. Flood and other types of irrigation (e.g. border) are used in a small extent due to high demand for water while providing low water use efficiency.

The type of hydraulic structures for irrigation is related with irrigation method. In sub-irrigation systems on permanent grasslands, there are various structures for water damming and distribution (damming structures on streams, water damming culverts, weirs, gates, dammed lakes, water intakes and discharge points), to increase water level in streams and canals as well as to supply water to the network of ditches. Grasslands, predominantly those located in river valleys, have been drained by open ditches. It amounts to 25% of the drained grasslands that may be irrigated—having been equipped with proper hydraulic structures. All in all, the subirrigation systems are the only method used for irrigating grasslands. Nowadays, only part of these systems is utilized by farmers. In pressure irrigation systems (sprinkling, micro-irrigation) there are different facilities for water intake from surface and ground waters (wells, pumping stations), network of pipelines to supply water to fields, facilities for water filtration and for preparing fertigation solutions. In those systems, there is no special hydraulic infrastructure for damming water, but for drainage systems, covered plastic or ceramic drains have been installed for desiccation of arable soils.

The future of irrigation lies in adaptable devices, that can be adjusted properly to field conditions and land use type (arable lands, horticulture and orchards). The potential is visible for micro and drip irrigation, but generally variable rate irrigation approach will gain significance due to varying environment and adaptation to climate changes.

5.4   Problems, Challenges, Opportunities and Constraints to Future Development of Irrigation and Drainage

5.4.1   Technical State, Maintenance, Reconstruction and Modernization of Melioration Systems

Having acknowledged the need to analyse the status of melioration equipment and systems, an insight into reclaimed agricultural lands discovers usually their poor technical condition. The study performed by Kaca [15] proved that in Poland, in 2013, the youngest facilities not demanding reconstruction, at the age of 21 were present on 3.5 thousand hectares of permanent grassland, while significant numbers, reaching the age of 64, cover the area of 37.8 thousand ha. Prevailing devices are at the age of 50, operating on the area of 55.1 thousand ha (Fig. 5.1). On arable lands, the youngest facilities not requiring reconstruction at the age of 20 were on the area of 17.5 thousand ha, and the oldest ones, around the age of 64, pertain to the area of 3.7 thousand ha. Devices prevailing at the age of 45, refer to a surface of 159.3 thousand ha.

No secondary melioration equipment was maintained on more than half of the reclaimed area of agricultural land (53% of agricultural land, including about 49% of arable land and 65% of grassland) (Table 5.3). In 2016, drainage and irrigation systems were maintained in working order on 51.0% (2.3 million hectares) of the drained and irrigated area of arable land, and 34.9% (0.6 million hectares) of permanent grasslands (meadows and pastures).

As given in the work of Kaca [16], a vital increase in this area should be planned, up to 64% (3.0 million hectares) and 89% (4.1 million hectares) in arable lands and 38% (0.7 million hectares) and 46% (0.8 million hectares) in permanent grasslands by years 2020 and 2030, respectively (Fig. 5.2).

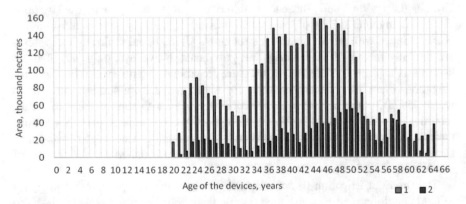

Fig. 5.1 Area of reclaimed arable land (1) and permanent grassland (2) with facilities recognized as not requiring reconstruction in a given age (as at the end of 2013) (acc. to [15])

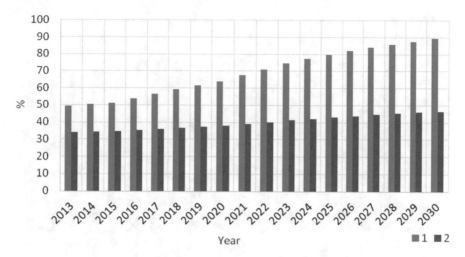

Fig. 5.2 Planned share (%) of arable land (1) and permanent grassland (2) with equipment maintained (acc. to [16])

According to the Ministry of Agriculture and Rural Development, devices on 33% (592 thousand hectares) of the area of reclaimed grasslands required reconstruction or modernization. The equipment on arable land was said to be in a much better condition. However, also arable lands demand the reconstruction or modernization of such equipment on 19% (861 thousand hectares) of the area of meliorated arable land (Table 5.3). According to the data elaborated by Kaca [17], a significant decrease in this area should be planned, going down to 32% and 28% in permanent grasslands and 18 and 15% in arable lands until years 2020 and 2030, respectively (Fig. 5.3).

Kaca [18] introduced the concept of valley water-melioration systems (SWM), which consists of water systems SW (i.e. primary melioration) and melioration systems SM (i.e. secondary melioration) (see Sect. 1.4). It should be stressed, first of all, that SWM systems are directed to be balanced. The measure of this balance is the WZ index. It is assumed that water and melioration systems (SWM) are maintained balanced (in harmony) when WZ = 1. In other cases, these systems are thought to show imbalance. The degree and direction of this imbalance result from the value of WZ—melioration systems dominate when WZ > 1 and water systems predominate when WZ < 1. The degree of balancing of valley water and melioration systems in the provinces can be assessed using a five-point order scale (Table 5.5).

On the basis of the WZ index, it can be demonstrated that the valley water-melioration systems (SWM) in individual provinces are usually not balanced (Fig. 5.4). Water systems (SW) dominate over melioration systems (SM), and this domination becomes more intense. It may suggest that water facilities with water-courses in the province can only be used partly for agriculture. A particular case concerns the Zachodniopomorskie (Western Pomeranian) province, within which the WZ decreased to 0.13. In such cases, one should strive to accelerate the develop-ment of melioration systems all over the province. The situation in Śląskie (Silesia)

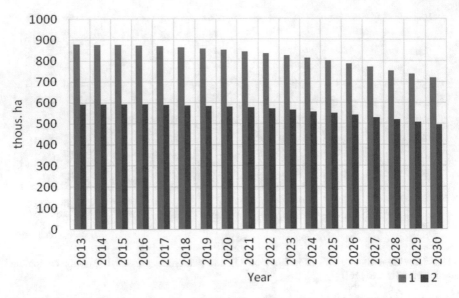

Fig. 5.3 Planned area (thous. ha) of arable land (1) and permanent grassland (2) with equipment for reconstruction or modernization (acc. to [17])

Table 5.5 The degree of balancing of valley water and melioration systems (SWM) (acc. to [14])

Degree of balancing	Balance index WZ
Very small with a predominance of melioration system SM	>2
Small with a predominance of melioration system SM	1.2–2
Satisfactory	0.8–1.2
Small with a predominance of water system SW	0.5–0.8
Very small with a predominance of water system SW	<0.5

and Wielkopolskie provinces seems to be different. Water systems are in deficiency in relation to melioration systems. In these cases, the development of water systems ought to be accelerated eventually.

5.4.2 Irrigation Problems

Potentially high demand for irrigation concerns a large area of Poland in the form of the final agro-technical operation, that permits to increase crop yields and guarantee

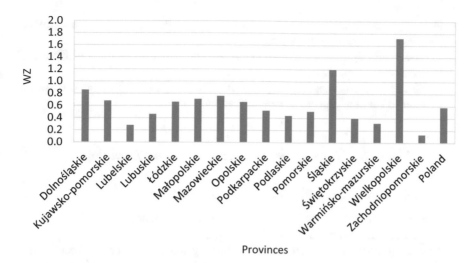

Fig. 5.4 WZ index—the degree of balancing water-melioration systems in the provinces (based on [16])

the stability and quality of crops. Five basic goals for irrigation can be specified for polish agriculture:

- mitigating the effects of drought in agriculture
- ensuring the stability of good quality crops
- increasing plant production efficiency
- increasing the competetiveness of farms
- protection of organic soils and permanent grassland habitats in river valleys.

It is in consent to substantial rainfall shortages, creating the need for irrigation. Contemporary irrigation practices in Poland, as it was stressed before, seem to be only supportive and in the wet years play a marginal role. The requirement of irrigation every single year exists only for vegetable and fruit plants. Even though, irrigation is considered indispensable to maintain agricultural production at a higher level [19].

The main problems in irrigation, that need urgent solution or elimination, involve:

- too small irrigated area; plant production in Poland is based mainly on rainwater and soil post-winter retention. The irrigation aimed at reducing water shortages is performed in a limited extent,
- insufficient disposable water resources—incurred as a result of irrational and ineffective use of water,
- lack of adjustment of irrigation to current meteorological and soil-water conditions,
- application of water-consuming irrigation technologies,
- noticed low efficiency of irrigation water,
- insufficient irrigation infrastructure that could provide for sustainable and wise regulation of outflow and retention of water in the basin,

– lack of preparation (investment, exploitation) and the scope of irrigation water sources as well as supply systems,
– a constant and significant increase in the uncontrolled abstraction of groundwater for agricultural irrigation,
– grasslands located on peat soils (about 2 million ha) cannot be denied irrigation in order to stop the process of peat degradation.

5.4.3   Needs for Development of Irrigation and Drainage

With regard to irrigation, some basic questions could be addressed:

– how many agricultural lands should be irrigated in Poland;
– to what extent irrigation is needed in Poland;
– how much and what (what crops) need to be irrigated, also taking into account, and perhaps first of all, the demand for agricultural products and food (domestic and export needs), which is associated with the strategy of agricultural development in Poland;
– how much food we want to produce for the internal and external market;
– whether irrigation is needed to obtain this production;
– how farmers look at irrigation, whether they need it and in what production departments;
– whether irrigation will be economically justified;
– do farmers receive compensation for losses in crops as a result of a drought or support the improvement of irrigation financially
– etc.

Similar unresponded questions and unsolved problems can be formulated with regard to drainage.

Out of 16 Polish voivodships (Fig. 5.5), due to climatic conditions described by CV index [19], the following ones urgently need development of irrigation (Fig. 5.6): Wielkopolskie (CV = 1.00), Lubuskie (CV = 0.96) and Kujawsko-Pomorskie (CV = 0.83), while the needs seem to be the lowest for: Małopolskie (CV = 0.00), Podkarpackie (CV = 0.09) and Śląskie (CV = 0.10).

The similar analysis, realized for the improvement of drainage, proves that due to climatic conditions the Zachodniopomorskie and Pomorskie provinces have the highest validity (need) of drainage solutions implementation (Fig. 5.7), whereas the lowest pertains to Wielkopolskie and Kujawsko-Pomorskie [19]. When soil-water criteria are concerned, Podlaskie seems generally characterized by the highest validity of developing drainage, in terms of hydrological conditions—Małopolskie, while from the environmental-ecological point of view Łódzkie province gains the highest importance.

Fig. 5.5 Voivodships in Poland (according to geographical atlas)

5.4.4 Recapitulation—Challenges and Constraints

The overall result of the analysis, performed in this paper, on the qualitative, quantitative and technical state of melioration leads to the following summary and concluding remarks.

There is no doubt, that at present moment meliorations have passed into regression phase. With respect to the importance of irrigation, it will increase along with the intensification of agriculture and unfavorable climate changes, especially in vegetable and fruit farming. An increase in the irrigated area, especially vegetables and orchards, is observed with the help of modern water-saving micro-solvent systems (drip, micro-coverage, root, etc.). In orcharding, drip installations are employed for approx. 80% of the irrigated area.

Among the emerging problems, water shortage is already a factor limiting the intensification of agricultural production. This unfavorable situation may even

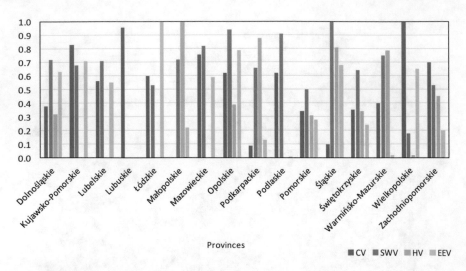

Fig. 5.6 Values of indices of the climatic (CV), soil-water (SWV), hydrological (HV) and environmental-ecological (EEV) validity of developing irrigation in the individual provinces (acc. to [19])

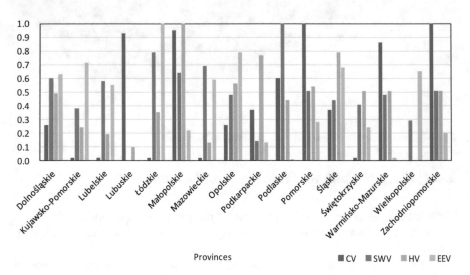

Fig. 5.7 Values of indices of the climatic (CV), soil-water (SWV), hydrological and environmental-ecological validity of developing drainage in the individual provinces (acc. to [19])

become worse in the light of observed and predicted climate changes. The accomplishment of irrigation and ensuring the required amount of water is treated as a great challenge, and prequisite for maintaining current high level of achieved agricultural produce, as well as its intensification. Irrigation seems necessary so much, so it maintains agricultural production at a desired level.

All the improvement and realistic effects of irrigation will depend primarily on the economic conditions of agriculture, investment costs and operating costs of the equipment, the quality of irrigation systems management so as the administration. Lastly, the availability of irrigation water should be taken into account [19].

Modern, energy-saving and water-saving methods and techniques of irrigation should be implemented: in gravity irrigation—irrigation with a constant water table and regulated outflow, in pressure irrigation—microirrigation (drip, root, mini-sprinklers).

It is necessary to vitally increase the available water resources for irrigation, focusing on the following remarks:

- It needs the form of concepts, directions of actions, legal provisions, etc. solutions in order to increase the possibilities and the broader use of groundwater resources for irrigation
- To a certain degree it develops concepts of regional (collective) water supply systems for irrigated lands
- Seems indispensable to continue the implementation of small retention programs, with a focus on retention for the sake of irrigated facilities.

The development of irrigation, taking into account the intensive development of agricultural production focused on vegetables production, should find its place in agricultural development strategies. Water resources protection is stressed here, because it needs the basis for further organizational and technical activities for the creation of modern irrigation systems. Implementation of such a program will require many instructional and training activities for future users of those systems.

The effects of irrigation and drainage are associated with the quality of exploitation (O&M—operation and maintenance) of irrigation/drainage systems. Once exploitation of facilities has reached its' goal, water supply is secured and then the adverse effects of droughts and water excess could be avoided, leading even to higher productivity score. To achieve a high efficiency of these systems, systematic conservation of the network of ditches and facilities is necessary. A lack of upgraded, but improper exploitation of the systems and facilities restrict competent water management and take effect in crop production decline. The proper use allows effective water management on irrigated and drained areas and greater yield stability. A decision support system can also play a considerable role in effective irrigation/drainage performance.

In recent years, increase in the area of applied microirrigation, especially drip irrigation in vegetable and fruit farming has been observed. Estimates said that the development of this type of irrigation would run in line with the development and intensification of agriculture. Undoubtedly, emphasis should be laid on negative consequence of climate changes here. It has already been a matter of dispute, microirrigation systems becoming more popular owing to their high efficiency. It is expected that this trend will dominate in Polish agriculture. On permanent valley grasslands, gravitational subirrigation systems will remain in the future as a source of fodder and a way of safe feeding base for cattle. They also exist for the preservation of biodiversity and safeguard the state of organic soils in river valleys.

The reason seems not to be the technique itself, that will hinder the progress of irrigation in Poland, but rather the economics as well as water availability in sufficient quantities and acceptable quality. The emerging problem tends to be the improvement and accomplishment of irrigation systems to make possible the use of modern energy- and water-saving solutions.

Long-term irrigation research and gained practical abilities exhibited water use as much more efficient in pressurized irrigation than in surface irrigation. For this reason, the inevitable search for more advanced techniques and accessories has revealed following solutions:

– micro sprinkling with low-discharge emitters and minisprinklers,
– drip irrigation (surface and subsurface directly into the root zone),
– compensated drippers,
– automatic valves and controllers,
– media and automatic filtration.

The above mentioned solutions and devices should be enhanced for a wide use. Besides, a routine of fertigation procedure has to be introduced to the most of irrigated areas. Highly soluble and liquid fertilizers ought to be applied as compatible to the actual technology. As a fully controlled process, irrigation by automatic valves and computerized steering units deserves attention. Modern irrigation and drainage equipment will enable better control and monitoring of water status of agricultural land, which will be translated into higher effectiveness of water intake. A regional network of agro-meteorological stations should strive to deliver real-time weather data to farmers along with the guidance regarding operational irrigation/drainage scheduling. Besides weather monitoring, soil water content control utilizes tensiometers, pressure chamber patterns, electrical resistance sensors, to secure more precise, site- specific local adaptation. Vegetation indices, such as leaf water potential, stomatal resistance, infrared canopy temperature deserve to be used to achieve better accuracy in water dosage.

In the contemporary times, the advanced irrigation technologies and equipment are available in Poland and can be purchased and implemented in a farm by many private companies, either domestic or the representatives of foreign companies (e.g. from Israel, USA, Italy). They provide a complete service for investment, initially from advisory and design consultation and ending with the maintenance and service. The only barrier to acknowledge in the common implementation of modern irrigation devices is their cost.

Currently, as it was already mentioned, under the economic circumstances and tendencies in Poland, the irrigation of most field crops is considered to be unprofitable measure. It should be noted, however, that the irrigation of potatoes, vegetables and orchards proved to enhance profits to some degree [12, 14, 19]. It can be foreseen that the link between the investment and operation costs of irrigation and farther profits will gain new connotation. From one side, the lowering of prices of irrigation gear could be expected; from the other side, the market prices of agricultural products will cause irrigation bringing more profits to farmers. One should acknowledge such developments as: productivity effectiveness, the use of different types of irrigation technology and their prices, operation and maintenance costs, prices for water drawn for irrigation, application of fertilizers, etc.

Increased frequency and dangerous character of droughts, the encouragement of agricultural production, being forced by the internal domestic and all-European free-market competition aspects, cause more rapid irrigation improvement. Moreover, the importance of reaching sufficiently high quality of almost all agricultural products is found to be accelerating the development of irrigation. This will gain new significance in coherence with the intensification of agriculture (e.g. in horticulture, orchards, seed crops) and major effects of climate changes, as it was already stressed. It has already been discussed, that many regions of Poland would face serious irrigation restrictions because of the availability of water resources. From the viewpoint of hydrologic conditions, a limit to irrigation is created by the lack of water due to the co-existence of drought and lowered river water levels as well as the decreased storage of lakes and retention reservoirs. Then, the water management is assigned a role to regulate the outflow, meaning proper conditions for water retention and adjusted discharge in the cases of water excess. This seems to be crucial in the spring season and particularly after the occurrence of high precipitation, when storage in the soil profile and the network of ditches should take place. It may be also supported by implementation of local water harvesting reservoirs and dams as well as across-field structures to restrict water outflow directly at local field-plots.

5.4.5   Recommendations

There exists a variety of measures that could be recommended in order to accomplish the strategic goal—controlling water management in irrigation/drainage systems. All of them, individually, should be aimed at improving:

- conditions for water retention in the periods of its excess—in the spring and after abundant precipitation (construction of retention reservoirs and water structures to restrict water outflow from fields, erosion control measures, soil technical management, water harvesting etc.),
- the efficiency of water utilization including irrigation.

Finally, they should lead to:

- an increase in the standing and flowing water resources,
- a growth in agricultural water use effectiveness,
- a decrease in crop water requirements.

These actions trigger the water unit productivity, at the same time reduce water losses and pointless water discharges. Moreover, they are directed at the decrease of droughts risk by collecting water beforehand or reduce all the major consequences in periods of water deficit. Controlled water management in a watershed that contains retention reservoirs and dammed lakes also limits the low water stage periods in running waters and modifies the frequency of their occurrence. A point should be made, for particular actions associated with agriculture to be acknowledged by the provincial centers of the agricultural advisory, agricultural chambers, research institutes, agricultural schools to a degree etc.

In recent years, as well as in all years coming onwards, three main factors drive the development of irrigation and drainage [18]:

- increased frequency and intensity of extreme climate events that conditions the deficit or excess of water in the soil profile, which, as a result, leads to reduced crops and causes uncertainty in economic enterprises of agricultural farms,
- the intensification of possible agricultural produce, being stimulated by the inner domestic and all-European merchandise competition; its stability is needed every year because, among other things, only then helps to conclude different agreements with the users of agricultural crops.
- the demand for high quality of the majority of agricultural products; one of the examples depicts consumable potatoes, especially for collective nutrition units; their indispensable qualitative standard (size; shape, equality) could be realistic only under the conditions of sprinkler irrigation, i.e. in the conditions of prolonged, optimum moisture content of soils.

Irrigation and drainage are to be considered as elements of integrated water management approach in a watershed [20]. Apart from the problems, general objectives, options and future fate of irrigation and drainage deserves insight from the viewpoint of water management planning. The scale of problems to be addressed by the wise management of water resources (respecting the principles of sustainable development) has been recently recognized in Poland as it was already discussed in previous sections of this paper.

Due to foreseen climate changes and related extreme meteorological and hydrological events (droughts, floods), it is essential to revise planning guidelines, design criteria, operation assumptions, emergency plans and management policies for water infrastructure and to outline the role of irrigation, drainage systems and agricultural water storage so as to hamper utmost events. In order to restrict the aftermath of whether extremes, the appropriate methods and adaptation strategies should be elaborated and implemented to existing irrigation/drainage systems.

Knowledge—based and consistent activities are directed at the refinement of operation, maintenance, administrative aspects and decision making on water problems. Proposed measures include:

- remodelling of irrigation/drainage and water supply systems to improve their effectiveness for infow and outflow of water,
- revision of O&M of irrigation/drainage schemes and water systems,
- utilization of current energy- and water-efficient approach in irrigation,
- refinement of irrigation efficiency,
- increase of crop water use capability,
- renovation of existing infrastructure for storage and distribution of water,
- increasing available water resources (in soils, streams, reservoirs),

- adjustment of water system control algorithms to changing climate conditions and extreme weather events,
- development of regional (local) systems of monitoring climate for the need of water system management,
- possible development of wireless telecommunication systems,
- application of remote-sensing methods and GIS in water management systems control.

All considered activities towards the refinement of contemporary systems need to be coupled with other activities undertaken for counteracting adverse effects of climate change in agriculture. As a joint result, the possibilities of adaptation of hydrologic systems will be substantial and their vulnerability to extreme events would become minor.

At present and in the near future, there seem to be several emerging challenges in the area of irrigation and drainage for the polish government. They also face self-governments, local councils, local administration of amelioration and water management, farmers' associations and farmers themselves in the following order:

- restrain from decapitalization of irrigation/drainage systems,
- improvement and upgrade of irrigation/drainage networks,
- creation of desired economic conditions in agriculture for investment, upkeep and better performance of irrigation/drainage devices,
- implementation of modern energy- and water-efficient approach and techniques of irrigation,
- improvement in the efficiency of irrigation,
- optimization of water distribution and water management in irrigation/drainage systems,
- implementation of different actions to increase current quality and availability of local water resources.
- management and adequate utilization of the so called "basic land reclamation facilities" like reservoirs and canals maintained by the state,
- organising courses for farmers to make public the upgraded irrigation systems and water-saving technologies,
- establishment of legal, financial and institutional basis for regional water supply systems for irrigation,
- stimulating farmers' participation in the planning of irrigation/drainage development and water resources management, emphasising the protection of water-dependent ecosystems and improvement of ecological status of surface waters,

5.4.6 The INOMEL Project

The above problems and constraints were identified in the framework of the project "Technological innovation and system of monitoring, forecasting and operational planning of drainage and irrigation for precise water management on the scale of drainage/irrigation facility (INOMEL)". The aim of the project is to develop and prepare for implementation a system of operational planning of controlled drainage and irrigation, including monitoring, forecasting and remote sensing. Irrigation/drainage facilities will be equipped with innovative devices designed for regulating (braking) the outflow of water from the facility and for regulation and measuring the water flow rate for irrigation. The products of the project may increase the efficiency of managing "own" water in drainage systems maintained in working order (drain systems and ditch system), and "foreign" water in subirrigation systems maintained in working order (two-way action). This means a radical improvement in planning the times and intensity of drainage and irrigation, as well as the amount of water removed from the system and taken up for irrigation. Firstly, systems should be equipped with these devices following their rebuilding and modernization. The rate of such reconstructions ought to increase. The medium and long term programmes for the development of water drainage and irrigation on the scale of the country and individual provinces, drainage and irrigation devices on a surfaces area of 8.8 thousand hectares of arable land in 2020 and 18.9 thousand hectares in 2030 will be subjected to rebuilding. In the case of permanent grasslands and pastures, these areas will be 10.6 thousand hectares and 22.4 thousand hectares in 2020 and 2030, respectively. The need for developing drainage and irrigation is indicated in provincial development strategies [15, 21]. Issues connected with agriculture, agro-food industry, and thus factors determining the development of water management and drainage and irrigation, are highlighted in provincial development strategies for Central Poland, have relatively high importance in provincial strategies in the east and west, and low in the development strategies of the provinces in Southern Poland.

The project products can improve water management on irrigation/drainage facilities and lead to a significant increase in the profitability and competitiveness of Polish agriculture, especially in years of high water shortages (occurring, on average, every 3–4 years). Years with high shortages are characterized by decreased crop yields or even their complete loss. Advantages stemming from the realization of the project will also have a social effect.

As a result of implementing the project, an increase in the efficiency (productivity) of water use will follow. Operational planning of regulated drainage and irrigation and decreased useless outflow of water from the system may lead to decreasing the outflow of biogenic compounds to rivers, especially nitrogen compounds from drained areas. This is especially important in areas at particular risk of contamination with nitrogen compounds from agricultural sources (NVZs). The results of the project can be particularly beneficial to irrigated (equipped with irrigation devices) permanent grasslands on hydrogenic soils, especially peat soils. As a result of maintaining moisture content at an adequate level, their degradation will be prevented, and the emissions of CO_2 to the atmosphere as a consequence of their mineralization reduced, thus contributing to a reduction in the emission of harmful gasses into the atmosphere.

It should be added here that production-oriented drainage and irrigation companies, as well as related companies, are interested in applying the project results in business activities. These are small and medium companies (SMEs), which number can be assessed at a few dozen. These businesses are interested in the production and installation, on systems of drainage using drains and ditches, regulators for stopping the outflow of water (for damming-up water in ditches and drainage wells), as well as the adaptation of intakes for the regulation and measurement of water flow rate. Another group are the few design service companies which, in cooperation with water users' associations, would undertake the remunerated operational planning of drainage and irrigation of land along with monitoring and short-term agro-hydro-meteorological forecasting. The most numerous group of entities expressing interest in the subject matter include groups of water users' associations. A specific group of entities expressing interest in the operational planning of drainage and subirrigation are agricultural advisory services as well as units of the local government. All entities express an interest in implementing the results of the project in business, indicating risk connected with the fact that the market of drainage/irrigation innovations is a rising market, the results of the project may have a positive influence on. There is an urgent need to establish contacts between the science and business communities, as well as other entities, in order to jointly devise innovative drainage and irrigation solutions, implemented in, e.g. systems undergoing rebuilding or modernization.

The results of the project will be of high importance on the scale of a farm, as well as to society. The direct end recipients (users) will be farmers, owners or users of drained and irrigated agricultural lands, mainly in lowlands (low land slopes). The number of end users of the project results is reflected by the area of agricultural land containing drainage/irrigation equipment (systems) maintained in working order. On the scale of the country, such systems function on 2.9 million hectares of agricultural land, and in the case of lowland provinces, benefitting most from the development of drainage/irrigation, on 1.7 million hectares of arable land and 0.4 million hectares of permanent grasslands. The end recipients of the project results also include water users' associations, which, by law, are responsible for the exploitation of drainage/irrigation devices. The number of such units is approx. 2300. Society, on the other hand, would be an indirect end user of the project results. The efficient management of water in irrigated agriculture signifies more water for its users.

The problem of regulating (stopping) the outflow of water from a system is taken up very rarely in practice, which results from, among others, the lack of appropriate outflow regulators. Some of these types of devices have been presented in the form of untested prototypes. The need for stopping the outflow of water from drainage systems is indicated by the research of Szejba et al. [22], who showed that, under conditions of unrestricted outflow from drained areas, significant amounts of biogenic compounds also flow out together with water.

The issue of operational planning stemmed from the need to improve water management in drained/irrigated agricultural areas, especially the management of "own" water on drained arable land, and own water and water taken in from the outside on permanent grasslands (meadows and pastures).

5.5 Conclusions

(1) Irrigation and drainage systems are in poor condition in Poland. Melioration equipment and systems on reclaimed agricultural land are usually in poor technical condition. No detailed melioration equipment is maintained for more than half of the meliorated area of agricultural land.

(2) Economic status of agriculture and shortage of sufficient funds by farmers, the associations of the land owners, and local government assemblages for drainage-irrigation practices in the catchment, are the main reasons for the negligence of the proper use of the land reclamation systems and facilities, for the blockage of irrigation/drainage system maintaining and conservation as well as controlled water management. Decreased interest in the use of hydraulic structure is observed.

(3) Nowadays the significance of irrigation in agriculture seems to be minor on account of a very small occupied area (0.5% of the total agricultural land area). The observed, actual frequency and intensity of droughts, the more intense agricultural production being shaped by the domestic and European merchandize competition and the need for high quality of the overall agricultural products will stimulate irrigation development.

(4) Under the current economic conditions of Polish agriculture irrigation is an unprofitable measure, except irrigation of potatoes, vegetables and orchards.

(5) Limited water supply in some regions constraints agriculture management in general. Surface and ground water storage may become strongly limited and environmental conditions require the implementation of water management strategies for the sustainability of agriculture and food security. Under conditions of increasing competition for water, there is a need for optimum use of limited water supplies.

(6) For many irrigation projects, water becomes a limiting factor for development. Proper water management would maximize the water use efficiency.

(7) There is a need for reconstruction and/or modernization of irrigation and drainage systems to inhibit and control the outflow of water from the systems. This will contribute to create favourable conditions for water retention, controlling water discharge, decreasing the outflow of biogenic compounds to rivers, especially nitrogen compounds from drained areas and effective water management in agricultural land.

(8) A variety of steps should be taken to stimulate the development of irrigation and drainage and to adapt the existing systems to extreme meteorological events resulting from climate change. A number of technological and organizational actions should be taken to enhance operation, maintenance and management.

(9) The risk of dangerous floods and droughts increases. Current water management systems are not efficient and cannot effectively counteract these phenomena. Losses caused by floods and droughts are enormous and therefore not socially and economically acceptable.

(10)   Water melioration is an element of the water cycle in the river basin, which is an integral part of water management. Water status factors (outflow, evaporation, retention) in river basins and the status of waters in rivers, including flood surges and low flows, depend largely on the condition of water facilities.

(11)   The unfavourable condition of water management is conditioned by the poor organization of water management entities. These entities are under-monitored and controlled. Their activity is underfunded and is based on water and drainage facilities with low technical condition. They have problems in implementing innovative procedures for organizing, planning, implementing and controlling water management.

(12)   There is an urgent need to start work on changing the current state. Contemporary requirements of modern precise agriculture, environmental protection and climate change point to the need for technical modernization of existing irrigation/drainage systems and devices and to improve the organization of their service and maintenance. It is necessary to improve the organization of water management and supervision over this management, improve the knowledge and skills of water management entities and increase the financial resources for maintenance of water melioration facilities and construction of new ones. Modernization of irrigation/drainage systems should take into account experience and forecasts of conditions in which they will operate. Modernization should take into account technical progress, economic conditions, management system and environmental protection requirements.

(13)   Water management should be systemic and take place within the hydrological catchment. Responsibility for organizing the water management system and for supervision and control over this management should rest with the bodies of the inter-municipal agreement or inter-municipal association.

(14)   Sustainable water management requires innovative recording, monitoring, operational planning, response and control procedures. These procedures may be established over a period of several years, as a result of extensive, interdisciplinary strategic research and development works, financially supported by the National Centre for Research and Development.

Acknowledgements This study was done within the project "Technological innovations and system of monitoring, forecasting and planning of irrigation and drainage for precise water management on the scale of drainage/irrigation system (INOMEL)" under the BIOSTRATEG3 program, funded by the Polish National Centre for Research and Development. Contract No. BIOSTRATEG3/347837/11/NCBR/2017.

References

1. Łabędzki L, Kuźniar A, Lipiński J, Mioduszewski W (2005) Irrigation Management and Transfer in European Countries of Transition (Polish Report). Polish National Committee,

International Commission on Irrigation and Drainage. Institute for Land Reclamation and Grassland Farming, Falenty, Aug. www.imuz.edu.pl

2. Łabędzki L (2009) Droughts in Poland—their impact on agriculture and mitigation measures. In: Leśny J. (eds) Climate change and agriculture in Poland—impacts, mitigation and adaptation measures. Acta Agropyhsica No. 169. Institue of Agrophysics, Polish Academy of Sciences, Treatises and Monographs (1)
3. GUS. Statistical Yearbook of the Republic of Poland (2018)
4. Fertilizer use by crop in Poland (2003) Chapter 1 "Introduction". Food and Agriculture Organization of the United Nations. www.fao.org/docrep/005/Y4620E/y4620e05.htm#bm05.1. 8 Oct 2018
5. Gleboznawstwo [Soil Sciences] (1999) Ed. S. Zawadzki. PWRiL, Warszawa
6. Woś A (1999) Klimat Polski [Climate of Poland]. Wydaw. Nauk. PWN, Warszawa
7. Klimada (2013) Polish National Strategy for Adaptation to Climate Change (NAS 2020) with the perspective by 2030. Ministry of the Environment Republic of Poland
8. Water Act (2017) Ustawa z dnia 20 lipca 2017 r. Prawo wodne. (Dz. U. poz. 1566 i 2180 z 2017 r. oraz z 2018 r. poz. 650)
9. Environment (2017) Central Statistical Office. Warsaw
10. Mioduszewski W, Szymczak T, Kowalewski Z (2011) Gospodarka wodna jako dyscyplina naukowa służącą rolnictwu [Water management as a scientific discipline serving agriculture]. Woda-Środowisko-Obszary Wiejskie. 11(1):179–202
11. Łabędzki L (2006) Susze rolnicze. Zarys problematyki oraz metody monitorowania i klasyfikacji [Agricultural droughts. An outline of problems and methods of monitoring and classification]. Rozprawy naukowe i monografie nr 17. Falenty. Wyd. IMUZ
12. Scheumann W, Freisem C (2002) The role of drainage for sustainable agriculture. J Appl Irrig Sci 37(1):33–61
13. Lipiński J (2016) Stan, potrzeby i uwarunkowania rozwoju melioracji w Polsce [State, needs and conditions of development of land reclamation in Poland]. In: Dembek W, Kuś J, Wiatkowski M, Żurek G (eds) Innowacyjne metody gospodarowania zasobami wody w rolnictwie [Innovative methods of water resources management in agriculture]. CDR Brwinów, pp 79–89
14. Brandyk T, Skąpski K, Szatyłowicz J (1993) Design and operation of drainage-subirrigation systems in Poland. Irrigat Drain Syst 7:173–187
15. Kaca E (2015) Podstawy metodyczne obliczeń w programowaniu rozwoju melioracji wodnych. Aspekty rzeczowo-kosztowe. [The methodological basis for calculations in programming of the land reclamation development. Material and cost aspects]. Woda-Środowisko-Obszary Wiejskie. Rozprawy naukowe i monografie nr 41. Falenty. Wyd. ITP
16. Kaca E (2015) Średnio- i długookresowe program rozwoju melioracji wodnych w skali kraju i województw. [Medium and long-term program for the development of water melioration on the national and provincial level]. Falenty. Wyd. ITP
17. Kaca E (2017) Methodology of assessing the relative environmental validity of developing drainage and irrigation on a regional scale. J Water Land Dev 35:101–112
18. Kaca E, Rek-Kaca G (2017) Względna przyrodnicza zasadność rozwoju odwodnień i nawodnień w skali województw [The relative validity of developing drainage and irrigation in the provinces scale]. Woda-Środowisko-Obszary Wiejskie 17(4):67–80
19. Łabędzki, L (2009) Expected development of irrigation in Poland in the context of climate change. J Water Land Dev 13b:17–29
20. Quevauviller P (2010) Is IWRM achievable in practice? Attempts to break disciplinary and sectoral walls through a science-policy interfacing framework in the context of the EU Water Framework Directive. Irrig Drain Syst 24:177–189
21. Kaca E (2015) Rozwój melioracji wodnych w świetle wojewódzkich opracowań strategicznych [Development of reclamation and water management in view of provincial strategic plans]. Woda-Środowisko-Obszary Wiejskie. Rozprawy naukowe i monografie nr 38. Falenty. Wyd. ITP
22. Szejba D, Papierowska E, Cymes I, Bańkowska A (2016) Nitrate nitrogen and phosphate concentrations in drainflow: an example of clay soil. J Elem 21(3):899–913

Part III
Water Quality Evaluation

Chapter 6
Quality Assessment of Water Resources of River in Poland

Adam Solarczyk

Abstract The chapter presents an assessment of water quality in Polish rivers as examined in 2010–2015 under the State Environmental Monitoring Programme. Using selected averaged physicochemical indicator values, an analysis was made of the spatial variability of mineral pollution (SEC, Total hardness), organic pollution (BOD$_5$, TOC) and biogenic pollution (TN and TP) in the rivers of the catchments and subcatchments of the Vistula, the Odra, and the Baltic Sea. Graded indexation of six quality characteristics (SEC, Total hardness, BOD$_5$, TOC, TN, TP) characterising particular types of pollution has shown that there are rivers with high levels of pollution in catchments in Wielkopolska, Kujawia and Górny Śląsk. There are very low and low levels of pollution in mountain and foothill rivers of the Carpathians and Sudetes, and in the Pomorze region of northern Poland. The longest rivers in Poland (the Vistula and the Oder) and their main tributaries have different water qualities that vary along their courses. Due to heavily mineralised drainage waters from bituminous coal mines in Górny Śląsk, the upper sections of the Vistula and Odra exceed norms in terms of mineral substances, in particular, chloride and sulphate ions. In over 63% of river Surface Water Bodies (SWBs) surveyed in 2010–2015 in Poland, the quality of water in terms of physicochemical characteristics correspond to class I and II standards according to the Water Framework Directive in force in Poland. The physicochemical state of the rivers does not correlate with their biological state. Only 36% of the surveyed river SWBs met the class I and II standards for biological elements. In terms of physicochemical and biological water quality, only 28% of the analysed Polish river SWBs are classified as being in a good or very good ecological state in 2010–2015.

Keywords River water quality · Water framework directive · Vistula basin · Oder basin · Poland

A. Solarczyk (✉)
Department of Hydrology and Water Management, Faculty of Earth Sciences and Spatial Management, Nicolaus Copernicus University, Lwowska 1, 87-100 Toruń, Poland
e-mail: adamsol@umk.pl

© Springer Nature Switzerland AG 2021
M. Zeleňáková et al. (eds.), *Quality of Water Resources in Poland*, Springer Water,
https://doi.org/10.1007/978-3-030-64892-3_6

121

6.1   Introduction

Poland's surface water resources are extremely limited compared to those of most European countries. *Per capita*, they amount to only approx. 1,600 m^3/year, while the European average is 4,560 m^3/person/year [1]. Its water resources are distributed very unevenly. In particular, the Kujawia region has a water deficit problem. Poland's problems with the amount of water resources are being aggravated by climate change. It is estimated that resources have been decreasing at a rate of 10 m^3/person/year [2]. The fact that Poland's water resources are very limited requires that care be taken to protect their quality because it is quality that determines the use of water for economic purposes. Knowing the quality of water on the macro scale makes it possible to develop water management plans for protecting it against pollution. On the micro-scale, it makes it possible to specify the use of water resources for economic purposes in river and lake catchments.

6.2   The Anthropogenic Factor in River Pollution in Poland

The quality of water in rivers is shaped by natural conditions in the catchment, and by the degree of anthropogenic pressure, which depends on how intensively natural resources are used. Agriculture is the most significant source of pollution. In Poland, agricultural-use areas occupy about 29.1 million ha, which is over 93% of the country's area. Catchments in central Poland have a high share of agricultural land. In terms of suitability for agricultural production, most of the areas with the lowest production value are found in the Podlaskie, Mazowieckie, and Pomorskie voivodeships [3]. Intensive agricultural management is carried out in most of the catchments in the following voivodeships: Opolskie, Dolnośląskie, Wielkopolskie, Lubuskie, Kujawsko-Pomorskie, Zachodnio-Pomorskie, Pomorskie, and Warmińsko-Mazurskie. The highest doses of mineral fertilisation with nitrogen, phosphorus, and calcium compounds are used in farms in the Opolskie, Dolnośląskie, Wielkopolskie and Kujawsko-Pomorskie voivodeships. In the rest of Poland, the agricultural economy is extensive. There, farms with an area of < 10 ha dominate.

A significant pollution load reaches the rivers from point sources. Changes to the economy after 1989 resulted in a decrease in sewage output, while the share of treated sewage increased. According to GUS [4] data, in total, 2,122,133,000 m^3 of sewage was discharged into surface waters (6,800 m^3/km^2, 56 m^3/inhabitant). Over 95% of wastewater was subjected to mechanical, chemical, and biological cleaning processes along with increased treatment of biogenic compounds. Over 51% of all wastewater was treated in a tertiary treatment process. The largest municipal and industrial point sources discharged sewage primarily to Poland's main rivers. The exception is wastewater discharged from Łódź to the Ner, from Lublin to the Bystrzyca, and from Białystok to the Supraśl. The largest number of pollution point sources is found in the Górny Śląsk conurbation.

6.3 River Water Quality in Poland, 2010–2015

The Water Framework Directive (WFD) introduced a systematic study of rivers in discrete Surface Water Bodies (SWBs). In the years 2010–2015, which covered the water management "Cycle 2", monitoring research was carried out under the State Environmental Monitoring Programme by Voivodeship Inspectorates for Environmental Protection at over 2,300 sites in water bodies of various abiotic types (27 abiotic river types have been designated in Poland). Depending on the type of monitoring performed in water bodies (diagnostic, operational, research), water samples are taken for physico-chemical indicators between 6 and 12 times a year, which gives a total of over 20,000 assays per indicator, creating a comprehensive database.

The study of river waters under the WFD covers a wide range of physico-chemical indicators. Some of the indicators are assayed obligatorily, regardless of the type of monitoring and abiotic type of water body. Based on these indicators, the degree of water salinity (general conductivity [SEC], and total hardness), organic compound load (BOD_5 and TOC) and biogenic compound content (Total Nitrogen, Total Phosphorus). These indicators were used to assess the spatial diversity of water quality in watercourses in the catchments of the main courses of Poland's largest rivers and their subcatchments. The assessment does not include the results of research in Poland's largest main rivers: the Vistula, Oder, San, Narew, Bug, and Warta. In the above-mentioned rivers, the variability of selected indicators along their course is described.

The physico-chemical properties of river waters and the hydromorphological features of individual river sections affect the growth of hydrobionts, whose status was assessed by biological indicators as the basis for ecological classification according to the WFD.

In 2010–2015 Polish rivers varied significantly in water salinity, organic compound loads, and biogenic substances. The ranges of values for specific electrical conductivity, total hardness, and total nitrogen concentration were particularly wide (Table 6.1). For this reason, the best measure of the average of the selected indicators was determined to be the median. To assess the spatial diversity of water quality in Polish rivers, a division into five groups was adopted. Group threshold values were determined based on the fractal distribution of average values for the catchment and subcatchments of the Polish rivers. The class limits were 20, 40, 60, and 80% [5].

6.3.1 Spatial Variability of River Water Quality

The extent of river water pollution by dissolved mineral compounds was diverse across the catchment system (Fig. 6.1). The highest specific electrical conductivity (SEC) was found in Martwa Wisła (7,556 µS/cm), which is exposed to the inflow of marine salt waters due to the current hydrographic system. An additional factor in the salinity increase in this former Vistula estuary section may be the inflow of

Table 6.1 Basic statistics of annual average values of selected quality indicators of Polish river waters examined in 2010–2015 (own calculations according to data of Voivodeship Inspectorates for Environmental Protection)

Indicator	Unit	Number	Min.	Max.	Median	Mean	95th percentile
SEC (20 °C)	μS/cm	2,362	46	33,249	476	636	1,119
Total hardness	mgCaCO$_3$/L	2,296	5.2	3,598	235	257	449
BOD$_5$	mgO$_2$/L	2,383	0.4	58.4	2.5	3.0	5.6
TOC	mgC/L	2,353	0.3	88.1	8.8	9.5	18.3
TP	mgP/L	2,374	0.008	5.350	0.160	0.222	0.560
TN	mgN/L	2,371	0.09	38.25	2.63	3.43	8.65

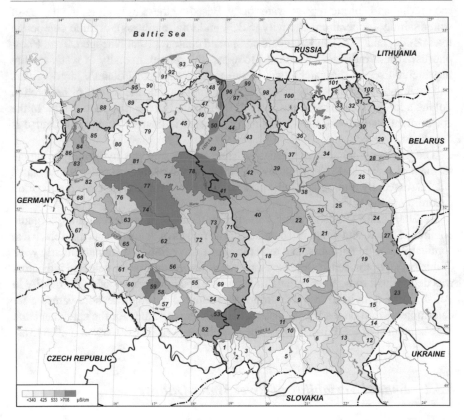

Fig. 6.1 Spatial diversity of average specific electrical conductivity (μS/cm) in Polish river catchments and subcatchments in the years 2010–2015 (own elaboration based on the data of Voivodeship Inspectorates for Environmental Protection). Explanations: catchments and subcatchments numbered as in Tables 6.2, 6.3, and 6.4

pollutants from the Gdańsk industrial zone. In the rivers of the Przemsza and Kłodnica catchments, the high SEC is caused by the inflow of mining discharge waters loaded above all with significant levels of chlorides and sulphates. These catchments contain the most industrialised areas in Poland, with mining, metallurgical and chemical industries. High contents of the main ions is also a feature of river catchments with intensive farming in the Wielkopolska and Kujawia regions. The lowest levels of dissolved mineral substances are seen in rivers in mountain and foothill catchments (Mała Wisła and Biała, Soła [170 μS/cm], Skawa, Dunajec, Nysa Kłodzka) and in sandy areas with high proportions of forest (e.g. Bóbr, Liswarta, Drawa, Gwda, Pisa, Tanew, Lubaczówka).

The vast majority of Polish rivers have high total hardness, at over 180 mgCaCO$_3$/L (Fig. 6.2). The catchments where rivers have the lowest content of calcium and magnesium ions are located in southern and south-eastern Poland (the Mała Wisła and Biała subcatchments; and the catchment of the Soła [87

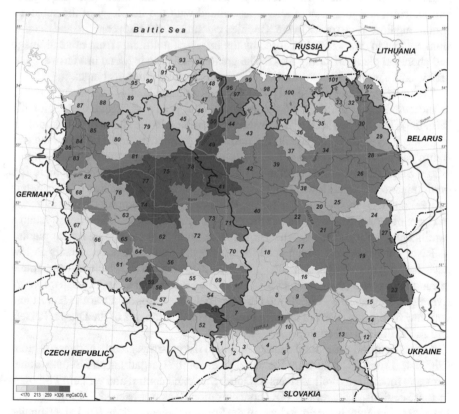

Fig. 6.2 Spatial diversity of average total hardness (mgCaCO$_3$/L) in Polish river catchments and subcatchments in 2010–2015 (own elaboration based on data of Voivodeship Inspectorates for Environmental Protection). Explanations: catchments and subcatchments numbered as in Tables 6.2, 6.3, and 6.4

mgCaCO$_3$/L], Skawa and Tanew). The exception is the rivers of the Wisłok river catchment, where the total hardness (259 mgCaCO$_3$/L) is higher, and this increase is likely caused by the high content of calcium and magnesium in flysch rocks in the Rzeszów region [6]. Low total hardness is also found in river catchments covering the Sudete region and forest complexes of Dolny Śląsk (Nysa Kłodzka, Bóbr [90 mgCaCO$_3$/L], Nysa Łużycka). Central Pomorze is an area that stands out among Poland's lowland areas; in the Łupawa, Słupia, and Wieprza river catchments, average total hardness did not exceed 170 mg of CaCO$_3$/L. Elsewhere, in the lowland regions of Wielkopolska, Kujawia, Mazowsze, Podlasie and Lubelszczyzna, total hardness in rivers is high, which is caused by the systematic liming of soil. Within these regions, there is a particularly high content of calcium and magnesium ions in the catchment of the Wełna and Zgłowiączka rivers and the Obra Canal, as well as in the subcatchments of the upper Noteć, middle Warta and lower Vistula. The high total hardness of the Ślęża river catchment (Dolny Śląsk) is probably caused by agrotechnical treatments. The enrichment of the rivers of the Huczwa river catchment (Zamość region) with calcium and magnesium ions is associated with the lithology of the substrate containing limestone rocks. In the Kłodnica catchment, the high calcium and magnesium content in rivers may be caused by the inflow of pollutants from mine drainage and chemical plants in Górny Śląsk. The greatest hardness is found in Martwa Wisła (Table 6.2). Gdańsk's chemical industry zone is located along its banks.

The average content of easily decomposed organic matter in Poland's rivers, as measured using the BOD$_5$ index, was low and poorly differentiated regionally (Fig. 6.3). The highest BOD$_5$ values were determined in the rivers of Wielkopolska (e.g., the Barycz river catchment, the middle Warta subcatchment), Mazowsza (the Bzura catchment, the subcatchment of left tributaries of the middle Vistula), and Górny Śląsk (the Kłodnica catchment). In Wielkopolska and Mazowsza, there is a predominance of small towns and villages, which may be the source of the organic matter contained in sewage. The higher loads of easily decomposed organic matter in rivers of Górny Śląsk are due to the region's very high population density, and a significant quantity of sewage discharged from municipal sewage treatment plants.

River catchments in which forest areas prevail and population density is low have low organic matter contents. The lowest average BOD$_5$ values were found in mountain rivers (the catchments of the Soła, Skawa, Raba, and Dunajec rivers), rivers of north-eastern Poland (the Nurzec, Biebrza, Rospuda, Czarna Hańcza and Omulew rivers) and rivers of western Pomorze (the catchments of the Drawa, Parsęta and Rega) (Tables 6.3 and 6.4).

Organic carbon compound content in Polish rivers varies greatly between regions (Fig. 6.4). Low TOC values are typical of rivers of the Carpathian and Sudete mountains and foothills, as well as rivers of Dolny Śląsk and central and western Pomorze. In these regions, the average TOC concentration does not exceed 7.8 mgC/L. The lowest TOC content is found in rivers of the Soła catchment (2.2 mgC/L), the Dunajec (2.4 mgC/L) and the Skawa (2.9 mgC/L). The highest TOC concentrations (> 9.9 mgC/L) occur in the rivers of eastern and north-eastern Poland (e.g., the Krzna, upper Narew and Węgorapa catchments), as well as in Wielkopolska and western and north-western Poland. Particularly high organic carbon contents are found in the

Table 6.2 Average (median) concentrations of selected physico-chemical indicators of water pollution in the catchments and subcatchments[1] of the Vistula basin in 2010–2015 (own calculations according to data from Voivodeship Inspectorates for Environmental Protection)

No.	Catchments/subcatchments	SEC (20 °C) µS/cm	Hardness mgCaCO₃/L	BOD₅ mgO₂/L	TOC mgC/L	TN mgP/L	TP mgN/L
1	Mała Wisła and Biała[1]	283	102	2.8	5.1	2.61	0.120
2	Soła	170	87	1.6	2.2	0.99	0.020
3	Skawa	274	129	1.7	2.9	1.80	0.050
4	Raba	378	184	1.8	3.2	1.47	0.061
5	Dunajec	332	175	1.8	2.4	1.16	0.047
6	Wisłoka	345	183	1.9	4.3	1.62	0.085
7	Przemsza	724	321	2.9	8.9	4.10	0.180
8	Nida	464	205	2.1	8.0	3.42	0.120
9	Czarna	438	214	2.6	8.9	2.52	0.130
10	Small right tributaries of the upper Vistula[1]	470	204	2.6	9.0	2.91	0.137
11	Small left tributaries of the upper Vistula[1]	627	315	2.4	6.6	3.79	0.158
12	Small tributaries of the San[1]	372	197	2.2	6.2	1.69	0.097
13	Wisłok	511	259	2.6	5.2	2.48	0.122
14	Lubaczówka	330	176	2.2	7.0	1.52	0.098
15	Tanew	320	169	3.4	8.1	1.80	0.120
16	Kamienna	404	160	2.4	7.7	2.86	0.130
17	Radomka	455	233	2.9	10.2	2.90	0.165
18	Pilica	345	172	2.5	9.4	2.20	0.120
19	Wieprz	530	279	3.2	9.2	2.50	0.160
20	Świder	472	229	3.5	9.3	3.03	0.149
21	Small right tributaries of the middle Vistula[1]	530	261	3.4	8.7	3.24	0.181
22	Small left tributaries of the middle Vistula[1]	586	273	4.3	8.9	3.40	0.224
23	Huczwa	725	385	2.9	10.0	2.35	0.110
24	Krzna	446	206	1.9	14.8	2.07	0.155
25	Liwiec	447	225	3.0	10.0	2.27	0.145
26	Nurzec	391	300	1.8	9.7	4.80	0.180
27	Small tributaries of the Bug[1]	594	227	3.5	10.5	2.58	0.170
28	Small tributaries of the upper Narwa (to Biebrza)[1]	475	263	2.7	12.8	3.34	0.180

(continued)

Table 6.2 (continued)

No.	Catchments/subcatchments	SEC (20 °C) µS/cm	Hardness mgCaCO$_3$/L	BOD$_5$ mgO$_2$/L	TOC mgC/L	TN mgP/L	TP mgN/L
29	Supraśl	390	233	2.3	11.6	2.57	0.200
30	Biebrza, excluding the Ełk, Rospuda-Netta and Jegrznia[1]	480	281	1.8	11.3	2.60	0.080
31	Rospuda-Netta	481	262	1.6	11.2	1.72	0.068
32	Jegrznia	378	253	2.2	11.6	2.10	0.078
33	Ełk	355	246	2.0	11.1	2.00	0.090
34	Small tributaries of the middle Narwa (to Pułtusk)[1]	484	274	2.2	9.1	3.30	0.170
35	Pisa	274	176	2.6	10.5	1.60	0.150
36	Omulew	351	184	1.8	11.8	1.37	0.169
37	Orzyc	518	250	2.0	8.2	2.89	0.240
38	Small tributaries of the lower Narwa[1]	530	219	3.6	9.8	2.54	0.184
39	Wkra	644	301	2.2	8.6	3.38	0.205
40	Bzura	577	270	4.1	9.5	3.92	0.250
41	Zgłowiączka	792	391	2.3	13.3	2.24	0.260
42	Skrwa Prawa	571	305	3.4	10.1	3.30	0.131
43	Drwęca	480	253	2.1	8.5	2.42	0.200
44	Osa	599	314	2.0	9.2	1.23	0.215
45	Brda	370	190	1.9	6.3	1.51	0.120
46	Wda	366	188	1.5	5.8	1.23	0.215
47	Wierzyca	423	221	2.4	6.5	2.61	0.320
48	Radunia-Motława	388	192	2.0	5.2	1.39	0.130
49	Small right tributaries of the lower Vistula[1]	648	353	2.5	10.8	3.19	0.230
50	Small left tributaries of the lower Vistula[1]	783	419	2.5	10.0	3.56	0.280
51	Martwa Wisła[1]	7,556	954	2.3	6.0	0.94	0.140

rivers of the lower Warta subcatchment (22.9 mgC/L), the Pliszka river catchment (19.5 mgC/L) and the lower Oder subcatchment (15.3 mgC/L). In these hydrographic units, there are significant areas of wetlands, which are a source of organic carbon.

The content of total nitrogen in rivers of Poland varied considerably between regions (Fig. 6.5). The lowest concentrations of nitrogen compounds occurred in rivers of: mountain and foothill catchments of the Soła, Raby, Dunajca and Wisłoka; the Pomorksie Lakeland (the Wda, Brda, Radunia-Motława and Drawa catchments);

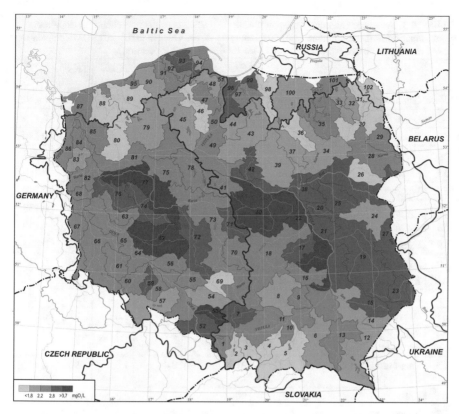

Fig. 6.3 Spatial diversity of average BOD$_5$ (mgO$_2$/L) in Polish river catchments and subcatchments in 2010–2015 (own elaboration based on data from Voivodeship Inspectorates for Environmental Protection). Explanations: catchments and subcatchments numbered as in Tables 6.2, 6.3, and 6.4

the Mazurskie Lakeland (the Pisa and Omulew catchments); and the Suwalskie Lakeland (the Czarna Hańcza catchment). Average total nitrogen in the rivers of these catchments did not exceed 1.64 mgN/L. The region with the highest concentrations of total nitrogen is Wielkopolska. The high content of nitrogen compounds in the rivers of the Obra Canal, middle Warta, Prosna and Wełna catchments is mainly caused by intensive farming. The impact of agriculture on nitrogen compound contents in the rivers is confirmed by high nitrate concentrations in drain water and drainage ditch water [5].

In the catchments of the Kłodnica, Ner and Nurzec rivers, elevated nitrogen compound concentrations mainly result from their inflowing from point sources of pollution.

In terms of content in Polish rivers, phosphorus compounds were spatially distributed similarly to nitrogen compounds (Fig. 6.6). The lowest total phosphorus concentrations (< 0.099 mgP/L) were found in: the mountain and foothill rivers of the Carpathians (the Soła, Skawa, Raba, Dunajec, Wisłoka and San river catchments); the

Table 6.3 Average (median) concentrations of selected physico-chemical indicators of contamination in the catchments and subcatchments of the Oder in 2010–2015 (own calculations based on data of Voivodeship Inspectorates for Environmental Protection)

No.	Catchments/subcatchments	SEC (20 °C) μS/cm	Hardness mgCaCO₃/L	BOD₅ mgO₂/L	TOC mgC/L	TN mgP/L	TP mgN/L
52	Upper Oder and Olza[1]	672	233	3.0	6.7	3.75	0.179
53	Kłodnica	2,087	390	4.0	10.0	5.24	0.280
54	Mała Panew	362	183	2.2	8.2	3.90	0.135
55	Stobrawa	362	152	2.7	10.5	2.65	0.180
56	Widawa	628	322	2.6	8.2	3.83	0.180
57	Nysa Kłodzka	280	132	2.1	3.9	2.74	0.158
58	Oława	689	298	2.8	5.6	3.15	0.140
59	Ślęza	1,105	381	3.3	7.8	4.05	0.425
60	Bystrzyca	451	207	2.8	6.0	3.66	0.215
61	Kaczawa	483	214	2.5	6.1	3.71	0.124
62	Barycz	545	267	3.9	11.7	3.57	0.255
63	Obrzyca	588	250	2.9	9.0	1.82	0.188
64	Small right tributaries of the middle Oder[1]	531	254	2.7	9.3	2.08	0.130
65	Small left tributaries of the middle Oder[1]	583	272	2.4	7.6	2.90	0.153
66	Bóbr	239	90	2.7	4.5	2.32	0.125
67	Nysa Łużycka	393	164	2.7	6.0	2.5	0.162
68	Pliszka	345	200	2.5	19.5	0.73	0.140
69	Liswarta	331	145	1.6	9.4	3.79	0.130
70	Widawka	466	199	2.3	10.5	3.37	0.145
71	Ner	529	266	3.7	10.5	5.00	0.260
72	Prosna	485	230	3.2	8.7	5.17	0.216
73	Small tributaries of the upper Warta (to the Prosna)[1]	605	277	2.7	12.0	4.00	0.197
74	Obra Canal	918	423	3.5	13.7	10.1	0.255
75	Wełna	626	328	3.2	12.5	4.45	0.226
76	Obra	505	245	2.9	12.7	2.18	0.210
77	Small tributaries of the middle Warta (to the Noteć)[1]	832	362	4.3	13.0	5.85	0.296
78	Upper Noteć (to the Bydgoszcz Canal)[1]	1,106	470	3.3	10.3	3.26	0.245
79	Gwda	295	173	2.1	6.8	2.02	0.140

(continued)

Table 6.3 (continued)

No.	Catchments/subcatchments	SEC (20 °C) μS/cm	Hardness mgCaCO₃/L	BOD₅ mgO₂/L	TOC mgC/L	TN mgP/L	TP mgN/L
80	Drawa	311	176	1.6	7.2	1.27	0.089
81	Lower Noteć, excluding the Gwda and Drawa[1)]	536	268	2.3	11.8	2.50	0.175
82	Small tributaries of the lower Warta[1]	451	234	2.3	22.9	1.20	0.140
83	Myśla	558	298	2.1	18.0	2.33	0.070
84	Płonia	579	292	2.6	11.7	2.39	0.127
85	Ina	521	269	2.8	13.8	3.04	0.190
86	Small tributaries of the lower Oder[1)]	532	267	2.7	15.3	2.22	0.135

Table 6.4 Average (median) concentrations of selected physico-chemical indicators of water pollution in catchments and subcatchments[1] of the Baltic Coast, Szczecin Lagoon, the Vistula Lagoon (including the Pregola[2] catchment) and the Neman[3] in 2010–2015 (own calculations according to data from Voivodeship Inspectorates for Environmental Protection)

No.	Catchments/subcatchments	SEC (20 °C) μS/cm	Hardness mgCaCO₃/L	BOD₅ mgO₂/L	TOC mgC/L	TN mgP/L	TP mgN/L
87	Small tributaries of the Szczecin Lagoon[1]	448	213	2.5	14.9	2.38	0.140
88	Rega	433	215	1.7	9.1	2.36	0.108
89	Parsęta	372	188	1.6	6.9	1.95	0.120
90	Wieprza	316	167	1.9	6.0	1.67	0.120
91	Słupia	308	164	2.7	4.9	1.73	0.115
92	Łupawa	313	161	2.6	5.4	2.33	0.130
93	Łeba	340	178	3.0	6.2	2.20	0.105
94	Reda	383	195	2.2	4.9	1.62	0.170
95	Small coastal rivers[1]	447	205	2.6	8.9	2.18	0.202
96	Nogat[1]	602	305	3.5	11.4	3.55	0.360
97	Elbląg	693	275	2.3	10.6	2.14	0.190
98	Pasłęka	477	246	1.8	10.8	2.06	0.190
99	Small tributaries of the Vistula Lagoon[1]	553	245	3.0	11.8	2.50	0.200
100	Łyna[2]	470	243	2.2	13.5	2.35	0.146
101	Węgorapa[2]	321	240	2.7	12.4	2.20	0.100
102	Czarna Hańcza[3]	415	232	1.8	9.3	1.13	0.028

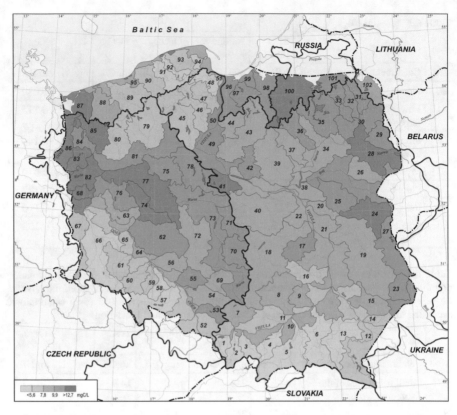

Fig. 6.4 Spatial diversity of average TOC (mgC/L) in Polish river catchments and subcatchments in 2010–2015 (own elaboration based on data from Voivodeship Inspectorates for Environmental Protection). Explanations: catchments and subcatchments numbered as in Tables 6.2, 6.3, and 6.4

lakeland and lowland rivers of north-eastern Poland (the Pisa, Rospuda, Jegrznia, Ełk, Czarna Hańcza and Biebrza river catchments); and the Drawa and Myśli river catchments in the Pomorskie Lakeland. The highest concentrations of total phosphorus were found in the rivers of Wielkopolska, Kujawia, Mazowsza and north-central Poland (the Nogat, Osa, Drwęca, Wierzyca, Wda, and Pasłęka river catchments), as well as in Dolny Śląsk (the Ślęża and Bystrzyca river catchments). These regions' elevated phosphorus concentrations as compared to other areas of Poland are caused by influx from diffuse sources (intensive farming) and point sources (mechanical and biological sewage treatment plants).

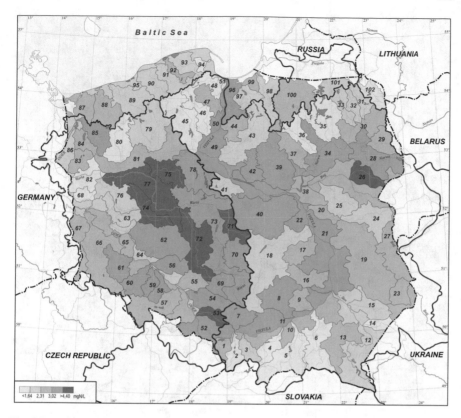

Fig. 6.5 Spatial diversity of average total nitrogen (mgN/L) in Polish river catchments and subcatchments in 2010–2015 (own elaboration based on data from Voivodeship Inspectorates for Environmental Protection). Explanations: catchments and subcatchments numbered as in Tables 6.2, 6.3, and 6.4

6.3.2 The Extent of Water Pollution in Polish River Catchments, by Region

Polish rivers are very diverse in terms of mineral, organic, and biogenic pollution. This is determined by the diversity of Poland's natural conditions and the varying intensity of anthropogenic pressure in different regions. Based on analysis of the degree of river pollution based on six quality characteristics (SEC, Total hardness, BOD_5, TOC, TN and TP), Polish river catchments and subcatchments were indexed. Ten classes were established for each indicator. The thresholds for these classes were established at every 10th fractal. Each class was assigned from 1 to 10 points. Based on the average score, five classes were then distinguished (every 2 points), and then assigned the following river pollution grades: Class I—very low; class II—low; class III—moderate; class IV—moderately high; class V—high.

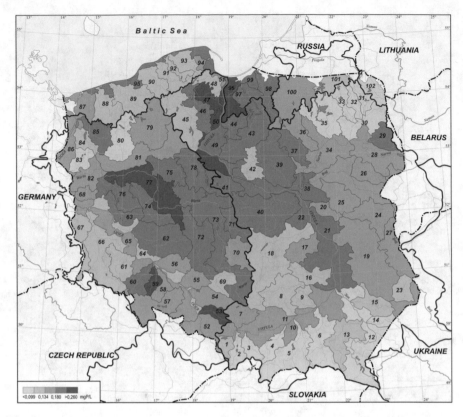

Fig. 6.6 Spatial diversity of average total phosphorus (mgP/L) in Polish river catchments and subcatchments in 2010–2015 (own elaboration based on data from Voivodeship Inspectorates for Environmental Protection). Explanations: catchments and subcatchments numbered as in Tables 6.2, 6.3, and 6.4

Spatial analysis of the six selected physico-chemical indicators of water quality shows that most of Poland has catchments where rivers have a moderate to a moderately high degree of pollution. Rivers with elevated pollution levels occur in central Poland (Fig. 6.7). The rivers in Wielkopolska in the Obra and Wełna catchments, and the middle Warta subcatchment, as well as in Kujawia (the upper Noteć subcatchment), are heavily polluted. The main significant sources of pollution are primarily intensive farming and heavy settlement. In the Ślęza catchment, the reduced quality of river water is primarily caused by pressure from the agricultural sector. In Górny Śląsk, including the Kłodnica and Przemsza river catchments, as well as in the subcatchment of the lower Vistula's left tributaries, the low water quality is primarily caused by pollution from the industrial activities and built-up areas of the Górny Śląsk conurbation.

The environmental conditions most conducive to maintaining good water quality occur in mountainous and foothill areas of the Carpathians, and parts of the Sudetes.

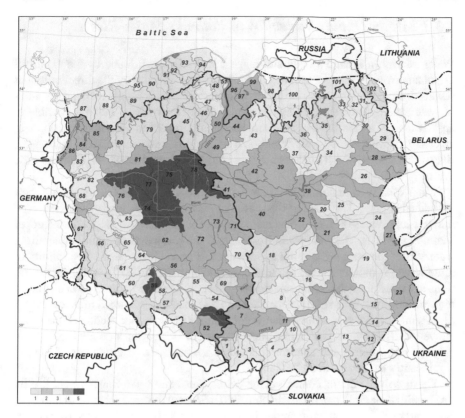

Fig. 6.7 Regionalisation of Polish river catchments and subcatchments in terms of the extent of water pollution in 2010–2015 (own elaboration based on data from Voivodeship Inspectorates for Environmental Protection) Explanation: Degree of water pollution: 1—very low; 2—low; 3—moderate; 4—moderately high; 5—high. Catchments and subcatchments numbered as in Tables 6.2, 6.3, and 6.4

The Soła, Skawa, Raby, and Dunajec catchment rivers have particularly low pollution levels for Poland. Lowland rivers with low pollution levels occur in the Pomorskie Lakeland (the Gwda, Drawa, Brda, Wda, Wierzyca, Parsęta, Wieprza, Słupia, Łupawa, Łeba, and Reda catchments) and the central part of the Mazurskie Lakeland (the Pisa and Omulew catchments). Within these catchments, significant areas are forested, and population density is among the lowest in Poland (< 50 person/km^2). In the remaining catchments with low river pollution levels (the Bóbr, Nysa Łużycka, Liswarta, Pilica, Wisłoka, Tanwia and Lubaczówka catchments, and the subcatchments of small tributaries of the San), forests and extensive farming play an important role in limiting the influx of mineral, organic and biogenic substances.

6.3.3 Water Quality Variability in Poland's Major Rivers

Water quality in Poland's main and longest rivers is conditioned by the export of pollutants from the largest tributary catchments and affected by the direct inflow of sewage from point sources in the largest cities or near major industrial plants. Studies of water quality indicators along the course of the main rivers make it possible to approximately assess the extent of anthropogenic impact in individual sections. In 2010–2015, water quality was monitored in various years in Poland's six longest rivers: the Vistula and its tributaries (the San, Narew, and Bug), the Oder and the Warta.

Water quality varies markedly between Poland's largest rivers, and along their individual courses. The Vistula is one of the rivers most heavily loaded with dissolved mineral substances (Fig. 6.8). The main reason for the increase in SEC in the upper Vistula, below the Goczałkowicki Reservoir is the discharge of highly mineralised waters from the draining of a bituminous coal mine in Górny Śląsk. The salinity of mine waters is mainly caused by chlorides and sulphates. In 2013–2015, concentrations of these ions in the Vistula upstream of the confluence with the Skawa displayed high variability, which strictly depended on their load in discharged mine waters [7, 8]. A similar process also increased water salinity in the upper section of the Oder. However, discharge waters from a mine in western Górny Śląsk do not increase the chloride and sulphate ion concentrations that are the main determinants of SEC as significantly as they do in the Vistula.

The dissolved mineral contents in the Warta, San, Narew and Bug are significantly lower, with a maximum of approximately 750 μS/cm in the upper section of the upper Bug. Part of the mineral pollutant load on the cross-border section of the Bug comes from Ukraine.

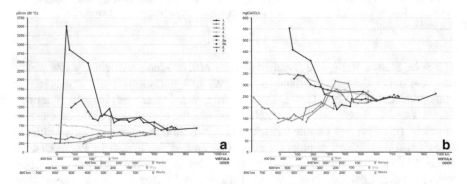

Fig. 6.8 Changes in average (median) SEC and total hardness in major Polish rivers in 2010–2015 (own elaboration based on data from Voivodeship Inspectorates for Environmental Protection) Explanations: (1) Vistula (2012); (2) San (2015); (3) Narew (2012); (4) Bug (2014); (5) Oder (2011); (5a) Western Oder (2011); (5b) Eastern Oder (2011) (6) Warta (2014); (7) estuary to the main river

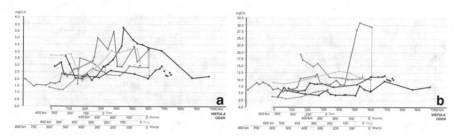

Fig. 6.9 Changes in average (median) BOD₅ and TOC in Poland's main rivers in 2010–2015 (own elaboration based on data from Voivodeship Inspectorates of Environmental Protection). Explanations as in Fig. 6.8

The content of readily decomposable organic matter, as expressed by the BOD₅ indicator, varies greatly along the course of Poland's main rivers (Fig. 6.9A). The fluctuations in organic compound loads in rivers are associated with the inflow of sewage from large cities and industrial plants on rivers and their tributaries. Anthropogenic organic matter content in rivers is generally low, not exceeding 6.0 mgO₂/L (class II according to WFD). The current low organic substance loads in rivers are the result of a long-term programme of constructing sewage treatment plants.

Total organic carbon concentrations in Poland's main rivers vary considerably (Fig. 6.9B). The lowest TOC occurs in a mountain river—the San. TOC loads in the San, as well as in the Vistula, Bug, Oder, and Warta, generally increase along their courses. In the case of the Warta, a significant increase in the TOC concentration is caused by the inflow of organic carbon from wetlands in the ice marginal valley section of the river. In the Narew, the highest TOC concentration occurs in the upper and middle sections of the catchment. This is associated with the increased share of peat soils in this part of the catchment [9].

The total organic carbon concentrations in the major rivers—the Vistula and the Oder—are lower than their lowland tributaries, the Narew, Bug, and Warta. TOC in the Vistula and Oder gradually increases along their courses. In the Vistula, the increase in the TOC concentration caused by its confluence with the Narew is reduced by up to 20% in the Włocławek Reservoir [9].

Biogenic substance concentrations varied significantly between the analysed main rivers, and along their courses (Fig. 6.10A, B). The Vistula's and Oder's nitrogen compound loads generally decreased towards their Baltic Sea estuaries. Only near Puławy on the Vistula was there a clear temporary increase in total nitrogen concentration, which is caused by the operation of large nitrogen plants. The average total nitrogen concentration in this section of the Vistula did not exceed the standards of class I according to WFD. The Warta's very significant nitrogen compound load in central Wielkopolska is mainly due to inflow from areas of intensive agriculture. The lowest nitrogen compound concentrations occurred in the San, whose catchment has a significant share of forests and a low population density. Detailed studies on the emission of nitrogen compounds from the Vistula and Oder catchments have shown that the inflow of this bioelement into the rivers and then the Baltic Sea in 1995–2015

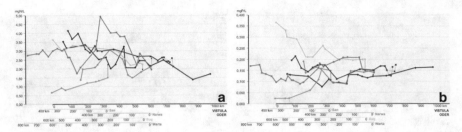

Fig. 6.10 Changes in average (median) total nitrogen and total phosphorus content in Poland's main rivers in 2010–2015 (own elaboration based on data from Voivodeship Inspectorates for Environmental Protection). Explanations as in Fig. 6.8

is mainly from drainage discharge (50% in the Oder catchment), groundwater (40% in the Vistula catchment) and discharge from sewage treatment plants [10].

Total phosphorus concentrations in Poland's main rivers ranged widely, from 0.020 mgP/L in the upper San to over 0.350 mgP/L in the Bug in the Kryłów profile—the first site tested in Poland. The high phosphorus compound concentrations in the upper Bug are therefore associated with their inflow from Ukraine. In the Bug's lower section, total phosphorus concentrations fell to 0.200 mgP/L. Phosphorus compound contents in the San increased significantly along the river's course. The San river catchment's middle and lower parts have large urban centres (Przemyśl, Rzeszów, Stalowa Wola) that discharge treated sewage into the San and its main tributaries (Wisłok). In the Vistula, Narew, Oder, and Warta, the highest total phosphorus concentrations occur in the rivers' upper and lower sections. In the case of the Warta, the increase in phosphorus compound concentrations in the Wielkopolska region is caused by the inflow of treated sewage from urban centres and the export of this bioelement from areas of intensive agricultural use. These phosphorus outflow sources account for approximately 80% of the bioelement's total emission from the Vistula and Oder catchments [10].

6.4 River Ecological Status Assessment According to the Water Framework Directive

The rational management of water resources requires knowledge of ecological status, which involves assessing physico-chemical and biological elements. The ecological status of Polish rivers examined under the WFD's Water Cycle 2 in 2010–2015 was classified according to the standards published in the Regulation of the Minister of the Environment regarding the classification method for surface water bodies and environmental quality standards for priority substances [11].

6.4.1 Assessment of Status of Physico-Chemical Elements

In physico-chemical terms, river water quality is classified according to the standards defining the threshold values for classes I and II. Due to the general assumption of the WFD, that the ecological status of surface waters is considered appropriate if it corresponds to two states: high and good, no standards are set for the other three classes of physico-chemical indicators. The standards for physico-chemical quality class I and II waters also takes into account rivers' abiotic types. The threshold values for physico-chemical indicators of river water quality are given in Table 6.5. According to WFD rules, the state of physico-chemical elements is determined by the least favourable indicator.

In physico-chemical terms, water quality is highest in the catchments and subcatchments of the Baltic Sea coast, Szczecin Lagoon, Vistula Lagoon (including the Pregola) and Neman (Fig. 6.11). Over 74% of water bodies tested in 2010–2015 were assessed as meeting the standards for cleanliness classes I and II. In the Vistula catchment, 66% of surface water bodies met WFD requirements. The lowest percentage of surface water bodies to belong to cleanliness classes I and II was in the Oder catchment, at 56%. The most frequently exceeded standards concerned indicators of biogenic pollution, and, to a lesser extent, organic.

The river water bodies of the top water quality class I occur mainly in the catchments of mountain and foothill rivers of the Carpathians (the Soła, Skawa, Raba, Dunajec, Wisłoka) and the Sudetes (the Nysa Kłodzka, Bóbr, Nysa Łużycka,

Table 6.5 Threshold values of selected of water quality in rivers designated as discrete Surface Water Bodies (according to the Regulation of the Minister of the Environment 2014)

Indicator	Unit	Water quality indicator threshold	
		Class I	Class II
BOD$_5$	mgO$_2$/L	≤ 3	≤ 6
TOC	mgC/L	≤ 10	≤ 15
TOC[1]	mgC/L	≤ 15	≤ 20
SEC	μS/cm	$\leq 1,000$	$\leq 1,500$
Hardness[2]	mgCaCO$_3$/L	≤ 200	≤ 300
Hardness[3]	mgCaCO$_3$/L	≤ 300	≤ 500
Ammonium	mgN-NH$_4$/L	≤ 0.78	≤ 1.56
Total Kjeldahl Nitrogen	mgN/L	≤ 1.00	≤ 2.00
Nitrate as nitrogen	mgN-NO$_3$/L	≤ 2.20	≤ 5.00
Total nitrogen	mgN/L	≤ 5.00	≤ 10.00
Phosphates	mgPO$_4$/L	≤ 0.20	≤ 0.31
Total phosphorus	mgP/L	≤ 0.20	≤ 0.40

[1] for type rivers 23 and 24
[2] for type rivers: 1.3–5, 8,10
[3] for type rivers 2, 6, 7, 9, 12, 14–26

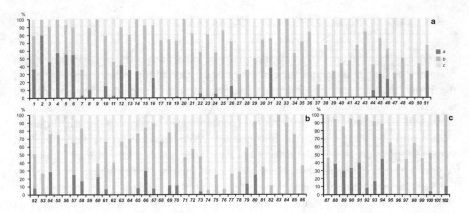

Fig. 6.11 Assessment of physico-chemical elements of the ecological status of Polish rivers according to the WFD in 2010–2015 (own elaboration based on data from Voivodeship Inspectorates for Environmental Protection). Explanations: A—Vistula catchment; B—Oder catchment; C—rivers of the Baltic Coast, Szczecin Lagoon, Vistula Lagoon, and Neman, a—class I; b—class II; c—below class II. Catchments and subcatchments numbered as in Tables 6.2, 6.3, and 6.4

Bystrzyca). There is a high proportion of the highest quality waters in riverine water bodies in Pomorze (the Wieprza, Parsęta, Słupia, Reda, Rega, Drawa and Gwda).

6.4.2 Assessment of the State of Biological Elements

The current system for assessing the ecological status of surface water bodies prioritises biological elements. Depending on the river's abiotic type, different groups of aquatic organisms are assessed (indicators characterising phytoplankton, phytobenthos, macrophytes, benthic macroinvertebrates, and ichthyofauna) using appropriate indicators. Five quality classes were adopted for indicators describing rivers' biological environments. As in the case of the physico-chemical characteristics of water, the most favourable biological statuses occur in rivers that feed directly into the Baltic Sea and its lagoons (Fig. 6.12). In biological terms, over 62% of water bodies meet requirements for classes I and II. In the Vistula and Oder catchments, the favourable status of the assessed biological elements in SWBs is significantly lower and amounts to 29% and 39% respectively. In the Vistula catchment, which is subject to almost relentless anthropogenic pressure, the studied biotic environmental elements occurred mainly in the rivers of the Carpathian mountains and foothills and in catchments with an increased share of forest areas (e.g., Brda, Lubaczówka). In the Oder catchment, there was a favourable biological status in rivers in catchments with larger forest complexes (e.g., the Bóbr, Nysa Łużycka, Gwda, and Liswarta). The percentage of rivers in the Vistula and Oder catchments that had satisfactory biological characteristics was lower than for rivers that directly discharge into the

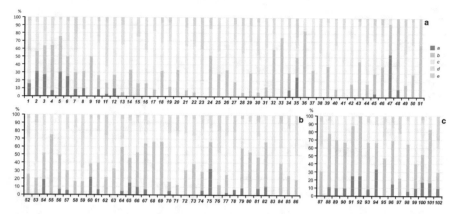

Fig. 6.12 Assessment of biological elements of the ecological status of Polish rivers according to the WFD in 2010–2015 (own elaboration based on data from Voivodeship Inspectorates for Environmental Protection). Explanations: A—Vistula catchment; B—Oder catchment; C—rivers of the Baltic Coast, Szczecin Lagoon, Vistula Lagoo, and Neman, a—class I; b—class II; c—class III; d—class IV; e—class V. Catchments and subcatchments numbered as in Tables 6.2, 6.3, and 6.4

Baltic Sea, and this may be caused by changes related to heavier regulation of river channels.

6.4.3   Assessment of Ecological Status

The water quality in the water bodies surveyed in 2010–2015 proved to be generally satisfactory. The ecological status of most Polish rivers is, however, not satisfactory. Only 28% of the riverine water bodies surveyed in 2010–2015 had a good or high ecological status (Fig. 6.13). The most favourable ecological status occurred in the rivers of the Baltic Coast, Szczecin Lagoon, Vistula Lagoon and Neman catchments (over 52%). In the Vistula and Oder catchments, about three quarters of water bodies failed to meet the requirements of the Water Framework Directive. The percentage of riverine water bodies that had a good or high status within the Vistula catchment was 23%, and 26% in the Oder catchment.

The low percentage of water bodies with a favourable ecological status is the results of the unsatisfactory state of the biotic environment elements in the rivers.

6.5   Conclusions

The water quality in surface water bodies surveyed in Poland in 2010–2015 was satisfactory. Over 63% of rivers had a good or high status according to the WFD's physico-chemical indicators. In surface water bodies, standards for biogenic and

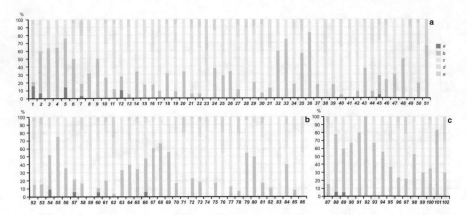

Fig. 6.13 Assessment of the ecological status of Polish rivers according to the WFD in 2010–2015 (own elaboration based on data from Voivodeship Inspectorates for Environmental Protection). Explanations: A—Vistula catchment; B—Oder catchment; C—rivers of the Baltic Coast, Szczecin Lagoon, Vistula Lagoon, and Neman, a—high; b—good; c—moderate; d—weak; e—bad. Catchments and subcatchments numbered as in Tables 6.2, 6.3, and 6.4

organic indicators were usually not met. By region, in terms of pollution, the Polish river catchments and subcatchments with the least favourable surface water quality occur in Wielkopolska, Kujawia and Górny Śląsk, and in the Ślęża catchment (Dolny Śląsk). There is a very low level of river water pollution in rivers of the Carpathian and Sudete mountains and foothills, as well as in central Pomorze. A factor contributing to maintaining good water quality in the rivers of these areas is the high share of forest areas in catchments (> 40%).

In 2010–2015, Poland's main rivers varied significantly in pollutant contents along their courses. At the majority of study sites on the main rivers, the values of indicators of mineral, organic, and biogenic pollutants did not exceed the water quality standards. The exceptions are the upper sections of the Vistula and the Oder, whose mineral substance loads exceed threshold limits, and the mouth section of the Warta, whose total organic carbon concentrations exceed norms.

Favourable features of physico-chemical water quality do not correlate with the status of rivers' biotic environments. Only 36% of river bodies in Poland were in classes I or II in terms of biological elements. In terms of the physico-chemical quality of surface waters, only 28% of the Polish rivers surveyed in 2010–2015 were considered to be of high or good ecological status.

Detailed analyses of changes in the content of biogenic and organic compounds have shown that the physico-chemical quality of Polish river waters is improving. Throughout 1985–2015, the Vistula and Oder catchments' discharge of phosphorus and nitrogen compounds to the Baltic decreased significantly, and one reason for this positive trend was the increase in the amount of treated municipal sewage [10]. The increase in the number of sewage treatment plants also reduced total organic content in the Vistula in 1960–2014 [9]. In the upper sections of the Vistula and Oder, the discharge of saline mining waters is a major problem. The highest chloride ion concentrations occurred in the early 1990s. As a result of introducing a hydrotechnical method for discharging mining wastewater, the chloride content in the Vistula fell to a level that was, relatively, stable, but still with significant fluctuations of 500–1,300 mgCl/L [7].

6.6 Recommendations

Due to Poland's water resources being very limited, and the possibility of them falling further as a result of climate change, further protection of waters against pollution is extremely important. Water quality can be improved by, among others, increasing the number of sewage treatment plants and by improving the performance of existing facilities. At present, in Poland, about 52% of wastewater requiring treatment is treated using enhanced nutrient removal processes. The increased effectiveness of treatment is also important because of the potential for more frequent occurrence of severe and long-term low waters. In some rivers, treated wastewater can be a significant component of their discharge and can significantly influence water quality during low water conditions.

Activities aimed at improving water quality are particularly important in areas with the smallest water resources, which include Wielkopolska and Kujawy. These regions also have higher levels of water pollution. For this reason, it is advisable to undertake measures to protect the waters of the Wielkopolska and Kujawy river catchments, due to the planned long-term governmental "Retention Development Programme". Reducing the export of pollutants from catchments, primarily including biogenic compounds, should limit the negative effects (water blooms) of water storage in the planned reservoirs, dammed lakes and other small retention facilities.

Acknowledgements The author thanks the employees of the Voivodeship Inspectorates for Environmental Protection for preparing and making available data from monitoring of the state of river cleanliness according to the Water Framework Directive.

References

1. Kundzewicz ZW, Zalewski M, Kędziora A, Pierzgalski E (2010) Water-related threats. Nauka 4(2010):87–96 (in polish)
2. Gutry-Korycka M, Jokiel P (2017) The balance and water resources of Poland: Surface water. In: Jokiel P, Marszelewski W, Pociask-Karteczka J (eds) Chapter 13.1. Hydrology of Poland. Wydawnictwo Naukowe PWN SA, pp 281–287 (in polish)
3. Krasowicz S, Górski T, Budzyńska K, Kopiński J (2009) Agricultural description of Polish territory. In: Igras, J, Pastuszak M (eds) Contribution of Polish agriculture to emission of nitrogen and phosphorus compounds to the Baltic Sea. IUNG-PIB Puławy, pp 41–108 (in polish)
4. GUS (Central Statistical Office) (2015) Environment 2015, Warszawa (in polish)

5. Igras J (2004) Mineral element concentrations in drainage water from agricultural area in Poland. *Monografie i rozprawy naukowe IUNG Puławy*, 13 (in polish)
6. Lis J, Pasieczna A (1995) Geochemical atlas of Poland 1: 2 500 000. PIG Warszawa (in polish)
7. Motyka J, Juśko K, Kasprzak A (2015) Changes of chloride concentration in the Vistula River in Kraków. Przegl Geol 63(10/2):963–966 (in polish)
8. Kasprzak A, Juśko K, Motyka J (2016) Changes of chloride and sulfates concentration in the Vistula River form the Goczałkowice reservoir to the Skawa River. Zesz Nauk Inst Gospodarki Surowcami Mineralnymi i Energią Polskiej Akad Nauk 94:197–204 (in polish)
9. Górniak A (2017) Spatial and temporal patterns of Total organic carbon along the Vistula River course (Central Europe). Appl Geochem 87:93–101. https://doi.org/10.1016/j.apgeochem.2017.10.006
10. Pastuszak M, Kowalkowski T, Kopiński J, Doroszewski A, Jurga B, Buszewski B (2018) Long-term changes in nitrogen and phosphorus emission into the Vistula and Oder catchments (Poland)—modelling (MONERIS) studies. Environ Sci Pollut Res. https://doi.org/10.1007/s11356-018-2945-7
11. Regulation of the Minister of the Environment regarding the classification method for surface water bodies and environmental quality standards for priority substances (Dz.U. 2014, item 1482) (in polish). http://prawo.sejm.gov.pl/isap.nsf/download.xsp/WDU20140001482/O/D20141482.pdf

Chapter 7
Water Quality in Main Dam Reservoirs in Poland

Damian Absalon, Magdalena Matysik, and Michał Habel

Abstract This chapter presents the analysis of data on water quality for 22 selected dam reservoirs in Poland. The conditions of water quality have been described based on available data. The characteristics of physical, chemical, and biological parameters have been presented, and ecological status, chemical status, and the overall status of reservoirs was assessed. As a result, good ecological status was established for 13 reservoirs and good chemical status for 8 reservoirs. However, the overall classification points to good water quality of only 6 dam reservoirs located in the Carpathians, in upper sections of rivers. Other studied reservoirs have a bad quality of water. The possibilities of modern continuous monitoring on one of the dam reservoirs have also been presented.

Keywords Water quality · Dam reservoirs · Ecological status · Chemical status · Continuous (real-time) monitoring · Poland

7.1 Introduction

Dam reservoirs are formed when a river valley is artificially divided with a dam [1]. So far in Poland, 101 dam reservoirs have been built with the storage capacity of over 1 mln^3 [2, 3], and their total capacity constitutes only about 5–6% of annual discharge to the Baltic Sea. [4]. It can thus be assumed that Poland does not have favourable conditions for creating big water reservoirs, which is the result of low flow in rivers, high flow changeability, and unfavourable natural conditions for their construction.

D. Absalon (✉) · M. Matysik
University of Silesia in Katowice, Katowice, Poland
e-mail: damian.absalon@us.edu.pl

M. Matysik
e-mail: magdalena.matysik@us.edu.pl

M. Habel
Kazimierz Wielki University in Bydgoszcz, Bydgoszcz, Poland
e-mail: mihab@ukw.edu.pl

© Springer Nature Switzerland AG 2021
M. Zeleňáková et al. (eds.), *Quality of Water Resources in Poland*, Springer Water,
https://doi.org/10.1007/978-3-030-64892-3_7

According to the Water Framework Directive (WFD, Directive 2000/60/EC), dam reservoirs are treated as heavily modified water bodies, so until 2021, they are required to reach at least good ecological status. As reservoirs are considered to be heavily modified sections of rivers, they are mostly not very similar to rivers. Each river, as regards ecology, is an open ecosystem, which is in a state of dynamic equilibrium [5, 6]. According to the *river continuum concept (RCC)* by [7] breaking its continuity undoubtedly influences changing hydrological conditions, which can significantly shape physical, chemical and biological processes determining the quality of water in the reservoir and thus in the river downstream of the dam [8–11]. In general, particularly in reservoirs with long retention time, the flow velocity is minimum, what makes them more similar to lakes. Like lakes, dam reservoirs should be characterised with primary production reflecting their processing capacity, which means that biomass production in a reservoir or lake should not exceed its destruction capacity in the course of natural processes. Imbalance of this equilibrium by increasing the content of compounds of allochtonic or autochthonic origin in water as a consequence leads to the changes triggering the deterioration of the chemical status and ecological status of the reservoir [12]. Dam reservoirs, despite being man-made structures, have developed rich ecosystems with a high diversity of living organisms [13–16]. The functioning of reservoirs significantly modifies balance and flow of energy [17] and matter in the World Ocean. The evidence for that is many times smaller transport of all kinds of load in the river downstream of the dam [18]. This is directly connected with reservoir accumulation [4, 19–23]. Most of the bed load remains in the reservoir. Suspended load and wash load is retained only in part. Effectiveness of dam reservoirs in trapping sediments is in some cases significant and amounts to 70–90% of the reservoir's storage capacity [24]. It is assumed that about 30–40% of matter transported in the form of suspension by rivers in the world does not reach seas, oceans, and some big lakes and is trapped in man-made dam reservoirs at least for the period of existence of this infrastructure [25].

For the purposes of implementing the Water Framework Directive, a distinction is made between heavily modified and artificial surface water bodies, in accordance with the applied parameters, into such category of surface water bodies, which is the closest to the given artificial or heavily modified water body. Not all dam reservoirs may be treated as lakes. Elements differentiating dam reservoirs from lakes depend on parameters characteristic of reservoirs, the most important of which is regarded as the rate of water exchange. On this basis, dam reservoirs may be divided into reolimnic, transitory, and limnic [12]. Reolimnic reservoirs—with the water retention period shorter than 20 days—are reservoirs, which in extreme cases are not much different than rivers with a low flow velocity. Transitory reservoirs—with the retention period of 20 do 40 days—are reservoirs which in the river part are more similar to rivers and in the lake part (close to the dam) to lakes. Limnic reservoirs—with the retention period of over 40 days—are more similar to lakes.

7.2 Research Methods and the Used Data

The analysis of water quality comprised 22 dam reservoirs (Fig. 7.1), constituting 20 surface water bodies. Input data come from monitoring carried out by the Voivodship Inspectorates of Environmental Protection. The analysed reservoirs were subject to diagnostic and operational monitoring. Monitoring is carried out in four-year cycles, the present one within the framework of the third water management cycle lasting from 2016 to 2021. Diagnostic monitoring comprises measurements of all groups of water quality indexes: biological, hydromorphological, physicochemical character-ising physical and oxygen conditions as well as organic pollutants, salinity, acidity, biogenic indexes, specific synthetic and non-synthetic pollutants and also chemical indexes characterising the presence of substances particularly harmful to the water environment, including priority substances. Operational monitoring program encom-passes flowing water bodies, including dam reservoirs threatened with not meeting the environmental objectives. This kind of monitoring encompasses selected biolog-ical indexes, physicochemical indexes characterising physical, oxygen conditions and organic pollutants, salinity, acidity, biogenic indexes as well as substances from

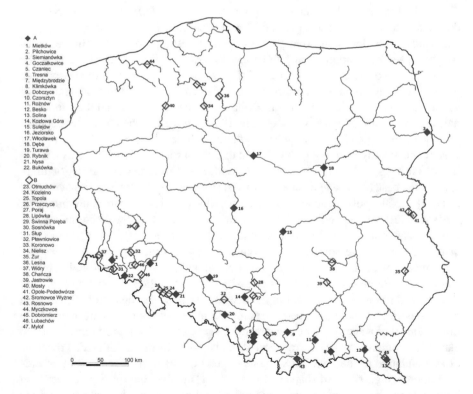

Fig. 7.1 Distribution of the main dam reservoirs in Poland A—studied reservoirs B—other reservoirs (by M. Matysik and D. Absalon on the basis [26])

the group of specific synthetic and non-synthetic pollutants and also chemical indexes characterising the presence of substances particularly harmful to water environment, which have been identified in the catchment.

The aim of monitoring carried out by the State Inspection for Environmental Protection is providing knowledge about the status of water, necessary for taking actions for the improvement of the condition and protection of water from pollution. Research results obtained based on the monitoring enable classification of elements of water quality, ecological status or chemical status as well as the assessment of the status of surface water bodies that are dam reservoirs.

The analysis encompassed only these reservoirs for which operational or diagnostic monitoring was carried out in the years 2012–2017, and the scope of tasks was comparable for them.

7.3 Conditions of Water Quality in Dam Reservoirs—Selected Issues

The studied reservoirs are located in different regions of Poland, and they perform different functions, e.g., flood protection, power engineering, water supply, tourism, and recreation, aiding water transport, and fish harvesting. Some of them are multi-purpose reservoirs serving many functions at the same time. An exception is the Czaniec reservoir (the last one in the Soła Cascade) which only supplies potable water. The analysed dam reservoirs have a very diversified time of water exchange and ability to accumulate river load. Water exchange conditions and their morphometric parameters influence water quality. Pollutant load in the reservoirs depends mainly on the degree of human impact in their catchment [27]. Substances polluting water come mainly from point sources and dispersed sources [28]. Nowadays, rivers in Poland are still heavily polluted with pollutants from sewage and agriculture [29]. In the case of sewage discharged to water, these may by toxic substances such as heavy metals, polycyclic aromatic hydrocarbons (PAHs) or petroleum products. Surface runoff from agricultural areas contaminates water with fertilising substances (organic and mineral) and plant protection sprays (pesticides). Point sources and dispersed sources are sources of biogenic substances (phosphorus and nitrogen). The process of eutrophication caused by human activity is called cultural eutrophication [30]. Pollutants reaching water easily connect to the silt and clay sediment fraction and are thus transported in the deep water as suspended load. The chance of these pollutants being trapped in sediments in reservoirs may contribute to the improvement of water quality downstream of the reservoir [31]. Dam reservoirs trap clastic and organic material, transported in the solid form and diluted. Processes of accumulation of material in reservoirs depend on many factors constant or variable in time, e.g.: size and shape of the reservoir, number and quality of river load brought from the catchment, and mainly the level of water in the reservoir and the course of hydrological phenomena in the river on which the reservoir was formed.

For example, the Solina reservoir on the upper San traps as much as 99% of load transported by this river (average from the period of dam functioning), and other deep reservoirs located in the Carpathians and Sudetes over 80% [32]. In shallow reservoirs located in different parts of Poland (Czaniec, Goczałkowice, Włocławek, Sulejów, Dębe, Jeziorsko), the multiannual retention capacity of these reservoirs ranges from 18% to about 50% [32]. Downstream of the barrage in Włocławek the transport of suspended load in the Vistula is lower by 41%, and bed load by nearly 100% [20, 33]. Material accumulating in reservoirs has a huge impact on its ecological status because along with material trapped in the reservoir pollutants accumulate in bottom sediments, including also toxic compounds and heavy metals [34]. The suspended load is attributed to an important role in the transport of metals along the course of the river [35]. Besides trapping the suspended load, the functioning of dam reservoirs may favourably impact the reduction of selected pollutants in water due to the process of self-purification, which is observed in retention reservoirs. In the case of a big lowland reservoir, comparing quality parameters of water entering the reservoir and exiting it, we may count on the reduction of heavy metal load by 70–90%, organic pollutants by about 50% and overall suspended load by 30%, what was proven by [36] on the basis of studies of the Włocławek reservoir. Reactivation of polluted sediments trapped in reservoirs, e.g. by waving, floods, dredging bottom sediments or by shipping may lead to secondary water pollution. The capacity of the reservoir to accumulate sediments also depends on hydrodynamic conditions.

An important factor influencing the concentration of mineral substances in reservoirs is their storing capacity and intensity of water exchange. As regards water retention the following dam reservoirs have been included into the limnic group: Solina, Czorsztyn, Jeziorsko, Goczałkowice, Dobczyce, Nysa, Turawa, Tresna, Siemianówka, Mietków, Klimkówka, Rybnik, Kozłowa Góra, Bukówka, and Besko. Average water retention time in the analysed reservoirs of the limnic type is 138.13 days. The longest retention period was recorded for Kozłowa Góra, where it was as long as 307 days. Transitory reservoirs in the presented group are: Rożnów, Sulejów, Pilchowice, and Porąbka—average retention time in them is 32 days. Reolimnic reservoirs include Dębe, Włocławek, and Czaniec. Average retention time, in this case, is 4.5 days. In the Włocławek reservoir, it is 5.6 days [36].

Usually water quality is potentially the best in reservoirs with the greatest depth and volume of water mass and small total area of catchment, without point pollution sources. However, different trends apply to water quality in shallow reservoirs with scarce water resources, with vast catchment areas that are agriculturally used and densely populated. An important factor that may influence the quality of surface layers of water is the possibility of export of the so-called internal biogenic load, which increases with growing trophism and is caused by the release of phosphates from bottom sediments, particularly in anaerobic conditions. Additional phosphate load in the summer period stimulates algae development leading to algal bloom [30].

7.4 Characteristics of Physical, Chemical and Biological Parameters of the Studied Dam Reservoirs

The content of dissolved oxygen in dam reservoirs depends on the trophic state and morphometric conditions [30]. Average concentrations of dissolved oxygen in most of the analysed reservoirs exceed limit values for class I water quality (values ≥ 7.0 mg O_2/L). One exception is the Pilchowice reservoir with the lowest average dissolved oxygen content of 6.83 mg O_2/L—which makes it fall into water quality class II (values 5.0–7.0 mg O_2/L). The Dobczyce reservoir has the highest concentration of dissolved oxygen, with an average value of 14.0 mg O_2/L (Fig. 7.2).

Biochemical Oxygen Demand (five days—BOD_5) is one of the indexes characterising oxygen conditions. In most reservoirs, average BOD_5 values do not exceed the limit value for class I (≤ 3.0 mg O_2/L) and the lowest value was recorded for the Klimkówka reservoir—0.97 mg O_2/L. Average BOD_5 concentrations in reservoirs: Siemianówka, Dębe, Sulejów, Jeziorsko, Rybnik do not exceed the limit value for class II (3.0–6.0 mg O_2/L), with the highest average value for the Siemianówka reservoir—5.40 mg O_2/L (Fig. 7.3).

The content of organic matter is one of the parameters defining the degree of environmental pollution. Total organic carbon (TOC) provides full data about the sum of all organic compounds found in water. Organic matter suspended in water may be both natural and anthropogenic origin [37]. The lowest average TOC concentration was recorded for reservoirs located in mountainous areas, with the highest TOC value of 4.70 mg C/L for the Besko reservoir and the lowest 2.06 mg C/L for the

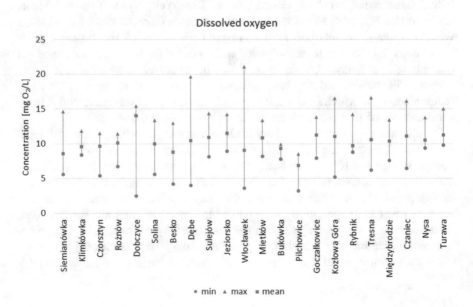

Fig. 7.2 Distribution of concentrations of dissolved oxygen in analysed reservoirs

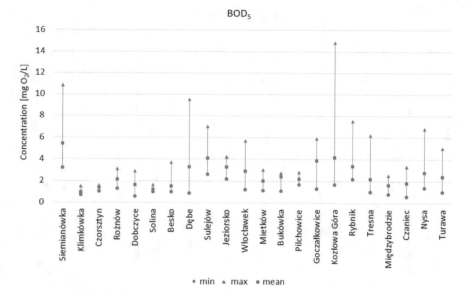

Fig. 7.3 The distribution of BOD₅ in analysed reservoirs

Czorsztyn reservoir. Siemianówka has the highest TOC value, with an average of 19.09 mg C/L. TOC limit value for class I of water quality is ≤10.0 mg C/L— and most reservoirs meet it. The following reservoirs have been classified as class II (10.0–15.0.0 mg C/L): Dębe, Sulejów and Kozłowa Góra (Fig. 7.4). High TOC concentration in waters of the Siemianówka reservoir may also be the result of a large number of wetland areas in the catchment. Bog and post-bog ecosystems trigger the enrichment of underground and surface water in organic matter and dissolved carbon [38, 39].

Specific conductance is one of the indexes in the classification of water quality characterising salinity. The lowest conductance value was found for the Bukówka reservoir—132.8 µS/cm. In the water of reservoirs located in mountainous areas, conductance does not exceed 300 µS/cm (Fig. 7.5). Higher conductance is found in reservoirs with large catchments and located in lowland areas. As an example: the conductance value for the water of the Włocławek reservoir is 620 µS/cm. Classification of water quality takes ≤1000 µS/cm as the limit value for class I. The highest specific conductance of 1075.5 µS/cm is characteristic only of the Rybnik reservoir. The main factor causing the rise in salinity in this reservoir is the inflow from the Nacyna and the Ruda rivers. The water of the Nacyna river, which is the main recipient of saline mine water and municipal sewage from the city of Rybnik is discharged downstream of the reservoir but only to the value of 1.25 m³/s. Above this value (due to the capacity of the pumping station) the water from the Nacyna enters the reservoir [40]. The water of the reservoir also has high values of chlorides and sulphates—which are pollutants characteristic of water coming from draining

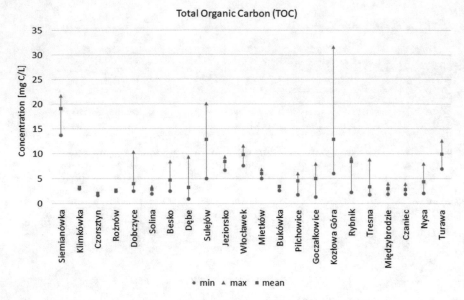

Fig. 7.4 The distribution of TOC in the analysed reservoirs

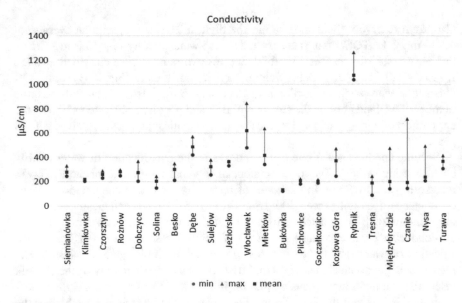

Fig. 7.5 Distribution of conductance in analysed reservoirs

coal mines. Concentrations of chlorides and sulphates are a few times higher than in other reservoirs.

Total nitrogen and total phosphorus are indexes characterising biogenic conditions. High values of biogenic substances in reservoirs, particularly phosphorus, lead to their eutrophication, often triggering algal bloom in the water of dam reservoirs. Poland-wide the main factor causing nitrogen and phosphorus pollution from dispersed (or nonpoint) sources is agriculture, producing 79.24% of nitrogen load and 96.7% of phosphorus load. The limit value for total nitrogen for class I of water quality in dam reservoirs is ≤ 5.0 mg N/L. Average concentrations of total nitrogen in analysed reservoirs may be classified as class I water quality. The lowest values are found in reservoirs located in the Carpathians and in waters of the Włocławek reservoir. The lowest average values are 0.67 mg N/L for Włocławek and 0.68 mg N/L for Solina. The highest average value of total nitrogen is recorded for the Turawa reservoir—2.98 mg N/L (Fig. 7.6).

The limit value of total phosphorus for class I water quality is ≤ 0.20 mg/L. The water of all analysed reservoirs excluding Siemianówka does not exceed this value. Average concentrations of total phosphorus reflect distinctive differences—in Carpathian reservoirs average concentrations are more than 6 times lower than in reservoirs with the highest concentrations—Dębe, Rybnik and Turawa with values above 0.180 mg P/L (Fig. 7.7).

A different distribution can be found as regards average nitrate nitrogen and phosphates. The limit value for class I water quality for dam reservoirs is 2.2 mg N-NO$_3$/L. Only in water of the Czaniec reservoir, this value was exceeded—2.96 mg N-NO$_3$/L.

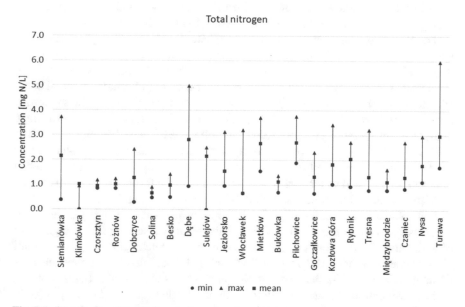

Fig. 7.6 Distribution of total nitrogen in analysed reservoirs

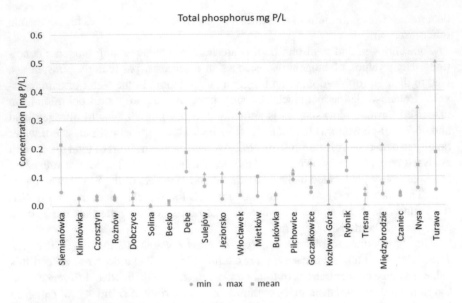

Fig. 7.7 Distribution of total phosphorus in analysed reservoirs

The lowest average concentration of nitrate nitrogen is found in the Siemianówka reservoir, with an average value of 0.20 mg N-NO₃/L. In most of the analysed reservoirs, average concentration of nitrate nitrogen did not exceed 1.0 mg N-NO₃/L. Only in Dębe, Mietków, Pilchowice, Kozłowa Góra, Nysa and Turawa reservoirs average concentrations of nitrate nitrogen exceed 1.0 mg N-NO₃/L (Fig. 7.8).

The highest average concentration of phosphates is found in the Rybnik reservoir, where this value is 0.247 mg P-PO₄/L, and the limit value for class II water quality is ≤0.130 P-PO₄/L. The main impact in the reservoir catchment is industry and municipal economy. The lowest average concentration of phosphates is found in the Włocławek reservoir—0.03 mg P-PO₄/L, and it is comparable with values recorded for the Solina reservoir—0.05 mg P-PO₄/L. In most of the analysed reservoirs average phosphate concentrations do not exceed the limit value for class I quality ≤0.065 mg P-PO₄/L. Reservoirs with average concentrations reflecting class II water quality for dam reservoirs include Dębe, Sulejów, Jeziorsko, Mietków, and Nysa. Exceeded limit value for class II water quality was found only in the Rybnik reservoir 0.247 mg P-PO₄/L and the Pilchowice reservoir 0.197 mg P-PO₄/L (Fig. 7.9).

Phytoplankton and chlorophyl (a) well reflect impacts causing oversupply of nutrients in water in the whole reservoir catchment (caused e.g., by a high share of agricultural lands, high population density in the catchment, etc.). The values of phytoplankton index IFPL vary in the analysed reservoirs. The highest index is found in Solina—0.950, the lowest in Siemianówka—0.179 (Fig. 7.10). According to the classification of water quality [41], the limit value for class I quality is ≥0.8. The following reservoirs were classified as class I quality: Klimkówka, Czorsztyn, Rożnów, Solina, Besko, Goczałkowice. Quality class II reservoirs (0.6–0.8) include

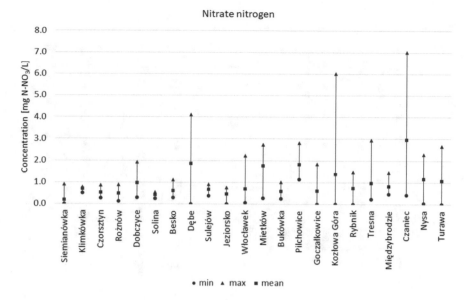

Fig. 7.8 Distribution of nitrate nitrogen N-NO₃ in analysed reservoirs

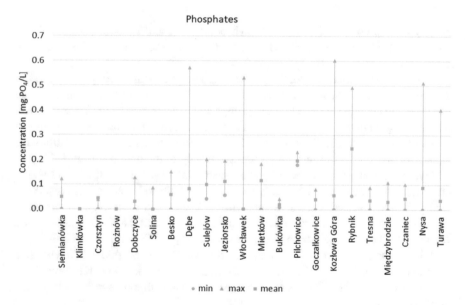

Fig. 7.9 Distribution of phosphates in analysed reservoirs

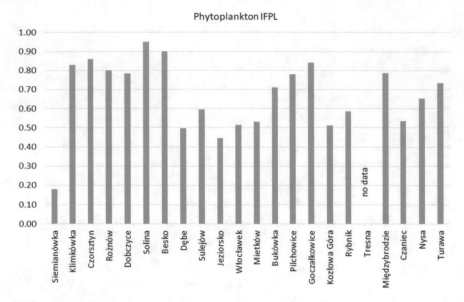

Fig. 7.10 Distribution of phytoplankton indexes in analysed reservoirs

Dobczyce, Bukówka, Pilchowice, Międzybrodzie, Nysa and Turawa. The other reservoirs are considered to be class III quality (0.4–0.6), except for the Siemianówka reservoir, for which the phytoplankton index is in class V ≥0.2.

After the reservoir is created, the plankton grows quickly as it has quite a short life cycle and it proliferates rapidly. In the littoral zone, it is much richer and characterised by a differing abundance of individual groups and species than in the limnetic zone. The growth of plankton in artificial lakes depends mainly on the nutrient content and on retention time. In small, and particularly fast flowing rivers, the conditions for this are usually not favourable, so growth starts only in the dam reservoir [42].

Phytobenthic diatoms reflect impacts causing increased nutrient level in water, but they also react to local impacts (the nature of littoral zone of the reservoir, local pollution sources, excessive recreational use of the reservoir). In the analysed reservoirs, phytobenthos values (IO) are in the range of 0.860 (Solina) to 0.482 (Jeziorsko). Class I as regards phytobenthos content (≥0.75) comprises: Klimkówka, Rożnów, Czorsztyn, Rożnów, Dobczyce, Solina, Bukówka, and Międzybrodzie whereas class II (0.65–0.75) includes the following reservoirs: Besko, Kozłowa Góra, and Czaniec (Fig. 7.11). Phytobenthos values (IO) in the other reservoirs reflect class III (0.45–0.65).

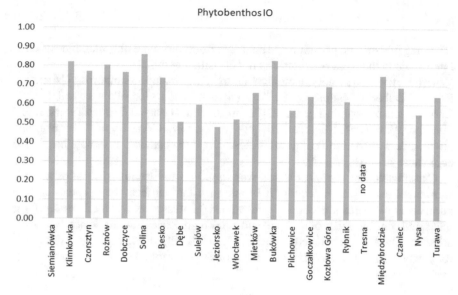

Fig. 7.11 Distribution of phytobenthos indexes in analysed reservoirs

7.5 Classification and Assessment of the Status of Dam Reservoirs

Ecological status of water is based on the results of tests of biological elements such as the indexes of phytoplankton (IFPL), macrophytes (MIR), phytobenthos (IO), benthic macroinvertebrates (MMI) and ichthyofauna and supporting parameters—hydromorphological and physicochemical. Ecological status is assessed for artificial or heavily modified water bodies. The assessment is made on the basis of results of classification of the tested biological, physicochemical, and hydromorphological elements. Ecological status is assessed as maximum, good, moderate, poor, bad.

Only two studied reservoirs are in class I of biological elements: Czorsztyn on the Dunajec and Besko on the Wisłok. Eleven reservoirs were classified as class II as regards biological elements: Klimkówka, Rożnów, Dobczyce, Solina, Sulejów, Jeziorsko, Goczałkowice, Turawa, Nysa, Tresna, Międzybrodzie (Porąbka) and Czaniec. In the following reservoirs, biological elements fall into class III: Dębe, Włocławek, Mietków, Pilchowice, Kozłowa Góra, and Rybnik. The Siemianówka reservoir was classified as class V as regards biological elements.

Classification of hydromorphological elements of the studied dam reservoirs points to class I and II. Class I of physicochemical elements characterised the following reservoirs: Klimkówka, Rożnów, Czorsztyn, Solina, Besko, Jeziorsko, Pilchowice and the reservoirs in the Soła Cascade (Tresna, Międzybrodzie, Czaniec). Class II of physicochemical elements includes the reservoirs: Dobczyce, Sulejów, Włocławek, Mietków, Goczałkowice, Rybnik, Turawa, and Nysa. The Kozłowa Góra reservoir is in class III. Siemianówka and Dębe have the worst quality class—below good.

As regards specific pollution, the analysed reservoirs were only classified as class I and II. Only the Nysa reservoir was not classified in this respect.

12 reservoirs have a good ecological status. These include Klimkówka, Czorsztyn, Rożnów, Dobczyce, Solina, Besko, Jeziorsko, Bukówka, Goczałkowice and the Soła Cascade dam reservoirs: Tresna, Międzybrodzie, and Czaniec. The reservoirs: Dębe, Sulejów, Włocławek, Mietków, Pilchowice, Kozłowa Góra, Rybnik, Turawa, and Nysa have a moderate ecological status. The Siemianówka reservoir has a bad ecological status (Fig. 7.12).

Classification of chemical status was based on the results of tests of priority substances and other pollutants. It is assumed that surface water bodies have a good chemical status if average annual values (expressed as arithmetic mean of the measured concentrations of indexes) and maximum concentrations do not exceed the allowed values of environmental quality standards (EQS). Exceeding relevant environmental quality standards for at least one positively verified value of concentrations

Fig. 7.12 Ecological status of the studied dam reservoirs (by M. Matysik and D. Absalon on the basis [26] and data from Table 7.1)

Table 7.1 Classification of water quality in analysed dam reservoirs

Dam reservoirs	Surface water body code (SWB code)	Type of reservoirs	Class Biological Quality Elements	Class Hydromorphological Quality Elements	Class Physico Chemical Quality Elements	Class Specific Pollutants	Classification status		
							Ecological	Chemical	Overall
Siemianówka	PLRW200017261449	L	V	I	Moderate	II	Bad	Fail	Bad
Klimkówka	PLRW20000218239	L	II	II	I	II	Good	Good	Good
Czorsztyn	PLRW20000214179	L	I	II	I	I	Good	Good	Good
Rożnów	PLRW20000214739	P	II	I	I	II	Good	Good	Good
Dobczyce	PLRW200002138599	L	II	II	II	I	Good	Good	Good
Solina	PLRW20000221559	L	II	I	I	I	Good	Good	Good
Besko	PLRW20000226159	L	I	I	I	II	Good	Good	Good
Dębe (Zalew Zegrzyński)	PLRW200002671999	R	III	II	Moderate	II	Moderate	Fail	Bad
Sulejów	PLRW200002545399	P	II	I	II	II	Moderate	Fail	Bad
Jeziorsko	PLRW60000183179	L	II	II	I	II	Good	Fail	Bad
Włocławek	PLRW20000275999	R	III	II	II	II	Moderate	Good	Bad
Mietków	PLRW6000013455	L	III	I	II	I	Moderate	Fail	Bad
Bukówka	PLRW60000161159	L	II	I	I	I	Good	Fail	Bad
Pilchowice	PLRW6000016333	P	III	I	I	I	Moderate	Fail	Bad
Goczałkowice	PLRW20000211179	L	II	II	II	II	Good	Fail	Bad
Kozłowa Góra	PLRW20000212639	L	III	I	III	II	Moderate	Fail	Bad
Rybnik	PLRW600001156539	L	III	II	II	II	Moderate	Fail	Bad
Tresna, Międzybrodzie, Czaniec	PLRW2000021329553	L/ P/ R	II	II	I	II	Good	Fail	Bad
Turawa	PLRW6000011859	L	II	II	II	III	Moderate	Fail	Bad
Nysa	PLRW6000012599	L	II	II	II	n.d.	Moderate	Good	Bad

Explanations: L - limnic type; P – transitory type; R – reolimnic type; n.d. – no data

of a priority substance tested in water or biota leads to lowering the classification of chemical state to "below good" (Table 7.1).

8 reservoirs have a good chemical status: Klimkówka, Czorsztyn, Rożnów, Dobczyce, Solina, Besko, Włocławek and Nysa. Other reservoirs have a bad chemical status (Fig. 7.13). In most reservoirs bad chemical status was due to exceeded concentrations of polycyclic aromatic hydrocarbons (PAH): Benzo(ghi)perylene, Indeno(1,2,3-cd)pyrene (Table 7.2).

Overall status classification points to a good status for reservoirs located in the Carpathians: Klimkówka, Czorsztyn, Rożnów, Solina, Besko, and Dobczyce. Other reservoirs have a bad status of water (Fig. 7.14).

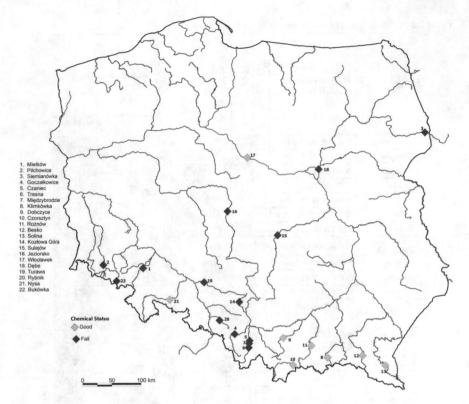

1. Mietków
2. Pilchowice
3. Siemianówka
4. Goczałkowice
5. Czaniec
6. Tresna
7. Międzybrodzie
8. Klimkówka
9. Dobczyce
10. Czorsztyn
11. Rożnów
12. Besko
13. Solina
14. Kozłowa Góra
15. Sulejów
16. Jeziorsko
17. Włocławek
18. Dębe
19. Turawa
20. Rybnik
21. Nysa
22. Bukówka

Chemical Status
◇ Good
◆ Fail

0   50   100 km

Fig. 7.13 Chemical status of the studied dam reservoirs (by M. Matysik and D. Absalon on the basis [26] and data from Table 7.1)

7.6 Continuous Monitoring of a Dam Reservoir for Management and Water Quality Protection—The Goczałkowice Reservoir Case Study

Implementing the latest technical and technological developments for the analysis of quality of water resources is connected with the need to meet the requirements of the EU Water Framework Directive and state regulations. In the face of this challenge constant monitoring of physical and chemical parameters of water gives new possibilities and perspectives of a better assessment of water quality [43, 44].

The subject of research carried out as part of the project Integrated System Supporting Management and Protection of a Dam Reservoir (ZiZOZap) within the Operational Program Innovative Economy was the Goczałkowice Reservoir on the Vistula. Out of over 100 Polish reservoirs and barrages of a capacity bigger than 1 hm^3 the "Goczałkowice" reservoir is one of the biggest ones. As regards total storing capacity (165.6 hm^3) and area at the maximum water level (32 km^2) it takes the fifth

Table 7.2 Priority hazardous substances leading to fail status

Dam reservoir	SWB code	Name of water body	Priority hazardous substances
Siemianówka	PLRW200017261449	Narew-zb. Siemianówka	Brominated diphenylethers, mercury and its compounds benzo(a)pyrene
Dębe (Zalew Zegrzyński)	PLRW200002671999	Zalew Zegrzyński	Benzo(a)pyrene
Sulejów	PLRW200002545399	Zbiornik Sulejów	Brominated diphenylethers, benzo(a)pyrene, heptachlor
Jeziorsko	PLRW60000183179	Warta ze Zb. Jeziorsko	Brominated diphenylethers, mercury and its compounds, benzo(a)pyrene, heptachlor
Mietków	PLRW6000013455	Bystrzyca, zb. Mietków	Benzo(g,h,i)perylene, indeno(1,2,3-cd)pirene
Bukówka	PLRW60000161159	Bóbr, zb. Bukówka	Benzo(g,h,i)perylene, indeno(1,2,3-cd)pirene
Pilchowice	PLRW6000016333	Bóbr, zb. Pilchowice	Benzo(g,h,i)perylene, indeno(1,2,3-cd)pirene
Goczałkowice	PLRW20000211179	Zbiornik Goczałkowice	Benzo(a)pyrene, brominated diphenylethers, mercury, and its compounds, heptachlor
Kozłowa Góra	PLRW20000212639	Zbiornik Kozłowa Góra	Benzo(a)pyrene, fluoranthene
Rybnicki	PLRW600001156539	Ruda w obrębie zbiornika Rybnik	Hexachlorocyclohexane (HCH), nickel and its compounds, benzo(a)pyrene
Tresna, Międzybrodzie, Czaniec	PLRW2000021329553	Kaskada Soły (Soła od zb. Tresna do zb. Czaniec)	Benzo(a)pyrene
Turawa	PLRW6000011859	Mała Panew, Turawa	Cadmium and its compounds, benzo(a)pyrene

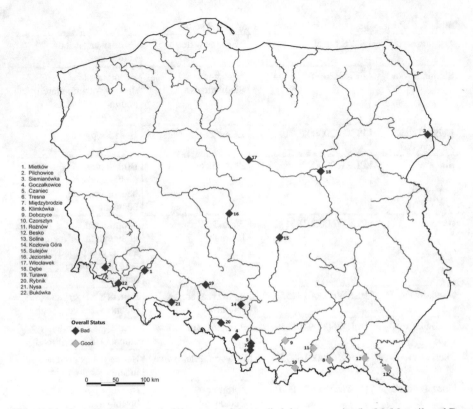

1. Miełków
2. Pilchowice
3. Siemianówka
4. Goczałkowice
5. Czaniec
6. Tresna
7. Międzybrodzie
8. Klimkówka
9. Dobczyce
10. Czorsztyn
11. Rożnów
12. Besko
13. Solina
14. Kozłowa Góra
15. Sulejów
16. Jeziorsko
17. Włocławek
18. Dębe
19. Turawa
20. Rybnik
21. Nysa
22. Bukówka

Overall Status
◆ Bad
◆ Good

0   50   100 km

Fig. 7.14 Overall classification of the status of the studied dam reservoirs (by M. Matysik and D. Absalon on the basis [26] and data from Table 7.1)

place. Moreover, as regards maximum depth (14 metres) it is one of the most shallow ones.

Goczałkowice is a multi-purpose reservoir. Its basic functions are potable water supply for the Silesia conurbation, flood control, compensation of low-flow during the drought, fisheries management, nature conservation, and recreation [45].

Location of continuous monitoring points of physical and chemical parameters of water was preceded by detailed field research, desk research (analysis of the bathymetric plan of the Goczałkowice Reservoir, analysis of historical data and the Vistula river discharge dynamics) that were to ensure maximum representativeness of tested parameters. Three research and measurement points were selected (Fig. 7.15):

1. the Vistula river—weir in Strumień—inflow to the reservoir;
2. Pelagic zone of the reservoir in the deepest point, in the old Vistula riverbed about 400 m southwest of the bottom drain valve of the dam;
3. the Vistula river—the outflow of the reservoir.

Continuous monitoring of physical and chemical parameters of water was made with three automatic multi parameter probes DS5X (HYDROLAB) by

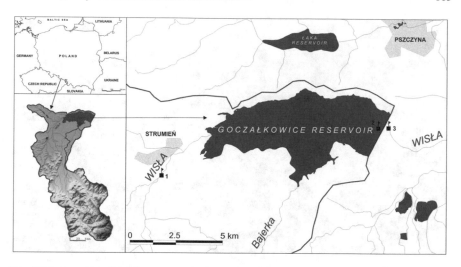

Fig. 7.15 Location of continuous monitoring points for the Goczałkowice Reservoir: 1—water inflow, weir in Strumień village on the Vistula (Wisła) River; 2—lacustrine zone; 3—water outflow below dam on the Vistula (Wisła)

OTT Messtechnik GmbH, enabling measurement of such parameters as: water temperature, dissolved oxygen, depth, pH, REDOX potential, specific conductance, chlorophyl, turbidity, ammonia, nitrate, chloride concentrations, sunlight intensity.

Building the monitoring system required the preparation of installation of two probes on the inflow and outflow of the Goczałkowice Reservoir, in the riverbed of the Vistula in order to carry out tests of the quality of water flowing into the reservoir and impact of the reservoir on the quality of water downstream of the facility. The most complicated element of the system was installation in the depths of the reservoir of a fully autonomous measuring buoy (Fig. 7.16). This buoy, besides monitoring physical and chemical parameters of water, measures basic meteorological parameters—air temperature, its relative humidity and force and direction of the wind. It also has an autonomous power supply in the form of solar panels. Transmission of data from the system is carried out remotely thanks to GSM-GPRS modems and sending it to the server making it possible to analyse the monitored parameters. This enables not only instant access to the collected measurements but also the remote configuration of selected parameters.

The range of measured parameters not only makes research monitoring (potamic and limnic) possible, but it also provides a number of information about the quality of water flowing in and out of the reservoir. This enables evaluation of the impact of the reservoir on the quality of river water and provides information about the impact on the microclimate (meteorological monitoring). This also provides important data about the development of processes of eutrophication and conditions of life of flora and fauna (biological monitoring). All measured parameters add to the model of the reservoir's functioning. Continuous monitoring of physical and chemical parameters of water shows the dynamics of the observed indexes, not only seasonal but also dial,

Fig. 7.16 Equipment of the water quality monitoring station on the Goczałkowice Reservoir: A—Buoy equipped with measuring multiparametre probe and meteorological sensors; B—multi-parametre probe immediately before mounting; C—the appearance of the probe installation site on the Vistula river below dam—outflow of the reservoir. Photos by D. Absalon

what improves the possibilities of the proper interpretation of the obtained values and the speed of reaction to threats.

An important innovation of continuous monitoring is the possibility to observe up to date changes of thermal conditions of water in the reservoir, which is not possible with traditional measurements. Already in the initial period of measurements, it was proven that in the reservoir there occurs periodic thermal stratification, despite the small maximum depth of the reservoir.

Continuous research monitoring of physical and chemical parameters of water also enables the analysis of the impact of the Goczałkowice Reservoir on water quality in the Vistula downstream of the reservoir.

Research carried out in the period 2010–2013 proved that the reservoir exhibits significant chemical retention and most quality parameters improve. An exception is the pH value, which goes slightly up, what is caused by eutrophication processes taking place in the reservoir. Water temperature also increases, what is mainly connected with the area of the reservoir and the retention time. On the other hand, seasonal analyses carried out in the period 2016/2017 showed that in the summer the reservoir exhibits chemical retention of nitrates. The concentration of chlorides on the outflow of the reservoir goes up both in winter and summer. In winter, also water temperature increases on the outflow (Figs. 7.17 and 7.18).

Monitoring physical and chemical parameters of water quality with the use of multi-parameter sondes with continuous data transmission through GSM-GPRS provides new possibilities as regards water quality assessment. Traditional monitoring, based on infrequent measurements in an annual cycle, in specific conditions may lead to incorrect conclusions concerning the ecological status of the subject of

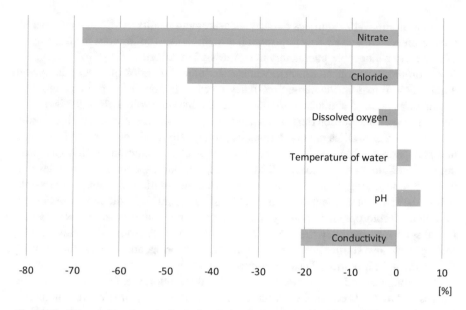

Fig. 7.17 Changes in selected physical and chemical parameters of water of the Vistula river flowing out of the Goczałkowice Reservoir as compared to inflow in the period from June 2010 to January 2013

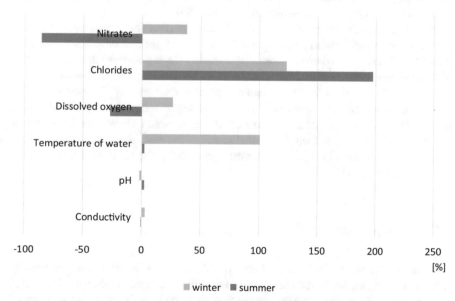

Fig. 7.18 Changes in selected physical and chemical parameters of water of the Vistula river flowing out of the Goczałkowice Reservoir as compared to inflow in the summer and winter period of 2016/2017: summer: June, July, and August 2016; winter: December 2016, January and February 2017

research. In Fig. 7.19 examples of daily measurement results of selected parameters are presented. They show how important it is to implement continuous measurements of relatively changeable parameters into "classic" monitoring.

Chlorophyll (a) concentration in water is an important index of water quality and algal bloom threat, at the same time carrying information about primary biomass production and food abundance for consumers in the reservoir. This index is dependent on oxygen and carbon dioxide concentration, pH, water temperature, insolation, and turbidity, as well as concentrations of biologically available forms of nitrogen and phosphorus. Chlorophyll (a) concentration is also a function of phytoplankton count. Determination of chlorophyll by counting plankton specimens is troublesome and time-consuming, moreover, at the stage of condensation of tests, it is possible to lose the smallest nanoplankton forms. Classical monitoring with laboratory determination of chlorophyll is equally labour-consuming. Metagenomic methods that reveal semi-quantitatively the presence of ribosomal DNA from chloroplasts of the main groups have similar limitations as manual monitoring, and also require creating databases and calibration. These limitations do not apply to the monitoring with the use of automatic sondes detecting chlorophyll with the fluorescent method. Currently, fluorescent methods enable in situ measurements of photosynthetic dyes characteristic of 4–5 different taxonomic phytoplankton groups (cyanobacteria, chlorophytes, diatoms with dinoflagellates and chrysophytes, cryptophytes) [46, 47]. Continuous monitoring makes it possible to observe diel vertical migration of phytoplankton and correlation of oxygen concentration and pH [48–50].

Although continuous monitoring is not yet so perfect for, excluding the necessity of measurements with traditional methods and expensive laboratory analyses, it shows the direction where modern water quality testing methodology is heading. Measurements with the use of automatic sondes, besides constant access to results—to a great extent independent of weather, season, time of day, or meteorological conditions—point to new possibilities of monitoring water quality [51]. Continuous monitoring of physical and chemical parameters of water shows the dynamics of the observed indexes, not only seasonal changes but also daily fluctuations. This improves the possibilities of the proper interpretation of the obtained values and the speed of reaction to threats.

7.7 Conclusions

Pursuant to the Water Framework Directive, dam reservoirs are treated as heavily modified water bodies, so they are required to achieve at least good ecological status by year 2021. Reservoirs, which are considered to be heavily modified sections of rivers, are mostly not very similar to rivers.

The analysis of water quality of 22 selected dam reservoirs showed the follow-ing regularities:

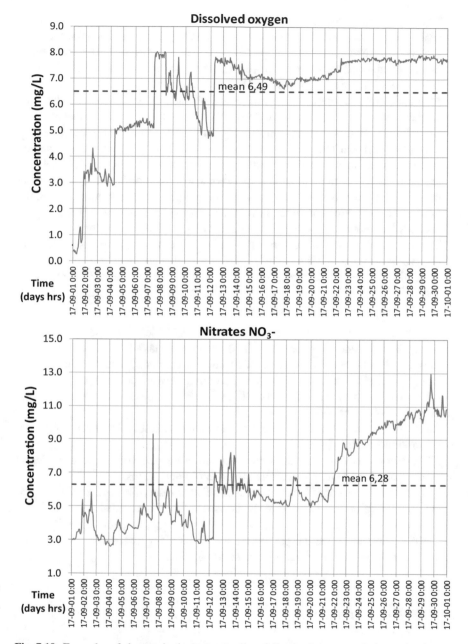

Fig. 7.19 Examples of changes in the concentration of dissolved oxygen and nitrates in water of the Goczałkowice Reservoir in September 2017 (time step of the horizontal axis—1 day)

- The average concentrations of dissolved oxygen and BOD5 values in most of the analysed reservoirs place them in class I of water quality. On the other hand, the reservoirs located in mountainous areas have the lowest average TOC concentration and conductance not exceeding 300 μS/cm. Higher conductance is found in reservoirs with large catchments and located in lowland areas;
- Average concentrations of total nitrogen in analysed reservoirs may be classified as class I water quality. The lowest values are found in reservoirs located in the Carpathians. Analyses of phosphorus concentra-tions in all analysed reservoirs (except one) reveal that they do not ex-ceed the limit value for class I quality. However, significant differences in the average concentrations of total phosphorus are noticeable—in the Carpathian reservoirs average concentrations are more than 6 times lower than in reservoirs with the highest concentrations. Also, the average concentrations of nitrate nitrogen and phosphates in most of the analysed reservoirs do not exceed the limit values for class I of water quality;
- The values of phytoplankton index IFPL vary in the analysed reservoirs. Six reservoirs were classified as class I quality and also six as class II quality. The other nine reservoirs are considered to be class III quality, and one is class V. In terms of phytobenthos content, eight reservoirs were in class I, three in class II, and the remaining 11 were in class III.

Only two of the analysed reservoirs have class I of biological elements, eleven reservoirs are considered to be class II, 6 reservoirs class III, and one reservoir is considered to be class V in terms of biological elements. The classification of hydromorphological elements of the studied dam reservoirs points to class I and II. One reservoir is characterised by class III, whereas two reservoirs have the worst quality class—below good.

As regards specific pollution, most of the analysed reservoirs (except for one) were classified as class I and II.

12 reservoirs have a good ecological status. Nine reservoirs have a moderate ecological status, and one reservoir has a bad ecological status.

8 reservoirs have a good chemical status. The other reservoirs have a bad chemical status. In most reservoirs bad chemical status was due to exceeded values of poly-cyclic aromatic hydrocarbons (PAH): Benzo(ghi)perylene, Indeno(1,2,3-cd)pyrene.

Overall status classification points to a good status for reservoirs located in the Carpathians. Other reservoirs have a bad status of water.

7.8 Recommendations

Implementing the latest technical and technological developments for the analysis of quality of water resources is associated with the need to meet the requirements of the EU Water Framework Directive and state regulations. In the face of this challenge, reservoirs important from the point of view of water supply should aim to develop continuous monitoring of physical and chemical parameters of water, which gives new possibilities and prospects for a better assessment of water quality.

Constant monitoring of the physical and chemical properties of water can be carried out using automatic multi-parameter probes. Continuous monitoring of physical and chemical parameters of water shows seasonal and even daily fluctuations of the observed indicators.

Measurements with the use of automatic probes provide continuous access to results that are largely independent of weather, season, time of day or meteorological conditions. They also point to new possibilities in the area of water quality monitoring—they improve the possibility of proper interpretation of the values.

References

1. Głodek J (1985) Jeziora zaporowe świata. PWN, Warszawa
2. Choiński A (2007) Limnologia fizyczna Polski. Wydawnictwo Naukowe UAM
3. Statistical Yearbook 2014, Główny Urząd Statystyczny w Warszawie (2015)
4. Łajczak A (1995) Studium nad zamulaniem wybranych zbiorników zaporowych w dorzeczu Wisły. Monografie Komitetu Gospodarki Wodnej PAN 8:106
5. Allan JD (1998) Ekologia wód płynących. PWN, Warszawa
6. Kajak Z (1998) Hydrobiologia – limnologia. Ekosystemy wód śródlądowych. PWN, Warszawa
7. Vannote RL, Minshall GW, Cummins KW, Sedell JR, Cushing CE (1980) The River Continuum Concept. Can J Fish Aquat Sci 37(1):130–137
8. Znamenski VA (1975) The role of hydrological factors with respect to water quality in reservoirs. The effects of storage on water quality. Water Research Centre, Medmenham, pp 567–573
9. Ward JS, Stanford JA (1983) The intermediate—disturbance hypothesis: an explanation for biotic diversity patterns in lotic ecosystems. In: Fontaine TD, Bartell SM (eds) Dynamic of lotic ecosystems. Ann Arbor Science Publishers, The Butterworth Group, Ann Arbor, MI, pp 347–356
10. Kajak Z (1984) Changes in river water quality in reservoirs, exemplified by studies in Poland. In: Lillehamer A, Saltveit SJ Regulated rivers. Publ. Universitetsforlanget AS, Oslo, pp 521–531
11. Galicka W, Kruk A, Zięba G (2007) Bilans azotu i fosforu w Zbiorniku Jeziorsko. Wyd. Akademii Rolniczej im. A, Cieszkowskiego, Poznań
12. Picińska-Fałtynowicz J, Błachuta J (2012) Wytyczne metodyczne do przeprowadzenia monitoringu i oceny potencjału ekologicznego zbiorników zaporowych w Polsce. Inspekcja Ochrony Środowiska, Warszawa
13. Giziński J (2013) Hydroelectricity and ecological considerations. Falsification of the environmental reality by the opponents of hydropowe. Acta Energetica 3:32–44
14. Kentzer A, Dembowska E, Giziński A, Napiórkowski P (2010) Influence of the Włocławek Reservoir on hydrochemistry and plankton of a large, lowland river (the Lower Vistula River, Poland). Ecol Engineer 36(12):1747–1753
15. Napiórkowski P, Kentzer A, Dembowska E (2006) Zoo-Zooplankton of the lower Vistula River: the effect of Włocławek Dam Reservoir (Poland) on community structure. Verh Int Ver Limnol 29:2109–2114
16. Giziński A, Kentzer A, Żytkowicz R (1993) Ekologiczne skutki kaskadowej zabudowy dolnej Wisły (prognoza oparta na wynikach badań zbiornika włocławskiego) [Ecological Consequences of Cascade Development of the Lower Vistula River (Forecast Based on Findings of Research of the Włocławek Reservoir], In: Churski Z (ed) Uwarunkowania przyrodnicze i społeczno-ekonomiczne zagospodarowania dolnej Wisły [Environmental and Socio-Economic Considerations of Development of the lower Vistula River], Pub. TNT, pp 179–188

17. Gierszewski P, Habel M, Szmańda J, Luc M (2020) Evaluating effects of dam operation on flow regimes and riverbed adaptation to those changes. Sci Total Environ 710:136202
18. Obodovskyi O, Habel M, Szatten D, Rozlach Z, Babiński Z, Maerker M (2020) Assessment of the Dnieper Alluvial Riverbed Stability Affected by Intervention Discharge Downstream of Kaniv Dam. Water 12:1104
19. Cheng X (1992) Reservoir sedimentation at Chinese hydro systems. Water Power Dam Constr, 44
20. Babiński Z, Habel M (2017) Impact of a single dam on sediment transport continuity in large lowland rivers. In: Wieprecht S, Haun S, Weber K, Noack M, Terheiden K (eds) River Sedimentation. Taylor & Francis Group, London, UK, pp 975–982
21. Morris GL (1995) Reservoir serimentation and sustiainable development in India, Proceeding of the 6th International Symposium on River Sedimentaion, New Delhi, India
22. Gierszewski P, Zakonnov V, Kaszubski M, Kordowski J (2017) Transformacja właściwości wody i osadów w profilu podłużnym zbiorników zaporowych Kaskady Górnej Wołgi. Przegląd Geograficzny 89(3):391–412
23. Szatten D, Habel M, Pellegrini L, Maerker M (2018) Assessment of siltation processes of the Koronowski Reservoir in the Northern Polish Lowland based on bathymetry and empirical formulas. Water 10(11):1681–1694
24. Wildi W (2010) Environmental hazards of dams and reservoirs. Nat Environ Sci 88:187–197
25. Vörosmarty CJ, Sharma KP, Fekete BM, Copeland AH, Holden J, Marble J, Lough JA (1997) The storage and aging of continental runoff in large reservoir systems of the world. Ambio 26:269–278
26. MPHP—Digital Map of Poland's Hydrographic Division. https://dane.gov.pl/dataset/869,kom puterowa-mapa-podziau-hydrograficznego-polski. Accessed on 30 Sep 2020
27. Absalon D, Matysik M, Woźnica A, Łozowski B, Jarosz W, Ulańczyk R, Babczyńska A, Pasierbiński A (2020) Multi-faceted environmental analysis to improve the quality of anthropogenic water reservoirs (Paprocany Reservoir Case Study). Sensors 20:2626
28. Matysik M, Absalon D, Habel M, Maerker M (2020) Surface water quality analysis using CORINE data: an application to assess reservoirs in Poland. Remote Sens 12:979
29. Krengel F, Bernhofer Ch, Chalov S, Efimov E, Efimova L, Gorbachova L, Habel M, Helm B, Kruhlov I, Nabyvanets Y, Osadcha N, Osadchyi V, Pluntke T, Reeh T, Terskii P, Karthe D (2018) Challenges for transboundary river management in Eastern Europe—three case studies. Erde 149:1–16
30. Solarczyk A (2017) Jakość wody oraz stan ekologiczny jezior i zbiorników zaporowych. In: Jokiel P, Marszelewski W, Pociask-Karteczka J (eds) Hydrologia Polski. PWN, Warszawa, pp 247–255
31. Dojlido JR (1995) Chemia wód powierzchniowych. Wyd. Ekonomia i Środowisko, Białystok
32. Ciupa T, Łajczak A, Babiński Z (2017) Rumowisko klastyczne. In: Jokiel P, Marszelewski W, Pociask-Karteczka J (eds) Hydrologia Polski. PWN, Warszawa, pp 146–152
33. Szatten D, Babiński Z, Habel M (2018) Reducing of water turbidity by hydrotechnical structures on the example of the Wloclawek Reservoir. J Ecol Eng 19(3):197–205
34. Achrem E, Gierszewski P (2007) Zbiornik Włocławski, Biblioteka Monitoringu Środowiska WIOŚ w Bydgoszczy, p 146
35. Barbusiński K, Nocoń M, Nocoń K, Kernert J (2012) Rola zawiesin w transporcie metali ciężkich w wodach powierzchniowych na przykładzie Kłodnicy. Ochrona Środowiska 33(1):13–17
36. Gierszewski P (2008) Koncentracja metali ciężkich w osadach zbiornika włocławskiego jako wskaźnik hydrodynamicznych warunków depozycji. Land Anal 9:79–82
37. Pietrzyk A, Papciak D (2016) Materia organiczna w wodach naturalnych – formy występowania i metody oznaczania. JCEEA, t. XXXIII, z. 63(2/I/16):241–252
38. Kiryluk A (2005) Stężenia biogenów i węgla organicznego w wodach pochodzących z różnie użytkowanych torfowisk niskich. Mongrafie Komitetu Inżynierii Środowiska PAN, vol. 30
39. Kiryluk A (2006) Stężenie ogólnego węgla organicznego w wodzie ekosystemów pobagiennych różnie użytkowanych. Woda–Środowisko–Obszary Wiejskie 6(1):173–181

40. Absalon D (1998) Antropogeniczne zmiany odpływu rzecznego w zlewni Rudy. Wydawnictwo Uniwersytetu Śląskiego, Katowice
41. Dziennik Ustaw (Journal of Laws) of 2016 item 1187
42. Traczewska T (2012) Problemy ekologiczne zbiorników retencyjnych w aspekcie ich wielo-funkcyjności. Conference proceedings, European Symposium "Anti-flood Defences—Today's Problems", Paris–Orléans, 28–30 Mar 2012. Accessed 11 Dec 2018
43. Absalon D, Ruman M, Matysik M, Kozioł K, Polkowska Ż (2014) Innovative solutions in surface water quality monitoring. APCBEE Procedia 10:26–30
44. Absalon D, Matysik M, Ruman M (2015) Novel methods and solutions in hydrology and water management. Papers Glob Change 22:137–138
45. Absalon D, Matysik M, Ruman M (2011) Location, hydrological conditions and factors influencing water quality of Goczałkowice reservoir and its catchment. Anthropogenic Nat Transformations Lakes 5:7–15
46. Leeuw T, Bossemail ES, Wrightemail DL (2013) In situ measurements of phytoplankton fluorescence using low cost electronics. Sensors 13(6):7872–7883
47. Submersible Spectrofluorometer with Automatic Algae Class and Chlorophyll Analysis® (2013) bbe moldaenke, http://www.ppsystems.com/Literature/FluoroProbe.pdf. Accessed 20 Jan 2014
48. Łaszczyca P, Augustyniak M, Absalon D, Długosz J, Gwiazda R, Kliś C, Kostecki M, Migula P, Nachlik E, Palowski B, Pasierbiński A, Strzelec M, Wilk-Woźniak E, Woźnica A (2013) Monitoring hydrologiczny i ekologiczny jako podstawa modelowania zbiornika zaporowego i predykcji zjawisk w nim zachodzących (Projekt PO IG ZiZOZap). In Wiśniewski R (ed) Ochrona i rekultywacja jezior, Toruń, pp 123–134
49. Absalon D, Kostecki M, Łaszczyca P, Matysik M, Ruman M (2014) Ciągły monitoring automatyczny a monitoring klasyczny – alternatywa czy dopełnienie metod oceny jakości wody. Gospodarka Wodna 8:296–299
50. Absalon D, Ruman M, Matysik M, (2014) Monitoring ciągły – możliwości i perspektywy lepszej oceny zmian jakości wód. In: Magnuszewski A (ed) Hydrologia w ochronie i kształtowaniu środowiska. Monografie Komitetu Gospodarki Wodnej PAN, vol 2, 20, pp 329–340
51. Ziemińska-Stolarska A, Adamiec J, Imbierowicz M, Imbierowicz E, Jaskulski M, Aleksander Szmidt A, Zbiciński I (2018) Online Measurement Method of Water Quality in The Sulejow Reservoir. Ecol Chem Eng S 25(1):89–100

Chapter 8
Water Quality and Ecosystem Modelling: Practical Application on Lakes and Reservoirs

Rafał Ulańczyk, Bartosz Łozowski, Andrzej Woźnica, Damian Absalon, and Agnieszka Kolada

Abstract Impacts of anthropogenic and other pressures on lakes and reservoirs are often a subject of detailed analyses, which can be aimed, e.g. to identify causes of the water quality deterioration or changes in ecosystems and to predict effects of projects likely to have significant effects on the environment. Since the mid-nineteenth century, such analyses were increasingly supported by an application of mathematical models, which are nowadays capable of 3-dimensional, high-resolution simulation of complex interactions in water bodies, including fluid dynamics, heat transfer and biogeochemical reactions. Mathematical models, until recently perceived as an innovative approach or a good practice, quickly became a legally required element of water management procedures. This chapter presents five examples of the application of aquatic ecosystem models in Poland. Applications were selected in such a way, as to represent various case studies differentiated not only by characteristics of the water body and its catchment area but also by the goal of the model application. Examples include reservoirs with the surface area ranging from 0.1 to 32 km^2 and catchment areas ranging from 13 to more than 500 km^2. Presented applications were either a part of research projects or studies commissioned by local authorities. Aims of presented studies include among others a presentation of real-time status of the water body, analyses of impacts of reservoir and catchment management scenarios, an

R. Ulańczyk (✉) · A. Kolada
Institute of Environmental Protection, National Research Institute, Krucza
5/11D, 00-548 Warsaw, Poland
e-mail: rafal.ulanczyk@ios.edu.pl

A. Kolada
e-mail: agnieszka.kolada@ios.edu.pl

B. Łozowski · A. Woźnica · D. Absalon
Silesian Water Centre, Faculty of Natural Sciences, University of Silesia in Katowice, Bankowa
Str. 9, 40-007 Katowice, Poland
e-mail: bartosz.lozowski@us.edu.pl

A. Woźnica
e-mail: andrzej.woznica@us.edu.pl

D. Absalon
e-mail: damian.absalon@us.edu.pl

© Springer Nature Switzerland AG 2021
M. Zeleňáková et al. (eds.), *Quality of Water Resources in Poland*, Springer Water,
https://doi.org/10.1007/978-3-030-64892-3_8

evaluation of the recreational potential, an evaluation of planned remedial measures and support to the state environmental monitoring system.

Keywords Lake · Reservoir · Ecosystem · Model · Case study · Impact assessment · Poland

8.1 Introduction to Models and Model-Based Decision Support Systems

An increase in anthropogenic pressures results in the deterioration of water quality in rivers, lakes, and reservoirs [1]. Because of the severity of the problem, numerous actions aimed at the improvement of the water quality are undertaken to present various methods applied to the assessment of the water quality status and the prediction of effects of corrective measures [2–5].

Water quality is potentially the best in lakes and reservoirs of relatively large depths and volumes and of small catchment areas with no sources of pollution. On the contrary, water bodies which are shallow, contain a smaller amount of water and have large catchment areas, including agriculture and dense residential areas are much more sensitive to pressures [6]. In flow-through lakes and reservoirs characterised by a short retention time physical, chemical and biological processes depend to a large extent on the status of contributing streams. By contrast, large lakes and reservoirs that have low water exchange rates develop own ecosystems. Features mentioned above, as well as other forms of water dynamics (such as waves, horizontal and vertical currents, stratifications and mixing events), have a definitive influence on the water quality.

To cope with the problem of lack of readily available information on the status of water resources and factors affecting these resources more and more complex models and decision support systems (DSS) are being developed and applied at various scales (local, national, international). These tools should support the expert who analyses processes occurring in a reservoir based on clearly defined objectives and based on the knowledge of the expert and of the model developer. One of the model types is a physical model that represents a reservoir, a lake, or a hydraulic structure in a defined scale. Such models are not widely applied nowadays because of the cost and the duration of construction and limited capabilities of parameterization (adjustments) and scope of simulations. Physical models are also prone to ineffectively represent force ratios which are adequate to the model scale; another issue is the problem of scaling of various dimensions of such models (linear, square, cubic) [7]. Physical models were used mostly before the sufficiently fast computers were developed and available. An example of such a model is the Vistula River model constructed to simulate the mixing of cooling water discharged from the Kozienice power plant and reported in 1978 [8]. At present, physical models are still in use [e.g., 9, 10], however, their application is aimed primarily to validate mathematical models [11–13]. The latter is the most widely used type of models. Mathematical models supporting the protection of water resources date back to the 1950s and were rapidly developed thanks to advances in information and communication technology [14]. However, origins of mathematical modelling in the hydrology can be traced to works like the analysis of the biological oxygen demand in the Ohio river by Streeter and Phelps in 1925 [15]. The model was run with the use of computers in the 1960s, and it included a first-order linear differential equation and 1-dimensional flow equation based on the water level and the flow rate. There are various classifications of mathematical models, for example, stochastic and deterministic, steady-state and unsteady, 0-dimensional to 3-dimensional, however, in this chapter we focus on a division of models into two groups based on the scope of simulated processes, i.e., hydrodynamic models and aquatic ecosystem models. The first type allows simulating the water flow and the transport of variables taking into account a wide scope of driving forces and features of the analysed water body (e.g., wind energy, heat transfer, solar radiation, bathymetry, inflows/outflows, damming level, etc.). Models of aquatic ecosystems are usually coupled to hydrodynamic models in order to simulate both the water flow and quality. These models compute equations representing chemical and biological processes, thus enabling the simulation of cycles of individual elements (e.g., N, P, Si, O_2, C, metals), chemical compounds and biomass of plankton, bacteria, fish, macrophytes and other organisms. Taking into account the variability and complexity of aims, various legal regulations, socio-economic conditions and unique characteristics of analysed areas (geography, hydrology, geology, climate, etc.) it is not possible or at least it is unreasonable to develop one, all-purpose and widely applicable model [16]. At present, there are thousands of mathematical models dedicated to various components of the environment, for example, Janssen et al. [17] reported over 1500 aquatic ecosystem models that varied in their complexity. There are widely used one-dimensional models, like CE-QUAL-R1, DYRESM, DUFLOW, GLM, GOTM, LIMNMOD, MINLAKE, Mylake, PROTECH, SIMSTRAT, which are usually applied for the simulation of water mixing and quality in a water column [17]. Based on such tools, two-dimensional models are often developed (e.g., CE-QUAL-W2), which are capable of simulating processes along rivers and streams or narrow reservoirs/lakes. Two-dimensional models can be used to simulate floods taking into account the elevation of bed and floodplains. In these models, however, it is assumed, that the variation of analysed parameters (e.g., water temperature, concentrations) is negligible in a horizontal plane. It is a considerable limitation of the use of 2-D models in the analysis of local problems, as the algal blooms [18, 19]. The most advanced tools have a form of three-dimensional models (e.g., AEM3D, ELCOM-CAEDYM, GEMSS, GETM) or systems allowing the user to choose the number of dimensions appropriately to the goal of application (e.g., Delft3D, EFDC, WASP) [17]. Three-dimensional models usually allow to simulate changes in multiple water quality parameters (even teens or hundreds of parameters simultaneously) and includes non-linear interrelated equations. The water body is represented as a set of cuboids or polyhedrons, for which output variables are calculated. Advantages and disadvantages of various model types were a subject of numerous studies, e.g., [17, 20, 21].

The Aquatic Ecosystem Model (AEM3D) is an example of a three-dimensional coupled hydrodynamic and ecosystems model. It is provided and supported by the HydroNumerics company (www.hydronumerics.com.au) and is based on two models, i.e. ELCOM (Estuary, Lake, and Coastal Ocean Model) and CAEDYM (Computation Aquatic Ecosystem DYnamics Model) developed earlier by the Centre for Water Research—University of Western Australia [22–24]. The AEM3D allows to simulate among others: water flow, water temperature and density, cycles of nitrogen, phosphorus, oxygen, silica, carbon, sediments, metals and organic matter, the biomass of bacteria, plankton and macrophytes [24].

8.2   Mathematical Models of Lakes and Dammed Reservoirs—Examples of Practical Applications

This section presents five applications of the AEM3D model and coupled ELCOM-CAEDYM models. Applications were selected in such way, as to represent various case studies differentiated not only by characteristics of the water body and its catchment area but also (and primarily) by the goal of the model application.

8.2.1   Goczałkowice Reservoir

The Goczałkowice Reservoir is the biggest dam reservoir in the south of Poland. It covers over 32 km² and has two main inflows: Small Vistula (approx. 80% of the inflow) and Bajerka. The reservoir was created in 1955 and it protects downstream areas from floods and droughts, serves as a part of system supplying the Upper Silesian agglomeration with potable water (~ 3.4 million inhabitants) and helps to protect a wide range of habitats and species as a part of the Natura 2000 system [25–27]. Average water depth is 5.5 m, and the maximum is 13 m for the normal damming level. The reservoir drains the area of 530 km² that includes high plains and mountains. Model of the reservoir was prepared as one of the tasks in the research project "Integrated system supporting management and protection of water dam reservoir" (Polish acronym: ZiZOZap) funded from the European Regional Development Fund under the Operational Programme Innovative Economy. Goals of the model application included a presentation of real-time status of the reservoir and provision of a tool capable of analyses of impacts of various reservoir management scenarios. The model that was used for the real-time analyses was of 100 m horizontal resolution and included 20 water layers with thickness ranging from 0.5 to 1.25 m. The total number of calculation cells was 41,683. However, for detailed analyses of reservoir hydrodynamics, a model of horizontal resolution equal to 20 m was used. The model was parameterized in order to enable the simulation of hydrodynamics and thermodynamics based on inputs regarding two main surface inflows, five inflows from pumps dewatering surrounding areas and the meteorology (temperature, wind speed, and gust, relative humidity, pressure and cloud cover). The 100 m resolution model includes additionally a water quality module (CAEDYM) which was used to simulate nutrients, oxygen and silica cycle, two fractions of inorganic sediments, four groups of phytoplankton, three groups of zooplankton and three groups of fish. The model has been validated based on hourly observed water temperature in a water profile (9 intervals) and based on hourly measurements of the chlorophyll a. Validation of the calculated water temperature resulted in the coefficient of determination (R^2) 0.96 and the Nash Sutcliffe Efficiency (NSE) coefficient 0.95. For the chlorophyll, these coefficients were lower but still satisfactory ($R^2 = 0.52$ and NSE $= 0.51$). An example of model outputs prepared for the online presentation is shown in Fig. 8.1.

The model of higher spatial resolution was used among others for the analysis of impacts of dredging the reservoir near to the inflow of the Vistula River. The dredging was planned in order to restore the original geometry of the Vistula River channel which was blocked by sediments transported with surface waters. The sediment deposition impeded water flow to the east (central part of the reservoir) and hindered the navigation. The analysis included three scenarios: 0) current status, (1) dredging the Vistula channel to the elevation of 253.1 m amsl. (length of the dredged channel: 700 m; maximum depth: 1.5 m; average depth: 0.46 m). Moreover, (2) dredging the Vistula channel to the elevation of 252.0 m amsl (length of the dredged channel: 1470 m; maximum depth: 3.4 m; average depth: 0.87 m). In both cases, the restored channel bottom was 40 m wide with 20 m wide slopes (Fig. 8.2). The goal of the analysis was to assess if the planned dredging will result in increased concentrations of mineral particles and dissolved pollutants in the north-eastern part of the reservoir, where the water intake is located, and if it results in a shorter pollutants travel time. Such risk was considered by the operator of the reservoir, as a consequence of facilitated water flow in the direction of the water intake. Simulations included the transport of two fractions of sediments (0.45 and 1 μm) and a virtual tracer representing dissolved, conservative pollution. Model outputs indicated that after the dredging, the pollutants travel time to the intake will not decrease and, what is more, the calculated concentration of sediments in the location of intake will be reduced (Fig. 8.3). The projected, positive effect of the dredging results from the increased water flow velocity and water mixing rate after the restoration of the Vistula River channel.

8.2.2 Rogoźnik I Reservoir

The Rogoźnik I is a relatively small, dammed reservoir (12.6 ha) located in the municipality of Bobrowniki, in southern Poland. Maximum and average depths of the reservoir are 2.4 and 1.28 m, respectively. Its catchment (15 km^2) is covered primarily by agricultural areas and includes two other reservoirs connected in a series. Outflow from these reservoirs represents nearly 100% of the total surface inflow to the Rogoźnik I. To exploit the recreational potential of the reservoir and its surroundings,

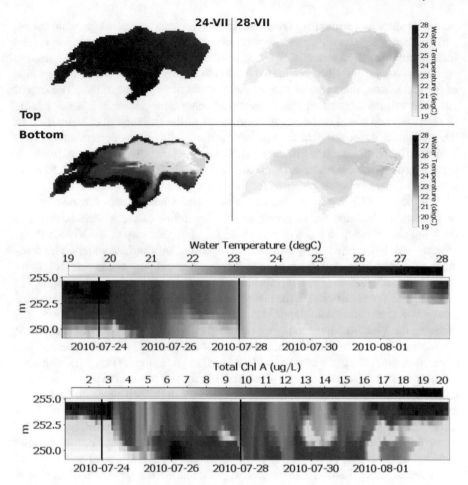

Fig. 8.1 An example of ELCOM-CAEDYM outputs for the Goczałkowice Reservoir: Spatial distribution of water temperature in the top and the bottom water layer on the 24 and the 28 of July 2010 and the water temperature and chlorophyll *a* in a water column located near the main outflow (after [28])

an analysis of the status of the Rogoźnik I reservoir and a proposition of remedial measures were done at a request of the Municipality Office. The study included the preparation of a mathematical model capable of the simulation of impacts of the damming level, various inflow rates (including extreme) and the reduction of nutrients load on the retention time and the water quality (nutrients and phytoplankton). The model was prepared using the AEM3D. It had the horizontal resolution of 10 m, and 7 water layers of thickness ranging from 0.2 m near the surface to 0.5 m in the deepest part. The simulated period included one year and a set of scenarios, that enabled the assessment of impacts of proposed remedial measures. According to the model outputs, the water mixing was wind-driven, and the largest flow velocity

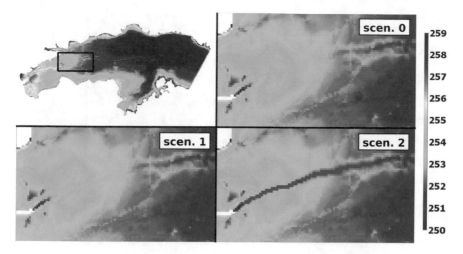

Fig. 8.2 Scenarios of the Goczałkowice Reservoir dredging [m amsl] (after [28])

was in the central part of the reservoir, along central-northern and central-southern water edges. This pattern was characteristic for both the high inflow scenario and the low inflow scenario. The water retention time was 15 days on average in the analysed period and varied from 3 to 29 days during high and low inflow rates, respectively. The main inflow from the upstream reservoir had a minor impact on the water flow velocity and directions limited to periods of high inflow and low wind energy. Nonetheless, it was a key factor affecting the water quality in the Rogoźnik I. Simulated reduction of the nutrient load in inflows by 10 and 20% resulted in almost the same reduction in the outflow from the reservoir. The simulated reduction also caused a significant decrease in the chlorophyll *a* concentration in the reservoir and the outflow (decrease by 8 and 17.5%). The largest decrease in the chlorophyll *a* concentration was estimated for the north-eastern part of the reservoir, close to the main inflow. This is the area where the concentration of nutrients and the abundance of phytoplankton are the highest in the current status and all other scenarios. Another measure analysed with the use of the model was the temporary decrease in the damming level to the allowed minimum, that enables cleaning the reservoir bottom. It was estimated that the decreased damming level could result in increased concentration of nutrients and chlorophyll *a* for most of a year and especially in spring and summer (Fig. 8.4). The adverse effect mainly concerned the phytoplankton in the north-eastern part of the reservoir, where the calculated concentrations were several times greater than for the normal damming level. In case of nutrients, the estimated concentrations were more uniformly distributed, and the increase was usually less than 10% (see Fig. 8.4)

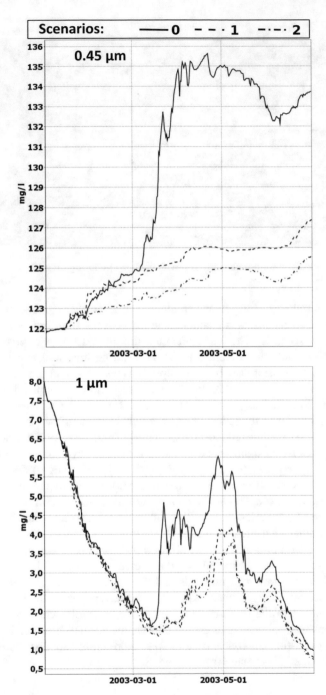

Fig. 8.3 Concentrations of sediments in the location of the water intake, calculated for three scenarios of the dredging of the Goczałkowice Reservoir

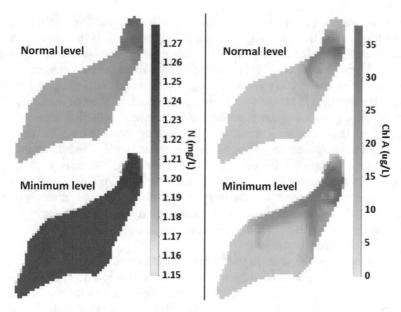

Fig. 8.4 Simulated impact of the decreased damming level on the concentrations of nitrogen and chlorophyll a in the reservoir Rogoźnik I

8.2.3   Paprocany Lake

The lake Paprocany is located in the City of Tychy, in southern Poland. Constructed in the first half of the eighteenth century, it is one of the oldest dammed reservoirs in Poland [29]. Currently, the lake is used for recreational purposes and fish farming. The lake covers the area of 105.1 ha, and its mean and maximum depths are 1.34 and 2.51 m, respectively. The natural catchment area is 130.6 km^2. However most of the surface water is transferred currently downstream the lake limiting the catchment area to 17.94 km^2 and increasing considerably the water residence time. Despite the efforts made to improve the water quality, the use of the lake for recreational purposes was limited. Therefore, the "Programme of the restoration of the Lake Paprocany" was commissioned by the Municipal Office in Tychy and the model of the lake was prepared to support analyses of proposed measures [30]. Main objectives of the model application included the assessment of dynamics in the water quality, the assessment of impacts of the restoration of the natural catchment area and the evaluation of the performance of the solid barrier partially isolating the bathing area. The horizontal resolution of the model prepared using the coupled ELCOM-CAEDYM was 10 × 10 m, while the vertical resolution ranged from 0.1 m near the surface to 1.3 m in the deepest part.

In order to assess the impact of the restoration of the catchment area, two scenarios were analysed. The first one assumed that the lake, in addition to current inflows, will receive water from the Gostynia River basin upstream the S1 ditch but only if the flow exceeds the minimum measured in a dry season. The second scenario assumed the restoration to a larger extent, including the same as the scenario (1) and additionally transfer of the Żwakowski stream back to the lake's catchment area. The model simulation allowed to identify locations where the peak concentration of chlorophyll *a* exceeded 200 µg/l suggesting algal blooms. According to the model outputs, the increase in inflow (restoration of the catchment area) should increase the flow velocity and the water mixing rate. The water retention time increased from 60 days in the eastern part including the bathing area and from 77 days in the northern part (maximum) to 40 and 58 days in the scenario 1 and 25 and 47 days in the scenario 2. Both restoration scenarios resulted in the decreased chlorophyll *a* concentration and a number of locations, where the chlorophyll *a* concentration was much greater than the average in the lake (see Fig. 8.5). It is worth noticing that the increased catchment area can improve the water quality considerably, but the proper land management has to be ensured to prevent the pollution (e.g., increased loads of nutrients from agricultural areas).

Among measures proposed in the "Programme" to protect the bathing area, was the isolation of bathing area from the western part of the lake, where main inflow and sources of nutrients are located. This option was analysed with the use of a model, which included a 250 m curtain (Fig. 8.6). Simulations indicated that the curtain will not affect the average concentration of chlorophyll *a* in the bathing area, because the main force driving the transport of nutrients and plankton is the wind and not currents caused by surface inflows. What is more, the curtain can increase the water retention time in the bathing area in a longer-term perspective, thus posing a risk of water quality deterioration. However, the curtain can have a significant impact during events of a large inflow to the lake, especially when the quality of inflowing

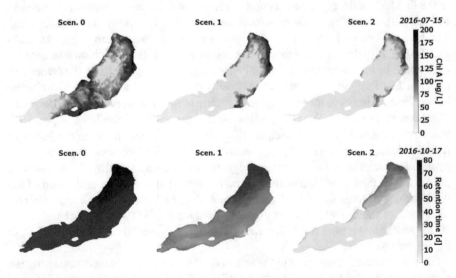

Fig. 8.5 Simulated chlorophyll *a* (top) and water retention time (bottom) for three scenarios of the inflow to the lake Paprocany

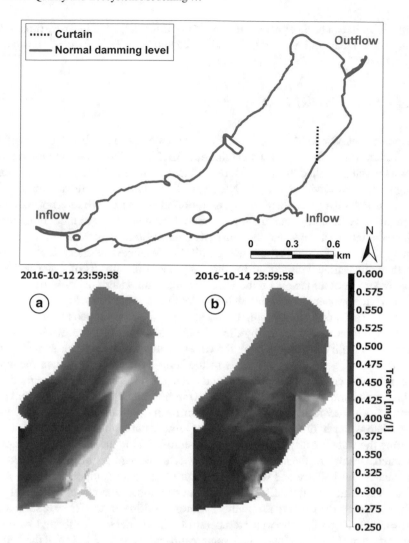

Fig. 8.6 Simulated impacts of the curtain installed near the bathing area of the lake Paprocany: (A) a negative impact of the curtain and (B) a positive impact

water differs from the quality in the lake. In the case of the inflow of polluted water, the curtain will protect the bathing area until the water is mixed by wind (Fig. 8.6). On the other hand, when inflows are of good quality, the curtain can slow down the improvement of the water quality in the bathing area (limited dilution, limited oxygen input, higher temperature). Such events accompanied by no wind or south/east winds only can lead to water stagnation and conditions favourable to algal blooms and other adverse effects. Taking into consideration the results of simulations, it was suggested, that the curtain can be an effective measure only if it is adjusted in a real-time to the quality of two main inflows and the wind speed and gust.

8.2.4 Łękuk Wielki Lake

The lake Łękuk Wielki has an area of 21.3 ha and mean and maximum depths of 5.2 and 12.5 m, respectively [31]. Its catchment (13.2 km^2) has been monitored since 1994 as a station called "Puszcza Borecka" and included in the Integrated Monitoring of Natural Environment—IMNE (part of the state environmental monitoring system in Poland). Since 2018, analyses reported by the IMNE, are not based on the monitoring data only but are supplemented by mathematical modelling aimed at the calculation of the water and nutrients balance in monitored catchments. The primary tool used for this purpose is the Soil And Water Assessment Tool (SWAT), a mathematical model developed for the assessment of impacts of water and land management or other factors on the water balance and water quality in river basins [32]. SWAT, however, does not allow for in-depth analyses of processes occurring in lakes and reservoirs, and therefore, the modelling system developed for the "Puszcza Borecka" station was extended by the AEM3D. The aim of the AEM3D application was to identify factors affecting the water quality in the lake Łękuk Wielki and to enable predictions based on trends in the long-term monitoring data including meteorological conditions, deposition, and changes in the land cover. The model of the lake was composed of 33,622 cells of 10 m horizontal resolution and thickness ranging from 0.25 m in the top layer to 1 m in the bottom. The model was used to simulate water hydrodynamics and water temperature considered as key factors affecting the water quality and aquatic ecosystems. Taking into account the scope of simulated parameters, the model calibration was aimed at matching the observed and simulated water balance and water temperature. The only component of the water balance calculated by the AEM3D was the evaporation. Therefore, the model was calibrated using the observed (monthly) values resulting in the $R^2 = 0.83$ and the NSE $= 0.88$. Regarding the water temperature, the model was calibrated based on temperature profiles (including 8 intervals) resulting in the $R^2 = 0.72$ and the NSE $= 0.28$. Simulations indicated that the average, yearly difference between bottom and surface water layers ranged from 11.8 to 18.2°C in the period of May–November. In remaining months the difference decreased to less than 1 °C. The calculated average water retention time was 280 days in the entire lake except for the areas close to surface inflows. During the warm season, the retention time was different at different depths. The difference between top and bottom reached 70 days (Fig. 8.7).

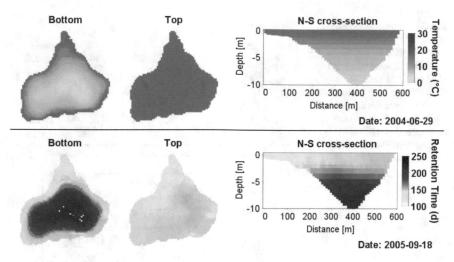

Fig. 8.7 Simulated water temperature and water retention time in the lake Łękuk Wielki

8.2.5   Kozłowa Góra Reservoir

The reservoir is located in southern Poland, in Świerklaniec. Its catchment area is of nearly 200 km² and is of mixed land use, with the domination of forests, but also with agricultural areas and urban areas (mainly residential and transportation— international airport). The area of the reservoir is 5 km², and its average and maximum depths are 2.4 and 4.5 m, respectively for the normal damming level. The reservoir is operated by the Silesian Waterworks PLC and serves for water supply and flood protection purposes. The reservoir and its catchment serve as one of the pilot sites of the PROLINE-CE project aimed at the improvement the protection of drinking water resources and protection measures against floods and droughts in an integrated land use management approach (funded by the European Regional Development Fund— Interreg Central Europe Programme). One of the tools applied in this pilot site in order to support water management is the model of the reservoir prepared using the AEM3D. The model is of horizontal resolution equal to 50 m and has 11 water layers of thickness ranging from 0.25 to 1.15 m in the deepest part. The model includes inputs regarding surface inflows and the quality of inflowing waters, groundwater inflow/outflow, meteorological conditions (precipitation, wind, humidity, pressure, cloud cover and solar radiation), water intake and discharge from the reservoir. An initial application of the model included assessment of impacts of inflows and wind on the water retention time and the transport of pollutants. The retention time is primarily dependent on the surface water inflows and the outflow from the reservoir. Spatial distribution of the retention time is, however, driven by wind energy. The retention time varied from 100 to 270 days in the period of 2014–2016 for the lowest and highest observed inflows, respectively. The average retention time was 165 days, and it was almost uniformly distributed over the reservoir except for areas close to

Fig. 8.8 Spatial distribution of tracer introduced to the reservoir with all inflows at the concentration of 1 mg/l after a year of simulation (top) and a concentration of the tracer in the outflow of the reservoir (bottom) for four scenarios: (Scen. 0) observed inflows, (1) SNQ (2) SSQ and (3) SWQ

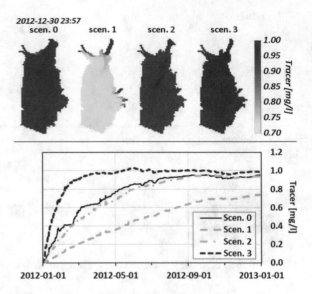

the main inflows. The difference between the retention time in the surface and the bottom layers did not exceed one day except for areas close to the main inflows. The average retention time calculated for characteristic inflow rates, i.e., the average low flow rate—SNQ, the average flow rate—SSQ and the average high flow rate—SWQ was 317, 141 and 37 days, respectively. Even though the flow direction in the reservoir was wind-driven mainly, it was estimated that the dominating flow direction was to the south in the central part of the reservoir and to the north along eastern and western banks. Potential paths of the pollutants transport were assessed based on virtual tracers introduced to the reservoir individually with separate, seven inflows and a tracer introduced with all inflows all together. Figure 8.8 presents the spatial distribution of the tracer after a year of simulation and changes in the tracer concentration in the outflow from the reservoir. Both outcomes are presented for four scenarios: (0) observed inflows, (1) SNQ, (2) SSQ, and (3) SWQ. According to the model outputs, dissolved substances that enter the reservoir with surface waters are uniformly distributed over the entire reservoir except for the locations next to main inflows in the northern and eastern parts, and except for periods of the average low inflow (scenario 1). The concentration in inflows and in the outflow is almost equal in all scenarios but the scenario 1, however, the travel time of the pollution (tracer) varied considerably in analysed scenarios (see Fig. 8.8).

8.3 Conclusions

Mathematical models are extensively used worldwide and are more and more commonly used in Poland as well. The application of mathematical models as tools supporting the water management is recommended and advised or even required by numerous international organisations which formulate the environmental protection policy and legal regulations on various scales. Models can be applied for virtually all lakes and reservoirs, provided that a proper conceptual design is prepared, tools are chosen based on the goal of the application, and sufficient data of appropriate quality are collected for model parameterisation, calibration, and validation. Application of models can be time-consuming and expensive (sometimes unreasonably), therefore, it is crucial to define exact objectives and clear expectations regarding the scope of model outputs.

In this chapter, only a small set of examples was presented, but it confirms, that models are applied to lakes/reservoirs of different sizes, types, and geographic features. Applications are aimed among others at the identification of processes occurring in the water body and factors affecting these processes, assessment of impacts of specified management practices including land management practices in catchment areas, etc. Mathematical models of lakes and reservoirs are no longer tools used for the research purpose only but became a common practice used and required by authorities/units responsible for the management of water resources, flood protection, water supplies the state monitoring system.

8.4 Recommendations

Mathematical models for lakes and reservoirs are widely available together with an extensive theoretical documentation and user manuals. Moreover, more and more experts (usually universities, research institutions and consulting companies) offer services and support in the application of models, what can be useful for authorities and other units responsible for the water management, the assessment of water status and the water supply.

Mathematical models, even if sometimes not required in the water management procedures, can be considered as an optional supporting tool. In such considerations, the objective should be precisely defined for the model in order to choose the appropriate tool (capable of providing the desired information) and minimise the cost of the application.

Even though mathematical models are used for decades and are proven, recommended and increasingly required tools, the application in some sectors, regions and countries is not at the level allowing for the exploitation of the full potential of such tools. Therefore, it is still important task to increase the public awareness of capacities of existing tools (models) and acceptance of models as a source of information about factors which affect, affected or can affect in future water resources and aquatic ecosystems.

Acknowledgements The research leading to a part of presented results has received funding from the European Regional Development Fund under the Operational Programme Innovative

Economy under grant agreement n° POIG 01.01.02-24-078/09 and from the Interreg Central Europe Programme—project n° CE110.

References

1. Cosgrove WJ, Loucks DP (2015) Water management: Current and future challenges and research directions. Water Resour Res 51:4823–4839. https://doi.org/10.1002/2014WR016869
2. Daoliang L, Shuangyin L (2019) Water quality evaluation. In Water quality monitoring and management basis, technology and case studies. Academic Press, London, pp 113–159
3. Xua C, Zhangb J, Bia X, Xua Z, Hea Y, Gin KY-H (2017) Developing an integrated 3D-hydrodynamic and emerging contaminant model for assessing water quality in a Yangtze Estuary Reservoir. Chemosphere 188:218–230
4. Kaçıkoç M, Beyhan M (2014) Hydrodynamic and Water Quality Modeling of Lake Eğirdir. CLEAN Soil Air Water 42:1573–1582. https://doi.org/10.1002/clen.201300455
5. Molina JP, Rodríguez-Gonzálvez P, Molina C, González-Aguilera D, Espejo F (2014) Geomatic methods at the service of water resources modelling. J Hydrol 509:150–162
6. Matysik M, Absalon D, Habel M, Maerker M (2020) Surface water quality analysis using CORINE data: an application to assess reservoirs in Poland. Remote Sens 2020(12):979. https://doi.org/10.3390/rs12060979
7. Heller V (2011) Scale effects in physical hydraulic engineering models. J Hydraul Res 49(3):293–306. https://doi.org/10.1080/00221686.2011.578914
8. Plewa S, Góralczyk M (1978) Rzecz o Wiśle. Program Wisła. The National Film Archive's Digital Repository 36/78B, Release Date: 9 Sep 1978. http://www.repozytorium.fn.org.pl. Accessed 6 Mar 2019
9. Gallisdorfer MS, Bennett SJ, Atkinson JF, Ghaneeizad SM, Brooks AP, Simon A, Langendoen EJ (2014) Physical-scale model designs for engineered log jams in rivers. J Hydro-environ Res 8(2):115–128. https://doi.org/10.1016/j.jher.2013.10.002
10. Bousmar D, Courtois E, Van Audenhaege L, Rollin X (2018) Performance of a fish pass for multiple species: Scale model investigation. E3S Web of Conferences 40, 03010. https://doi.org/10.1051/e3sconf/20184003010
11. Solov'eva DA, Nigmatulin RI (2010) Investigation of the spring thermal bar phenomenon based on mathematical and laboratory modeling. Dokl. Earth Sci 434:1346. https://doi.org/10.1134/S1028334X10100120
12. Ashraf M, Soliman AH, El-Ghorab E, El Zawahry A (2018) Assessment of embankment dams breaching using large scale physical modeling and statistical methods. Water Sci 32(2):362–379. https://doi.org/10.1016/j.wsj.2018.05.002
13. Begam S, Sen D, Dey S (2018) Moraine dam breach and glacial lake outburst flood generation by physical and numerical models. J Hydrol 563:694–710. https://doi.org/10.1016/j.jhydrol.2018.06.038
14. Orlob GT (1983) Introduction. In: Orlob GT (ed) Mathematical modelling of water quality: Streams, lakes and reservoirs. International series on applied system analysis. Wiley, Chichester, pp 1–10
15. Streeter HW, Phelps EB (1925) A study of the pollution and natural purification of the Ohio Rivers. US Public Health Service Bulletin No. 146
16. Mauersberger P (1983) General principles in deterministic water quality modeling. In: Orlob GT (ed) Mathematical modelling of water quality: Streams, lakes and reservoirs. International series on applied system analysis. Wiley, Chichester, pp 42–115
17. Janssen ABG, Arhonditsis GB, Beusen A, Bolding K, Bruce L, Bruggeman J, Couture R-M, Downing AS, Elliott JA, Frassl MA, Gal G, Gerla DJ, Hipsey MR, Hu F, Ives SC, Janse JH, Jeppesen E, Johnk KD, Kneis D, Kong X, Kuiper JJ, Lehmann MK, Lemmen C, Ozkundakci

D, Petzoldt T, Rinke K, Robson BJ, Sachse R, Schep SA, Schmid M, Scholten H, Teurlincx S, Trolle D, Troost TA, Van Dam AA, Van Gerven LPA, Weijerman M, Wells SA, Mooij WM (2015) Exploring, exploiting and evolving diversity of aquatic ecosystem models: a community perspective. Aquat Ecol 2015(49):513–548

18. Romero JR, Antenucci JP, Imberger J (2004) One- and three-dimensional biogeochemical simulations of two differing reservoirs. Ecol Model 174(2004):143–160

19. Lee HS, Chung SW, Ryu I, Choi J (2013) Three-dimensional modeling of thermal stratification of a deep and dendritic reservoir using ELCOM model. J Hydro-environ Res 7(2):124–133

20. Saloranta TM, Malve O, T. Bakken TH, Ibrekk AS, Moe J (2004) Lake water quality models and benchmark criteria. NIVA, SYKE. Delivery Report from the Lake Model Work Package (WP6) of the BMW-project

21. Gao L, Li D (2014) A review of hydrological/water-quality models. Front Agr Sci Eng 1(4):267–276

22. Hipsey MR, Antenucci JP, Hamilton D (2012) Computational aquatic ecosystem dynamics model: CAEDYM v3 v3.2 Science Manual (DRAFT). Centre for Water Research, University of Western Australia, 17 July 2012

23. Hodges B, Dallimore C (2013) Estuary, lake and coastal ocean model: ELCOM v3.0 user manual. Centre for Water Research, University of Western Australia, 23 May 2013

24. Hodges B, Dallimore C (2016) Aquatic ecosystem model: AEM3D v1.0 user manual. HydroNumerics, 7 Oct 2016

25. Polak J, Bartoszek M, Żądło M, Kos A, Sułkowski WW (2011) The spectroscopic studies of humic acid extracted from sediment collected at different seasons. Chemosphere 84(11):1548–1555. ISSN 0045-6535, http://dx.doi.org/10.1016/j.chemosphere.2011.05.046

26. Dabioch M, Kita K, Zerzucha P, Pytlakowska K (2013) Assessment of elemental contamination in the bottom sediments from a dam reservoir using a sequential extraction technique and chemometric analysis. Cent Eur J Chem 11(12):1981–1995

27. Młynarczyk N, Bartoszek M, Polak J, Sułkowski WW (2013) Forms of phosphorus in sediments from the Goczałkowice Reservoir. Appl Geochem 37:87–93

28. Łozowski B, Ulańczyk R (2018) Digital models of reservoirs—a tool to manage waters and ecosystem. In: Absalon D (ed) Current problems in water management. Monografie Śląskiego Centrum Wody, vol 1, pp 83–102

29. Absalon D, Czekaj J, Jarosz W, Łozowski B, Małkowski E, Migula P, Pasierbinski A, Pszczeliński Ł, Sitek S, Siudy A, Ulańczyk R, Woźnica A, Zarychta A (2017) Comprehensive restoration of the lake and its catchment area (in Polish). Komunalny Plus 4(2017):12–16

30. Absalon D, Matysik M, Woznica A, Łozowski B, Jarosz W, Ulanczyk R, Babczynska A, Pasierbinski A (2020) Multi-faceted environmental analysis to improve the quality of anthropogenic water reservoirs (Paprocany reservoir case study). Sensors 20:2626. https://doi.org/10.3390/s20092626

31. Siuta J (ed) (1994) Stacja Kompleksowego Monitoringu Środowiska Puszcza Borecka (Puszcza Borecka (the Borki Forest) Integrated Monitoring Station). Instytut Ochrony Środowiska, Warszawa 130p (in Polish with English summary)

32. Neitsch SL, Arnold JG, Kiniry JR, Williams JR (2011) Soil and water assessment tool theoretical documentation, grassland, soil and water research laboratory, Blackland Research Centre, Texas Water Resources Institute Technical Report No. 406, Texas A&M University System

Chapter 9
Assessment of Pollution of Water Resources and Process of Pollution Spreading

Jacek Kubiak and Sylwia Machula

Abstract Excessive fertilization of surface waters is an extremely important problem in their protection. Urbanization, industrialization, intensive agricultural production in the catchment area, as well as significant development of tourism, in the absence of effective methods of lake protection, affect their high nutrient loads. Therefore, in recent decades we have observed rapid eutrophication of lakes. The multidimensional importance of lakes (e.g. natural, economic) is indisputable, hence the need to protect them. The paper presents the results of long-term studies enabling to trace the changes taking place in the lakes of West Pomerania. The susceptibility of the studied lakes to degradation was assessed, as well as the role of the catchment area in this process. Analyzing the ecological systems of the catchment—lake, the intensity of eutrophication of each reservoir was determined; trophy level of the examined lakes was assessed. It was found that farming is the basic source of biogens for the studied waters, and the degradation of water resources under the influence of agricultural area pollution is a fundamental problem for their protection. The results of lake research clearly indicate the need to develop plans for their protection, which should include examining 7–8 times each year of each water body, which will allow to determine the functioning of lake ecosystems and take appropriate protection measures.

Keywords Eutrophication · Phosphorus · Nitrogen · Lakes · Poland

J. Kubiak (✉) · S. Machula
Department of Commodity Science, Quality Assessment, Process Engineering and Human Nutrition, West Pomeranian University of Technology in Szczecin, K. Królewicza, 71-550 Szczecin, Poland
e-mail: jacek.kubiak@zut.edu.pl

S. Machula
e-mail: sylwia.machula@zut.edu.pl

M. Zeleňáková et al. (eds.), *Quality of Water Resources in Poland*, Springer Water,
https://doi.org/10.1007/978-3-030-64892-3_9

9.1 Introduction

Large fertilization of surface waters is an extremely important problem in their protection [1–4]. Urbanization, industrialization, intensive agricultural production in the catchment area, as well as a significant development of tourism, with the lack of effective methods of lake protection, affect their high nutrient load, especially phosphorus and nitrogen [5]. Loads of nutrients often exceed the level considered dangerous by Vollenweider [6], which is why in the last decades we have observed fast eutrophication of lakes. The pace of this process and the high degree of advancement of surface water trophies, especially lakes, have become a global problem. The adverse effects caused by man intensified in the 1930s. Currently, in many countries, including Poland, they have assumed undesirable sizes [2, 7]. Recently, many publications inform about the deterioration of the quality of our lake waters, and the term "eutrophication of waters" is one of the most commonly used in limnological literature [2, 3, 8–10]. In England, Wales, Spain, Romania, Denmark, and the Netherlands, the condition of lakes is similar to that of lakes in Poland. Over 85% of respondents in recent years have a concentration of total phosphorus and water transparency characteristic of eutrophy [7, 11–14]. The state of cleanliness of lakes is primarily the result of progressive eutrophication [15].

In Western Pomerania, as in other parts of Poland, excessive runoff to surface waters of biogens and organic matter has been noted for years [16–24]. In the 1970s, only 5 tanks showed poor fertility—mesotrophy (lakes: Miedwie, Morzycko, Krzemień, Ińsko, and Zajezierze), and two lakes (Jelenin and Woświn) were located on the borderline of β-mesotrophy and eutrophy [18].

The progressing eutrophication process, leading to a decrease in the quality and disturbances of the biocenotic balance of waters, even covered particularly protected waters, i.e. located within national parks [25, 26], as well as estuaries and the sea with limited contact with ocean waters. Particular examples include the Gulf of Gdańsk, the Pomeranian Bay, the Szczecin Lagoon, and even the entire Baltic Sea [27], in which 30% of phosphorus and 40% of nitrogen discharged from the entire catchment comes from Poland [7].

Currently, it is considered that eutrophication is the biggest threat to the biodiversity of freshwater habitats [11]. This process is based on an excessive or disrupted increase in water fertility, caused by uncontrolled or insufficiently controlled inflow of biogenic compounds from external sources, as well as on anthropogenic, direct interference in the internal structure of the ecosystem [1, 2, 4, 12, 28]. The causes and effects of eutrophication are known and comprehensively described in the limnological literature [3, 4, 29–31]. It is believed that in the vast majority of cases phosphorus is the factor limiting primary production in inland water reservoirs [2–4, 30]. A similar situation occurs in estuarine waters [27]. Eutrophication also intensifies excessive amounts of organic matter, both allo- and indigenous, because its destruction causes the release of nutrients [2–4, 30].

The lake ecosystem is fed by matter in various ways; its productivity depends on the intensity of the flow of matter and the amount of biogenic compounds taking part in it. Elements accumulated in the ecosystem return to circulation in various amounts, at different times and using various mechanisms [3, 4, 28, 30, 32].

The basic phosphorus and nitrogen resources are accumulated in bottom sediments [2–4, 33–35]. In the surface 10-cm layer of bottom sediments, there is about 90% of phosphorus that occurs in the whole ecosystem [3]. The most important processes in the phosphorus retention in bottom sediments are sorption and binding of this element by iron hydroxides and addition by calcium carbonates and autogenous formation of minerals. The water supply of biogeols contained in bottom sediments plays the most important role during summer stagnation, mainly due to bacteria and release mechanisms to the solution, under annealed conditions, Fe-P and Al-P [4, 30, 33, 35] joints. This process is intensified by the mixing of bottom waters with the ones closer to the surface, which is why the dynamics of water masses play a fundamental role in the nutrient cycle. The average rate of vertical transport of fosfon from hypolimnion to the photonic zone is 4 mg·m-2 per day. Swiss research shows that this transport can take place faster and ranges from 20 to 33 mg·m-2 per day [4].

The release of phosphorus from bottom sediments located within the epilimnion is much more important than the same process carried out in hypolimnion on the border of sediments and water under anaerobic conditions [4]. The phosphorus exchange rate between water and bottom sediments varies in direction and pace; depends on the trophy level of the lake. In oligotrophic and mesotrophic tanks, phosphorus is released into the water regardless of the aerobic conditions prevailing in the bottom waters; in the absence of oxygen, the rate of this exchange is faster. In eutrophic lakes in the presence of oxygen, phosphorus is absorbed by bottom sediments [4, 33].

Much smaller phosphorus reserves are found in aquatic organisms (containing a small percentage of total nutrient resources), but due to the rapid rate of recirculation of these resources, especially in the summer, they play an important role in the ecosystem. From consumption and the distribution of life and from the secretion of various forms of matter by aquatic organisms comes its main part that circulates and is subject to transformation in the ecosystem. Separation of mineral nitrogen and phosphorus compounds by invertebrate animals is the basic mechanism of their circulation in the water depth, as well as significant, in addition to bacterial degradation, in bottom sediments [33]. This applies even more to bacteria. Their biomass is small, while the role in supplying the ecosystem with soluble forms of nutrients is fundamental [3]. All processes together mean that during the summer in lakes' epilimnion, the phosphorus circulation time is several minutes [30].

The basic source of water pollution is agriculture, and the degradation of water resources under the influence of agricultural area pollution is a significant problem [1, 4, 7, 28, 30, 32]. Agroecosystems are characterized by open biogeochemical cycles, significantly simplified. Therefore the losses of nutrients, mainly in the process of water migration, are higher within them. Among pollutants discharged from agricultural areas, organic and trophic impurities predominate, lakes are thus exposed to constantly progressing eutrophication, resulting in increased fertility [31].

The transformation of the primary landscape into the agricultural landscape has affected the deformation of water bodies, mainly consisting of shortening and simplifying the small water cycle. There was a decrease in soil retention, deterioration of infiltration conditions, and facilitation of surface runoff of rainwater [28]. In addition, the accelerated outflow of groundwater intensifies the process of soil leaching [36, 37]. As a result, the export of nutrients from the catchment to surface waters is increasing [38]. The drainage of agricultural land, due to the increased mineralization of organic matter, especially on organic soils, increases the load of nutrients discharged into the water bodies.

Agricultural land use is a source of nitrogen and phosphorus for waters. In Finland, it was found that N and P losses from agricultural areas are 10–20 times higher than from non-human areas [39]. The losses of 0.05 kg P ⁙ ha^{-1} during the year [40] are considered natural for the catchment of the Swiss lakes, and for the basin of southern Sweden—losses amounting to 0.06 kg P ha^{-1} during the year [41]. Kajak [12] described the export of nutrients from the Wisła basin at the level of 5 kg N ha^{-1} during the year and 0.3 kg P ha^{-1} during the year. In agricultural catchments, no more than 0.5 kg of P ha^{-1} are leached to surface waters over the course of the year; in the case of nitrogen, losses amount to several kilograms per 1 ha [40]. The average annual loss of fertilizer components from the catchment of coastal rivers—in the case of mineral nitrogen at 1.00–1.30 kg $_{NNH4+NO3}$·ha^{-1}, and in the case of phosphorus at 0.20 kg P_2O_5·ha^{-1}. Annually, soil losses in the bottom moraine landscape in Western Pomerania amounted to 23.9 kg P_2O_5·km^{-2} [42]. Durkowski [43] described them at the level of 4% of the applied fertilizer dose, while every year from intensely fertilized agricultural areas it amounts to a maximum of 5.5 kg N·ha^{-1} and 0.6 kg P·ha^{-1}. According to Lossowa [10], 1–5% of phosphorus and 10–20% of nitrogen introduced into the soil with mineral fertilizers is washed away into the surrounding waters. In the Wisła basin from area sources (mainly from agricultural areas), 52% of nitrogen and 45% of phosphorus come from the Odra catchment—43 and 30% respectively [44].

The catchments of the studied lakes were intensively used for agriculture. In the vast majority, these areas belonged to state-owned farms, using very intensive fertilization of up to 750 kg NPK·ha^{-1} [36], exceeding the national average (180–200 kg·ha^{-1}) [40]. In the period from 1979/1980 to 1992/1993. There was a decline in the country scale, mineral fertilization (on average from 193 to 66 kg NPK·ha^{-1}, and on individual farms even up to 61 kg·ha^{-1} agricultural land) [45]. A similar tendency was observed in Western Pomerania—in the mid-1970s, fertilization with mineral fertilizers averaged about 280 kg·ha^{-1}, by the end of the eighties it ranged between 250–280, and in 1992–1993 there was a sharp decrease in fertilization to 63, 1 kg·ha^{-1} [46–48]. Since the mid-1970s, the use of lands did not change in this part of the country—arable lands usually occupied about 400, meadows—from 85 to 90, pastures—from 37 to 38, and forests—from 265 to 270 thousand. ha. In the 1990s, the eutrophication of lakes could inhibit a significant reduction in fertilization [15]. It should be noted, however, that large point sources, including municipal sewage, may outweigh the impact of agriculture on a given basin [2, 3].

Inadequate water and sewage management in rural areas is an important cause of water eutrophication. At the same time, along with the expansion of water supply systems, sewage must be taken over and treated. Otherwise, not only rapid degradation of lakes will occur, but also deterioration of groundwater quality [7, 10].

It is unacceptable to run fish-breeding in lakes, which is a significant source of biogens and organic matter for the reservoir, which significantly accelerates their eutrophication [49].

The breeding of slaughter animals and the food industry, as well as tourism increasing anthropopressure, especially in the summer months, are less important in the process of eutrophication of lakes [2, 4, 6].

The municipal, industrial, and recreational suitability of water from pierced lakes decreases, resulting in certain economic effects, e.g. increased costs of water treatment and losses in tourism.

Sheries are also exposed to the adverse effects of eutrophication. This process, apart from the exploitation and introduction of new species, is one of the important factors affecting fish populations [50]. Exploitation and introduction directly affect ichtiofauna. Eutrophication exerts indirect effects on the fish, preceded by changes in environmental conditions [50–52]. These changes include: reduction of water transparency, disappearance of submerged vegetation, quantitative and qualitative changes in zooplankton and benthic fauna, algae blooms, increase in the amount of bottom sediments, deterioration of aerobic conditions, especially in the deepest water layers, as well as periodic drastic changes. changes in pH [2, 4, 30, 53]. The habitat conditions under the influence of these changes are unfavorable for some species of fish, and at the same time favorable for the development of species with different requirements. For example, the emergence and deepening of oxygen deficiencies in the pelagic can eliminate planktonic monsters from this zone, while the deoxidation of profundal causes the exclusion of benthophages from feeding grounds. Such phenomena cause the gathering of previously dispersed fish of different species and age groups in the littoral. Intra- and interspecies tensions increase, mainly in the field of food competition; the pressure of predators and the ability to control fish by parasites and invasive diseases also increase [50]. Other changes determine changes in fish breeding conditions and survival. With the increase in trophies can also be a threat to eggs from the developing, sometimes very strong, predatory invertebrates and fish specializing in its spawning, such as the stickleback *Gasterosteus aculeatus L.* [50, 53].

Transformations in fish communities are a sensitive indicator of changes in trophies of lakes [51–53]. In oligotrophic lakes, ichtiofauna is initially dominated by species from the *Salmonidae* family, e.g. (*Coregonus lavaretus L.*), vendace (*Coregonus albula L.*) and lake trout (*Salmo trutta m. lacustris L.*), which, as eutrophication progresses, are replaced by species of the carp family *Cyprinidae*. The latter is very diverse, abounds in species with different tolerance ranges to environmental and biocenotic factors. The period of occurrence of cyprinids is the longest—there are significant changes in the species composition of fish. Changes within the *Cyprinidae* family consist of displacement of larger, not too numerous species, playing an important role in fishing and angling sport, through so-called fish weed, i.e., fish species of small size, but many occurring. In the early stages of this period, cyprinids coexist with predatory fish such as pike (*Esox lucius L.*) and later zander (*Sander lucioperca L.*). With further progress of eutrophication, predatory fish disappear. Along with the qualitative transformations of ichthyofauna assemblies resulting from the increase in water trophies, there are also changes in the size of fishing production, which initially increases quickly, then undergoes inhibition, and in highly eutrophicated tanks decreases [15, 50–52].

The tendency to change the ichthyofauna syndrome along with the progressive eutrophication has a univocal nature, regardless of climatic conditions. Although species composition of fish is different, changes in subtropical lakes are similar to those observed in temperate lakes [51, 52].

Analyzes of quantitative and qualitative changes of ichthyofauna caused by the eutrophication of lakes progressing in Poland, among others Iwaszkiewicz [54] and Leopold et al. [55]. The vast majority of domestic lakes were dominated by families of cyprinidae and perchidae (*Percidae*) whose requirements correspond to the environmental conditions existing in these reservoirs (mesotrophy, and above all eutrophy). These families, although poorer in economically valuable species, develop well even in shallow, strongly political tanks, in which there are often mass algae blooms as well as summer and zoning oxygen deficits. This means that eutrophication of lakes, which does not lead to saprophy, does not threaten the existence of most species included in these families. On the contrary, in some mesotrophic reservoirs, the increase in fertility, even by the subsequent intensified development of macrophytes, may create better conditions for reproduction and living of a significant number of fish from these families. Most of our mesotrophic lakes are approaching or have already reached the level of eutrophication, after which there will no longer be an increase in the total biomass of fish [50].

The increase in the level of trophic waters and its effects cause the need to look for ways to slow down, inhibit, or even reverse this unfavorable process or remove its negative consequences [7, 8, 34]. Water protection is based on a comprehensive knowledge of ecosystems, which are open structures exchanging substances and energy with their surroundings. There is a need to analyze these interrelations and interactions, taking into account both biotic and abiotic factors. It is appropriate to assess the quality of the ecosystem based on physical-chemical, biological, and ecomorphological criteria [11]. At present, in the protection of lakes, internal reclamation methods, used within the lake basin, and external—protective ones, concerning their catchments are distinguished. The basic, most effective way to protect lakes is to reduce or eliminate feed sources in biogas [5, 7, 8]. Therefore it is necessary to eliminate point sources of pollution and maximum reduction of sewage disposal, without removing biogenic compounds.

Necessary for the protection of lake ecosystems and their rational use is the recognition of the natural resistance of lakes to degradation and the role of the basin in accelerating or inhibiting this process [2, 32, 57–59]. Depending on the natural physical and geographical features, the catchment can accelerate or inhibit the supply of matter (including nutrients) to the lake. The effects of the supply of matter can be different in the tank, depending on the natural resistance, resulting from morphometric and hydrological features [4, 59]. This requires individual consideration of the use and protection of the reservoir, taking into account not only the current water quality but also the features that condition the lake's susceptibility to external influences [2]. The lakes with favorable natural conditions, more resistant to degradation, have greater possibilities of use [31, 32].

Before taking action to protect lakes, it is advisable to learn about the pace and direction of changes in the trophies of each of the reservoirs. Such a procedure makes it possible to take appropriate protective measures, as well as possible withdrawal of the effects of excessive eutrophication [31, 58, 59]. Long-term research is needed to determine whether changes in water quality in the lake are the result of temporary changes in environmental conditions, the result of local conditions associated with the catchment or perhaps reflect the general deterioration of the environment associated with increased anthropopressure [58].

Water dynamics, natural catchment conditions, and susceptibility to degradation determine the quality of water in lakes and their trophic level [2, 4].

A recognized indicator of the trophy of lakes is aerobic conditions. They are shaped by the mutual ratio of intensity of assimilation and dissimilation, with some effect of oxygen exchange with the atmosphere [4, 30, 57]. There is a diversification of the water depth of the reservoir into layers: trophogenic, in which predominance prevails, and trophologic, in which dissimilation takes place more intensively. The ratio of the thickness of the two layers depends on the state of the trophy of the lake; less fertile has a more extensive trophogenic layer [29]. Oxygen systems defining the nature, extent, and intensity of basic processes that form biocenosis, mainly in hypolimnion, vary depending on: the seasons (geographical location), the extent of vertical mixing of water and the impact of tank morphology on them [4].

Lange and Maślanka [60] believe that the analysis of oxygen systems during the summer stagnation period is more useful to assess the level of trophic lakes than the multivariate analysis used in their monitoring. The conclusions regarding the susceptibility of the reservoir to degradation arising from the analysis of aerobic conditions relate to the actual course of the eutrophication process and not to the potential conditions. The susceptibility of the lake to degradation determined based on the physiographic indicators of the catchment may not have a direct relation to the quality of the waters. The most degraded lakes do not have to be characterized by a particularly low resistance, because their condition results primarily from strong anthropopressure (from intensive agriculture, point sources of pollution).

An alternative to the analysis of the level of trophic water based on one parameter is multi-indicator analysis. Methods proposed by Carlson [61] and Vollenweider [6] are widely used, and in the country—by Kajak [12] and Zdanowski [13]. The first one is based on measurements of concentrations in the surface waters of total phosphorus and chlorophyll "a" in summer and the visibility of the Secchi disc. The results of these studies, transformed using the proposed equations in numerical values, are indicators of the state of the TSI trophism (trophic state indicators). The second method, adopted by the OECD (Organization for Economic Cooperation and Development), is based on the average annual and maximum concentrations of total phosphorus and chlorophyll "a" in surface water, Secchi disc visibility and oxygenation of the waters in summer. The basis for the classification of trophies of lakes, proposed by Kajak [12] and Zdanowski [13] are basic indicators, ie summer concentrations of total phosphorus (TP) and chlorophyll "a" in surface waters and summer visibility, as well as additional indicators, i.e. summer concentration of total nitrogen in surface waters (TN), TN to TP ratio, algae biomass and percentage of cyanobacteria in biomass.

The necessity of research on the state of surface waters in our country also results from the fact of integration with the European Union, which requires harmonization of the law, also in the field of water protection. Legislative measures have been taken in the Union to create programs for the renewal of degraded aquatic ecosystems [62]. The associated countries decided to develop programs for specific catchments that would allow, among others, to counteract ecological degradation and to restore degraded surface waters. Until the end of 2010, these activities were to achieve home-ostasis [62]. At the same time, members of the Union were obliged to: elaborate the characteristics of the river basin, review the impact of human activities on the status of surface and groundwater, identify all relevant water resources used as drinking water sources, to develop river basin management plans [63]. In Poland, legislative measures have been taken to adapt Polish law in the field of environmental protec-tion to EU law, which were to lead to water management, in accordance with the principle of sustainable development, and the solutions it contains serve to achieve good ecological status [64].

Lakes are dynamic ecosystems that change over time and aim to enrich and intensify biological production. The occurrence of biogenic elements in lake waters depends on many factors that affect their complex bio-geochemical cycles, defining the circulation between biotic and abiotic elements [3, 65]. Transport, exchange, and redistribution of mineral matter originating from biogeochemical processes, organic matter arising in the ecosystem and pollutants introduced into the environment are processes determining the direction and rate of lakes evolution [28, 32]. In local conditions, these processes depend on a whole set of factors: climate, hydrological, hydrogeological, and soil in the catchment area. Additionally, they are conditioned by morphometric traits as well as hydrochemical relations of the reservoir [3, 4, 31, 65]. The dynamics of water masses [66], the functioning of its biocenosis and the exchange of elements between the solid phase and the solution [4, 30, 65] also play an important role in the lake. Rational management of lake resources requires learning all the aforementioned features of each of the reservoirs, as well as identifying the sources of their pollution.

The rate of matter circulation in the lake, including the rate of reactivation of biogens from the bottom layers, depends on the dynamics of water masses [66]. In the case of similar abundance in nitrogen and phosphorus, the primary production is characterized by lakes with faster circulation of matter, where mixing is more intense. Olszewski [29] acknowledges that its exponent is thermal systems. They are subject to modification through the influence of wind, the intensity of which depends on the morphological features of the bowls [67]. These dependencies determine the level of tank stratification, the intensity of exchange between epilimnion and hypolimnion, and between water and bottom; they are widely discussed by many authors [4, 30, 57].

The condition of Polish lakes is mainly a consequence of their progressive eutrophication. The causes and effects of eutrophication are now known and comprehensively described in the limnological literature [99]. The lake, being an important link in water retention, with deteriorating quality and contraction of groundwater reserves, becomes a major source of high-quality water [2].

9.2   State of Hydrochemical Tests in Lakes Western Pomerania

At the turn of the nineteenth and twentieth centuries and in the pre-war period, the hydrography of the lakes of Western Pomorze was handled by Halbfass [68, 69], Jentsch [70, 71] and Wahnschaffe [72], who examined the more important reservoirs of the region [73]. After the Second World War, limnology research significantly expanded, just after the war, a study by Młodziejewski [75] about Lake Miedwie appeared. At the end of the 1940s, an inventory of Polish lakes was carried out, including those located in West Pomerania. As a result of these works, in the mid-fifties, the "Catalog of Lakes of Poland" was created [76]. In the sixties, the Inland Fisheries Institute carried out morphometric and partly hydrochemical research, among others 200 lakes located in the West Pomeranian Voivodeship. The measurements made at that time are the basic source of knowledge about the morphometry of these tanks. Prawdzic [73] presented in the physiographical, hydrographic, and economic-tourism terms the largest lakes of Western Pomerania. In terms of natural history, the Zachodniopomorskie Lake District was described in detail by Jasnowska and Jasnowski [77], in geographical terms, it was done by Mikołajski [78]. Later, the morphometric data of reservoirs larger than 1 ha were reported in the catalog of lakes located in the Pomeranian Lake District [79]. In recent years, a publication has been published regarding the facts of Szczecin lakes [80] and West Pomeranian islands [81]. Brodzińska et al. [82, 83] are the authors of the "Atlas of Lakes of Poland," a very valuable publication, which contains a lot of morphometric and hydrochemical data concerning, among others, lakes with an area larger than 10 ha, located in the area of the West Pomeranian.

There are few comprehensive hydrochemical studies of the lakes of Western Pomerania. With increased intensity, such research began in the early seventies, at which time Nguyen [84] performed a two-year study of lakes: Ińsko Duże, Ińsko Małe, Wisala and Starzyc, tanks of various trophic degrees. At that time (1970–1976), the Center for Environmental Research and Control researched the larger lakes of this part of the country (lakes: Dąbie, Liwia Łuża, Miedwie, Morzycko, Płoń, and Woświn), the results of which were published in the Atlas of Lake Purity [85]. Many Szczecin lakes were then examined as part of works carried out by the University of Agriculture in Szczecin; the results are in the form of pens or publications [17, 18].

In Western Pomerania, as well as in other regions of Poland, for a long time, there was an excessive runoff to surface waters of biogens and organic matter, as described by Tadajewski [17], Tadajewski, Kubiak [16, 17], Tadajewski and others [18]. In the 1970s, only 5 tanks showed poor fertility—mesotrophy (lakes: Miedwie, Morzycko, Krzemień, Ińsko, and Zajezierze), two lakes (Jelenin and Woświn) remained on the borderline of β-mesotrophy and eutrophy [18].

In the second half of the 1990s, many publications about the hydrochemistry of West Pomeranian lakes were published. Studies on the lakes of the Woliński National Park were published by Poleszczuk [86, 87], Tadajewski and others. [19] and Kubiak [88], the susceptibility of these reservoirs to degradation was described by Doliński et al. [89]. Kubiak [90] determined the hydrochemical conditions of the lakes of this national park based on many years of research. Systematic reports of the PIOŚ appeared on the state of the Westpomeranian Voivodeship. They contained hydrochemical data for many lakes studied in accordance with the guidelines of lake monitoring [91].

In later years, publications regarding the hydrochemical conditions of the lakes: Ińsko [92], Woświn [93], Dąbie [19], and Lake Chłop [25] were published. The coastal lakes [22, 94, 95], river basins: Wołczenica [21], Płonia [96], and Rurzyca [97] were described in the field of hydrochemistry and susceptibility to degradation. Kubiak et al. [93] gave a hydrochemical description of the lakes of the Cedynia Landscape Park, a year later, the lakes of the Iński Landscape Park were described [98]. Changes in hydrochemical conditions in selected lakes of Western Pomerania (lake Ińsko, Morzycko, Chłop, Narost) for Kubiak's other twenty years were presented by [20]. Kubiak et al. [26] they undertook research on lakes around Szczecin.

Kubiak [99] presented the results of many years of research on changes in the trophys of the group of the largest dimictic lakes of Western Pomerania. Such studies in shorter periods were conducted in the Masurian Lake District [13, 14, 99]. Kubiak [99] published a work on the level of trophies of coastal lakes of Western Pomerania, and Tórz, Kubiak, and Chojnacki [100] presented an assessment of the quality of Miedwie lake water for the period 1998–2001.

In 2009, work was undertaken to assess the degree of degradation and/or degree of susceptibility to lakes and reservoirs demarcation, and to designate lakes and reservoirs degraded as a result of improper use of their water resources, the use of the catchment together with the assessment of technical and temporal possibilities to restore them to good state. Permanent proposed measures to improve the condition of lakes and reservoirs, as well as limitations in the use of their resources and use of the basin, are possible under the conditions for water use. These activities were to support the process of reconstruction of degraded lakes and reservoirs, as well as to prevent possible deterioration of the reservoirs susceptible to degradation.

The works included the following activities: proposing a method for assessing the degree of degradation and/ or the degree of susceptibility to degradation of lakes and water reservoirs, identification of water bodies of lakes and water reservoirs, constituting uniform surface water bodies that were considered degraded and susceptible to degradation together with determining the causes of degradation, determining for each JCWP lakes and water reservoirs, identified as degraded or susceptible to degradation, specific measures to improve the condition, specific conditions for using its water resources and restrictions on the use of the catchment, assessing the time possibilities to restore them to good the state [101].

In later years, research was undertaken to review and analyze, including in particular determining the temporal perspective of the possibility of achieving environmental objectives, uniform water bodies of the lakes of the Lower Oder and Western Pomerania water regions, which were indicated for derogation. The study carried out as part of this study was part of the task within the framework of the Task List and activities for the water management planning process in accordance with the requirements of the EU Water Framework Directive and the Water Law Act in 2010–2015 [102].

9.3   Lakes' Resistance to Degradations and the Role of the Catchment in the Eutrophication Process

Protection of lake ecosystems and their rational use require knowledge of the natural resistance of lakes to degradation and the role of the catchment in accelerating or inhibiting this process [5, 32, 57, 59, 103–106]. Natural physiogeographical features cause that the catchment (especially direct) may accelerate or inhibit the supply of materials (including biogenes) to the lake, which is why the reservoir due to its natural immunity, resulting from morphometric and hydrological features, may limit the catchment's influence or be more susceptible to outside influences. Each lake, therefore, has individual characteristics, which makes the susceptibility to degradation of individual bodies diversified [4, 59]. This requires individual consideration of the use and protection of the tank, taking into account not only the current water quality, but also the characteristics that determine the lake's vulnerability to external influences. Wider use possibilities have lakes with favorable natural conditions, resistant to degradation [31, 32].

In the catchment area, the water movement is the basic means of transport and exchange of various forms of matter between ecosystems. Water in the presence of living components of eco-systems is also the center of transformation of this matter in the landscape [105]. Bajkiewicz-Grabowska [107] proposed a valorising system, which on the one hand assesses the catchment as a source of matter for the reservoir, and on the other hand, determines the lake's vulnerability to the impact of cargo arriving from its area. The extent of the catchment impact on the lake is assessed on the basis of its features, which in the case of the direct catchment are the lake coefficient and the balance type of the lake, while in relation to the total basin the density of the river network in its area, its average fall, the area of outflow areas, geological structure and the way use (Table 9.1). In principle, indicators assessing a catchment as a supplier of biogenic matter can only slightly change over time.

From the point of view of lake protection, it is important to determine the susceptibility of tanks to de-gradation. There are two systems for assessing this property in Polish literature. The first of them is proposed by Bajkiewicz-Grabowska [58,

Table 9.1 Factors for estimation of lake catchment as biogenic matter suppliers by Bajkiewicz-Grabowska [58, 104, 107] and Bajkiewicz-Grabowska et al. [108]

Indicator	Punctation (groups)			
	0	1	2	3
Lake density coefficient	<10	10–40	40–150	>150
Balance type		Drain	Outflow	Flow-through
Density of river network	<0.5	0.5–1.0	1.0–1.5	>1.5
Mean slope of direct catchment	<5	5–10	10–20	>20
Closed drainage area of direct catchment	>60	45–60	20–45	<20
Geological structure of direct catchment	Clay	Sandy-clay	Clay-sandy	Sandy
Catchment use Direct catchment use	Forest, marsh, agriculture-forest, pasture-agricultural-forestry	Forestry-agricultural, pasture-agricultural	Agricultural, pasture-forestry-agricultural with buildings, forest with buildings	Forest—agricultural with buildings, pasture—agricultural with buildings, agriculture with buildings

104, 107]. The second system is provided by Kudelska et al. [59]. The assessment of lakes' susceptibility to degradation is based in both systems on similar indicators, a significant part of them are morphometric indices, eg the average depth, lake volume, the length of the shoreline or the lake surface (Table 9.2). They were determined in the sixties of the last century and in the majority of them they have become demeanorized, for example due to lake overgrowing. It is now necessary to update this data, using modern numerical techniques and echosounding in combination with GPS. Such surveys will also allow assessment of lake water resources.

9.4 "Internal Supply" in Biogenes

Elements accumulated in the ecosystem come back into circulation in various amounts, at different times and through various mechanisms. The state of trophies of lakes depends on the intensity of the circulation of matter and on the amount of biogenic compounds taking part in it [4, 28, 32]. The basic resources of phosphorus and nitrogen are accumulated in bottom sediments, a much smaller part in organisms, but the circulation in this ecosystem of this part of resources plays a significant role. Consumption and the distribution of life and the secretion of various forms of matter by aquatic organisms are the main ways of its circulation and transformation in the ecosystem. Separation of mineral nitrogen and phosphorus compounds by invertebrate animals is the basic mechanism of their circulation in the water depth, and also significant (in addition to bacterial degradation) in bottom sediments (benthos) [33]. The most important processes in the phosphorus retention in bottom sediments are the sorption and binding of this element by iron hydroxides, the addition of calcium carbonates and the autogenous formation of minerals. Feeding of biogenic water contained in bottom sediments plays the most important role during summer stagnation, mainly due to bacteria and release of Fe-P and Al-P connections in anaerobic conditions [35]. This process is intensified by mixing of near-surface waters closer to the surface. The release of phosphorus from bottom sediments located within the epilimnion is much more important than the same process taking place at the border of sediments and water under anaerobic conditions in hypolimnion [4].

Among the factors determining the rate of matter cycling in the lake, water mixing, affecting the rate of reactivation of nutrient reserves [66], is of great importance. Olszewski [29] recognizes that the thermal relations of the lake, subject to climatic factors of seasonal variability, are an indicator of the dynamics of water masses. They are modified by the effects of wind, which in turn depends on the morphological characteristics of the bowls. These dependencies are discussed extensively by many authors [4, 29, 57]. Especially in shallow reservoirs, they determine the epilimnion range (stratification level), the exchange intensity between epilimnion and hypolimnion, and between water and bottom [66]. Usually, tanks with high water mass dynamics are more susceptible to eutrophication. Research conducted by Kubiak [99] on the largest lakes of Western Pomerania confirmed the principle that those with intense water mixing are characterized by a higher level of trophies.

Table 9.2 Factors for lakes susceptibility to degradations based on data presented in Kudelska et al. [59], Kudelska and Soszka [109] and Bajkiewicz-Grabowska [58, 104, 107]

Index	Model	Category	Punctation			
			1	2	3	4
Mean depth (m)	Bajkiewicz-Grabowska [58, 104, 107]	0-3	>10	5-10	3-5	<3
	Kudelska et al. [59], Kudelska, Soszka [109]	1-4	≥ 10	≥ 5	≥ 3	<3
V—lake volume/L—length of shoreline	Bajkiewicz-Grabowska [58, 104, 107]	0-3	>5	3-5	1-3	<1
	Kudelska et al. [59], Kudelska, Soszka [109]	1-4	≥ 4	≥ 2	≥ 0.8	<0.8
Stratification %	Bajkiewicz-Grabowska [58, 104, 107]	0-3	>35	20-35	10-20	<10
	Kudelska et al. [59], Kudelska, Soszka [109]	1-4	≥ 35	≥ 20	≥ 10	<10
Active bottom area/V epilimnion (m^2/m^3)	Bajkiewicz-Grabowska [58, 104, 107]	0-3	<0,10	0.10-0.15	0.15-0.30	>0.30
	Kudelska et al. [59], Kudelska and Soszka [109]	1-4	≤ 0.10	≤ 0.15	≤ 0.30	>0.30
% water exchange	Bajkiewicz-Grabowska [58, 104, 107]	0-3	>10	5-10	1-5	<1
	Kudelska et al. [59], Kudelska and Soszka [109]	1-4	$\leq 30\%$	$\leq 200\%$	$\leq 1000\%$	>1000
Schindler index	Bajkiewicz-Grabowska [58, 104, 107]	0-3	<10	10-30	30-100	>100
	Kudelska et al. [59], Kudelska and Soszka [109]	1-4	≤ 2	≤ 10	≤ 50	>50
Catchment use	Kudelska et al. [59], Kudelska and Soszka [109]		$\geq 60\%$ of forests	<60% of forests <60% of arable land	$\geq 60\%$ of arable land	
Susceptibility class	Bajkiewicz-Grabowska [58, 104, 107]	od 0 do 3	cat. 0 ≤ 0.80	cat. 1 ≤ 1.60	cat. 2 ≤ 2.40	cat. 3 >2.40
	Kudelska et al. [59], Kudelska and Soszka [109]	od I do IV	cat. I ≤ 1.50	cat. II ≤ 2.50	cat. III ≤ 3.25	cat. IV >3.25

When assessing the susceptibility of a lake to degradation, it is necessary to determine the thermal systems occurring in it.

From the point of view of lake protection, it is unacceptable to keep fish in the lakes [86]. For example, in the years 1977–1982, as a result of the large-scale rainbow trout rearing (*Oncorhynchus mykiss Walbaum*) in Ińsko, water quality deteriorated significantly. About 27 t of fish were produced at that time, while the load of organic pollutants was estimated at 10,000–15,000—equivalent residents. The trophic level of this reservoir was then defined as the initial stage of eutrophy, whereas it was previously a mesotrophic reservoir [92]. After stopping the trout breeding, the lake returned to the a-mesotrophy state at the beginning of the nineties [92]. In Western Pomerania, the practice of carbon black fish farming was abandoned.

9.5 Assessment Systems for Landscape Water Trophy

Water dynamics, natural catchment conditions, and susceptibility to degradation determine the quality of water in lakes and the level of their trophies. In Poland, a method for determining the quality of their waters has been developed for lakes monitoring purposes [59, 109]. Research in this method is carried out only during the spring circulation and summer stagnation, and mainly includes surface waters, and to a lesser extent, near bottom waters (Table 9.3). When assessing the trophy of lakes, other methods are also used; adopted by the Organization for Economic Cooperation and Development (OECD), based on: measurements of percentage oxygenation at the summit of the heterothermal summer of the bottom layer, average annual and maximum values of measurements: secchi disc visibility (SD), content in surface waters of total phosphorus and chlorophyll "a" [5]. The method proposed by Zdanowski [13, 14] is similar, taking into account summer measurements of total phosphorus and chlorophyll "a" in surface waters and water transparency (SD), as well as additional indicators (total nitrogen content, N to P ratio, algae biomass and share of cyanobacteria in biomass). The Carlson method [61] is commonly used, based on the results of measurements in summer in surface waters of the total phosphorus, chlorophyll "a" and water visibility, transformed into numerical indicators (lake state trophies—WST), which are a kind of comparable "measure" the level of advancement of the eutrophication process. The description of these methods of trophic assessment is presented in Table 9.4. Lange, Maślanka [60] take the position that for the assessment of the level of trophic lakes, the most appropriate is the analysis of aerobic conditions during the summer stagnation. According to these authors, oxygen conditions determine the nature, extent, and severity of the basic processes that form biocenosis. In their analysis, the conclusions as to the susceptibility of the reservoir to degradation concern the actual course of the eutrophication process and not the potential determinants. The susceptibility of the lake to degradation determined on the basis of physiographic indicators and catchment may not be directly related to water quality. The most degraded lakes do not have to be distinguished by particularly low resilience, because their condition is primarily due to strong

Table 9.3 Factors for estimation of lake water quality by Kudelska et al. [59] and Kudelska and Soszka [109]

Index	Unit	Season	Type of lake	Layer of water	I class	II class	III class	IV class
Mean hypolimnetic oxigenation	$\%O_2$	Summer	S		≥ 40	≥ 2	≥ 1	<1
O_2 content	mg O_2 dm^{-3}	Summer	NS	bottom	≥ 40	≥ 20	≥ 5	<5
ChZTcr	mg O_2/dm^3	Summer	NS	surface	≤ 20	≤ 30	≤ 50	>50
BZT$_5$	mg O_2/dm^3	Summer	S;NS	surface	≤ 2	≤ 4	≤ 8	>8
BZT$_5$	mg O_2/dm^3	Summer	S	bottom	≤ 2	≤ 5	≤ 10	>10
Phosphates	mgP$_{PO4}$/dm^3	Spring	S;NS	surface	≤ 0.02	≤ 0.04	≤ 0.08	>0.08
Phosphates	mgP$_{PO4}$/dm^3	Summer	S	bottom	≤ 0.02	≤ 0.04	≤ 0.08	>0.08
Total phosphorus	mgP/ dm^3	Summer	S	bottom	≤ 0.06	≤ 0.15	≤ 0.60	>0.60
Total phosphorus	mgP/ dm^3	Average (spring+summer)	S;NS	surface	≤ 0.05	≤ 0.10	≤ 0.20	>0.20
Mineral nitrogen	mgN/dm^3	Spring	S;NS	surface	≤ 0.20	≤ 0.40	≤ 0.80	>0.80
Ammonia nitrogen (N_{NH4})	mgN$_{NH4}$/dm^3	Summer	S	bottom	≤ 0.20	≤ 1.0	≤ 5.0	>5.0
Total nitrogen	mgN/dm^3	Average (spring+summer)	S;NS	surface	≤ 1.0	≤ 1.5	≤ 2.0	>2.0
Conductivity	μS/cm	Spring	S;NS	surface	≤ 250	≤ 300	≤ 350	>350
Chlorophyll	mg/m^3	Average (spring+summer)	S;NS	surface	≤ 8.0	≤ 15.0	≤ 25.0	>25.0
Seston dry mass	mg/dm^3	Average (Spring+summer)	S;NS	surface	≤ 4.0	≤ 8.0	≤ 12.0	>12.0
Visibility	m	Average (Spring+summer)	S;NS	surface	≥ 4.0	≥ 2.0	≥ 1.0	<1.0

anthropopressure (intensive agriculture, point sources of pollution). Occurring in the surface layers of high-oxygenation, it is considered an indicator of a high level of trophic lakes. The lack of oxygen in the bottom layers of waters, also in winter, confirms the high level of trophies [4, 57].

9.6 Conclusion

Poland's membership in the European Union means that the necessary measures to assess the quality and size of lake water resources, as well as to protect them. It is necessary to determine the level of trophies and the impact of the catchment on lakes and their natural susceptibility to degradation, which will allow to determine

Table 9.4 Factors for estimation of lake trophic state

Method	Index	WST-SD (summer)[1]	WST-Chl (summer)[2]	WST-TP (summer)[3]	WST-TP (summer)[4]	WST-Chlorophyll "a" (summer)[5]	basic TP- (summer) (surface)	basic SD- (summer) (surface)	basic Chlorophyll "a" (summer) (surface)	additional TN (summer) (surface)	additional TN/TP (summer) (surface)	TP average (spring + summer)	Chl "a" max summer	Chl "a" average (spring + summer)	SD max (summer)	SD average (spring + summer)	%O2 – summer bottom
Carlson [6]	oligotrophy	≤40	≤40	≤40													
	mesotrophy	≤60	≤60	≤60													
	eutrophy	>60	>60	>60													
Walker [10]	oligotrophy				<30	<30											
	mesotrophy				<45	<45											
	eutrophy				<65	<65											
	hipereutrophy				>65	>67											
Hilbricht–Ilkowska[11], Kajak [12], Zadnowski [13]	mesotrophy						≤0.05	≥3	≤10	≤1.5	>30						
	mesoeutrophy						≤0.10	≥2	≤30	≤1.5	>20						
	eutrophy						>0.10	<2	>30	>1.5	<20						
Vollenweider [6]	ultraoligotrophy											≤0.004	≤2.5	≤1.0	≥6	≥12	≥90
	oligotrophy											≤0.01	≤8.0	≤2.5	≥3	≥6	≥80
	mesotrophy											≤0.035	≤25	≤8	≥1.5	≥3	≥40
	eutrophy											≤0.10	≤75	≤25	≥0.7	≥1.5	≥10
	hipertrophy											>0.10	>75	>25	<0.7	<1.5	<10

SD secchi disc visibility, Chl – chlorophyll "a", TP – total phosphorus, TN - total nitrogen, %O2 – oxygenation, WST - status indicator of trophies

1/ calculated according to the formula $WST_{SD} = 10(6-(\ln SD/\ln 2)$
2/ calculated according to the formula $WST_{Chl} = 10(6-(2.04-0.68Chl/\ln 2)$
3/ calculated according to the formula $WST_{TP} = 10(6-(\ln 48/TP/\ln 2)$
4/ calculated according to the formula $WST_{TP} = -15,6+48\log TP$
5/ calculated according to the formula $WST\ Chl = 20+33,2\ \log Chl$

the scope of activities in the catchment area (especially when it is susceptible to release of cargoes), protection of individual lakes (including the appropriate fishing industry) and determining the ways of their reclamation.

It is necessary to undertake research on the development of a concept for the protection of the waters of the West Pomeranian Lake District. This approach requires comprehensive research, including determination of water resources based on current morphometric data of individual bodies. Currently applied in Poland systems for assessing lake susceptibility to degradation are largely based on morphometric data of reservoirs, however, since they originate mainly from the 1960s, they should be updated. Currently, the assessment of lake water quality for monitoring purposes includes only selected hydrochemical indicators, measurements of which are made during the spring circulation and summer stagnation. Water quality tests in lakes should be conducted in annual cycles, in all limnological seasons, at a frequency of at least 7–8 times a year. On the basis of the research, the assessment of the functioning of lake ecosystems in the scope of physical, chemical, and biological factors should be made, taking into account their seasonal variability. Determined on the basis of physiographic indicators and catchment, lake susceptibility to degradation may not be directly related to water quality. The most degraded lakes do not have to be distinguished by particularly low resilience, because their condition is primarily the result of strong anthropopression (intensive agriculture, point sources of pollution).

9.7 Recommendations

In the longer term, protective measures should lead to immediate measures eliminating or reducing the size of area sources of pollution. Therefore, the use of the drainage basin should be fundamentally changed, the use of artificial fertilizers should be rationalized, pre- and catch crops should be used and all actions aimed at creating long-term soil cover of soils in the catchment area should be taken. It is necessary to regulate and control the sewage management of towns and recreation centers located around lakes. A sustainable strip of vegetation should be created as a barrier that limits the migration of nutrients towards lakes. It is advisable to undertake reclamation measures aimed at reducing the "internal" supply of biogens by their inactivation in bottom sediments.

References

1. Vollenweider RA (1971) Scientific fundamentals of the eutrophication of lakes and following waters, with reference to nitrogen and phosphorous as factors in eutrophication. OECD, Environment Directorate, Paris 27, 1–61
2. Kajak Z (1998) Hydrobiologia – limnologia. PWN, Warszawa, pp 1–355
3. Kajak Z (1979) Eutrofizacja jezior. Warszawa
4. Kalff J (2001) Limnology. Upper Saddle River, NJ

5. Lossow K (1996) Rekultywacja jezior i zbiorników wodnych – dotychczasowe osiągnięcia, możliwości i perspektywy. W: Ochrona i rekultywacja jezior i zbiorników wodnych, Materiały konferencyjne, Międzyzdroje, 7–8 marca 1996, Biuro Inf. Nauk. Szczecin, 47–56

6. Vollenweider RA (1989) Global problems of eutrophication and its control. Symp Biol Hung 38:19–41

7. Lossow K. (1998) Ochrona i rekultywacja jezior – teoria i praktyka. Idee Ekolog. Ser. Szkice 13(7):55–71. Poznań

8. Gawrońska H (1994) Wymiana fosforu i azotu miedzy osadami a wodą w jeziorze sztucznie napowietrzanym. Zesz. Nauk. ART Olszt. Ser. Ochr. Wód Ryb. Śródl. E 19 (supl.):1–50

9. Lossow K (1995) Odnowa jezior. Ekoprofit 5:11–15

10. Lossow K (1995) Zanikające jeziora. Ekoprofit 7(8):40–45

11. Hillbricht-Ilkowska A (1989) Różnorodność biologiczna siedlisk słodkowodnych – problemy, potrzeby, działania. Idee Ekol Ser Szkice 13(7):13–55

12. Kajak Z (1983) Dependences of chosen indices of structure and functioning of ecosystems of different trophic status and mictic type for 42 lakes. Ecological characteristics of lakes in north-eastern Poland versus their trophic gradient. Ekol Pol 31:495–530

13. Zdanowski B (1983) Chemistry of the water in 41 lakes. W: Ecological characteristics of lakes in north-eastern Poland versus their trophic gradient. Ekol Pol 31:287–308

14. Zdanowski B (1983) Chlorophyll content and visibility of Secchi's disc in 46 lakes. Ekol Pol 31:333–352

15. Zdanowski B (1996) Czystość jezior a możliwość rybackiego użytkowania. W: Rybactwo jeziorowe. Stan, uwarunkowania, perspektywy. I Krajowa Konferencja Użytkowników Jezior. Uroczysko Waszeta. Red. A. Wołos. Olsztyn, pp 7–14

16. Tadajewski A, Kubiak J (1976) Wstępna ocena stopnia zeutrofizowania wód strefy przy-brzeżnej w rejonie ujść niektórych rzek Pomorza Zachodniego. Stud Mater Oceanolog 15:91–108

17. Tadajewski A (1979) Zanieczyszczenie wód dopływowych Zatoki Pomorskiej. W: Bałtyk i jego wody dopływowe. Materiały sesyjne, Szczecin 2 lutego 1979. Warszawa, pp 7–13

18. Tadajewski A, Mutko T, Faberski Z, Górka-Niwińska E, Kubiak J (1980) Aktualny stan zanieczyszczenia i eutrofizacji jezior Pomorza Zachodniego. W: Stosunki wodne w zlewniach rzek Pomorza i dorzecza dolnej Wisły ze szczególnym uwzględnieniem gospodarki wodnej jezior. Materiały sesji nauk.-tech. Słupsk, 23–24 października 1980, Cz. 3. Słupsk, pp 46–63

19. Tadajewski A., Kubiak J., Sitkowska M., Knasiak M. (1996). Hydrochemia jezior Wolińskiego Parku Narodowego. W: Ochrona i rekultywacja jezior i zbiorników wodnych, Materiały konferencyjne, Międzyzdroje, 7–8 marca 1996, Biuro Inf. Nauk. Szczecin, pp 135–137

20. Kubiak J, Mutko T, Tórz A (1997) Trends of water chemistry changes in lakes of the West Pomeranian region during the last twenty years. In: Proceedings 3rd Meeting of Internat. Center of Ecology PAS, Szczecin, 9-12 December, pp 1–3

21. Kubiak J, Nędzarek A, Żurawska J (1996) Hydrochemia jezior zlewni rzeki Wołczenicy. W: Ochrona i rekultywacja jezior i zbiorników wodnych, Materiały z konferencji Międzyzdroje, 7–8 marca 1996, Biuro Inf. Nauk. Szczecin, pp 157–159

22. Kubiak J, Nędzarek A, Tórz A (1997) Charakterystyka limnologiczna jezior przymors-kich Pomorza Zachodniego. Materiały 17 Zajazdu Hydrobiologów Polskich Poznań, 8–11 września 1997. Poznań, pp 40–41

23. Szyper H, Kraska M (1998) Ocena oddziaływań zlewni i odporność na degradację 16 jezior Drawieńskiego Parku Narodowego. Poznań

24. Szyper H, Kraska M (1999) Ocena zewnętrznego obciążenia związkami biogennymi jezior Drawieńskiego Parku Narodowego. W: Mat. Konf. Limnolog. „Naturalne i antropogeniczne przemiany jezior". Radzyń, 20–22 września 1999. Warszawa, pp 255–265

25. Kubiak J. (2000). Evaluation of natural tolerance of anthropogenic impacts of the lakes at Wolin National Park. In: Proceedings 4th Limnological Conference „Naturalne i antro-pogeniczne przemiany jezior", 18–20 September 2000, Olsztyn, pp 133–147

26. Kubiak J (2001) Hydrochemistry of Wolin Island Lakes. Folia Univ Agric Stettin 218(28):63–76. Piscaria

27. Kubiak J (1980) Studia nad eutrofizacja Zatoki Pomorskiej w rejonie oddziaływania wód Świny i Dziwnej. Rozprawa doktorska. Akademia Rolnicza w Szczecinie (maszynopis)
28. Gotkiewicz J, Hutorowicz H, Lossow K, Mosiej J, Pawłat H, Szymczak T, Traczyk T (1990) Czynniki kształtujące obieg wody i biogenów w krajobrazie młodoglacialnym. Poznań, pp 105–126
29. Olszewski P (1971) Trofia a saprobia. Zesz. Nauk WSR Olszt. Ser. C, Supl. 3:5–14
30. Wetzel RG (1975) Limnology. W. B. Saunders Company, Philadelphia, pp 1–743
31. Lossow K (1996) Znaczenie jezior w krajobrazie młodoglacialnym Pojezierza Mazurskiego. Zesz Probl Post Nauk Rol 431:47–59
32. Lossow K, Więcławski F (1991) Migracja podstawowych pierwiastków pożywkowych z gleb, użytkowanych rolniczo do wód powierzchniowych. Biul Inform ART Olsztyn 31:123–133
33. Górski T, Rybak JI (1971) Czynniki wpływające na wymianę substancji pomiędzy mułem a wodą. Wiad Ekol 21(2):104–122
34. Lossow K, Gawrońska H (1998) External input to Lake Wadąg—Effective and estimate loadings. Pol J Envir Stud 7(2):95–98
35. Pettersson K, Amiard J (1998) Mechanisms for internal loading of phosphorus in lakes. Hydrobiologia 373–374(1–3):21–25
36. Borowiec S, Zabłocki Z (1988) Rolnicze zanieczyszczenia obszarowe wód odpływowych ze zlewni rolniczych i działów drenarskich północno-zachodniej Polski. Zesz Nauk AR Szczec Rolnictwo 134(45):25–39
37. Borowiec S, Zabłocki Z (1990) Czynniki kształtujące chemizm wód powierzchniowych i odcieków drenarskich obszarów rolniczych Polski północno-zachodniej. Materiały Seminaryjne IMUZ, Falenty 27:95–108
38. Wilgat T (1987) Ochrona zasobów wodnych Polski. PWN, Warszawa, pp 1–196
39. Kauppi L, Pietilainen OP, Knuuttila S (1993) Impacts of agricultural nutrient loading on Finnish water sources. Water Sci Tech 28:461–471
40. Stachowicz K (1995) Migracja wodna składników pokarmowych ze zlewni rolniczych. Człow Śr 19(1):125–141
41. Ahl T (1975) Effects of man-induced and natural loading of phosphorus and nitrogen on the large Swedish lakes. Verh Internat Verein Limnol 19:1125–1132
42. Chudecki Z, Duda L (1971) Annual losses of chemical components of the soil in the Płonia river basin. Pol Soil Sci 4(2):145–154
43. Durkowski T (1998) Chemizm wód drenarskich obiektów Pomorza Zachodniego. Zesz Probl Post Nauk Rol 458:349–356
44. Niemrycz E, Taylor R, Makowski Z (1993) Zagrożenie substancjami biogennymi wód powierzchniowych. Bibl Monit Śr PIOŚ, Warszawa, pp 1–50
45. Rajda W, Ostrowski K, Kowalik T, Marzec J (1995) Stężenia i ładunki niektórych składników chemicznych wynoszonych z opadem i odpływających z mikrozlewni rolniczej. Zesz Nauk AR Krak 298:45–57
46. Rocznik Statystyczny Województwa Szczecińskiego (1974) Wydaw. WUS, Szczecin, pp 1–324
47. Rocznik Statystyczny Województwa Szczecińskiego (1980) Wydaw. WUS, Szczecin, pp 1–356
48. Rocznik Statystyczny Województwa Szczecińskiego (1995) Wydaw. WUS, Szczecin, pp 1–385
49. Trzebiatowski R (1986) Chów karpi w sadzach a zanieczyszczenie wód. Aura 7:15–17
50. Prejs A (1978) Eutrofizacja jezior a ichtiofauna. Wiad Ekol 14(3):201–208
51. Opuszyński K (1983) Podstawy biologii ryb. PWRiL, Warszawa, pp 1–591
52. Opuszyński K (1997) Wpływ gospodarki rybackiej, szczególnie ryb roślinożernych, na jakość wody w jeziorach. Bibl Monitor Śr, PIOŚ, Zielona Góra, pp 1–156
53. Zdanowski B (1993) Eutrofizacja wód [w: Rybactwo śródlądowe]. In: Szczerbowski JA (ed) Red. IRŚ, Olsztyn, pp 124–135
54. Iwaszkiewicz M (1976) Wpływ eutrofizacji wód otwartych na ichtiofaunę [w: Nawożenie a eutrofizacja wód]. Materiały konferencyjne, Zielona Góra, 17–18 maja 1976. Komitet Ekologii PAN, Zielona Góra, pp 163–177

55. Leopold M, Bnińska M, Nowak W (1986) Commercial fish catches as an index of lake eutrophication. Arch Hydrobiol 106:513–524
56. Hillbricht-Ilkowska A, Wiśniewski R (1994) Zrónicowanie troficzne jezior Suwalskiego Parku Krajobrazowego i jego otuliny – Stan obecny, zmienność wieloletnia, miejsce w klasyfikacji troficznej jezior. W: Jeziora Suwalskiego Parku Krajobrazowego. Związki z krajobrazem, stan eutrofizacji i kierunki ochrony. In: A. Hillbricht-Ilkowska, R. J. Wiśniewski (red) Zesz. Nauk. Kom. "Człowiek i Środowisko" PAN, 7:181–200
57. Hutchinson GE (1957) A treatise on limnology. Vol. I. geography, physics and chemistry. London, pp 1–885
58. Bajkiewicz-Grabowska E (1990) Stopień naturalnej podatności jezior na eutrofizację na przykładzie wybranych jezior Polski. Gospod Wod 12:270–272
59. Kudelska D, Cydzik D, Soszka H (1994) Wytyczne monitoringu podstawowego jezior. PIOŚ, Warszawa, pp 1–42
60. Lange W, Maślanka W (1994) Próba oceny naturalnej tendencji jezior Pojezierza Mazurskiego [Problemy hydrologii regionalnej]. Materiały Ogólnopolskiej Konferencji Hydrograficznej, Karpacz, 26–28 września 1994. [b.w.], Karpacz, pp 35–38
61. Carlson RF (1977) A trophic state index for lakes. Limnol Oceanogr 22(2):361–369
62. Roman M (1998) Standardy jakości i ochrony śródlądowych wód powierzchniowych w przepisach Unii Europejskiej i w przepisach polskich. Monogr PZIiTS Ser Wod Kanal 1:1–94
63. Gromiec MJ (1998) Polityka wodna Unia Europejskiej i jej implikacje dla Polski. Monogr PZIiTS Ser Wod Kanal 2:1–62
64. Górski M (2002) Gospodarowanie wodami śródlądowymi w prawie polskim i Unii Europejskiej [w: Zaopatrzenie w wodę i jakość wód]. V Międzynarodowa Konferencja Naukowo-Techniczna, Poznań – Gdańsk, 26–28 czerwca 2002. [b.w.]
65. Stumm W, Morgan J (1996). Aquatic chemistry. Chemical equilibrium and rates in natural waters. Wiley, New York, pp 1–1022
66. Patalas K (1960) Mieszanie wiatrowe jako czynnik określający intensywność krążenia materii w różnych morfologicznie jeziorach okolic Węgorzewa. Rocz Nauk Rol ser B 77(1):224–241
67. Choiński A (1995) Zarys limnologii fizycznej Polski. Wydaw. Nauk. UAM, Poznań, pp 1–256
68. Halbfass W (1901) Beitrag zur Kenntnis der Pommerschen Seen. Gotha-Justus Perths
69. Halbfass W (1901) Ergebnisse seiner Seenforschung in Pommern. Verh. Der Gesellschaft für Erdkunde, p 28
70. Jentsch A (1912) Studien an Seen der Inseln Usedom und Wollin. Abh D Klage Pr Geol Land N. F. H. 51, Berlin
71. Jentsch A (1922) Über einige Seen Westpreußens, Beitrag zur Seenkunde. T. 3. Abh. Geol. Landesanstalt, Berlin
72. Wahnschaffe F (1921) Geologie und Oberflachengestaltung des Norddeutschen Flachlandes. Stuttgart, pp 1–265
73. Prawdzic K (1966) Jeziora Pomorza Zachodniego. Wiad. Zach. Szczecin, 1–127
74. Młodziejewski J (1948) Miedwie – największe jezioro Pomorza Kaszubskiego. Prz Zach 2:22–32
75. Czekańska M (1948) Obszar ujściowy Odry. Monografia Odry. Poznań, 68–98
76. Majdanowski S (1954) Jeziora Polski. Prz Geogr 26(2):22–42
77. Jasnowska J, Jasnowski M (1982) Pojezierze Zachodniopomorskie. Warszawa
78. Mikołajski J (1966) Geografia województwa szczecińskiego. Szczec Tow Nauk 12:1–156
79. Choiński A (1991) Katalog jezior Polski – Pojezierze Pomorskie. Poznań
80. Filipiak J, Sadowski J (1994) Jeziora szczecińskie – zarys faktografii. Szczecin
81. Filipiak J, Raczyński M (2000) Jeziora zachodniopomorskie (zarys faktografii). Szczecin
82. Brodzińska B, Jańczak J, Kowalik A, Sziwa R (1996) Atlas jezior Polski. T. 1. Poznań
83. Brodzińska B, Jańczak J, Kowalik A, Sziwa R (1997) Atlas jezior Polski. T. 2. Poznań
84. Nguyen Van T (1972) Studia na chemizmem wód jezior o różnym stopniu troficznym. Praca doktorska AR, Szczecin
85. Atlas czystości jezior województwa szczecińskiego 1970–1976 (1977) OBiKŚ, Szczecin, Maszynopis

86. Poleszczuk G (1996) Jeziora Wolińskiego Parku Narodowego – status troficzny, tendencje zmian i możliwości przeciwdziałania degradacji. W: Mat. Konfer. 19, Uniw. Szczec. Szczecin, pp 117–141
87. Poleszczuk G, (1994) Ekosystemy wodne Wolińskiego Parku Narodowego. Klify 1:99–119
88. Kubiak J, Wechterowicz Z, Tadajewski A. (2000) Jakość wód i podatność na degradację jeziora Chłop w latach 1980–1997. W: Materiały 18. Zjazdu Hydrobiologów Polskich, Białystok 4–8 września 2000. Białystok, pp 141–142
89. Doliński A, Godowac D, Sz Rolle, Wrzesiński D (1995) Ocena podatności jezior Wolińskiego Parku Narodowego na degradację i znaczenie warunków przyrodniczych zlewni w tym procesie. Klify 2:33–43
90. Kubiak J, Raczyńska M, Tórz A (2001) Wstępna charakterystyka hydrochemiczna jeziora Binowskiego. Mat Sesji z okazji 50-lecia Wydz Ryb Morsk AR w Szczecinie, Szczecin, p 37
91. Raporty o stanie środowiska w województwie szczecińskim: Bibl. Monit. Śr., PIOŚ, Szczecin, 1–126, 1994, 1995, 1997, 1999, 2000
92. Kubiak J, Knasiak M (1996) Jezioro Ińsko zmiany chemizmu wód. Ochrona i rekultywacja jezior i zbiorników wodnych. Materiały z konferencji Międzyzdroje 7–8 marca 1996, Biuro Inf. Nauk. Szczecin, pp 143–145
93. Kubiak J, Tórz A, Knasiak M (1996) Warunki hydrochemiczne wybranych jezior rejonu Cedyńskiego Parku Krajobrazowego. Ochrona i rekultywacja jezior i zbiorników wodnych. Materiały Konferencyjne Międzyzdroje 7–8 marca 1996, Biuro Inf. Nauk. Szczecin, pp 141–142
94. Tórz A, Kubiak J, Nędzarek A (2000) Naturalna podatność na degradację jezior Koprowo, Liwia Łuża i Resko Przymorskie. „Naturalne i antropogeniczne przemiany jezior", Materiały 4 Konferencji Limnologicznej, Zalesie, 18–20 września 2000. Olsztyn, pp 161–173
95. Kubiak J (2003) Eutrophication rate and trophic state of Western Pomeranian coastal lakes. Acta Sci Pol Piscaria 2(1):141–158
96. Kubiak J, Tórz A, Nędzarek A (1997) Charakterystyka limnologiczna wybranych jezior zlewni rzeki Płoni. Materiały 17 Zajazdu Hydrobiologów Polskich, 8–11 września 1997, Poznań, 43
97. Kubiak J, Żurawska J, Raczyńska M (1997) Hydrochemia wybranych jezior rzeki Rurzycy. W: Materiały 17 Zajazdu Hydrobiologów Polskich, 8–11 września 1997, Poznań, 44, Kudelska D, Cydzik D, Soszka H (1981) Propozycja systemu oceny jakości jezior. Wiad Ekol 27(2):149–173
98. Kubiak J, Żurawska J, Knasiak M (1996) Zmiany trofii jeziora Woświn. W: Ochrona i rekultywacja jezior i zbiorników wodnych. Materiały konferencyjne Międzyzdroje, 7–8 marca 1996, Biuro Inf. Nauk. Szczecin, pp 155–157
99. Kubiak J (2003) Największe dimiktyczne jeziora Pomorza zachodniego. Poziom trofii, podatność na degradację oraz warunki siedliskowe ichtiofauny. Rozprawy AR Szczecin 214:1–92
100. Tórz A, Kubiak J, Chojnacki J (2003) Assessment of lake Miedwie water quality in 1998–2001. Acta Sci Pol 2(1):279–290
101. Projekt „Sformułowanie w warunkach korzystania z wód regionu wodnego ograniczeń w korzystaniu z wód jezior lub zbiorników oraz w użytkowaniu ich zlewni". Pr. Zbior. (2010). Konsorcjum MGGP, IOŚ, Kraków, Warszawa, pp 1–1499
102. Przegląd i analiza jezior wskazanych do derogacji w obszarze regionu wodnego Dolnej Odry i Przymorza Zachodniego PTLim, Gdańsk (2011)
103. Bajkiewicz-Grabowska E (1981) The influence of the physical geographic environment on the biogenous matter delivery to the lake. J Hydrol Sci 8(1–2):63–73
104. Bajkiewicz-Grabowska E (1985) Struktura fizyczno-geograficzne zlewni jako podstawa oceny dostawy materii biogennej do jezior. Pr Stud Geogr 7:65–89
105. Hillbricht-Ilkowska A, Kostrzewska-Szalkowska I, Wiśniewski R (1996) Zróżnicowanie troficzne jezior rzeki Krutyni (Pojezierze Mazurskie) – Stan obecny, zmienność wieloletnia, zależności troficzne. W: Funkcjonowanie systemów rzeczno-jeziornych w krajobrazie pojeziernym: rzeka Krutynia (Pojezierze Mazurskie). In: Hillbricht-Ilkowska A, Wiśniewski, RJ (red). Zesz. Nauk. Kom. „Człowiek i Środowisko". PAN 13:125–153

106. Hillbricht-Ilkowska A, Kostrzewska-Szalkowska I (1996) Ocena ładunku fosforu i stanu zagrożenia jezior rzeki Krutyni (Pojezierze Mazurskie) oraz zależności pomiędzy ładunkiem a stężeniem fosforu w jeziorach. W: Funkcjonowanie systemów rzeczno-jeziornych w krajobrazie pojeziernym: rzeka Krutynia (Pojezierze Mazurskie). In: Hillbricht-Ilkowska A, Wiśniewski, RJ (red). Zesz. Nauk. Kom. „Człowiek i Środowisko". PAN 13:97–123

107. Bajkiewicz-Grabowska E (1987) Ocena naturalnej podatności jezior na degradację i rola zlewni w tym procesie. Wiad Ekol 33(3):279–289

108. Bajkiewicz-Grabowska E, Hillbricht-Ilkowska A, Kajak Z, Kufel L (1989) Metodyka oceny odporności i obciążenia jezior, stanu eutrofizacji i czystości ich wód oraz wpływu zlewni. Zesz. Nauk. Kom. „Człowiek i Środowisko", PAN. Jeziora Mazurskiego Parku Krajobrazowego 1:21–45

109. Kudelska D, Soszka H (1996) Przegląd stosowanych w różnych krajach sposobów oceny i klasyfikacji wód powierzchniowych. PIOŚ, Warszawa, pp 1–83

110. Walker W (1979) Use of hypolimnetic oxygen depletion as a trophic index for lakes. Water Resour Res 15(6):1463–1470

Chapter 10
Ingression and Ascension of Saline Waters on the Polish Coast of the Baltic Sea

Arkadiusz Krawiec

Abstract The sea coast zone of the Southern Baltic Sea is one of the regions in which the risk assessment of both fresh groundwater resources and intake of admissible volumes, defined as a maximum safe yield, is a complex problem, requiring a separate methodological approach. This results from the coexistence of two hydrogeochemical environments—fresh and saline waters and fluctuations of a border surface between the saline (sea) and freshwater. Salt and brackish waters that occur along the coastal Baltic lowlands in Poland originate from the seawater intrusion or brines ascending from the deep Mesozoic strata. The study area covers the Polish part of the southern Baltic coast from Uznam Island to the Vistula Lagoon. These areas are located in different geological units and different hydrogeological regions. The problem of groundwater salinity is presented for the following areas: islands, sandspits, coastal lowlands, cliff coasts, and the Vistula delta (Żuławy).

Keywords Groundwater · Ascension and ingression process · Groundwaters genesis and ages · Baltic coast · Poland

10.1 Introduction

The sea coast is a transitional zone between two hydrogeochemical environments important for life conditions and groundwater resources. On the one hand, there are saline seawaters not suitable for drinking, and on the other hand—freshwater resources sensitive to changes. On the borderline between these two environments, a zone has developed that separates two phases: saline waters of the Cl–Na type and groundwaters mostly of the HCO_3–Ca type. Furthermore, saline waters and brines are common in the Mesozoic strata of the Southern Baltic coast, which induce an increased concentration of Cl^- ions in Cainozoic aquifers as a consequence of ascension (see Fig. 10.1).

A. Krawiec (✉)
Faculty of Earth Sciences and Spatial Management,
Nicolaus Copernicus University, Lwowska 1, 87-100 Toruń, Poland
e-mail: arkadiusz.krawiec@umk.pl

© Springer Nature Switzerland AG 2021   215
M. Zeleňáková et al. (eds.), *Quality of Water Resources in Poland*, Springer Water,
https://doi.org/10.1007/978-3-030-64892-3_10

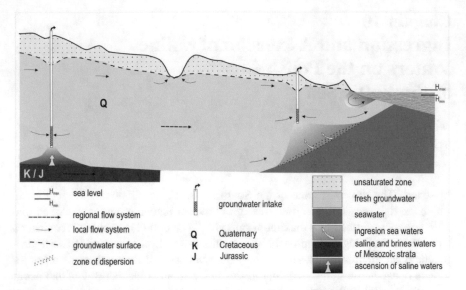

Fig. 10.1 Brackish or saltwater intrusion and ascension in the Southern Baltic coastal area

Natural geochemical anomalies in the coastal zone are mostly induced by seawater ingression or ascension of saline waters from the Mesozoic aquifers. The progressive increase in the chloride concentration in the groundwater used is referred to as salinity [1, 2]. The author understands the term „salinity" as a process of saline water encroachment into aquifers. As a consequence, the concentration of Cl^- ions increases, the chemical state of groundwater deteriorates, and the use of water for drinking is reduced.

The term ingression (encroachment) of saline waters into aquifers involves the process of saline seawater penetration observed along the coastline (Fig. 10.2). This process may also occur as a consequence of excessive exploitation of water intakes in the coastal zone. The phenomena responsible for the salinity on the Polish Baltic coast include ascension, i.e., ascending groundwater movements resulting from a difference in the hydraulic head between groundwater and confined groundwater levels at Mesozoic aquifers [3].

Salt and brackish waters that occur along the coastal Baltic lowlands in Poland originate from the seawater intrusion or brines ascending from the deep Mesozoic strata. Lowering the water table or piezometric pressures of groundwater caused by the exploitation of water intakes in the coastal zone results in an increasing risk of saline water intrusion from the lower bedrock, or the risk of seawater ingression. The issue of explaining the salinity of water in the intakes located within the coastal zone has been considered in many publications including [3–12], among others.

Fig. 10.2 Location of the study area on the background of geological units

10.2   Outline of the Geological Structure of the Coast

The Southern Baltic coast is situated within the range of several geological units (Fig. 10.3). The Cretaceous deposits in the Cainozoic substratum are developed mostly as marls, siltstone, and claystone, as well as sands and sandstone, and they are present mainly in the eastern part of the area. However, in the western part of the study area, Jurassic rocks dominate in the Cainozoic substratum, including mostly limestone, marl, claystone, and sandstone. The depth of the Mesozoic roof in the analysed section of the coast varies (Fig. 10.3): in the western part the Cainozoic substratum is located at the altitude of ca. −50 m amsl and is descending in the

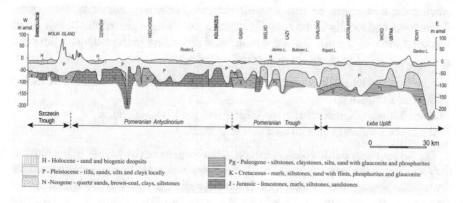

Fig. 10.3 Geological cross-section along the Polish Baltic coast (after [14] modified by the author)

easterly direction towards the altitude of ca. −100 m amsl In deep erosion depressions, the Mesozoic roof sometimes reaches the altitude of −250 m amsl—e.g., the Dreżewo channel, or even −340 m amsl—the region of Dąbki [13]. The Neogene and Paleogene deposits are present in the central and eastern parts of the Polish Baltic coast. Those are mainly Oligocene clay and sands, as well as sands, siltstone, and claystone with the interbedding of Miocene brown coal (Fig. 10.3) [14]. The Palaeocene, Eocene and Pliocene sediments are fragmentarily preserved in the form of isolated patches of mostly sand and gravel, with interlayers of clay and silt. In the western part of the coast, Neogene and Paleogene deposits do not occur, and Pleistocene glaciofluvial sediments are deposited directly on Mesozoic rocks.

The Quaternary sediments occur on the terrain surface all over the Baltic coast. They usually form alternate sequences of boulder clay and sand strata, sometimes also with muds and clays. The substratum and the thickness of the Quaternary deposits are varied; the latter ranges from 30 m in the region of Kołobrzeg, Darłowo, and Gdynia to over 200 m in the area with deep fossil valleys near the town of Żarnowiec.

In the estuarial parts of the river valleys and ice-marginal valleys, in the areas of coastal lowlands and the Żuławy region, the Holocene sandy and organic sediments can be found—peat and alluvial muds, while sand barriers and peninsulas formed in the other parts of the Baltic Coast. The southern Baltic Coast is mostly of a sedimentational type, being levelled with sand barriers since the Littorina transgression, i.e., during the last 7000 years. The cliff type coast, in which the Pleistocene and locally the Neogene sediments are exposed, accounts for about 15% of the coastal line.

10.3   Hydrogeological Conditions After the Review

The Baltic Sea forms a discharge area for all the multi-aquifer formations and groundwater flow systems recharged from the moraine uplands within the coastal belt of Lakelands, where the elevation exceeds 200 m amsl, and locally even 300 m amsl. Fresh groundwater can be found mainly in the Quaternary, Neogene, Paleogene, Cretaceous, and locally in the Jurassic strata in the western part of the Polish coast. The conditions of the first surface aquifer vary according to the extent of the main geomorphological-geological units, which consist of:

– dune barriers that separate lakes or coastal lagoons from the Baltic Sea, including the Hel Peninsula,
– coastal lowlands, including river valleys and the Vistula delta,
– the highest upland hills and cliff coasts (Fig. 10.4).

On the southern Baltic coast, common groundwaters occur mainly in the Quaternary aquifer. In the western part of the coast, freshwater is extracted locally from the Cretaceous aquifer in the syncline of Trzebiatów and occasionally from the Jurassic aquifer e.g. in the towns of Mrzeżyno, Podczele and Pogorzelica (Figs. 10.2 and 10.3). On the eastern coast, apart from the generally exploited Quaternary aquifer, also the Miocene, Oligocene and Cretaceous aquifers are exploited (especially in the

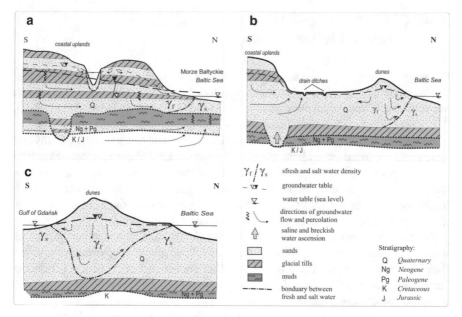

Fig. 10.4 Groundwater occurrence in main geomorphological and geological units of the southern Baltic coast. A—Aquifers at moraine upland and cliff coast; B—Example of an aquifer at coastal lowland with a dune barrier at the shoreline; C—freshwater lens at a sand dune

region of Gdańsk). Freshwaters are mostly of the HCO_3–Ca or HCO_3–Ca–Mg types and they occur mainly in the Quaternary aquifers. Locally, they are also present in the roof parts of the Cretaceous and Jurassic deposits. In the Neogene, Paleogene and Cretaceous beds, there are mostly waters with higher diversity of chemical types: HCO_3–Na, HCO_3–Cl–Na, HCO_3–Cl–Na–Ca and Cl–HCO_3–Na. Brackish and saline waters in the Cretaceous and Jurassic aquifers are usually of the Cl–Na type [3].

A distinct seasonal diversity in the amount of groundwater extractions can be observed on the Baltic coast. In summer time this extraction is locally 3 times bigger than the annual average. The highest amount of water every year is extracted in Świnoujście—about 7500 m^3/day and Międzyzdroje—around 1700 m^3/day. In other coastal towns, the multiyear average is Kamień Pomorski—1200 m^3/day, Pogorzelica—2000 m^3/day (summer), Darłowo—2000 m^3/day, Dąbki 1500 m^3/day and Ustka—2800 m^3/day, Krynica Morska—1600 m^3/day.

The biggest groundwater intakes that are in use in big cities are moved away from the coastal zone by a distance of several or more than 20 km from the shoreline. These intakes are located in: Kołobrzeg—17 000 m^3/day, Koszalin—around 19 000 m^3/day, Władysławowo-Cetniewo—above 5000 m^3/day, Gdynia—above 35,000 m^3/day, Gdańsk—around 50,000 m^3/day and the central waterworks in the Vistula Delta area around few thousands m^3/day.

10.4 Chloride Anomalies in the Groundwaters of the Polish Baltic Coast

The paper presents results of the origin of chloride anomalies in the groundwaters of the Polish Baltic coast. In the selected exploratory boreholes and wells, water samples were collected for detailed physicochemical analysis, isotope determination (^3H, ^{14}C ^{18}O, ^2H) and concentrations of noble gases (^4He, ^{40}Ar, and ^{21}Ne). In order to determine the extent of ingression and ascension of saline and brine waters, geophysical surveys were carried out with the use of electrical resistivity tomography—Fig. 10.5.

The paper presents comprehensive results of the hydrogeochemical and isotope analysis of waters in terms of their genesis and "ages." The water exchange rate was determined in each aquifer systems within the study areas on the basis of the results of isotope analysis, the concentrations of ^4He and ^3H (see Table 10.1).

Frequent discontinuities in the substratum of Cainozoic deposits affect the groundwater circulation system and are often the cause of merging among different aquifers. Such connections, particularly between saline waters occurring in the Jurassic or Cretaceous strata and waters from the Quaternary beds, may result in chloride anomalies in shallower strata.

Mostly chloride anomalies were found on the Polish Baltic coast. The processes of ascension and ingression of saline and brackish waters are the main reasons for their development. Several factors determining the occurrence of chloride anomalies in the study area were identified (circulation systems, hydrogeological evolution, water storage capacity of strata).

Groundwater circulation systems are largely dependent on the geological structure, morphology of the coast and tectonics—dislocation zones. Waters from the Mesozoic strata usually have higher hydraulic heads compared to waters from the Quaternary strata, which contributes to ascension of saline waters into shallower aquifers. In the western part of the coast, there are no Neogene and Paleogene impermeable clay deposits—aquiclude, which greatly facilitates the flow of saline waters from the Mesozoic substratum to the Quaternary beds (Fig. 10.6). On the coastal lowlands, islands, and sandspits, low hydraulic heads in the exploited aquifers (frequently induced by excessive exploitation) are the cause of seawater ingression and ascension of brine/saline water from the Mesozoic substratum [11, 15, 16].

The fastest changes in the water chemical composition occur in aquifers of low thickness. Intensive water extraction from an aquifer with a low thickness in adverse conditions may be followed by fast degradation of its resources through the inflow of saline waters from the substratum or the shoreline. If the aquifer or the aquifer system is thick (high water capacity), and hence is abundant in groundwater resources, all the changes are much slower. The total thickness of the Cainozoic aquifers in the eastern part of the Polish Baltic coast is usually 2–3 times larger compared to the western part.

Hydrogeological evolution of the study area, particularly in the late Pleistocene and the early Holocene had a major effect on the formation of chemical composition and genesis of groundwater. During the glaciation, in places where the ground

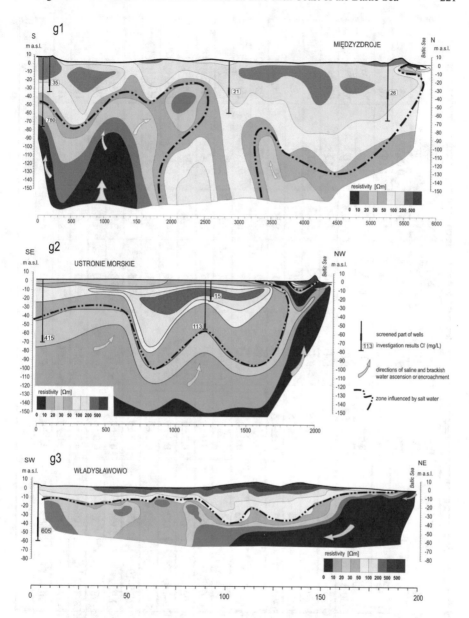

Fig. 10.5 Geoelectrical cross-sections (g1—Międzyzdroje, g2—Ustronie Morskie, g3—Władysła-wowo)

Table 10.1 Hydrochemical data, available isotope and noble gas data from sampling sites in the Polish Baltic coast (after [3] modified by the author)

No.	Sampling site; Strat./depth [m]	Cl⁻ [mg/L]	Tryt (TU)	$\delta^{18}O$ (‰) V-SMOW	$\delta^{2}H$ (‰) V-SMOW	NGT [°C]	$^{4}He_{excess}$ $10^{-6}cm^{3}g^{-1}$	"Age"
1	Świnoujście 15a/29	45	2.9 ± 0.3	−9.22	−64.3	10.3	0.083	M+
2	Świnoujście 16; Q/30.5	700	0.9 ± 0.4	−9.30	−64.0	7.5	1.69	M+pQ
3	Wolin 10; Q/27.5	200	8.6 ± 0.5	−9.06	−63.2			M+SW
4	Świętoujść S-1; Q/30.5	240	16.7 ± 1.5	−9.83	−68.9	6.0	1.10	M+SW+pQ?
5	Jaroszewo H-1; Q/48	498		−9.41	−66.0	7.4	2.19	H+pQ
6	Pobierowo s.1; Q-J/70	200	1.2±0.5	−9.39	−64.9	5.9	2.31	H+pQ
7	Trzęsacz S-1; Q/17	70	8.6±0.5	−9.37	−63.8			M
8	Trzęsacz S-2a; J3/33	672	0.0±0.5	−9.87	−70.1	5.1	30.70	G+pQ
9	Niechorze 2; K/57	106	0.0±0.3	−8.90	−60.9	4.4	0.533	H+G+pQ?
10	Mrzeżyno S-1; J1/94	60	0.0±0.3	−9.15	−63.9	6.4	0.239	H+pQ
11	Dźwirzyno; J1/75	546		−10.40	−73.7	1.1	2.33	G+pQ?
12	Kołobrzeg 16A; Q/46	652	0.0±0.5	−9.80	−67.5	2.9	3.94	H+G+pQ?
13	Darłówko; Tr/83–102	226	0.2±0.4	−9.40	−65.4	1.6	2.29	H+G+pQ?
14	Jarosławiec; K/126–150	112	0.3±0.5	−9.35	−63.0	4.3	0.74	H+G+pQ?
15	Ustka; K/127–155	423	0.2±0.3	−9.63	−68.0	4.8	1.81	H+G+pQ?
16	Łeba Rąbka V-2; Q/15–19	188	0.7±0.3	−7.99	−58.5			M+SW
17	Łeba IMGW 1; Q-M/89	3354	0.0±0.5	−11.50	−85.0	2.9	380.00	G+pQ
18	Nowęcin 1; Tr/85–92	37	0.0±0.5	−9.60	−66.0	4.4	1.10	H+G
19	Łebieniec 1; Q/108–137	20	4.0±0.4	−9.89	−69.6		0.0083	H
20	Cetniewo 2; Q/87–124.5	12	2.6±0.3	−10.01	−70.4	8.3	0.22	H
21	Władysławowo 4; M/38–71	605	2.7±0.3	−9.71	−69.0	8.6	0.75	H+SW

(continued)

Table 10.1 (continued)

No.	Sampling site; Strat./depth [m]	Cl⁻ [mg/L]	Tryt. (TU)	$\delta^{18}O$ (‰) V-SMOW	$\delta^{2}H$ (‰) V-SMOW	NGT [°C]	$^{4}He_{excess}$ $10^{-6}cm^3g^{-1}$	"Age"
22	Jastarnia 9; Q-Tr/128	532	0.0±0.5	−13.75	−99.1			G+pQ?
23	Jurata 1/5; Q-K/148	97		−13.90	−102.0	0.0	2.01	G
24	Hel 3; Q-/173	116	0.2±0.3	−11.03	−79.9	0.9	3.03	G
25	Łaszka; Q/5	453	3.2±0.3	−8.89	−68.5			M
26	Elbląg K-4; M/120	278	0.0±1.0	−11.20	−81.0	3.8	10.40	G+H?
27	Suchy Dąb 2; Q/18-36	15.1	0.0±0.5	−10.10	−67.0			G+H

"Age" classification: M—modern (atomic bomb era), H—Holocene (pre-bomb Holocene), G—glacial, pQ—pre-Quaternary (ascension—salt water); SW—sea water; question mark means doubtful interpretation

Uncertainty (standard deviation) of: $\delta^{18}O$ is ±0.1‰, δD—±1‰, NGT—±0.7°C, ^{4}He excess—±10%

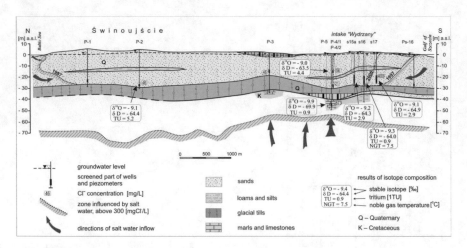

Fig. 10.6 Hydrogeological cross-section through the eastern part of Uznam Island (after [21] modified by the author)

was not frozen, or long-term permafrost was not continuous, the groundwater was recharged by waters from the melting continental ice-sheet. This involved an extensive groundwater exchange [3].

Based on ^{14}C analysis and determinations of noble gases, the so-called"glacial" waters infiltrating in the late Pleistocene were localized on the Hel Peninsula (Jastarnia, Hel—Table 10.1, nos. 22–24) and in the area of Żuławy Wiślane (e.g. Elbląg, Gozdawa, Suchy Dąb—Table 10.1, nos. 26 and 27). In Jurata (a seaside resort on the Hel Sandspit), there are extracted waters with delta values higher than the lower threshold typical of glacial recharge in the territory of Poland: $\delta^{18}O = -13.9\%o$ and $\delta^2H = -102.0\%o$ (Table 10.1). A significant contribution of infiltrating waters in cold climate is apparent also in samples taken from Ustka, Jarosławiec, Dąbki and Łeba (Fig. 10.7), for which NGT values range from 4.3 to 5.3°C (Table 10.1, nos.: 13–15 and 17–18). Waters with this genesis are found mainly in eastern parts of the Polish coast in the deep Pleistocene aquifers and Cretaceous roof layers at the depth ranging from 90 to over 180 m.

The results of the research proved the presence of the so-called „young relict" waters, which were reported by Schroedter [17] and further documented by Kozerski and Kwaterkiewicz [18], as well as Zuber et al. [6, 19]. In the central and northern parts of the Vistula Delta Plane (Żuławy), where hydraulic gradients are very low, there are zones with hindered water exchange and zones with stagnant waters occur. The so-called young relict waters or components of these waters could have survived in these regions, mixed with waters of different origin, e.g., present-day waters or older waters penetrating into more shallow aquifers as a result of ascension from the Mesozoic substratum [20].

Groundwater drainage from different water circulation systems occurs in the area of the southern Baltic coast. Water streams of varying "age" often mix with each other, contributing to the multicomponent genesis of water. Furthermore, changes in the

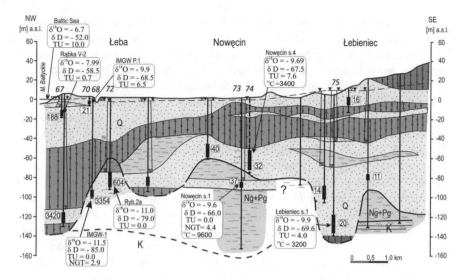

Fig. 10.7 Hydrogeological cross-section perpendicular to the shore-line in Łeba (Other explanations are given in Fig. 10.6)

chemical composition of waters in this area are strongly affected by human activity consisting of excessive exploitation of aquifers (mainly in the 1970s and 1980s). The process of saline water penetration into exploited aquifers was most extensive at that time. At present, it has been significantly inhibited as a result of reduced water extraction at the water intakes in this area. Water intakes with the continuously increased Cl^- concentrations have been closed. Some of the water intakes in the holiday resorts are used only seasonally, which contributes to better sustainability of resources. Basically, all water intakes are managed by well-organized enterprises/water supply companies, which seek to minimize the losses and to protect the groundwater resources. New water intakes are located at a considerable distance from the areas threatened with ingression or ascension of brine and saline water. In some areas where water intakes were excluded from exploitation or water extraction was significantly reduced (e.g., the region of Gdańsk), the process of groundwater resources restoration is observed together with slow freshening of aquifers degraded by saline waters. This process also continues on the Hel Peninsula where after ceasing the exploitation of the first aquifer, the concentration of chloride ions greatly decreased and at present fluctuates at the level below $100 \, mg/dm^3$. Locally the process of water freshening is also observed near the towns of Mielno and Unieście, where the presence of a freshwater lens was confirmed by geoelectric logging (Fig. 10.5).

The efficient groundwater extraction in the coastal zone is an important issue. In holiday and health resorts, the demand for freshwater greatly increases, especially in the summer season. In larger coastal towns, water extraction increased ca. 2-3 times in the summer season, and at some small water intakes—even several times.

Excessive exploitation of aquifers on the southern Baltic coast leads to the development of chloride anomalies and the closure of many groundwater intakes [3, 10, 21, 22]. Monitoring the volume and quality of freshwater extraction, as well as safely assessing the admissible volume of extracted water for all water management areas will prevent the future degradation of groundwater resources through salinity.

The presence of salt waters in shallower aquifers due to ascension was documented in the region of Kołobrzeg, Mielno, Międzywodzie, Kamień Pomorski, Łeba and Dąbki (Fig. 10.8), therefore, several wells were removed there. On the southern Baltic coast, the impact of mineralized waters flowing from the Mesozoic substratum is also observed in operating water intakes. Such a situation takes place, e.g. in the region of Kamień Pomorski, Pobierowo, Trzęsacz, Niechorze, Dąbki, Darłówek, Jarosławiec, Ustka, Hel, or Jurata [3]. As a result of reduced water extraction, the concentration of Cl^- ion has stabilized at those places at the level ranging from 100 to 300 mg/dm³. Seawater ingressions have a limited range in the study area, and at present, they are observed occasionally, usually in the first aquifer on islands and sandspits (Świnoujście, Wolin, Mielno, Dąbkowice, Rąbka, Dębki and Jurata— Fig. 10.8, Table 10.1—nos: 4, 16, 21). Studies have confirmed the presence of areas with chloride anomalies on the Polish Baltic coast, where the impact of seawaters and ascension of saline waters from the substratum overlap with each other causing the degradation of drinking water resources (Uznam Island, Międzywodzie-Dziwnów, Mielno-Unieście and the region of Dąbki and Dąbkowice—Fig. 10.8).

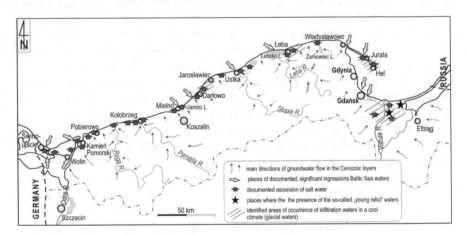

Fig. 10.8 The occurrence of ascension and ingressions of saline or brackish waters on the southern Baltic coast (after [3] modified by the author)

10.5 Conclusions

1. The characteristic feature of the Polish coast is the presence of brines of the Cl–Na type within the Mesozoic strata, from where they ascend into the Cainozoic aquifers (for example Kołobrzeg, Kamień Pomorski, Darłowo, Łeba, Dębki).
2. High piezometric pressure in the Mesozoic layers, deep erosional structures, and presence of tectonic discontinuities result in the saltwater ingress into the younger multi-aquifer complexes. Displacements and tectonic discontinuities within the Mesozoic layers influence the groundwater circulation system and often result in interferences between different aquifers. The brine ascension is supported by the groundwater extraction at intakes in coastal towns.
3. Changes in the chemical composition of waters in this area are strongly affected by human activity consisting of excessive exploitation of aquifers, mainly in the 1970s and 1980s. The process of saline water penetration into exploited aquifers was most extensive at that time. At present, it has been significantly inhibited as a result of reduced water extraction at the water intakes in this area.
4. Saline water ingressions from the Baltic Sea occur locally in shallow aquifers only and have limited extent. Until now, they have occurred mainly in the coastal regions, in holiday locations that are supplied with water from local groundwater intakes. During the last 20 years, as a result of the decreasing extraction and abandonment of many intakes located directly in the shore zone, as well as the construction of new intakes outside the influence of the sea zone, (most often from several to a dozen or so km from the shore), the phenomenon of ingression has been halted. Sometimes, restoration of the natural equilibrium between saline and freshwater is observed as it occurred in Gdańsk.
5. Young fossil waters, originating from the beginning of the Littorina transgression, have been found only locally, within the deltaic series of Żuławy. A clarification of the origin of saline groundwater along the Polish Baltic coast is required for water resource protection and safe yield calculations for water intakes, including the saltwater in baths.

10.6 Recommendations

Assessments of safe yields restricting degradation of fresh groundwater resources are fundamental for the continuous safe exploitation of groundwater intakes in the coastal zone. Therefore, issues of water quality are decisive in estimating safe exploitation volumes in the coastal zone.

All water intakes are managed by well-organized enterprises, which seek to protect the groundwater resources. New water intakes are located at a considerable distance from the areas threatened with ingression or ascension of brine and saline water. In some areas where water intakes were excluded from exploitation or water extraction was significantly reduced (e.g., the region of Gdańsk), the process of groundwater resources restoration is observed together with slow freshening of aquifers degraded by saline waters.

The efficient groundwater extraction in the coastal zone is an important issue. In holiday resorts, the demand for freshwater greatly increases, especially in the summer season. In larger coastal towns, water extraction increased ca. 2-3 times in the summer season, and at some small water intakes—even several times. Excessive exploitation of aquifers on the southern Baltic coast leads to the development of chloride anomalies and the closure of many groundwater intakes. Monitoring the volume and quality of freshwater extraction, as well as safely assessing the admissible volume of extracted water for all water management areas will prevent the future degradation of groundwater resources through salinity.

References

1. Kozerski B (1981) Saltwater intrusions into coastal aquifers of Gdansk region. In: Proceedings 7th Salt Water Intrusion Meeting, Uppsala, 27:83–89
2. Kozerski B (1983) Problems of the saltwater origin in the Vistula delta aquifers. In: Geologia Applicata e Idrogeologia. In: Proceedings 8th Salt Water Intrusion Meeting, Bari XVIII(II): 325–334. In: Proceedings 16th Salt Water Intrusion Meeting
3. Krawiec, A (2013) The origin of chloride anomalies in the groundwaters of the Polish Baltic coast (in polish, eng. Sum). Nicholas Copernicus Univ Publ Toruń, pp 1–143
4. Kleczkowski AS, Nguyen-Manh-Ha (1977) The Effect of the Baltic Water on the chemical composition of groundwater. Bull L'Academie Pol Sci 251:31–38
5. Kozerski B, Kwaterkiewicz A (1984) Zones of salinity of groundwater and its dynamics on the Vistula delta plane (in polish, eng.sum). Arch Hydrotech 31(3):231–255
6. Zuber A, Kozerski B, Sadurski A, Kwaterkiewicz A, Grabczak J (1990) Origin of brackish waters in the Quaternary aquifer of the Vistula delta. In: Proceedings 11th Salt Water Intrusion Meeting, Gdańsk, pp 249–262
7. Zuber A, Grabczak J (1991) Origin of saline waters in the Mesozoic of Central and Northern Poland (in Polish). Współczesne Problemy Hydrogeologii, SGGW-AR, Warsaw, pp 202–208
8. Burzyński K, Sadurski A (1990) The groundwater exchange rate of the Southern Baltic coastal lowland. Jour Hydrology 119:293–306
9. Burzyński K, Sadurski A (1991) Application of the Finite Element Method to Problems Related to the Origin of Mineralised Waters, as Exemplified by the Trough of Lake Żarnowieckie. Bull Pol Acad Sci 394: 389–397
10. Burzyński K, Krawiec A, Sadurski A (2005) The Origin of Groundwater in the Light of Circulation Systems on the Polish Western Coast of the Baltic Sea. In: Proceedings 18th Salt Water Intrusion Meeting, Cartagena, Madrid, pp 521–531
11. Krawiec A, Rübel A, Sadurski, A, Weise SM, Zuber A (2000) Preliminary hydrochemical, isotope, and noble gas investigations on the origin of salinity in coastal aquifers of western pomerania, Poland. In: Proceedings 16th Salt Water Intrusion Meeting, Toruń, pp 87–94
12. Przewłócka M, Jaworska-Szulc B, Kozerski B, Pruszkowska-Caceres M, Szymkiewicz A, Kwaterkiewicz A (2018) Freshening of salinized groundwater in Gdańsk Quaternary aquifer. In: Proceedings 25th Salt Water Intrusion Meeting, Gdańsk, pp 230–235
13. Mojski JE, Tomczak A (1994) Większe formy subglacjalne w rzeźbie podczwartorzędowej polskiego wybrzeża. Acta Universitatis Nicolai Copernici. Geografia 27:241–247
14. Tomczak A (1995) Geological structure and Holocene evolution of the Polish coastal zone. Prace Państwowego Instytutu Geologicznego 149:90–102

15. Sadurski A, Krawiec A (1998) The menace with groundwater resources in the coastal urban areas of Sothern Baltic. In: Hydrogeologia obszarów zurbanizowanych i uprzemysłowionych, Wyd. Uniwersytetu Śląskiego, Katowice, pp 176–184
16. Krawiec A (2015) Ascension and ingression of saline waters of the Pobrzeże Słowińskie Region (in Polish). Prz Geol 63(1):867–872
17. Schroedter M (1931) Die salzhaltigen grundwasser an der Weichselmundung. Geol. u. chem. Untersuchungen Danzig Statist Mitt, p 167
18. Kozerski B, Kwaterkiewicz A (1988) The reasons and the degree of salinity in Quaternary groundwater of the Gdańsk region (in Polish). *Aktualne Problemy Hydrogeologii*, Gdańsk, pp 93–104
19. Zuber A, Sadurski A, Weise SM, Rübel A, Osenbrück K, Grabczak J (2000) Isotope and noble gas data of the Gdańsk Cretaceous Aquifer, Northern Poland. In: Proceedings 16th Salt Water Intrusion Meeting, Toruń, pp 181–186
20. Krawiec A, Sadurski A (2018) Groundwater chemistry and origin of the Vistula delta plain. In: Proceedings 25th Salt Water Intrusion Meeting, Gdańsk, pp. 141–147
21. Krawiec A, Sadurski A (2010) Salinization of the Świnoujście groundwater body (Polish part of Uznam Island). In: 21th Salt Water Intrusion Meeting, pp 165–169
22. Kwaterkiewicz A, Sadurski A, Zuber A (2000) Origin of salinity in coastal aquifers of the Łeba region as indicated by environmental isotopes. In: Proceedings 16th Salt Water Intrusion Meeting, Toruń, pp 169–174

Chapter 11
Monitoring of Groundwater Quality in Poland

Izabela Jamorska

Abstract Monitoring of the groundwater quality in Poland is conducted by the representatives of the Polish Geological Institute—National Research Institute acting as Polish Hydrological Survey. The obligation to conduct the ongoing monitoring, both qualitative and quantitative, was imposed on all Member States of the European Union by the regulations of the Water Framework Directive. The specific scope of such operations is specified in the Ordinance of the Minister of Environment from November 15, 2011, regarding the forms and methods of conducting the monitoring of surface water and groundwater bodies (Journal of Laws of the Republic of Poland, 2011, No. 258, Item 1550). Within the framework of this publication the chemical properties of groundwater of the Quaternary, Palaeogene and Neogene, Cretaceous and Jurassic aquifers on a regional and point basis have been presented. Additionally, due to the size of the chapter, the assessment of the chemical properties of groundwater in Poland in the last five years has been conducted only for selected observation wells and springs. The source material has been provided by the results of chemical analyses published in Hydrogeological Annuals between 2013–2017. The main selection criterion for the assessment of changes was the completeness of the data, the aquifer mentioned in the sources as well as the location of the observation point.

Keywords Groundwater · Monitoring · Groundwater quality · Poland

11.1 Introduction

According to the definition by Water Framework Directive [1], groundwater bodies include groundwater occurring in the aquifers of sufficient porosity and permeability to allow either a significant water intake during the process of supplying local population with water or a flow of significant consequences for the formation of the desired state of surface water as well as terrestrial ecosystems.

I. Jamorska (✉)
Faculty of Earth Sciences and Spatial Management, Nicolaus Copernicus University, Lwowska 1, 87-100 Toruń, Poland
e-mail: izabela.jamorska@umk.pl

© Springer Nature Switzerland AG 2021
M. Zeleňáková et al. (eds.), *Quality of Water Resources in Poland*, Springer Water,
https://doi.org/10.1007/978-3-030-64892-3_11

Currently, the territory of Poland has been divided into 172 groundwater bodies (Fig. 11.1). The division of Poland into groundwater bodies.

The criteria of division included among others the following factors: types and extent of aquifers, interrelationships between ground waters and terrestrial ecosystems and surface waters as well as the possibility of water intake.

Monitoring of groundwater on the territory of Poland is carried out on a national, regional and local scale. This policy reflects the requirements included in the Water Framework Directive, Groundwater Directive as well as Nitrates Directive (91/676/EWG). Nationwide monitoring plays the most important role as it covers all usable water-bearing levels as well as groundwater levels. The groundwater-monitoring network currently comprises approximately 1252 observation wells and springs (as of 2018) [2, 3].

Groundwater chemical status is based on both diagnostic and operational monitoring. Diagnostic monitoring is conducted nationwide every three years. The operational monitoring concerns mainly the testing of groundwater bodies at risk of failing to achieve good groundwater chemical status for 2021 and beyond and is held once or twice a year.

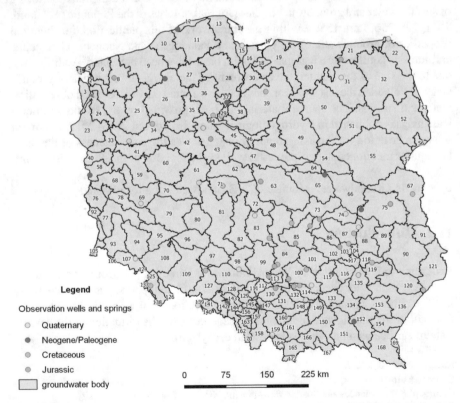

Fig. 11.1 Division of Poland into groundwater bodies (own elaboration based on Polish Geological Institute—National Research Institute data)

Fig. 11.2 Types of groundwater quality monitoring in Poland

11.2   Monitoring of the Chemical Status of Groundwater in Poland

The task of the groundwater monitoring, carried out in the Polish Geological Institute—National Research Institute, is to provide measurement results, tests in terms of quality of groundwater, necessary for chemical assessment of groundwater. The main role is carried out by national monitoring, which constitutes the groundwater monitoring network, which spans across the entire country and all usable aquifers and groundwaters. According to the Decree of Minister of Environment (Dz.U. 2016, Item 1178), the following types of monitoring of groundwater bodies (GWB) are defined: diagnostic, operational, research and protected area (Fig. 11.2).

11.3   Chemical Properties of Groundwater in Poland in the Aquifer System

The assessment of chemical properties of groundwater in Poland on particular water-bearing levels has been conducted on the basis of physico-chemical analyses from 2017. It includes the results interpretation of chemical composition of groundwater samples from 65 diagnostic and operational observation wells and springs provided by the Polish Hydrogeological Survey (PHS) including 18 Quaternary, 13 Paleogene and Neogene, 22 Cretaceous and 12 Jurassic observation wells and springs. In order to promote a more uniform distribution of monitoring points, references have been made to the territorial administrative division of the state, which has been marked both on the map (Fig. 11.3) and the tables (11.1–11.2, 11.3, and 11.4). The basis of the groundwater quality assessment constitutes the Regulation of the Minister of Environment of December 21, 2015 [4] concerning the criteria and methods of assessment of the condition of groundwater bodies. This document distinguishes five ranking water quality standards: class I—excellent quality water, class I good

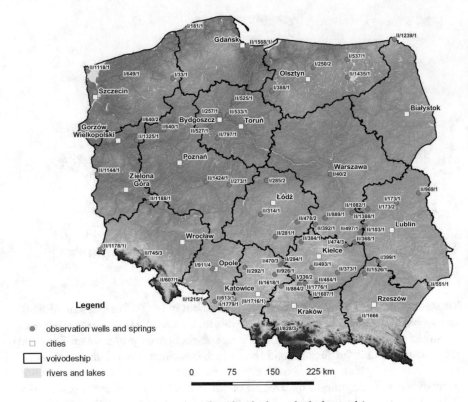

Fig. 11.3 Observation wells and springs (based on hydrogeological annuals)

quality water, class III—satisfactory quality water, class IV—unsatisfactory quality water, class V—poor water quality. Groundwater quality standards from class I to class III indicate that the chemical status of groundwater is at least good whereas groundwater quality standards from class IV to class V indicate that the chemical status of groundwater is rather poor. According to the aforementioned Regulation, the threshold values of physico-chemical elements for the good groundwater chemical status are the limit values of the physico-chemical elements defined for the class III of groundwater quality standards.

11.3.1 Quaternary Aquifer

In the Quaternary aquifer two-ionic and three-ionic waters of the HCO_3–Ca, HCO_3–Ca–Mg and HCO_3–SO_4–Ca type prevail. Apart from bicarbonates, calcium and magnesium, multi-ionic waters contain in their composition mainly chlorides. Exceedances in relation to drinking water generally concern Fe and Mn ions and locally NH_4 and NO_3 ions. With the exception of isolated cases, this is good quality

Table 11.1 Selected observation wells and springs of the quaternary aquifer

Observation well number	Voivodeship	Groundwater body	Monitoring point type	Lithology	Depth of the well [m]	Depth of the specified water table [m]	Groundwater chemical status	Quality class	Exceedances*
II/1118/1	ZPM	1	Well	fine-grained sands	21.00	1.6	HCO_3–SO_4–Ca	III	Mn, NH_4, Fe
II/314/1	ŁDZ	83	Well	sands	51.00	15.70	HCO_3–SO_4–Ca–Mg	III	Mn, Fe
II/1607/1	MŁP	132	Well	sands+gravels	27.00	9.00	HCO_3–SO_4–Ca–Mg	III	Mn, Fe
II/1388/1	MAZ	74	Well	sands+gravels	18.00	3.70	Cl–HCO_3–SO_4–Ca–Na	III	pH
II/1215/1	OPL	140	Well	gravels	37.00	9.80	HCO_3–Cl–SO_4–Ca–Mg	IV	Mn, Mn
II/1526/1	PKR	135	Piezometer	sands+gravels	19.00	3.07	SO_4–Ca	V	Fe, SO_4, Mn, Mn, NH_4, SO_4, Fe, pH
II/1239/1	PDL	22	Well	gravels	52.00	21.50	HCO_3–Ca–Mg	II	Mn, Fe
II/1568/1/	POM	15	Piezometer	sands	5.00	2.40	HCO_3–SO_4–Ca–Na	III	Mn, Fe
II/292/1	SLK	98	Well	sands	23.50	14.00	SO_4–NO_3–Ca	IV	NO_3, NO_3
II/484/1	SWK	100	Well	gravels	13.00	0.60	HCO_3–Ca	II	Mn, NH_4, Fe
II/1435/1	WMZ	31	Well	sands	34.50	4.20	HCO_3–Ca	II	Mn, Fe
II/1424/1	WKP	61	Piezometer	sands+gravels	9.00	2.70	HCO_3–SO_4–Cl–Ca	IV	Mn, Mn
II/1188/1	DLS	69	Piezometer	sands	25.00	10.10	HCO_3–SO_4–Ca	II	Mn, Fe

(continued)

Table 11.1 (continued)

II/527/1	KPM	43	Well	sands	43.00	4.00	Cl–HCO$_3$–Na	V	Na, Cl, TOC, Cl, Mn, Na, NH4, Fe
II/103/1	LBL	87	Piezometer	sands	52.00	32.40	HCO$_3$–Ca	II	
II/1325/1	LBU	34	Well	sands+gravels	13.00	0.50	HCO$_3$–Ca	II	Mn, Fe
II/745/3	DLS	107	Well	gravels	38.00	7.50	HCO$_3$–SO$_4$–Ca	III	
I/828/3	MŁP	159	Well	sands+gravels	8.00	1.85	HCO$_3$–Ca	II	

*Elements beyond the potable water quality standards issued by Decree of Minister of Health regarding the requirements water quality for human consumption, dated 7th December 2017 (Dz.U. 2017, Item 2294)

Table 11.2 Selected observation wells and springs of Palaeogene and Neogene aquifer

Observation well number	Voivodeship	Groundwater body	Monitoring point type	Lithology	Depth of the well [m]	Depth of the specified water table [m]	Groundwater chemical status	Quality class	Exceedances*
I/33/1	ZPM	26	Well	sands	220.00	0.77	HCO_3–Ca	II	Mn, Fe
II/1716/1	MŁP	147	Well	shales	19.00	5.60	HCO_3–Ca	III	Mn, Fe
I/40/2	MAZ	65	Well	sands	270.70	33.75	HCO_3–Cl–Na–Ca	II	Mn, NH4, Fe
II/1666/1	PKR	152	Spring	sandstones+shales	–	–	HCO_3–Ca–Mg	III	
I/181/1	POM	11	Well	sands	200	31.40	HCO_3–Ca	II	Mn, Fe
II/1779/1	SLK	144	Piezometer	gravels	58.00	44.52	HCO_3–SO_4–Ca	II	
II/373/1	SWK	115	Well	limestones+sandstones	42.00	17.00	HCO_3–Ca	III	
I/250/2	WMZ	20	Well	sands	205.00	27.02	HCO_3–Ca	IV	Mn, Fe
I/640/2	WKP	34	Well	sands	164.00	4.00	HCO_3–Ca	II	Mn, Fe
II/1178/1	DLS	105	Well	sands+gravels	36.00	5.30	HCO_3–SO_4–Ca–Mg	IV	pH, Fe, Mn Mn, NH4, Fe, pH
II/525/1	KPM	37	Well	sands	59.60	13.00	Cl–Ca	V	Cl Cl, Mn, NH4, Fe
II/1082/1	LBL	75	Well	sands	115.00	13.00	HCO_3–Ca	III	Mn, Fe
II/1144/1	LBU	58	Piezometer	fine-grained sands	171.00	8.60	Cl–HCO_3–Na	V	Na B, Cl, Na, NH4, Fe

*Elements beyond the potable water quality standards issued by Decree of Minister of Health regarding the requirements water quality for human consumption, dated 7th December 2017 (Dz.U. 2017, Item 2294)

Table 11.3 Selected Observation Wells and Springs of Createous Aquifer

Observation well number	Voivodeship	Groundwater body	Monitoring point type	Lithology	Depth of the well [m]	Depth of the specified water table [m]	Groundwater chemical status	Quality class	Exceedances*
I/173/2	LBL	75	Well	marls	50.00	15.40	HCO_3–Ca	II	Mn, NH_4, Fe
I/257/1	KPM	36	Well	sands	300.00	31.20	HCO_3–Ca–Mg–Na	II	Mn, NH_4, Fe
I/273/1	WKP	71	Well	marls	100.00	6.00	HCO_3–Ca	I	Mn, Fe
II/281/1	ŁDZ	83	Well	limestones	87.10	13.10	HCO_3–Cl–Ca	III	Mn, Fe
II/294/1	SLK	84	Well	marls	25.00	8.10	HCO_3–SO_4–Ca	III	
I/336/2	SWK	100	Well	sandstones	235.00	11,65+	HCO_3–Ca	II	Fe
II/368/1	MAZ	87	Well	marls	25.00	11.30	HCO_3–Ca	III	
I/388/1	WMZ	39	Well	sands	333.00	9.90	HCO_3–Ca–Mg	II	Mn, Fe
I/399/1	PKR	118	Well	limestones+conglomerates	100.30	11.60	HCO_3–Ca	II	Mn, NH_4, Fe
II/478/2	ŁDZ	84	Piezometer	sandstones	25.00	10.95	HCO_3–$SO4$–Ca	IV	Al,Al., Fe
II/493/1	SWK	100	Well	marls	25.00	4.00	HCO_3–Ca	III	
II/497/1	MAZ	87	Well	marls	150.00	16.30	HCO_3–Ca	II	Fe
II/533/1	KPM	29	Well	limestones	90.00	20.50	HCO_3–Ca–Na	II	Mn, NH_4, Fe
I/537/1	WMZ	21	Well	limestone+marls	301.00	7.40	Cl–Na	V	K, PEW, Na, Cl, B, B, Cl, Mn, Na, NH_4, Se, Fe, PEW
II/551/1	PKR	121	Well	limestones	30.00	4.00	HCO_3–Ca	II	
II/607/1	DLS	125	Well	marls	–	–	HCO_3–Ca	II	

(continued)

Table 11.3 (continued)

Observation well number	Voivodeship	Groundwater body	Monitoring point type	Lithology	Depth of the well [m]	Depth of the specified water table [m]	Groundwater chemical status	Quality class	Exceedances*
II/613/1	OPL	141	Dug well	limestones	14.20	6.50	HCO_3–Ca–K	V	K, NO_3
I/640/1	WKP	34	Well	sands	285.00	7.36	HCO_3–Cl–Na	II	NH_4, Fe
II/884/2	MŁP	114	Piezometer	marls	60.00	29.28	HCO_3–Ca	III	
I/911/4	OPL	127	Well	sandstones	200.00	20.00	HCO_3–SO_4–Ca–Na	II	Fe
II/969/1	LBL	67	Well	chalkstones	160.00	6.10	HCO_3–Ca	III	Mn, NH_4, Fe
II/1776/1	MŁP	114	Piezometer	marls	55.00	28.52	HCO_3–SO_4–Ca–Na	IV	NO_3, NO_3

*Elements beyond the potable water quality standards issued by Decree of Minister of Health regarding the requirements water quality for human consumption, dated 7th December 2017 (Dz.U. 2017, Item 2294)

Table 11.4 Selected Observation Wells and Springs of Createous Aquifer

Observation well number	Voivodeship	Groundwater body	Monitoring point type	Lithology	Depth of the well [m]	Depth of the specified water table [m]	Groundwater chemical status	Quality class	Exceedances*
I/173/1	LBL	75	Well	limestones	2355.50	10.00	HCO_3–Na–Mg–Ca	IV	Ba, Fe
I/285/2	ŁDZ	63	Well	limestone+marls	220.00	8.10		V	SO_4, Mg, B, Cl, F, Mg, Mn, Na, NH_4, SO_4
I/285/3	ŁDZ	63	Piezometer	limestones	130.00	10.70	HCO_3–Ca–Mg	III	Mn, Fe
II/384/1	SWK	85	Well	sandstones	25.00	4.20	HCO_3–Cl–SO_4–Ca–Na	V	K, Mn, Mn, Ni, Fe, pH
II/392/1	MAZ	74	Well	sandstones	25.00	4.00	HCO_3–Cl–SO_4–NO_3–Ca–Na	III	pH
I/470/3	SLK	84	Well	limestones	570.00	9.27	HCO_3–Ca–Mg	II	Fe
I/474/3	SWK	103	Well	sandstones	200.00	28.20	HCO_3–Ca	II	Mn, Fe
I/649/1	ZPM	8	Well	sandstones+mudstones	145.00	1.95	HCO_3–Ca	II	Mn, Fe
II/797/1	KPM	43	Well	sandstones	90.00	10.70	HCO_3–Ca–Mg	III	Mn, Fe
II/889/1	MAZ	74	Well	limestones	100.00	14.00	HCO_3–Ca–Mg	II	Fe
II/926/1	SLK	113	Well	limestones	40.00	22.00	HCO_3–Ca	II	
II/1618/1	MŁP	130	Piezometer	limestones	48.00	1.70	HCO_3–Ca	II	

*Elements beyond the potable water quality standards issued by Decree of Minister of Health regarding the requirements water quality for human consumption, dated 7th December 2017 (Dz.U. 2017, Item 2294)

water belonging to class II and class III (Table 11.1). Class IV water has been identified in three observation wells numbered II/1215/1, II/292/1 and II/1424/1, whereas class V water has been detected within the area of the observation well numbered II/1526/1. The chemical composition of water from selected observation wells and springs has been presented in the Piper diagram (Fig. 11.4).

The Piper diagram clearly shows the multi-ionic water from the observation well number II/527/1 located in the Kuyavian and Pomeranian voivodeship and from the observation well number II/1318/1 located in the Masovian voivodeship as well as the water of the SO_4–Ca type identified in the Subcarpathian voivodeship. The ions

2017 Quaternary

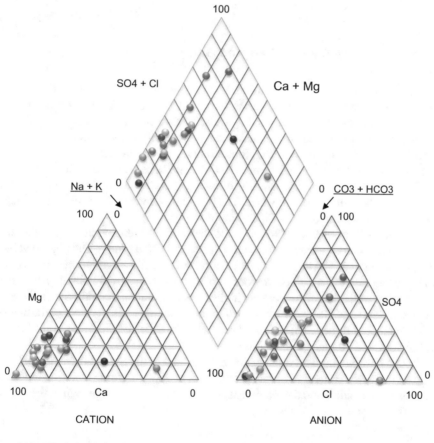

●II/1178/1 ●II/527/1 ●II/1325/1 ●II/314/1 ●II/1526/1 ●II/1568/1 ●II/1118/1 ●II/1607/1 ●II/1388/1
●II/1215/1 ●II/1239/1 ●II/292/1 ●II/484/1 ●II/1435/1 ●II/1424/1 ●II/1188/1 ●II/103/1 ●II/745/3

Fig. 11.4 Chemical composition of quaternary aquifer in the piper diagram (own elaboration based on Polish Geological Institute—National Research Institute data)

which alter the type of this water are sulphate and chloride ions, the concentrations of which are much higher than admissible in potable water.

Such high concentrations of sulphates have been recorded in the well number II/1526/1 amounting up to 640 mg/dm^3 occur pointwise. In the remaining area, their concentration values range from 6,4 (well number I/828/3) to 113 mg/dm^3 (well number II/314/1). The chloride concentrations range from 6,83 (well number I/745/3) to 92 mg/dm^3 (well number I/1526/1). The observation well number II/527/1 with the highest concentration of this ion on the level of 640 mg/dm^3 is located in the broad lowland (groundwater body number 43 in Table 11.1). Such a high result is a consequence of salt diapirs occurring in the soil. The maximum and minimum concentration thresholds of particular ions from the observation wells of the Quaternary aquifer level have been presented in Table 11.1.

11.3.2 Palaeogene and Neogene Aquifer

In the Palaeogene and Neogene aquifer level two-ionic waters of the HCO$_3$-Ca type belonging mainly to the quality class II prevail (Table 11.2). The occurrence of multi-ionic waters of the HCO$_3$–Cl–Na–Ca, HCO–SO–Ca–Mg and Cl–HCO$_3$–Na types have been reported in the Masovian voivodeship (observation well number I/40/2), Lower Silesian voivodeship (observation well number II/1178/1) and Lubusz voivodeship (observation well number II/1144/1). These are waters belonging to the quality classes IV and V, in which the following ions Cl, Na, NH$_4$ and Fe exceed the quality standards for drinking water. The occurrence range of the remaining major ions has been presented in the Piper diagram (Fig. 11.5).

On the territory of Poland in the Paleogene and Neogene aquifer, sulphate ions generally occur in concentrations ranging from 0,5 to 30 mg/dm^3. The highest values of concentration at 149 mg/dm^3 have been recorded in the observation well number II/1144/1 located in the Lubusz voivodeship in the western part of Poland. In the same observation well there occur also the highest concentrations of chlorides at the level of 408 mg/dm^3 and Na—433,2 mg/dm^3. In the remaining area, the concentrations of this ion amount to between 3,84 mg/dm^3 and 20 mg/dm^3 of chlorides and between 1,4 mg/dm^3 and 20 mg/dm^3 of sodium.

In one observation well number II/373/1 located in the Świętokrzyskie voivodeship an increased content of nitrates at the level of 20,75 mg/dm3 has been reported. Taking into consideration the typically agricultural character of the region, the excess of nitrates is of anthropogenic origin.

11.3.3 Cretaceous Aquifer Level

On account of the different depth of the Cretaceous formation placements in the area of Poland, 22 observation wells representing water-bearing levels of various lithology

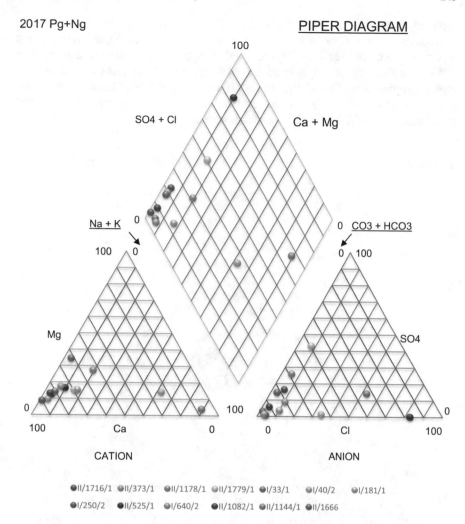

Fig. 11.5 Chemical composition of Paleogene and Neogene Aquifer in the Piper diagram (own elaboration based on Polish Geological Institute—National Research Institute data)

have been selected for the characterisation of groundwater chemical composition (Table 11.3).

The groundwater of the Cretaceous aquifer level is ordinary water, mainly of the HCO_3–Ca type, low-mineralised and of a good chemical status belonging to the quality class II and III. Unlike the groundwater of the Quaternary or Paleogene and Neogene aquifer level, it contains small concentrations of Fe and Mn ions and requires only simple treatment. In several observation wells of a great depth exceeding 200 m, multi-ionic waters of the HCO_3–SO_4–Ca–Na type (observation well number I/911/4, II/1776/1) and HCO_3–Ca–Mg–Na type (observation well number I/257/1) have been reported. In the observation well number I/537/1, located in the Warmian-Masurian voivodeship, the water of the Cl–Na type has been found. The chloride content reported in the well reached 2200 mg/dm^3, and the sodium content amounted to 1358.7 mg/dm^3. These ranges exceed considerably concentrations of such ions in typical groundwater of the Cretaceous aquifer level, which is noticeable in the Piper diagram below (Fig. 11.6). The reason for such results is the great depth of the observation well and the rising of brackish water.

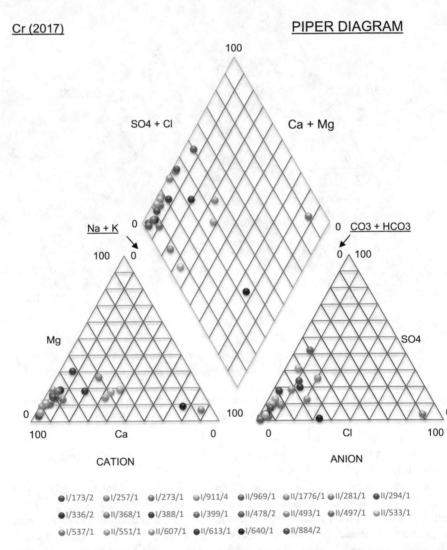

Fig. 11.6 Chemical composition of Cretaceous Aquifer in the Piper diagram (own elaboration based on Polish Geological Institute—National Research Institute data)

In the remaining area of Poland, concentrations of sodium ions occur within the range from 2,1–65,5 mg/dm^3 to 100 mg/dm^3 and occasionally exceeding this amount—observation wells number II/1776/1 and I/640/1, whereas the concentrations of chlorides range from 2 to 50 mg/dm^3, reaching the value over 100 mg/dm^3 only in one observation well number II/1776/1. The increased concentrations of nitrates ranging from 39,2 to 57 mg/dm^3 have been reported in the observation wells at a depth of 60 metres, where the groundwater table stabilises at a depth of around 10 metres (observation wells number II/294/1, II/613/1, II/1776/1). The concentrations of sulphates occur within the range from 1,41 mg/dm^3 (observation well number II/969/1) to 60 mg/dm^3 (observation wells I/388/1 and II/493/1). Concentrations of the remaining ions are not fundamentally different from the neutral hydrochemical background.

11.3.4   Jurassic Aquifer Level

The chemical properties of the groundwater of the Jurassic aquifer have been presented on the basis of analyses of major ion concentrations from 12 observation wells of different depth ranging from 25 to 2355 m (Table 11.4). Two-ionic waters of the HCO_3-Ca type and three-ionic waters of the HCO_3–Ca–Mg type prevail. In three observation wells the occurrence of multi-ionic waters of the HCO_3–Cl–SO_4–Ca–Na type (observation well number II/384/1), HCO_3–Cl–SO_4–NO_3–Ca–Na type (observation well number II/392/1) and HCO_3–Na–Mg–Ca type (observation well number I/173/1) has been reported. The reason for this state of affairs is anthropogenic pollution in the case of the first two wells on account of the shallowly seated groundwater table (approximately 4 m). The chemical status of the groundwater from the observation well number I/173/1 is affected by the depth of the borehole reaching over 2300 m, and it is most likely connected with geogenic factors. The different chemical status of the groundwater from the observation wells number II/384/1 and II/392/1 has been presented in the Piper diagram (Fig. 11.7).

The concentrations of sulphates reach the highest value in the observation well number II/384/1 and amount to 76,6 mg/dm^3. In the remaining monitoring points, their values range from 2,66 to 39,4 mg/dm^3. The concentrations of chlorides and sodium exceed the range of hydrogeochemical background typical for the area of Poland only in the borehole number II/384/1. In the remaining observation wells their values range from 2,46 (observation well number I/470/3) to 18 mg/dm^3 (observation well number I/173/1) as far as chlorides are concerned and from 1,9 (observation well number II/1618/1) to 40 mg/dm^3 (observation well number I/173/1). In two wells numbered II/384/1 and II/392/1 increased concentrations of nitrates have been reported. Exceedances of Fe and Mn which are widespread in the groundwater of the Quaternary aquifer and Palaeogene and Neogene aquifer occur here only in a single case, in the observation well number II/384/1.

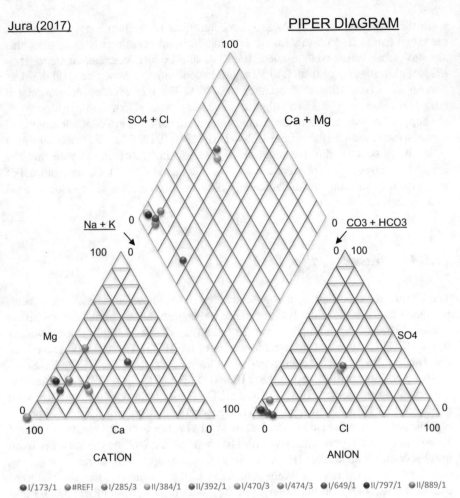

Fig. 11.7 Chemical Composition of Jurassic Aquifer in the Piper diagram (own elaboration based on Polish Geological Institute—National Research Institute data)

In order to present the diversification of ion concentrations on particular water-bearing levels in Table 11.5, minimum and maximum concentrations of major ions have been presented.

11.4 Chemical Properties of Groundwater Within Regional Configuration

Taking into account the hydrogeological regionalisation of Poland according to Paczyński and Sadurski [5], which makes allowances for groundwater bodies and

Table 11.5 Minimum and maximum concentrations of major ions

Ion [mg/dm³]	Quaternary aquifer		Palaeogene and Neogene aquifer		Cretaceous Aquifer		Jurassic aquifer	
	Min	Max	Min	Max	Min	Max	Min	Max
HCO_3	60	368.5	43	443	55	492	29	432
SO_4	6.42	640	<0.05	149	1.41	171	0.63	580
Cl	6.31	640	3.29	620	2.49	2200	2.46	290
Ca	32.8	184.6	20.2	111.5	23.00	155.3	20.7	107.9
Mg	4.3	38.8	2.68	35.9	2.2	54.6	1.0	159.5
Na	3.3	377.4	1.4	433.2	2.1	1358.7	1.9	229.6
K	0.7	10.7	1	9.3	0.9	83.3	<0.05	23.9
Fe	<0.01	93.98	<0.01	17.74	<0.01	8.16	<0.01	6.64
Mn	0	2.25	<0.001	1.25	0.0	0.17	<0.001	4.56
NO_3	0.17	69.9	0.09	26.75	0.09	81.9	<0.03	25.2
NO_2	<0.01	0.07	<0.01	<0.01	<0.01	<0.01	<0.01	0.04
NH_4	<0.05	1.48	<0.05	1.48	<0.05	2.48	<0.05	1.27

water regions of Vistula and Oder basins, the description of chemical properties of water in the following areas has been made (Fig. 11.8):

- Baltic Sea-Shore Zone
- Lake Districts Zone
- Lowlands Zone
- Uplands Zone
- Mountains Zone—The Sudetes
- Mountains Zone—The Carpathians

11.4.1 Baltic Sea-Shore Zone

In the marginal zone of Żuławy Wiślane, the Pleistocene and Holocene groundwater represents mainly the HCO_3–Ca type. The further away from the edge of the upland, the greater the amount of water flowing from a deeper level, locally altering the chemical composition of water to the Cl–Na type. At the Cretaceous level there occurs mainly groundwater of the Cl–HCO_3–Na type [6–8]. The ingredient which plays the most decisive role when it comes to water quality is the salinity reaching the level of 4000 mgCl/dm³ in the Cretaceous aquifer and 3100 mgCl/dm³ in the Pleistocene and Holocene aquifer [9].

The water quality here represents mainly class II and III, locally even class IV. Such a state of affairs results from high concentrations of Fe and Mn, which exceed the limits admissible in potable water. In the remaining area, in the eastern and western Pomerania region according to the regional division, groundwater occurs in

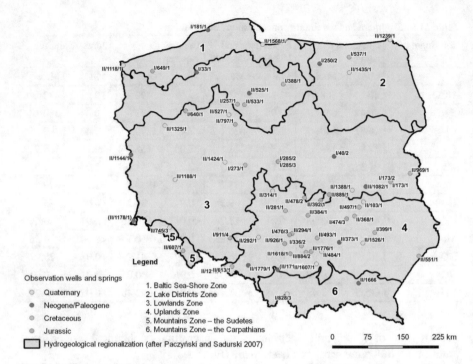

Fig. 11.8 Hydrogeological regionalization based on upper mentioned items after Paczyński and Sadurski with observation wells

the Quaternary, Paleogene and Neogene, Jurassic and Cretaceous formations. The two-ionic water of the HCO_3–Ca type and three-ionic water of the HCO_3–Ca–Na type prevail in the Quaternary aquifer. Groundwater occurring in this area is characterised by large variability in terms of the concentrations of Fe and Mn, as well as Cl and SO_4. Particularly high concentrations of chlorides (locally exceeding 1 g/dm³—Mielno) is typical of coastal zone groundwater. The coastal zone is the area affected by brackish water.

Groundwater occurring in the Paleogene and Neogene formations generally does not show an increased concentration of chlorides (except the Baltic sea-shore zone) and includes waters of the HCO_3–Ca and HCO_3–Ca–Na as well as HCO_3–Ca–Cl–Na.

In the case of the Mesozoic aquifer groundwater of the quality class II prevails. Similarly to the Quaternary and Paleogene and Neogene aquifers, the components which contribute to the lower quality of water are Fe and Mn, as well as Cl locally (in the coastal zone). Groundwater of the HCO_3–Cl–Ca or HCO_3–Cl–Ca–Na are predominant in the area [10–13, 15]

11.4.2   Lake District Zone

The western part of the lakeland area constitutes the Lower Oder region and Szczecin Lagoon. The two-ionic water of the HCO_3–Ca type and three-ionic water of the HCO_3–Ca–Mg type prevail here, and they represent high-quality class water. This is mainly well-insulated groundwater aquifer occurring in the upland area within the Quaternary formations. In the lowland areas and outwash plains with poor insulation of the water-bearing level two-ionic and three-ionic waters can be found where apart from the Ca and Mg ions, there are also chloride, sodium and sulphate ions. The Fe and Mn ions occur in excessive concentration in the whole area whereas the NH_4 ion occurs in excessive concentrations only locally [14].

In the central part of the lakeland area (the lakeland subregion represented by the observation wells number I/257/1, II/525/1, II/533/1 and I/388/1) the chemical composition of the Quaternary and Palaeogene and Neogene aquifers is similar. Ordinary waters of the HCO_3–Ca and HCO_3–Ca–Mg type occur in the greater part of the area. Locally, in the vicinity of the observation well number II/525/1, the increased concentration of the chloride ion becomes noticeable, which is probably connected with anthropogenic pollution. The water of these water-bearing levels needs simple treatment for the presence of Fe and Mn.

The chemical composition of the Cretaceous aquifer differs from the aforementioned water-bearing levels in terms of increased concentrations of Ca, Cl and Na ions. Two types of water prevail in the area: HCO_3–Ca and HCO_3–Ca–Mg. The chemical composition of groundwater in this area changes towards Żuławy Wiślane and the Vistula River Valley, where its mineralization and concentration of chlorides increase reaching its maximum in the area to the south of Bydgoszcz [5]. The eastern part of the lakeland zone is the basin of the Narew, Pregoła and the Neman River. Similarly to the remaining part of the Lake District zone, in the majority of observation wells (II/1239/1, II/1435/1, I/250/2, I/537/1) waters of the HCO_3–Ca and HCO_3–Ca–Mg types occur. These types of groundwater are typical of shallow areas of intensive exchange. The factors which negatively affect the groundwater quality are the exceedances of the Fe and Mn ions. In the Cretaceous formations water of the HCO_3–Na type locally occurs. According to Michalski [15] and Liszkowska [16], this type of water represents cryogenic transformed groundwater, by which they mean water the chemical composition of which has been transformed as a result of freezing. The occurrence of cryogenic water in this area is a consequence of the presence of permafrost, the thickness of which during the last glaciation reached up to 500 m [17].

11.4.3 Lowlands Zone

In the western part of the lowland zone, in the so-called subregion of the Wartanian lowland, groundwater occurs in the Quaternary, Palaeogene and Neogene, Cretaceous and Jurassic formations. The lowest quality class of groundwater can be ascribed to the Quaternary aquifer (observation wells number II/527/1, II/1424/1). The reason for this is the presence of excessive concentrations of Fe, Mn and NH_4. This type of water occurs in central parts of valleys, within the range of flood terraces, where the Holocene formations contain young reactive organic matter [18–21]. The hydrogeochemical types of this groundwater are $Cl–HCO_3–Na$ (observation well number II/527/1) or $HCO_3–SO_4–Cl–Ca$ (observation well number II/1424/1). The best quality water can be found in shallow reservoirs created by fluvioglacial formations (observation well number II/1325/1), where the groundwater type HCO_3-Ca of the quality class II prevails. The chemical properties of the Palaeogene and Neogene aquifer are diversified due to the inflow of saline water from the Mesozoic basement supplemented by the percolation of water from the Quaternary aquifer. As a consequence of this, both good quality water of the $HCO_3–Ca$ type and multi-ionic water of the $Cl–HCO_3–Na$ type with excessive concentrations of Cl, Na, NH4 and Fe ions occur here (observation well II/1144/1).

Groundwater of the Cretaceous aquifer is characterised by its very good quality, and locally they attain quality class I (monitoring point number I/273/1). On account of carbonate environment of this type of groundwater, the concentrations of Fe and Mn are on a very low level. The groundwater of the Jurassic aquifer, due to locally direct contact with Quaternary and Palaeogene and Neogene formations, has slightly lower quality and it belongs to quality class II.

In the eastern part of the lowland, in the subregion of the middle Vistula River, apart from the typical groundwater of the $HCO_3–Ca$ and $HCO_3–Ca–Mg$ types, multi-ionic waters, the chemical composition of which has been shaped by both natural processes and anthropogenic pollution can be found. In the central part of the area, in the isolated groundwater of the Paleogene and Neogene aquifer, among the main ions, the prevalence of the Na and Cl ions can be recorded. In the so-called transition zone groundwater of the $HCO_3–Ca–Na$, $HCO_3–Cl–Na$ or $HCO_3–Cl–Na-Ca$ types (observation well number I/40/2). Long-term observations indicate considerable sustainability of their chemical composition, constant and unchanged even in the zones of intensive consumption [22, 23]. The groundwater of the Jurassic aquifer in the eastern part of the lowlands constitutes good quality water of the $HCO_3–Ca$ and $HCO_3–Ca–Mg$ types.

11.4.4 Uplands Zone

The chemical composition of groundwater in the western part of the upland area (the subregion of the middle upland Vistula River) is extremely diverse. The Uplands

Zone includes within its boundaries the Silesian voivodeship, on the territory of which numerous mining areas are situated. The Quaternary aquifer, in its areas of anthropogenic activity (in the vicinity of mining areas), belongs to multi-ionic water of the SO_4–NO_3–Ca type (observation well number II/292/1) and HCO_3–SO_4–Ca–Mg type. Beyond the mining areas, ordinary water of the HCO_3–Ca and HCO_3–Ca–Mg types prevail. The groundwater of the early-Jurassic aquifer level represents good quality water, which in the greater part of its area belongs to the quality class II.

In the central part of the upland area, particularly in the region of the Świętokrzyskie Mountains, groundwater of the HCO_3–Ca and HCO_3–Ca–Mg types prevails at all water-bearing levels. This is mainly water of good quality in the whole area with local exceedances of Fe and Mn. Only locally, in the Quaternary and Paleogene and Neogene aquifers as well as shallow deposits of the Cretaceous aquifer, can one encounter multi-ionic water with excessive concentrations of the sulphate ion. It is a consequence of anthropogenic activity or, as it is in the case of the Nida Valley, the presence of gypsum deposits.

11.4.5 Mountains Zone—The Carpathians

The Uplands Zone borders with the mountain area include the Tatra Massif and the Carpathian Foredeep. The groundwater of the Tatra Massif belongs to ultra-fresh and fresh waters of the HCO_3–Ca and HCO_3–Ca–Mg types. This is water of high quality in which none of the ions usually exceeds the allowed concentrations for freshwater [24]. In the area of the Podhale Artesian aquifer groundwater of the HCO_3–Ca–Mg and HCO_3–SO_4–Ca–Mg types occurs. In the central and northern part of the aquifer, the mineralisation of groundwater increases, reaching the value of approximately 3000 mg/dm^3 whereas its temperature allows it to be assigned to the category of thermal waters. In the area of Outer Carpathians (flysch mountains) mineral waters commonly occur while ordinary freshwater occurs only in the Quaternary formations and surface parts of the flysch. This area is susceptible to ascending inflow of salt water from the deeper parts of the massif. There occur predominantly two-ionic waters of the HCO_3–Ca type as well as numerous multi-ionic waters of the HCO_3–Ca–Mg, HCO_3–Ca–Na, HCO_3–SO_4–Cl–Ca–Mg types.

11.4.6 Mountains Zone—The Sudetes

In the area of the Sudetes fresh water is extracted from the Quaternary, Palaeogene and Neogene, as well as from Cretaceous formations. In the case of the Quaternary aquifer, multi-ionic waters, mainly three-ionic, four-ionic and five-ionic waters of the HCO_3–SO_4–Ca type prevail (II/745/3). This is water of high quality which requires only ordinary treatment. The exceedances of the limit values concern mainly Fe and

Mn and NH_4 and NO_3 only locally, which is strictly connected with agricultural activity. The groundwater of the Neogene aquifer also belongs to mono-ionic waters where the ions such as Ca, Mg, SO_4 i HCO_3 prevail (II/1178/1). This water belongs to the quality class II and III on account of the increased concentrations of Fe and Mn as well as due to the colour of the water. The groundwater of the Cretaceous aquifer represents mainly the HCO_3–Ca–Mg and HCO_3–Ca types. This water frequently shows increased concentrations of free CO_2, Fe and F [25, 26].

11.5 Changes in Selected Physico-Chemical Parameters of Groundwater (Within the Past 5 Years)

In order to illustrate the variability of the chemical composition of groundwater in Poland on the basis of local sources, specific observation wells have been selected for particular water-bearing levels in which alterations in concentrations of the main ions between 2013 and 2017 have been analysed. (Table 11.6).

11.5.1 The Quaternary Aquifer

The changes in the Quaternary aquifer are represented by four observation wells located in the Kuyavian and Pomeranian (**II/527/1**), Łódź (**II/314/1**), Lubusz (**II/1325/1**) and Subcarpathian (**II/1526/1**) voivodeships. The well number II/527/1 has been selected for analysis on account of its high concentrations of chloride and sulphate ions in 2017. Taking into consideration concentrations of these ions since 2013, their continuous increase can be observed. The concentration of chloride ions increased from 595 mg/dm3 in 2013 up to 640 mg/dm3 in 2017, whereas that of sodium ions rose from 352,46 mg/dm3 to 377,40 mg/ dm3. The concentration of Fe and Mn in this aquifer remains on a continuously high level (Table 11.6). In the well located in the Subcarpathian voivodeship high concentrations of sulphates can be found, which have increased their contribution in the mineralisation of groundwater from 510 to 640 mg/dm3. The concentration of iron ions remains on a high level, which is steadily growing, and has risen by approximately 30 mg/dm3 since 2013. Taking into account the remaining ions, a decline in the concentrations of Cl and Na ions can be observed (Fig. 11.9).

The highest stability of main ion concentrations within the five-year period under analysis is shown by groundwater from the observation well number II/1325/1 in the Lubusz voivodeship in the western part of Poland. In comparison to 2013, a favourable change has occurred now in the concentration of SO_4, Cl and Na ions, which has had a considerable influence on raising water quality class from III to II (Table 11.6). In the Łódź voivodeship the concentrations of SO_4, Cl, Na and K ions in the observation well number II/314/1 had already fallen substantially by 2016,

Table 11.6 Changes in selected physico-chemical parameters of groundwater

Observation well number	Year	HCO$_3^-$	SO$_4^{2-}$	Cl$^-$	Ca^{2+}	Mg^{2+}	Na$^+$	K$^+$	Fe	Mn	NO$_3^-$	NO$_2^-$	NH$_4^-$
kujawsko-pomorskie II/527/1 Q	2013	301.95	5.48	595.00	81.44	16.85	352.46	4.33	7.89	0.25	0.23	0.05	0.84
	2014	324.52	4.12	565.00	77.72	15.96	329.03	4.62	8.31	0.24	0.26	0.16	0.81
	2015	328.18	2.54	600.00	78.65	16.69	366.45	3.72	8.70	0.22	0.57	<0.01	0.82
	2016	290.00	5.73	628.00	82.30	16.40	390.90	4.70	7.90	0.226	0.19	0.03	0.83
	2017	313.50	6.56	640.00	84.30	16.90	377.40	4.60	7.86	0.24	0.17	0.07	0.82
podkarpackie II/1526/1 Q	2013	125.05	510.00	151.50	186.33	30.74	70.17	5.29	64.86	1.95	0.32	0.03	0.94
	2014	141.52	545.00	169.50	192.18	31.89	73.30	5.51	63.56	2.07	0.21	0.01	0.96
	2015	157.38	580.00	211.00	173.60	35.14	87.03	5.50	80.25	1.97	3.20	<0.01	0.85
	2016	153.00	580.00	127.00	162.10	35.60	78.90	5.00	81.96	1900.00	0.95	<0.01	0.84
	2017	126.50	640.00	92.05	184.60	38.80	62.50	4.20	93.98	2.25	1.82	<0.01	0.79
lubuskie II/1325/1 Q	2013	157.38	38.55	12.00	59.54	3.89	6.93	0.78	0.35	0.12	0.41	0.04	0.10
	2014	152.50	37.10	8.92	61.59	3.99	6.35	0.67	0.24	0.08	0.23	0.01	0.18
	2015	165.92	33.50	9.03	66.61	4.06	5.88	0.64	0.67	0.16	0.13	<0.01	0.14
	2016	168.00	33.10	8.28	64.00	4.10	6.60	0.80	0.43	0.142	0.20	<0.01	0.15
	2017	168.00	31.75	7.98	63.50	4.30	6.30	0.70	0.36	0.11	0.50	<0.01	0.08
łódzkie II/314/1 Q	2013	366.00	207.50	76.40	160.56	26.92	49.94	6.20	1.27	0.06	0.29	0.02	0.03
	2014	309.88	67.50	25.70	114.92	18.28	15.62	2.94	1.28	0.05	0.43	<0.01	<0.05
	2015	294.02	18.90	21.40	87.52	12.81	5.96	1.70	2.22	0.15	0.41	<0.01	0.13
	2016	287.00	16.00	19.30	88.00	11.90	5.70	1.70	2.41	0.175	0.06	<0.01	0.11
	2017	352.50	113.05	18.75	128.70	22.90	15.80	2.60	1.83	0.10	0.66	<0.01	<0.05
małopolskie II/1716/1 Ng(m)	2013	273.28	47.20	17.20	92.63	9.05	15.00	5.70	0.88	0.77	9.68	0.04	0.06

(continued)

Table 11.6 (continued)

Observation well number	Year	HCO$_3$$^-$	SO$_4$$^{2-}$	Cl$^-$	Ca^{2+}	Mg^{2+}	Na$^+$	K$^+$	Fe	Mn	NO$_3$$^-$	NO$_2$$^-$	NH$_4$$^-$
	2014	285.48	48.10	11.05	102.19	9.87	8.31	5.26	0.12	0.05	4.45	0.01	0.03
	2015	234.24	63.80	24.60	92.56	9.52	7.95	4.29	0.09	0.72	3.80	<0.01	0.19
	2016	276.00	52.90	11.90	105.60	9.30	7.70	4.90	0.10	0.204	3.08	<0.01	0.13
	2017	303.50	50.70	14.82	111.50	10.00	7.10	4.30	0.31	0.17	0.90	<0.01	<0.05
świętokrzyskie II/373/1 Ng(m)	2013	286.09	23.80	20.75	102.17	6.94	5.45	3.28	0.01	0.00	28.15	0.02	0.03
	2014	267.18	24.85	20.75	107.61	6.92	5.44	3.44	0.01	0.00	27.30	0.01	0.04
	2015	283.04	25.50	19.60	101.77	6.68	5.21	3.15	<0.01	<0.001	27.40	<0.01	<0.05
	2016	266.00	24.00	18.20	102.00	6.90	5.50	3.90	0.01	0.001	27.40	<0.01	<0.05
	2017	263.50	23.25	20.15	103.50	6.90	5.50	3.50	<0.01	0.00	26.75	<0.01	<0.05
opolskie I/911/4 K	2013	320.86	118.00	18.00	70.23	18.75	73.73	9.63	0.73	0.031	0.07	0.02	0.22
	2014	303.78	124.00	20.30	79.98	20.27	72.78	11.26	0.51	0.03	0.29	0.02	0.25
	2015	303.78	130.00	28.00	71.19	19.31	67.97	10.44	0.61	0.030	0.84	<0.01	0.22
	2016	317.00	128.00	18.10	70.60	19.00	73.80	9.20	0.84	0.032	<0.01	<0.01	0.22
	2017	309.00	123.00	20.20	71.40	19.00	71.60	8.90	0.86	0.03	0.43	<0.01	0.23
warmińsko-mazurskie I/388/1 K	2013	401.38	97.90	10.50	122.43	28.28	12.15	2.25	<0.01	0.092	12.50	<0.01	1.86
	2014	419.68	3.91	95.70	39.99	14.66	146.56	10.31	3.55	0.08	0.20	0.33	1.37
	2015	407.48	<0.50	102.00	35.84	13.41	145.63	8.51	3.27	0.069	0.40	0.12	1.24
	2016	460.00	<0.50	89.30	33.80	13.60	154.00	8.50	3.21	0.063	0.10	<0.01	1,48
	2017	371.00	65.60	9.51	92.50	26.90	11.10	2.20	2.64	0.15	0.76	<0.01	0.50
kujawsko-pomorskie I/257/1 K	2013	273.28	2.34	5.26	43.72	14.54	23.37	9.28	3.10	0.110	0.19	0.01	1.06
	2014	272.06	7.10	8.27	45.74	14.62	24.76	12.42	4.18	0.07	0.26	0.14	0.88

(continued)

Table 11.6 (continued)

Observation well number	Year	HCO_3^-	SO_4^{2-}	Cl^-	Ca^{2+}	Mg^{2+}	Na^+	K^+	Fe	Mn	NO_3^-	NO_2^-	NH_4^-
	2015	283.04	20.10	10.70	55.24	17.04	23.81	9.22	5.68	0.088	0.42	<0.01	0.80
	2016	293.00	11.10	4.11	49.60	15.90	25.00	9.40	5.64	0.102	0.13	<0.01	0.92
	2017	289.00	9.48	5.00	48.90	15.50	24.10	8.80	4.57	0.10	0.33	<0.01	0.87
lubelskie I/173/1 J	2013	292.80	<0.50	17.70	30.09	19.32	39.10	11.00	0.73	0.010	0.06	0.01	0.51
	2014	275.72	0.73	18.40	31.79	19.30	42.25	10.98	0.78	0.01	0.10	<0.01	1.48
	2015	273.28	<0.50	20.10	29.40	19.27	40.94	11.27	0.42	0.009	0.11	<0.01	0.48
	2016	266.00	0.56	18.30	29.40	18.20	41.30	11.90	0.77	0.010	0.06	<0.01	0.46
	2017	254.00	0.63	18.00	29.30	19.30	40.00	11.90	0.80	0.02	0.10	0.13	0.48
śląskie I/470/3 J	2013	245.22	6.73	1.77	65.93	11.90	2.81	1.14	0.37	0.030	0.06	0.01	0.10
	2014	236.68	8.23	2.76	66.77	11.72	3.05	1.19	0.34	0.03	0.36	<0.01	0.10
	2015	250.10	7.54	2.69	62.32	11.80	2.61	1.08	0.37	0.030	0.29	<0.01	0.10
	2016	262.00	7.15	1.52	64.30	11.60	2.60	1.30	0.42	0.031	0.04	<0.01	0.11
	2017	243.00	8.80	2.46	61.70	11.30	2.50	1.10	0.41	0.03	0.42	<0.01	0.10

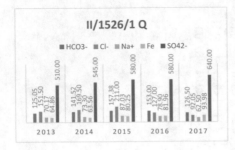 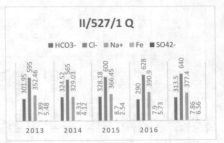

Fig. 11.9 Changes in concentrations of selected ions in time (own elaboration based on Polish Geological Institute—National Research Institute data)

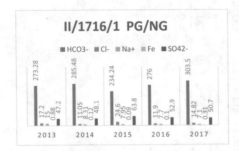 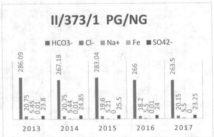

Fig. 11.10 Changes in concentrations of selected ions in time (own elaboration based on Polish Geological Institute—National Research Institute data)

which enabled the groundwater to reach the quality class II. The groundwater quality deteriorated again in 2017, particularly due to the repeated increase in the number of sulphate and sodium ions (Table 11.6).

11.5.2 The Neogene Aquifer

The groundwater of the Neogene aquifer is represented by observation wells from the Świętokrzyskie (II/737/1) and Lesser Poland (II/1716/1) voivodeships. These wells have been selected in order to analyse the content of nitrates on account of their increased concentrations in 2017 (Fig. 11.10). In the observation well number II/737/1, the concentration of nitrates slightly decreased since 2013 from 28,15 mg/dm^3—in 2013 to 26,75 mg/dm^3 in 2017. All the remaining main ions of the groundwater coming from this well reveal a stable range of concentrations.

It can be concluded then that the increased contents of nitrates has an anthropogenic background and probably constitutes a local exceedance. In the observation well from Lesser Poland, the nitrate concentration considerably decreased from

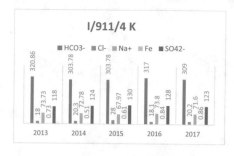
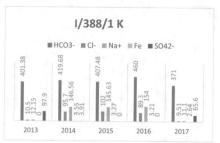

Fig. 11.11 Changes in concentrations of selected ions in time (own elaboration based on Polish Geological Institute—National Research Institute data)

9,68 mg/dm^3 in 2013 to 0,9 mg/dm3 in 2017. Furthermore, the chloride and sodium contents fell slightly as well (Table 11.6).

11.5.3 The Cretaceous Aquifer

The changes in main ions concentrations in the Cretaceous aquifers have been analysed on the basis of three monitoring locations situated in north-eastern (I/388/1), central (I/257/1) and southern Poland (I/911/4). In the first of the locations mentioned above (I/388/1) in 2013, a high concentration of sulphates was reported, which in subsequent years decreased to the level of < 0,05 mg/dm^3, and then in 2017 increased again reaching the concentration of 65,60 mg/dm^3 (Fig. 11.11). Within the analysed five-year period opposite were the concentrations of Cl and Na ions, which in 2013 amounted to 10,50 mgCl/dm^3 and 12,15 mg/dm3 respectively. In the subsequent years 2014-2016 the concentrations of these ions increased considerably to reach the value of 100 mg/dm^3, but the following year, in 2017, fell again to the value of 9,51 mgCl/dm^3 and 11,10 mgNa/dm^3 respectively. With regard to the depth of the observation well which exceeds 300 m, this phenomenon is strictly connected with the inflow of saline water. The chemical compositions of the boreholes number I/257/1 and I/911/4 have been stable for the past several years.

11.5.4 The Jurassic Aquifer

The time variability of the chemical composition of the Jurassic aquifer has been presented on the basis of the observation wells located in the Lublin voivodeship (I/173/1) and Silesian voivodeship (I/470/3). Within the analysed period, the concentrations of particular ions did not undergo major changes (Fig. 11.12).

In selected observation wells changes in the chemical types* and quality classes** of groundwater have also been analysed (Table 11.7)

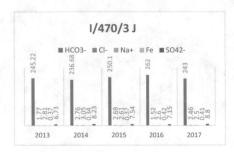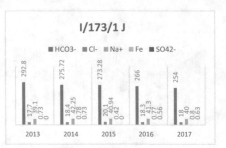

Fig. 11.12 Changes in concentrations of selected ions in time (own elaboration based on Polish Geological Institute—National Research Institute data)

11.6 Summary and Conclusions

As a result of the research conducted within the past five years, it can be concluded that the groundwater of the Jurassic, Cretaceous as well as Palaeogene and Neogene aquifers has had a stable chemical composition. The greatest variability is revealed by the groundwater of the Quaternary aquifer. It is a consequence of high concentrations of chloride and sulphate ions, which are of anthropogenic origin and frequently permeate into water-bearing levels. Taking into consideration the groundwater quality classes, it can be claimed that the quality of groundwater on all water-bearing levels either has increased or has remained stable. In some observation wells, for instance, in the well number I/388/1, extracting water from the Cretaceous aquifer groundwater quality class has changed from IV to II. Such a phenomenon indicates that agricultural standards have improved and raised social awareness concerning water pollution and contamination and protection of the water environment.

The improvement in the quality of groundwater in Poland is also visible in the comprehensive summary of quality classes concerning all water-bearing levels provided in the table below. It can be clearly discerned that in 2017 the percentage of observation wells containing water of quality class II increased (Table 11.8).

The evaluation of monitoring research from 2013 showed that 77% of groundwater examined in observation wells was included in the category of good chemical status (quality classes I–III) whereas 23% of groundwater examined in observation wells was classified as having poor chemical status (quality classes IV–V). In 2017, when a similar number of monitoring points was analysed, groundwater representing quality classes I–III accounted for 80% of all examined water, while the remaining 20% constituted poor quality groundwater—quality classes IV and V.

11.7 Recommendations

The analysis of Poland's groundwater quality conducted in the following study based on the points of network monitoring of the Polish Hydrogeological Survey

Table 11.7 Quality classes and chemical types of groundwater in Poland in selected observation wells

Well number / Year	2013		2014		2015		2016		2017	
	Type	Class	Type	Class	Type	Class	Type	Class	Type	Class
II/1568/1(Q)	HCO_3-Ca	V	HCO_3-Ca	III	HCO_3-Ca-Na	V	HCO_3-Ca-Na	V	HCO_3-SO_4-Ca-Na	III
II/527/1 (Q)	Cl-HCO_3-Na	V	Cl-HCO_3-Na	V	Cl-HCO_3-Na	V	Cl-HCO_3-Na	V	Cl-HCO_3-Na	V
II/1526/1 (Q)	SO_4-Cl-Ca	V	SO_4-Cl-Ca	V	SO_4-Cl-Ca-Na	V	SO_4-Ca	V	SO_4-Ca	V
II/1325/1 (Q)	HCO_3-SO_4-Ca	III	HCO_3-SO_4-Ca	II	HCO_3-Ca	II	HCO_3-Ca	II	HCO_3-Ca	II
II/314/1 (Q)	HCO_3-SO_4-Ca	III	HCO_3-Ca	III	HCO_3-Ca	II	HCO_3-Ca	II	HCO_3-SO_4-Ca-Mg	III
II/1178/1 (Pg/Ng)	HCO_3-SO_4-Ca	IV	HCO_3-SO_4-Ca-Mg	IV	HCO_3-SO_4-Ca-Mg	IV	HCO_3-SO_4-Ca-Mg	IV	HCO_3-SO_4-Ca-Mg	IV
II/1716/1 (Pg/Ng)	HCO_3-Ca	III	HCO_3-Ca	III	HCO_3-SO_4-Ca	II	HCO_3-Ca	III	HCO_3-Ca	III
II/373/1 (Pg/Ng)	HCO_3-Ca	III	HCO_3-Ca	III	HCO_3-Ca	III	HCO_3-Ca	III	HCO_3-Ca	III
I/911/4 (Cr)	HCO_3–SO_4–Ca–Na	III	HCO_3–SO_4–Ca–Na	III	HCO_3–SO_4–Ca–Na	III	HCO_3–SO_4–Ca–Na	II	HCO_3–SO_4–Ca–Na	II
I/388/1 (Cr)	HCO_3–SO_4–Ca–Mg	III	HCO_3–Cl–Na	IV	HCO_3–Cl–Na	IV	HCO_3–Cl–Na	IV	HCO_3–Ca–Mg	II
I/257/1 (Cr)	HCO_3–Ca–Mg–Na	III	HCO_3–Ca–Mg–Na	III	HCO_3–Ca–Mg	III	HCO_3–Ca–Mg–Na	III	HCO_3–Ca–Mg–Na	II
I/173/1 (J)	HCO_3–Na–Mg–Ca	IV	HCO_3–Na–Mg–Ca	V	HCO_3–Na–Mg–Ca	IV	HCO_3–Na–Mg–Ca	IV	HCO_3–Na–Mg–Ca	IV
I/470/3 (J)	HCO_3–Ca–Mg	III	HCO_3–Ca–Mg	II	HCO_3–Ca–Mg	II	HCO_3–Ca–Mg	II	HCO_3–Ca–Mg	II

* Chemical type of water according to Szczukariew-Prikłoński's classification
** Groundwater quality classes according to the Ministry of Environment regulation on surface and groundwater classification as far as the presentation of surface and groundwater state and the method of conducting the monitoring and interpretation of the results are concerned (21 December 2015, published in Dz.U. 2016, Item 85)

Table 11.8 Change in groundwater quality classes between 2013–2017

Quality class [%] Year (number of wells)	I	II	III	IV	V
2013 (589 pts)	1	15	61	13	11
2014 (601)	5	42	31	13	9
2015 (643)	4	42	32	9	13
2016 (1334)	3	42	33	15	7
2017 (593)	5	44	31	12	8

has revealed a visible improvement in its quality between 2013 and 2017. Thus, it can be stated that groundwater chemical status in particular regions of Poland has been well recognized and that the groundwater conservation measures have been effective. Bearing in mind the economic development of the country, it is strongly recommended that further monitoring of mainly the areas of intensive agricultural economy or mining should be continued. The maintenance and further development of the monitoring network is also dictated by prospective geogenic threats resulting from the geological structure of the area of Poland (for example, the occurrence of brackish water), climatic conditions (for instance, the occurrence of areas of hydrologic drought) and water resources.

References

1. Water Framework Directive, Groundwater Directive as well as Nitrates Directive (91/676/EWG)
2. Rocznik Hydrogeologiczny Państwowej Służby Hydrogeologicznej 2013. PIG-PIB, Warszawa
3. Rocznik Hydrogeologiczny Państwowej Służby Hydrogeologicznej 2017. PIG-PIB, Warszawa
4. Rozporządzenie Ministra Środowiska z dnia 21 grudnia 2015 r. w sprawie kryteriów i sposobu oceny stanu jednolitych części wód podziemnych
5. Paczyński B, Sadurski A (red.) (2006) Hydrogeologia regionalna Polski, tom I. Państ Inst Geol, Warszawa
6. Krawiec A, Sadurski A (2018) Groundwater chemistry and origin of the Vistula delta plain. In: Proceedings 25th Salt Water Intrusion Meeting, Gdańsk, pp 141–147
7. Kozerski B, Kwaterkiewicz A (1990) The origin and state of Quaternary groundwater salinization in Gdańsk region. W: Proceedings 11th Salt Water Intrusion Meeting (red.) B. Kozerski, A. Sadurski. Gdańsk, 14–17 May, pp 15–24
8. Sadurski A (1989) Górnokredowy system wód podziemnych Pomorza Wschodniego. Zesz. Nauk. AGH, Geol, 46
9. Kozerski B (1988) Warunki występowania i eksploatacji wód podziemnych w gdańskim systemie wodonośnym. Gdańsk
10. Krawiec A, Rübel A, Sadurski A, Weise SM, Zuber A (2000) Preliminary hydrochemical, isotope, and noble gas investigations on the origin of salinity in coastal aquifers of western pomerania, Poland. In: Proceedings of 16th Salt Water Intrusion Meeting, Toruń, pp 87–94
11. Jamorska I (2015) Hydrogeological conditions of southern Kujawy region (in Polish): Prz Geol, Vol. 63 nr 10 cz. 1: 756–761

12. Dąbrowski S, Rynarzewski W, Wojciechowska R (1996) System wodonośny Kotliny Żmigrodzkiej pradoliny Baryczy w świetle regionalnych badań hydrogeologicznych. W: Problemy hydrogeologiczne SW Polski. Wrocław, pp 177–185
13. Krawiec A (2015) Ascension and ingression of saline waters of the Pobrzeże Słowińskie Region. Prz. Geol., Vol. 63 nr 10, 1: 867–872
14. Jamorska I, Krawiec A (2017) Zmiany chemizmu wód podziemnych w województwie kujawsko-pomorskim w latach 2006–2015. Przegląd Geologiczny, vol. 65, nr 11/2: 1270–1275
15. Michalski T (1985) O genezie anomalii chemicznych w wodach podziemnych NE Polski. W: Aktualne problemy hydrogeologii 3: 501–511. AGH Kraków
16. Liszkowska E (1988) O wodach kriogenicznych zmetamorfizowanych Polski północno-wschodniej. W: Aktualne problemy hydrogeologii 4(2):81–91. Instytut Morski, Gdańsk
17. Nowicki Z, Szewczyk J (2003) Ocena miąższości wiecznej zmarzliny na podstawie danych geotermicznych jako element analizy paleohydrogeologicznej. W: Współczesne problemy hydrogeologii. 11:403–411. PGdań. WBWiIŚ. Gdańsk
18. Górski J, Przybyłek J (1994) Mikrozanieczyszczenia w wodach powierzchniowych i podziemnych w strefach ochronnych ujęcia mosińskiego. Mat. I Konferencji „Municipal and rural water supply and water quality". PZITS Poznań, pp 243–262.
19. Górski J, Przybyłek J (1995) Problematyka stref ochronnych infiltracyjnych ujęć wody w dolinach rzecznych. Mat. Symp. „ Strefy ochronne ujęć i głównych zbiorników wód podziemnych". Wyd. NOT. Częstochowa, pp 69–79
20. Górski J (1999) Wpływ powodzi na chemizm wód powierzchniowych oraz czerpanych z ujęć infiltracyjnych na przykładzie doliny Warty w rejonie Krajkowa. Mat. Symp. „Funkcjonowanie geoekosystemów zlewni rzecznych". Kołobrzeg, pp 24–27
21. Górski J (1981) Kształtowanie się jakości wód podziemnych utworów czwartorzędowych w warunkach naturalnych oraz wymuszonej eksploatacji. IKŚ. Warszawa
22. Płochniewski Z (1972) Żelazo i mangan w wodach podziemnych utworów plioceńskich, mioceńskich, oligoceńskich, kredowych i jurajskich na obszarze Niżu Polskiego. Biul Ins Geol 256:5–37
23. Macioszczykowa A (1979) Chemizm wód trzeciorzędowych i kredowych oraz jego geneza w zachodniej części niecki mazowieckiej. Prace Hydrogeol., seria spec. Z. 11. Inst. Geol. Warszawa
24. Małecka D, Nowicki Z (2002) Skład izotopowy wód podziemnych Tatr i Podhala. Biuletyn Państwowego Instytutu Geologicznego 404:67–83
25. Mroczkowska B (1989). Chemizm wód kredowego piętra wodonośnego w Sudetach Środkowych. Pr. Nauk.Inst Geotech PWroc, 58, Konferencje 29:463–466
26. Mroczkowska B (1995) Atlas hydrogeochemiczny kredy niecki śródsudeckiej i rowu Nysy Kłodzkiej. Państw Inst Geol, Warszawa
27. Rocznik Hydrogeologiczny Państwowej Służby Hydrogeologicznej 2014. PIG-PIB, Warszawa
28. Rocznik Hydrogeologiczny Państwowej Służby Hydrogeologicznej 2015. PIG-PIB, Warszawa
29. Rocznik Hydrogeologiczny Państwowej Służby Hydrogeologicznej 2016. PIG-PIB, Warszawa

Chapter 12
Sediment Management in River Basins: An Essential Element of the River Basin Management Plans

Michał Habel, Dawid Szatten, Zygmunt Babiński, and Grzegorz Nadolny

Abstract The Vistula River and the Oder River drain about 18% of the Baltic Sea Basin and drain 97.9% of the territory of Poland. This river basins serve as an example to show the kind of data, modelling, and analysis tools that are required for a qualitative and quantitative description of sediment dynamics. These methods are used to investigate sediment management options. First, a brief overview of the sediments types transported by Vistula and Oder and their tributaries is presented, anthropogenic impacts on sediment transport, followed by a description of the current methods of monitoring and modelling system used in this study. Lastly, the article looks at specific methods applications for sediment management tasks.

Keywords Sediment management plans · Poland · Vistula · Oder · River basins management · Sediment monitoring methods

12.1 Introduction

River sediment is a crucial, movable and sometimes very dynamic part of basins. Where human actions intervene with sediment balance change or quantity or quality, sediment management may become needful. One of the most essential for the EU Water Framework Directive (WFD) is the River Basin Management Plans (RBMP), which was produced and published by 2009. Into now, sediment linked quantity and quality questions have returned relatively minor attention in the national water

M. Habel (✉) · D. Szatten · Z. Babiński · G. Nadolny
Department of Inland Waterways Revitalization, Institute of Geography,
Kazimierz Wielki University, Kościelecki 8, 85-033 Bydgoszcz, Poland
e-mail: hydro.habel@ukw.edu.pl

D. Szatten
e-mail: szatten@ukw.edu.pl

Z. Babiński
e-mail: z.bab@ukw.edu.pl

G. Nadolny
e-mail: nadolnygrzegorz@ukw.edu.pl

© Springer Nature Switzerland AG 2021
M. Zeleňáková et al. (eds.), *Quality of Water Resources in Poland*, Springer Water,
https://doi.org/10.1007/978-3-030-64892-3_12

management policies. Sediment Management Plans are only carried out in a few selected European river basins, and they focus on selected problems, mainly estuaries [1–3]. A comprehension of the hydrodynamic and morphodynamic influences in sediment transport and knowledge about the relationships with hydrotechnical structures impacts are therefore an essential scientific reason for the optimum management of sediment. Namely, they need to be taken adequately into account when projects and fulfilmenting a sediment management strategy. Before the records in the WFD were formulated, it was known that rivers play a significant role in the global hydrological cycle, providing around 20 billion ton year^{-1} of river sediment to the world's ocean, reflecting climate change as a result of human activity [4]. Along with the amount of water, rivers transport debris, which hydrologists call the sediments. River debris in Russian literature, in terms of genesis, is often divided into riverbed and transit [5]. The first of them, with a mechanical composition exceeding 0.05 mm, enters the river as a result of deformation (erosion). While in Anglo-Saxon literature [6], three types of river sediments are most commonly found: dissolved load, suspended load and as bottom material—bed load. In the last two cases combined, it is a clastic load or solid sediment [7]. The suspended load is mainly clayey and mules. According to Knighton [6], it is a fraction smaller than 0.062 mm. The size of the suspension transport, as well as its intensity, depends on the rate of mechanical denudation of the basin and the supply from anthropogenic sources (including mining, sewage). Klimaszewski [8] or Bajkiewicz-Grabowska and Mikulski [9], also distinguish transport in suspension in the form of a lifted material (mineral and organic material whose specific weight is greater than the specific weight of water) and suspended (mostly organic material whose specific weight is lower than the specific weight of water). The first two types are discharged in the form of dissolved material and the remaining two as a clastic material, i.e., suspension and bed material [10]. The size of the transported particles is within a wide range of 10^{-6}–10^{-3} mm, i.e., from ions through colloids, clays, dusts, sands to boulders.

China has one among the best rates of continental erosion and is evaluated to be losing its land at a rate 57 times faster than it are often a substitute. At an equivalent time, its scaled infrastructures, like Three Gorges Dam, accumulated approximately 170 million tons of the Yangtze River's sediment per annum, reducing sediments reaching the Yangtze's delta by minimum as 82% [11]. Currently observed in the majority of large river systems in the world, the decrease in loads of transported sediments in the estuary sections is the result of its retention by various types of hydrotechnical operations carried out in the course of rivers [12, 13]. Particularly reservoirs contribute to reducing the delivery of river sediment to the world's ocean, contributing to the erosion of the estuary areas at a global scale [14–18]. Studies have shown that the efficiency of dam reservoirs in the capture of sediments is in some cases significant and amounts to 70–90% of the reservoir capacity [19, 20]. It is estimated that about 30–40% of a transported sediments in the form of a suspension through the global river network does not reach the seas, oceans and some great lakes, but is retained in reservoirs, at least for the time of existence of the infrastructure [21]. The calculations of hydrologists show that about 40% of the river outflow on a global scale is retained by artificial retention reservoirs [12, 22]. Since water is a factor affecting the intensity of sediment transport, partition of the riverbed by a dam affects the continuation of its transport. According to the theory of River Continuum Concept (RCC) by Vannote [23] interruption of its continuity, unquestionably affects the change of hydrological conditions. There are approximately 0.8 million dams in operation on the larger half of the world's rivers. Of these dams, approximately 40,000 are registered as large, meaning those that are at least 15 meters in height, or with a reservoir capacity greater than 3 million m^3 [24]. Almost all large reservoirs accumulate 98% of their river's sediment, thus holding limit delivery to seas and oceans.

Sedimentation of the debris in reservoirs creates many serious exploitation and ecological problems [25]. The most important ones include siltation of the reservoir, leading to the reduction of the reservoir capacity and eutrophication of its waters [18, 26–32]. The problem is successive filling the reservoir with sediments, which reduces its capacity. For example, the Kulekhani dam in Nepal, commissioned in 1981, whose lifespan was designed for 85 years, lost almost half of its dead capacity (i.e., approx. 12 million m^3) after just 8 years. The Cerron Grande reservoir in El Salvador has functioned for only 30 years, instead of the originally assumed viability, which was estimated at 350 years [33]. In order to reduce silting of large dam reservoirs, sludge cleaning is carried out by controlled discharge of water into the lower site by the bottom drains. This allows the release of liquefied bottom sediments from the reservoir [34]. In total, each year, 30 million tons of sediment accumulated in Xiaolangdi reservoir in Jiyuan created on the Yellow River are released in this way. In the last dozen or so years, 390 million tons of sediments collected in the reservoir were discharged along with the river's waters [35]. The variability of hydrological and operational conditions has a significant influence on changes in the backwater reach of the reservoir. Its length on the Cymlański Reservoir on the Volga is about 200 km, in the Dnieproges Reservoir on the Dnieper 70–80 km [36].

Undoubtedly, the accumulation of sediment in the reservoir affects the reduction of its operating volume and total capacity [37]. According to the Hartung criterion, the reduction of the reservoir volume to the level of 80% results in the loss of its functionality. In order to unblock the riverbeds (mainly waterways) and to release the reservoir's capacity, dredging works are carried out. In Germany, for example, more than 45 million m^3 of river sediment are dredging every year to provide safe waterway depths [1]. In addition to practical reasons, sediments extracted from rivers and reservoirs are a desirable raw material on the construction market (in particular sand-gravel fractions) [38]. The decisive economic factor in this type of mining branch is the cost of transporting the output to the customer, hence locating of the collection sites is related to the proximity of the markets and major transport routes. In addition, river gravel is usually of high enough quality to be classified as a product suitable for use in the production of concrete from Portland cement [39]. The material as a result of river transport is rounded, already granulated, well sorted and free of impurities. The exploitation of aggregates from the river bed requires small financial outlays [38]. In the case of mountain rivers, due to the ease of access to the aggregate exposed in the form of rapids, it is enough to use mainly excavators and vehicles used for transport. Alluvial sediments can also be exploited from neighbouring floodplains and older terraces. Another factor favouring the abstraction of river channel sediments is the reclamation of sediment, associated with periodic replenishment during floods [38], or in the case of the lower Vistula, excessive sediment accumulation caused by local hydromorphological conditions of the channel or hydrotechnical buildings [40, 41].

The material, transported by rivers in the form of clastic debris, originates mainly from mass movements, soil erosion, as well as from the erosion of the riverbed and the riverbanks itself [8]. Nowadays, cities are becoming more and more important in the supply of clastic material to riverbeds. They are an efficient source of anthropogenic material coming from many times larger surfaces than in natural conditions [42]. An important source of direct supply of clastic material for fluvial transport, under natural conditions, is a riverbed with variable bottom and intensively eroded riverbanks. The supply of clastic material transported also in the form of a suspended material increases many times when, under conditions of training of the channel (a regulated riverbed with river grones), it quickly deepens its bottom [36, 39, 43–48].

Currently, the rehabilitation of degraded and changed aquatic ecosystems has become an integral part of the river basin and river sediment management. For this reason, various sediment management projects have been initiated or fulfil in few countries [49]. Some examples include creation of lateral channels along the Rhine River [50], reconnecting the Danube old-arm system to the main channel [51], modernization and demolished of river groines to renew sediment dynamics in the Mur River [52], artificial replenishment of sediments the Ain River [53], and removing dams in US [36] to re-establish the river continuum. Sand and gravel are artificially replenishment to reinforce available spawning sediment supply below dams on a minimum of 13 rivers in California (Sacramento, Merced, Tuolumne, Mokelumne, Stanislaus, Russian, Dry, San Juan, San Luis Rey, Cache, Dry, Stony, and Tujunga Wash Rivers) [54]. The need for completing such hydrotechnical works arises from the necessity to implement the wants introduced by the WFD consisting within the implementation of procedures ensuring the minimization of negative impact of hydrotechnical structures oriented towards control and good conditions of surface waters and related aquatic ecosystems. One among the procedures recommended for the development of dam reservoirs requests maintaining the continuum of sediment transport or sediment replacement of the river downstream the dam. The term river sediment replenishment denotes the implementation of varied technical solutions that ensure constant replenishment of river load deficiency within the reach downstream of the dam, adapt to the sort of river competence and therefore the sort of flow operation by the dam. Such solutions include pumping systems, which move river sediments from the backwater zone of the reservoir to the reach below the dam, or directly shifting sediments downstream the reservoir by means of water transport e.g. employing a barge on plain rivers or carriage e.g. transport of gravel on mountain rivers. In western Europe and U.S. actions are undertaken to replenish the deficit of river sediments by employing point delivery of sediment material to the riverbed (Fig. 12.1), often directly below the dams [39, 55, 56]. The Ameri-cans, however, were the primary introducing this solution within the 90s, river sediment replenishing was already being conducted on 13 fragments of the Sacramento River in California

Fig. 12.1 A treatment on sediment replenishing below the Iffezheim dam on the upper Rhine. In the long-term, an average of 180,000 m^3 of gravel is dumped each year [55]. The implementation of the project is possible thanks to the cooperation German and French administrations and organisations responsible for the maintenance of the impounded section of the Upper Rhine since 1978 (photos prepared by M. Habel, Iffezheim in September 2016)

[54]. Investigation concerning the amount or dosage of river load downstream dams has been widely conducted in Germany and France—joint project on Rhine River below Iffezheim dam [55] and in Japan [49]. As it turns out, sustainable economic use of rivers requires extensive knowledge about the dynamics of sediments and their quality within river basins. Therefore, it is advisable to learn the natural hydrological conditions and to collect data on planned investments interfering with sediment circulation. The main goals of developing sediment management policies are to solve the following key problems in river basins:

- agricultural use of floodplains;
- dredging for navigation purposes;
- supply of drinking water;
- hydropower generation;
- nature protection and flood protection.

12.2   Sediment Dynamics in the Vistula and Oder River Basins

The Vistula and the Oder River drain about 18% of the Baltic Sea Basin and drain 97.9% of the territory of Poland: the Vistula (54% of the country's surface area) and the Oder (33.9% of the surface). Nowadays, Polish National Water Management Authority (Krajowy Zarząd Gospodarki Wodnej) classified more than 100 reservoirs in the Vistula and the Oder basin, but there are 58 main reservoirs with the storage capacity more than 2.2 million m^3: 36 in the Vistula and 22 in the Oder. Their estimated average total surface area is 944 ha, and average total storage capacity is 64.4 million m^3. The average flood storage capacity is 18.8 million m^3 while the total value for all reservoirs calculated for both basins is 0.864 km^3, which represents only

2% of the annual river flow of the Vistula River (33 km^3/year) and the Oder River (16.9 km^3/year) to the Baltic Sea. Most of the large reservoirs are located in lowlands. Hence the average height of the water reservoir is 18.5 m. Between 1945 and 1990 at the time of the communist party in Poland, 50% of present dams were built upstream, as the so-called 'Headwater Dams,' to generate hydropower, to prevent floods and to assure shipping. Up to one-third of the facilities in the Oder River basin was created before World War II. Only 4% of reservoirs were built before 1939 in the Vistula River basin. On the Vistula, there are three artificial reservoirs along its course: Czerniansk, Goczalkowice, and Włocławek. Oder River is an important waterway, navigable throughout most of its length. There are ten dams on the upper reach. The first dam is in Kędzierzyn Koźle, the last in Malczyce (commissioned in 2018).

It is assumed that the average annual share of the dissolved load discharged by rivers from Poland to the Baltic Sea constitutes approx. 85.6% of the total load and the suspended load and bed load is respectively: 8.3 and 6.1%. However, this clastic sediment (suspended and dragged in the channel) is a physical problem in the form of sediment, which should be rationally managed [57].

It is assumed that the largest two Polish rivers Vistula and Oder provide the sediment of the Baltic Sea sludge in the form of clastic sediment (suspended and bed load) of about 2.5 million tons per year [58]. During the culmination of the largest floods, Vistula River transferred around 9.55 thous. m^3·s^{-1} to the Baltic Sea., Oder River 3.3 thous. m^3·s^{-1} [59]. In 1960–2010, eleven great floods occurred in Poland. During the largest snowmelt flood, which covered a large part of northern Poland in March and April 1979, a flood wave that swept the Vistula River into the Baltic Sea had a volume of 18.2 km^3. It constituted approx. 53% of the average annual outflow from the Vistula basin [60]. Analysis of characterizing the dynamics of clastic sediment is carried out separately for the material transported in the form of a suspension and dragged in the channel.

12.2.1 Sediments Transported as a Suspended Load

The Suspended Sediment Concentration (SSC), also known as the turbidity, of the Polish rivers, shows regional diversity, determined by the morphogenetic conditions of the catchment area and the degree of anthropopressure. As it turns out, in the measurement cross-sections of rivers in the Vistula basin is much larger than in the rivers of the Oder basin [61]. In the Vistula basin, the greatest turbidity of waters (SSC >400 mg·l^{-1}) is characteristic in catchments basins of mountain rivers flowing out of the Carpathians [62]. The middle course of the Vistula River is characterized by a relatively high concentration of suspended load (SSC = 105 mg·l^{-1} in Puławy and 134 mg·l^{-1} in Warsaw) as a consequence of the supply of sediment from the Carpathian tributaries [63]. In the lower section, SSC decrease below 50 mg·l^{-1} (in Płock 43 mg·l^{-1} and Toruń 15 mg·l^{-1} [63]). The mountain tributaries of the Oder River are characterized by a lower mean concentration of suspended load (below 50 mg·l^{-1}) from the upper Vistula River tributaries. This is due to the higher resistance of the Sudetes surface to erosion. The Oder river turbidity increases in the middle part as a result of the supply of post-mining waters from Upper Silesia and the Ostrava region (in Miedonia 166 mg·l^{-1}, Brzeg 109 mg·l^{-1}), however, in the further course the turbidity is clearly decreasing, and in the lower section in the Gozdowice, the average value is about 6 mg·l^{-1} [61].

The Suspended Sediment Load (SSL) from the Vistula mountain catchment area is one order of magnitude higher than the upland and lowland basins. In the longitudinal profile of the upper Vistula, a systematic increase of the suspended load is observed [29, 61, 63]. In the upper section of the Vistula, from the profile in Skoczow— upper Vistula (SSL 10.4 thous. tons·year^{-1}) to Pulawy—middle reach (SSL 1.7 million tons·year^{-1}), there is a systematic increase in load associated with the supply of sediments from the Carpathian tributaries. In the middle part, from Pulawy to Płock, accumulation of the transported sediments is observed, as the SSL value is systematically decreasing, so in Płock profile SSL equals 0.7 million tons·year^{-1} [30, 64]. In the lower section, below the Włocławek Reservoir, suspended sediment load is about 0.35 million tons·year^{-1}, which is related to the reservoir accumulating about 42% of suspended sediments [27]. The systematic increase of suspended sediment load is strongly related to the erosion zone below the Włocławek Reservoir. The suspended sediment deficit in the waters of the Vistula River, which was previously accumulated in the reservoir, enforces increased erosion and transport. In literature, this phenomenon functions under the name "hungry water" [39]. However, the load of the sediments to the estuary of the Vistula is no longer rebuilt to the level which occurs above the water dam in Włocławek [40].

The Oder River, in terms of sediment transport, is characterized by not great variation in the suspended sediment load in the longitudinal profile. Despite a 20-fold increase in the catchment area and the 12-fold increase in water flow, the SSL is 0.33 million tons year^{-1} in the upper section (Chałupki profile) and 0.35 million tons year^{-1} in the estuary profile of the river [61].

Transport of suspended sediment in rivers of various regions of Poland shows high variability in the annual cycle, conditioned by climatic, hydrological and vegetative conditions. The most obvious regularities are observed in mountain rivers, where the maximum discharged loads of the suspension fall in June and July (50–70% of the annual sediment's mass). The second maximum in these rivers appears in March and April as a result of the thaw (Fig. 12.2). In the longitudinal profile of the Vistula, there is a large seasonal variation in the transport of the suspension, depending on the delivery of the sediment from the tributaries. In the upper and middle course, it dominates in summer, in the lower courses, the maximum can occur every month [61]. Both on the Oder and the Vistula River, the maximum transport of the suspension takes place during the maximum flood overflow or in the propagation phase of the flood wave [10]. During floods lasting up to several percents of days a year, over 95% of the annual load is carried, and during catastrophic floods—even 10 times more than in the average year. In the upland catchment areas, the share of flood in discharging the sediment reaches 70% of the annual load [65], and in urban basins exceeds 90% [66].

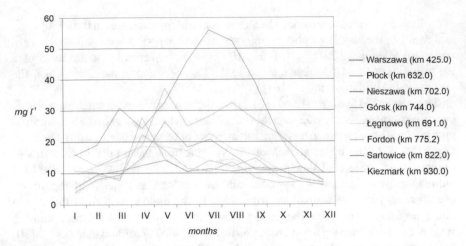

Fig. 12.2 The average monthly values of the sediment in the longitudinal profile of the Vistula River (km 694–822) in the years 2000–2010

12.2.2 Sediments Transported as a Bed Load

The lower sections of the two main rivers of Poland, Vistula and Oder River, are receivers of almost the entire load of bed load material from the territory of Poland. Accumulation of bed load material is most visible to the naked eye within the funnel-like artificial estuary of the Vistula. Research on this fragment of the Vistula's delta showed that the inflow cone, shaped from 1894 to 2000, reached the volume of approx. 130 million m^3, so the average growth rate of sediments over 105 years amounted to approx. 1.27 million m^3 per year, which calculated by weight means is more than 2.0 million tons per year [67, 68]. Based on the Skibiński formula [69], the determined size of bed load in the middle section of the Vistula River, in the Warsaw Nadwilanówka profile is about 504 thous. m^3 in the average year (about 0.9 million tons), while in the dry year 350 thous. m^3 (about 0.63 million tons). In the lower reach of the Vistula, from the estuary of the Narew to the backwater of the Włocławek Reservoir, the average annual bed load reaches approx. 2.0 million tons. In wet years, it is about 4 million tons, and in dry years—about 1.0 million tons [17]. Włocławek Reservoir interrupts the continuity of transport of bed load, but below the reservoir, there is its renewal as a result of channel incision [40, 41]. Regeneration of sediment transport is manifested in the river reach from Toruń (transport of bed load material 1.5 million tons) to Tczew (2.0 and over 0.6 million tons).

The Oder River, unlike the Vistula River, constitutes a more uniform river system (in the upper part built up by water dams, while the middle and lower parts are regulated), characterized by a smaller spatial differentiation of the bed load transport, with its constant increase downstream. Unfortunately, data on the transport of sediments in this river are either lacking or unsystematic. According to the studies of Pasławski [70], the average annual volume of bed load transported in the central

Oder (in Słubice cross-section in the period 1951–1969) was 137 thous. m^3 (about 0.22 tons per year), in the dry year 48 thous. m^3, and in the wet one—283.8 thous. m^3. Referring these values to the Vistula River in the Warsaw section (regulated) in the Nadwilanowka profile [69], it gives twice as much bed load sediment in 1 m^3 of water than the Oder in Słubice [70]. It should be added that the Slubice profile closes the Oder river basin area—53,580 km^2 (584.1 km of the river's course), and the Vistula Nadwilanowka 84,884 km^2 (504.1 km), so a comparison of the data is justified. This diversification of bed load transport results from differences in the management of the river channels of both rivers, where the Oder channel does not emerge point bars even at average low water levels, whereas in the Vistula riverbed they are common already at medium states. Also, data regarding the transport of bed load of both the Oder and especially Vistula should be considered as understated, as compared to those provided by Babiński [40], Babiński and Habel [71], which results from e.g., using different research methods. Because the amount of bed load sediment is affected, among others, by its feeding area, i.e. basin area and flow (transport capacity), and these parameters are respectively for the Vistula 194,424 km^2 and 1080 $m^3 \cdot s^{-1}$, and for the Oder 118,861 km^2 and 574 $m^3 \cdot s^{-1}$. Therefore, it can be concluded that Oder River has about 57% of the transport potential of the Vistula River. Assuming that the transport of the bed load sediment at the estuary of the Vistula is currently about 1.2 million tons per year, then the Oder transports almost 0.7 million tons/year. However, taking into account the fact that Vistula River has twice as much sediment in 1 m^3 of water than Oder [66], then the latter transports about 350 thousand tons of sand-gravel formations annually, which makes the Oder, in terms of morphodynamic conditions of the riverbed, friendlier for navigation.

12.2.3 Human Interference in the Dynamics of River Sediment Transport

Major changes in clastic sediment transport in the Polish rivers result from the impact of dams [27]. For example, the reservoir in Solina on the upper part of San captures as much as 99% of the suspension transported by this river (average from the dam operating period), and other deep reservoirs located in the Carpathians and Sudetes over 80% [61]. In shallow reservoirs located in different regions of Poland (Czaniec, Sromowce, Czchow, Myczkowce, Goczalkowice, Włocławek, Sulejow, Debe, Jeziorsko), erosion of sediments during rapid floods or quick release of water from the upper reservoir resulting in increased transport of the suspension in the river below the dam [18, 61]. Another effect cause by the impact of dams is a reduction in the variations of the transport of the suspension during the year, which results from the highest accumulation of material in reservoirs during numerous floods of low intensity. Significant changes in the sediment load in the longitudinal profiles of particularly smaller rivers may result from the impact of sewage discharge. Below the dam in Włocławek, the suspended sediment transport in Vistula is reduced by

40–44% and the bed load by 99% [27]. The Włocławek dam is located approx. 250 km from the estuary reach. Commissioned in 1968, it is the largest reservoir on the Vistula basin (area of 70 km², length of 55 km, storage 270 million m³). As the only one was completed in the plan of the cascade of the lower segment of the Vistula. It had an impact on the sediment transport balance. In consequence of the sediment transport deficit downstream the dam, and accumulation zone has formed (and continues to form) downstream the erosion zone. At this point, the river, being "saturated" with bed load, discharged it down its course to form, along with the "diminishing" training reach below, a braided channel, where bed load transport reaches up to 2.0 million tons per annum. Otherwise, the front of the erosion zone, moving down the river at a rate of 1.1–2.2 km year^{-1} [22, 40]. At present, the erosion zone front is located at—approx. 70 km below the dam [41]. According to long-term forecasts, this dam will visibly impact on the coastal zone around the year 2100 [17].

Many of the dredging works are the result of activities National Water Management Board "Polish Waters" (PGW Wody Polskie), which undertake a number of intervention activities on a regional and local scale aimed at clearing river channels. Flood safety in winter requires maintaining a proper depth on the lower Vistula for the operation of icebreakers. Every autumn, around 300–500 thous. m³ of sediments from the regulation of rivers and reservoirs are extracted in the lower section of the Vistula River. Generally, projects for the extraction of bed load sediments from rivers and reservoirs are conducted for commercial and non-commercial purposes. Non-commercial projects include the collection of dredges from the canals of river basins in order to maintain the necessary operational parameters and flood safety. Exploitation of aggregates from the Vistula River is usually of commercial nature, which enables a reliable assessment of the amount of material extracted from reservoirs. It is estimated that the current extraction of aggregates from these reservoirs is about 2.5 million m³ annually.

The most frequent is the dredging of sediment material of rivers deposited in the backwater zones of water reservoirs (e.g., at the estuary of the Dunajec River to the Czorsztyn Reservoir), which is used for the production of various types of construction aggregate. According to the Polish Association of Aggregate Producers, it is estimated that in Poland in 2012–2020, about 230 million tons of aggregates will be needed to build public (local and municipal) roads and rail routes, including 137 million tons of gravel aggregates and 240 million tons of earthen masses, rock masses and other materials for embankments, access roads, driveways [72]. Only on the lower Vistula commercial exploitation of sand and gravel sediments in 2005–2018 took place in at least seven locations: Toruń, Solec Kujawski, the area of Bydgoszcz—Fordon, Chełmno, Grudziądz, Korzeniewo, and Kiezmark. However, this collection collides with the areas covered by Nature 2000 program. The lower Vistula can transport during the wet hydrological year up to 4.0 million tons of debris above the Włocłwek Reservoir and over 1.5 million tons on the regulated section in the cross-section Toruń [36]. The dredging of more sediments than the given limit value may contribute to the disturbance of the ecosystem functioning of the river in the estuary section.

Anthropogenic sources of dredge origin are mainly wastewater. It is estimated that in Poland, the amount of industrial wastewater and municipal wastewater discharged into surface waters every year in 2005–2014 was from 8.9 to 9.2 km³ [73]. An example of an urban reservoir filled with wastewater dredge is an artificial urban reservoir called "Regatta Track" on the Brda River in Bydgoszcz. The research shows that between 350 and 500 thous. m³ of sewage sediment accumulate in the upstreamed section of the river [74]. In the past, Brda in Bydgoszcz was a receiver of untreated municipal, industrial and rainwater sewage. At the beginning of the present century, as much as 16.6% of the sewage discharged here was raw wastewater [75].

12.3 Sediment Transport Monitoring

Methods of research on suspended sediment transport were related to monitoring institutions in Poland. From the 1950s to the 1990s, the measurements of water turbidity were conducted by the National Hydrological and Meteorological Institute (PIHM)—from 1973 as the Research Institute of Meteorology and Water Management (IMGW). The monitoring network was connected with the stations of the basic hydrological observation network. Observations of water turbidity took place only on the part of the stations (Fig. 12.3). After Poland's accession to the EU in 2004 and the ratification of the records of the WFD, suspended sediment monitoring has become part of the "Państwowy Monitoring Środowiska" (SEM), constituting one of the elements determining the ecological status of waters [76]. At the local level, monitoring is carried out by Voivodship Inspectorates for Environmental Protection. The number of posts has increased significantly—1106 in the water cycle 2010–2012 [77].

Sampling by IMGW was carried out according to the methodology developed by Brański [62] with the help of the slowly filling bathometer. This method is one of those commonly used in sedimentology [78]. Depending on the type of river and the amount of water run, sampling took place at single-point or many-points in the cross-section of the river channel. A sample of water with a minimum volume of 2 l was collected in 15–30 s. The sample collection within the SEM with a volume of 1 l, takes place only at one point in the transverse profile of the channel in accordance with the records of the Polish/European Standard [79].

The laboratory markings for both systems consisted of the weight markings of the amount of the total suspended material in the water sample in mg·l⁻¹ by drying it. The differences in the applied methods resulted only in the volume of the filtered sample (IMGW—2 l, SEM—1 l) and the type of filters used (IMGW—paper filters, SEM—glass fiber filters with pore size: 0.7 μm). The marking method of suspension was done according to Polish/European Standard [79] used outside Poland also in: Austria, Belgium, Cyprus, Denmark, Estonia, Finland, France, Greece, Spain, Netherlands, Ireland, Iceland, Lithuania, Luxembourg, Latvia, Malta, Germany, Norway, Portugal, Czech Republic, Slovakia, Slovenia, Switzerland, Sweden, Hungary, Italy, and the UK.

The laboratory markings ... in the water sample in $mg \cdot l^{-1}$ by drying it.

Network stations:
- IMGW
- SEM

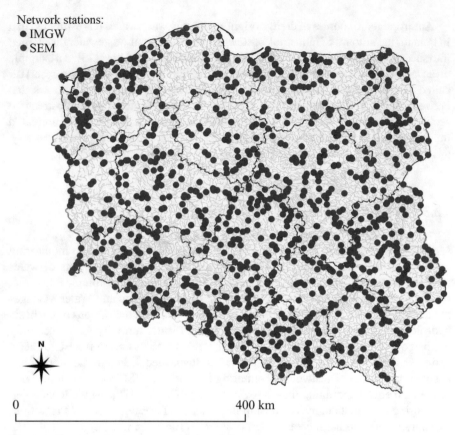

Fig. 12.3 Location of water turbidity stations in the IMGW and SEM networks (map prepared by D. Szatten)

Significant differences occur at the frequency of sampling in the described methods. Measurements within IMGW were dependent on the type of river (broken down into mountain and upland, lowland and watercourses linking lakes) and water levels. On the other hand, within the SEM, the frequency of sampling results from the type of program implemented for a given river water bodies—surveillance, operational or monitoring of protected areas [76]. Detailed data on the frequency of sampling are presented in Table 12.1.

It is worth noting that despite the clear increase in the density of the monitoring network, the quality of the data collected has deteriorated. The IMGW data series of almost continuous character has been replaced by almost seasonal measurements. Also, there is a lack of standardized methods for continuous monitoring of water turbidity in Polish rivers using turbidimeters.

Table 12.1 Frequency of suspended sediment concentration measurements IMGW and SEM

Water stages	IMGW			SEM		
	Mountain, Highlands	Lowland	Between lakes	Surveillance	Operational	Protected areas
<SSW	73 year^{-1} at 7 AM			12 year^{-1}	8 year^{-1}	4–8 year^{-1}
All	365 year^{-1} at 7 AM		182 year^{-1} at 7 AM			
SSW-SWW	365 year^{-1} at 7 AM		182 year^{-1} at 7 AM			
>SWW	730 year^{-1} at 7 AM and 6 PM		365 year^{-1} at 7 AM			

12.3.1 Mass, Volume and Intensity of Bed Load Sediment Transport

In alluvial rivers, such as the central and lower Vistula, the lower Oder bed load measurement is very difficult. However, it is often considerably affected by the subjective factor, which arises from the applied measurement method and its analysis. For instance, the same type of sampler used at a single sampling site provided results ranging from 5 to 725% [80]. Therefore, even at the initial stage of bed load transport analysis, the measurement error several times exceeds the absolute value. Hence, the available data related to the occurrence in question show considerable discrepancy and may be further distorted due to spatial variability of geological structure and riverbed morphology. This needs to be accounted for when selecting the sampling points. From the 1950s to the 1990s, measurements of bed load transport were conducted in selected sections of rivers by the National Hydrological and Meteorological Institute (PIHM)—since 1973 as the Research Institute of Meteorology and Water Management (IMGW). The research was carried out mainly with the use of net bathometers and catchers (traps) installed permanently in the whole cross-section of the channel (small watercourses) or temporarily functioning in various places of the channel [80, 81]. Field measurements of sediment marked with the use of indicator substances, so-called tracers: fluorescein, isotopes, magnetic and radioactive materials, magnetic detectors [10, 82]. An effective method on big rivers is the observation of the dynamics of sand forms (river bars) [27, 83, 84]. One of the simplest approaches for characterization of river's morphological changes is depths measurement and volumetric analysis of planimetric parameters. The features that are considered to be planimetric are those elements that are non-dependent on elevation and are represented on two—dimensional maps [85]. Among others, this method was used by Łomniewski [67] to calculate the sediment increments in the Vistula estuary to calculate the sediment load of the Vistula. In turn, the volumetric method is based on the use of bathymetric data developed in the form Digital Elevation Model (DEM). As it is created on the basis of depths measurements, DEM is considered to be a reliable and accurate mean of riverbed projection [48]. Application of DEMs into the assessment of riverbed's topographical variations provide insight on elevation changes due to erosion or deposition processes [86, 87]. To detect such changes, volumetric analysis using DEMs are based on comparison of two DEM data sets collected for two different periods, as only data sets captured for two different periods can result in the estimation of land loss or gain for a vast area [88]. Regarding analysis in GIS framework, there are two approaches commonly used for DEMs volumetric analysis. First one is related to the creation of DEM of Difference (DoD), whereas the second one is a comparison of cross-section profiles created based on developed DEMs. DoD approach is based on a mathematical algorithm for assessment volumetric changes of landforms on two different periods [89]. Creation of DoDs usually is related to subtraction of one elevation model from another to disclose a mosaic of morphological change.

The intensity of sediment and its load is also determined by indirect methods (empirical formulas), requiring knowledge of the hydraulic parameters of the channel and the granulometric characteristics of sediment grains [90]. The Skibiński formula [77] was developed and used on the Vistula in Warsaw. According to Skibiński, the amount of bed load is directly related to the water flow function:

$$G = 0.005\,Q^{1.286} \tag{1}$$

where

Q   water flow in m^3/s,
G   the size of bed load in dm^3/s.

12.4   Sediment Dynamics Investigation—The Vistula River Case Study

In the face of government plans for the development of the Vistula and Oder Rivers, there is a need for conducting pilot studies on the assessment of suspended and bed load sediment transport. On the Vistula River in the years 2020–2030, investments are to be implemented that affect the continuity of sediment transport. On the lower section, a new dam with a reservoir is to be built—a continuation of an unfinished cascade, and on the middle and lower Vistula, a governmental development program for inland waterways is planned [91]. In order to meet the planned investments, research on the dynamics of sediments in the Vistula channel is conducted mainly on the lower Vistula section [17, 64, 92]. As part of the research, the latest technical developments (measuring devices) and research approaches in the field of sediment monitoring are implemented.

12.4.1 Measurements of Suspended Load Concentration

The main part of the research is based on measurements of suspended load concentration in the longitudinal profile of the Vistula waters at the navigable reach of the river (on the distance more than 900 km). Research on total suspension was carried out in relation to two spatial aspects:

- heterogeneity of load in single cross-sectional profiles—to identify bottom and lateral erosion zones,
- heterogeneity of load on the longitudinal profile—to define the influence range of erosion processes/accumulation zones. We used mobile turbidimeter HACH 2100QIS (in FNU) and LISST 25X sensor (Laser In Situ Scattering Transmissometer manufactured by Sequoia Scientific Inc.) for suspended load measurement devices able to measure concentration in mg·l^{-1} and to differentiate specific grain size fractions. The instrument measures particles in two classes. In the range of 2.5–63 micron, and the other in the range of 63–500 micron. Turbidimeters are easy to use and most popular devices for suspended sediments monitoring [93].

From the research carried out in May 2018, in the longitudinal profile of the Vistula, several characteristic fragments can be distinguished, with a different course of the water turbidity. The recorded values reflect the sources of supply and the extent of the basin's transformation. The first section (0–92 km) is characterized by the presence of a cascade of retention reservoirs (Dwory, Smolice, Łączany, Kościuszko, Dąbie, and Przewóz), which affects the large variation of water turbidity (10–60 FNU, average 20 FNU). Then (up to approx. 420 km of the river's course) there is a systematic increase in water turbidity as a result of the supply of suspended sediment from the Carpathian tributaries, reaching a level of about 40 FNU. Between the estuary of the Wieprz and Pilica Rivers, there is a zone of increased accumulation of suspended load, which is marked by a clear decrease in water turbidity to about 30 FNU (Fig. 12.4). Below Warsaw, the influence of transformations of the Vistula river bed and local sources of sediment supply is noted. At km 551, the turbidity of the Vistula is once again drastically reduced (to about 20 FNU) as a result of the Narew River water supply—with slight turbidity (accumulation in the Zegrzyński Reservoir). Up to 624 km of the Vistula River (Płock), there is a systematic increase in the turbidity of waters because of the impact of the backwater zone of the Włocławek Reservoir. The turbidity of water in the reservoir (up to 672 km) drops even to the level below 10 FNU (Fig. 12.4), which reflects the accumulation role of hydrotechnical constructions in suspended sediment transport. Below the water dam of Włocławek, up to the estuary, the turbidity of waters undergoes systematic, though slight increase (up to about 15 FNU), which indicates the renewal of suspended sediment.

Investigations of the particle size distribution of suspended sediments in the longitudinal profile of the Vistula River may enable diversification of its supply sources, and in the cross profile determine the impact of hydrological and morphometric conditions on the dynamics of its transport. At selected sites in the longitudinal profile of the Vistula: Płock, Bydgoszcz, Kiezmark (Fig. 12.5) research was conducted at two

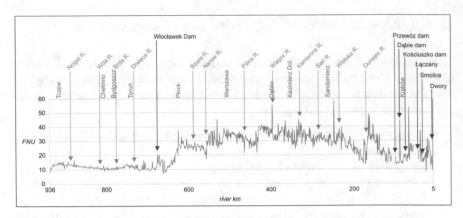

Fig. 12.4 Water turbidity in the longitudinal profile of the Vistula River. Measurements were taken during the rafting down the river on days: 2018 May 12–25

depths: in the surface layer and above the river bottom. For this purpose, the laser grain composition measurement sensor LISST-25X was used. It is a device, like the above mentioned turbidimeters, for measuring the transparency of water in real-time [94].

The volume concentration of suspended load in the longitudinal profile of the Vistula in the surface layer decreased along with the course of the river from the level of about 225 ml·l^{-1} (Płock) to about 40 ml·l^{-1} (Kiezmark). In the bottom layer, a higher volumetric concentration (in the range of 10–50 ml·l^{-1}) than in the surface layer was observed at all sites. This is related to the nature of the suspension grains transport in the depth profile of the channel. Also, the average grain size of the suspension in the longitudinal profile of the Vistula was reduced from about 30 μm in Płock to about 20 μm in Kiezmark (Fig. 12.6). The average grain size of the suspension in the zone above the river bottom also decreased along the Vistula— from over 30 μm (Płock) to about 23 μm (Kiezmark). This shows a decrease in the river's transport capacity. The greater variation in the average grain size of the suspension in the bottom area of the channel (Fig. 12.6) indicates a less homogeneous sediment nature.

In all transverse profiles studied, the volume concentration of the suspension was higher in the mainstream zone—both in the surface layer and above the river bottom. This shows the connection of the way of transport of the main sediment with the kinetic energy of flowing waters and the transit nature of transport of suspension in the lower Vistula. The occurrence of the bottom sediment resuspension process in the riverbed during the transport of the sediment shows the variation between the average size of the suspension grains in the range above 63 μm (Fig. 12.6). In the thalweg zone (deep water), there is not much variation in the course of this indicator between the surface and the bottom zone. Large discrepancies between the average size of the suspension grains in the range above 63 μm in these zones refer to the

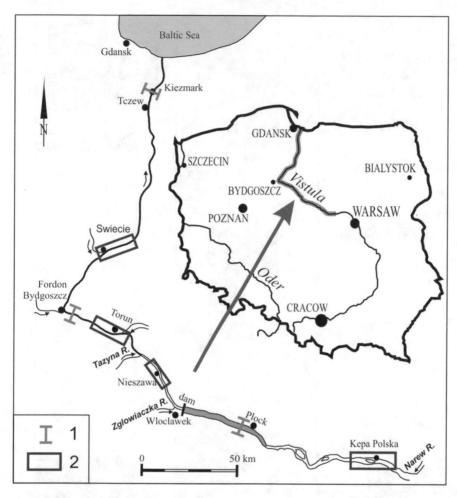

Fig. 12.5 The particle size distribution of suspended sediments (1) and dynamics of channel forms observation posts (2) on the Lower Vistula River

shallower parts of the channel. This proves the inclusion of the local material in the sediment transport, activated from the shallower parts of the channel.

12.4.2 Bead Load Transport—The Study on the Dynamics of Channel Forms

For large rivers, such as the Lower Vistula, it is possible to use the dynamics measurement methods to assess the size of the sediment load transported in the form of bed load. Bed load transport within the middle and lower Vistula being examples, is

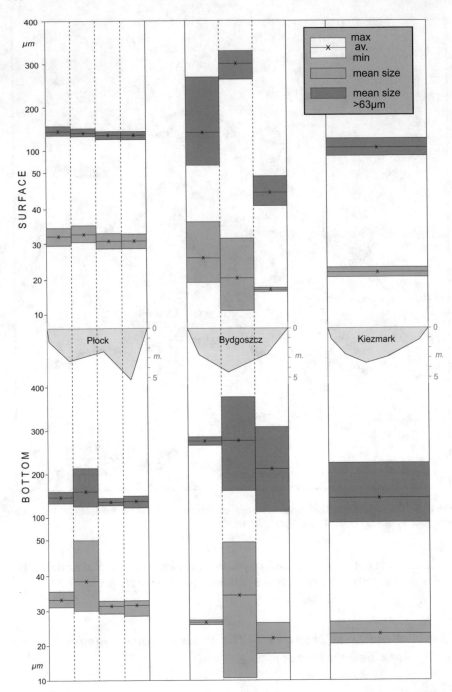

Fig. 12.6 The average grain size of the suspended sediment at selected stations in the longitudinal profile of the Vistula River

represented by two groups of riverbed forms: micro- and mesoforms [83]. The former constitutes a dynamic surface layer of mesoforms. Due to their small size (several centimetres high and wide, several meters long), and their tendency to permanently relocate and transform, they're inherently difficult to research outside of the lab. The said forms are commonly referred to as river bars, and their size usually corresponds to channel width or exceeds it, whereas their surface height coincides with average water levels—MWL [83]. They have a tendency to be stable and inert and may maintain their form over many seasons showing variable hydrological regime. They have a tendency to occur as individual, large sandy-gravel swells or bars and are typically accompanied by negative forms—pools.

The authors were ready to pursue the study objective by means of employing methods based on their experience [17, 27, 40]. In the case of big rivers, e.g. the Vistula, they involve performing measurements of sandbar dynamics at the selected segments of the river, as well as measurements of river load accumulation growth in delta zone of the Gdańsk Bay. The data were supported by depth maps of the selected segments and analysis of aerial pictures collected between 1989 and 2015. The method takes into a count within the monitoring of river bar dynamics and comparison of depth changes on bathymetric maps allows for estimating bed sediment transport capacity in cubic meters. In order to calculate the load weight, it was assumed that $1 \, m^3$ of sandy sediments weighs approx. 1.8 tons. Investigation on river bars dynamics was conducted in the years 1981–2014 over a 220 km-long fragment of the Vistula. The detailed research occurred at a characteristic part of the riverbed (Fig. 12.5), which is partially channel training and features a single dam. Movement of sandbars was made on the basis of own field methods (Fig. 12.7), including geodetic surveys carried out at the unregulated reach near the Warsaw (Kępa Polska section), as well as at the sediment deficit fragment downstream the dam in Włocławek. The second, 25 km far from Włocławek, a section in Nieszawa village (km 700.00). Nieszawa river section constitutes a maximum range of an erosive reach of channel. The third study fragment runs through the Toruń, 60 km downstream of the dam. It represents an initial, training section of the Vistula with a similar width and alternate diagonal river bars. The last fragment is located near the Swiecie, it is 140 km downstream of the dam. In this analysis, both cross—and longitudinal profiles of the channel proved to be greatly helpful in preparing the depths maps of the riverbed. The study was supplemented with sediment sampling and granulometric analysis composition of the river bars.

Systematic research on river bars led to the development of a model of dynamics of these forms and allowed for the estimation of river bedload transport. Field research and the analysis of available aerial pictures allowed for the identification of the following river bars in the Vistula: at the training section—lateral bars and mid-channel bars. The elevation of their surface corresponds with the average annual water stages. The linguoid bars, whose elevation coincides with average low water stage and alternate-diagonal bars that in terms of elevation fall between the abovementioned forms. Generally, each increase in water flows corresponds to a proportional increase in the river bars dynamics and vice versa. The analysis of riverbed fragments featuring different development levels shows that the front of river bars of the same kind moves

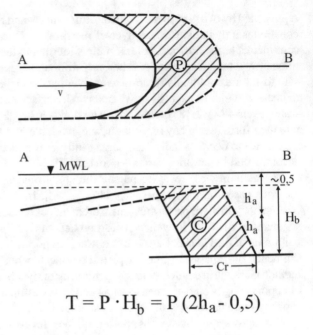

Fig. 12.7 Scheme of bottom transport value (T) based on measurements of alternate river bars dynamics at the training section of Vistula. Explanations: H_b—average sand-gravel thickness of river bar, C_r—rate of movement, P—the surface rate of movement (prepared by Z. Babiński)

$$T = P \cdot H_b = P\,(2h_a - 0{,}5)$$

at a similar rate, instead, this tends to change the more the bars differ in terms of, among others, the elevation of their body.

The conducted research [17, 40] shows that bed load transport rate at the training fragment, as a mean annual value for the years of 1971–1995, amounted more than 2.0 million tons, reaching up to approx. 4.0 million tons in the very wet year e.g. in Kępa Polska river section in 1975 and approx. 1.0 million tons in the dry year. These values were confirmed in the studies conducted at the estuary section of the Vistula, where according to Graniczny [68] the volume of the delta fan in the period spanning from the year 1894–1999 amounted to approx. 130 million m³. So, the mean sediment deposition rate over periods amounted to approx. 1.0–1.3 million m³ per annum, which explain into 2.0 million tons per year. Since the study in question also covered the period when the river estuary reach was affected by the training works performed in the nineteenth century, the value is bound to be undervalued. Therefore, it is believed that bed load transport over the entire study fragment of the Vistula oscillated between 2.2 and 2.3 million tons per annum. The study on the dynamics of river bars [40] indicates that bed load transport at the already training fragment of the Vistula ranges from approx. 1.0 million tons per annum in the Toruń up to 1.2 million tons per annum at the estuary reach. The approx. 20% increase in the supply of bed load over this 200 km-long reach arises from, to a lesser degree, (a) alimentation from the tributaries and, more importantly, (b) the fact that the river sediment transport is renewal downstream of the Włocławek dam. The said occurrence arises from the supply of material from the erosion of riverbanks [17].

12.4.3 Bead Load Transport—Volumetric Analysis

An even more accurate method of bed load transport assessment is a volumetric analysis of bathymetric maps. The date for the analysis was collected from 38 measurements on bed-elevation on the Vistula in Toruń (Fig. 12.5) within the period 2011–2019, within the section directly surrounded by the newly commissioned bridge. The measurements were collected from a motorboat equipped with a GPS rover for tracking the boat position and a single beam echosounder. Data was collected once by covering the water surface with dots every 5 m along a river section of 1900 m long, i.e. between 730.40 and 732.30 river kilometre. A digital sonar data record: x and y coordinates during a local PUWG-92 coordinate system and z depth were used to prepare a digital depths map for every single measurement within the sort of a TIN model, and then a DEM raster model—in reference to the average water level. Volumetric analysis was administrated on the example of an analysis of seven measurements carried out on July 2013 and July 2017 (Fig. 12.8). Volumetric analysis is a quantitative approach which involves measurement of the volume of specific factors.

The obtained result indicates the cyclicity of the appearance of bed forms in the studied section. The analyzes carried out showed that the full transition cycle of the river bar (sandy mesoform) in the researched section of the river is about 4 years. The volumetric analysis made it possible to calculate the total sediment transport of the bed load of Vistula River in this section of the river. In the period from June 2013 to June 2017, 2.35 million tons of sand and gravel sediments migrated here (annual transport of 0.585 million tons, which is about 330,000 m^3). It should be emphasized that the analyzed period was characterized by low flows (average for 4 years—853 $m^3 \cdot s^{-1}$). Hence the transport volume is relatively low.

12.5 Sediment Management Concept

Considering constant hydromorphological changes in river channels and water reservoirs caused by anthropogenic and natural influences, adaptation and optimization of sediment management strategies is an endless task. Therefore, in line with the WFD recommendations, measures are taken to manage the sediment on an ongoing basis. Sediment management in Poland is part of the tasks defined by River Basin Management Plans (RBMP) in river basin areas [95], which provide an overview of important water management problems. According to art. 315 of the Water Law Act [96] RBMP in the basin district is one of the main planning documents for programming and coordinating activities aimed at achieving or maintaining a good state Surface Water Bodies (SWB), sustainable use of water resources and improvement of flood protection and counteracting the effects of drought. Each RBMP includes, among others general description of the characteristics of the river basin district, summary of the identification of significant anthropogenic impacts and assessment of their

Fig. 12.8 An example of physical observation of the dynamics of sediment movement in the riverbed—successive bathymetric situations of the Vistula River for in Toruń (731.30 km). The brightest areas are sand bars (sand channel mesoforms), the darkest ones are bottom scours. The maps present the depth range from 0.0 to − 10.0 m

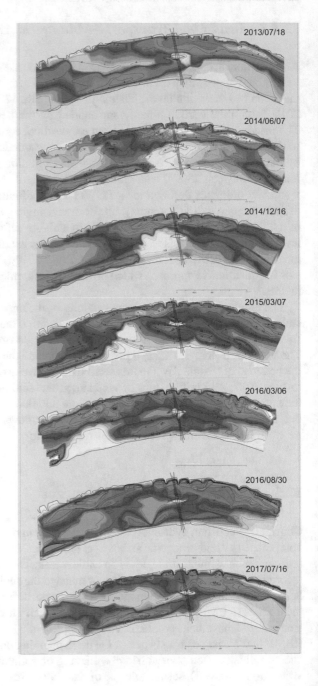

impact on the state of surface and ground waters, setting environmental objectives for surface water bodies and protected areas. Their also includes summary results of economic analysis related to the use of water and activities included in the country's water-environment program, a list of other detailed programs and management plans for the basin district, concerning the catchment area, economic sectors, problems or types of waters, including the content of these programs and plans. Each RBMP store information on methods and procedures for obtaining information and source documentation used to draw up the plan and information on expected results of the plan [97].

The records of the WFD require the development of RBMP in the river basin areas existing in a given country. Accurate identification of problems allows the application of appropriate action programs, implementation of which will be of key importance for achieving or maintaining at least a good condition of waters and ecosystems dependent on water. Important water management problems are the most important factors that hinder the achievement of environmental goals. When identifying important problems, the voice of the public is extremely important. This document, in accordance with the WFD assumptions and the Water Law Act [96], is subject to six-month public consultations for two years before the publication of subsequent (updated) RBMP.

In the first (2004–2009) and the second (2010–2015) planning cycle, the President of the National Water Management Board developed ten plans. Respectively for river basin districts: Oder, Vistula, Dniester, Danube, Jarft, Elbe, Niemen, Pregla, Swieza, Ücker. At present, Poland is in the third planning cycle (2016–2021) for which the President of the National Water Management "Polish Waters," is required to develop 9 plans—without the river basin Jarft and Ücker; with the Banowka River basin area. These documents make it possible to rationally manage water resources and to indicate actions aimed at keeping them unchanged for future generations. In 2016, the Council of Ministers adopted RBMP updates in the form of regulations, which contain provisions relating to, i.e. the two largest river basins in the country: Vistula [98] and Oder [99]. RBMP contains a detailed list of investments affecting the state of water, including activities directly or indirectly affecting the management of sediments in the catchment area. In addition to measures improving the quality of the aquatic environment, the planned investments also meet social or economic objectives. The most numerous group in the RBMP for the years 2010–2015 are flood protection projects that meet the requirements of art. 66 of the Water Law Act [96], i.e., "it is acceptable not to achieve (…) the failure to prevent deterioration of ecological status (…) as a result of new human activities consistent with the principle of sustainable development and necessary for the development of society". The types of investments, which due to their nature, are related to the direct impact on the dynamics of river sediment transport (Table 12.2), include:

- construction of dam reservoirs or damming constructions—activities that could adversely affect the SWB status,
- construction of a dry reservoir—neutral operation from the point of view of its impact on the SWB status.

Table 12.2 Number of planned investments, with the type of investments as construction of reservoirs, in individual water regions of River Basin Management Plans

Water regions	Number of investments
Small Vistula	2
Black Orawa	2
Upper Vistula	24
Middle Vistula	35
Lower Vistula	13
Lyna and Wilgorapa	3
Pregola	1
Niemen	4
Upper Oder	2
Middle Oder	12
Warta	16
Lower Oder and West Przymorze Rivers	18

The spatial distribution of investments planned for implementation in the second RBMP planning cycle in Poland is shown in Fig. 12.9. The most important investments affecting the continuity of river sediment transport are listed in Table 12.3.

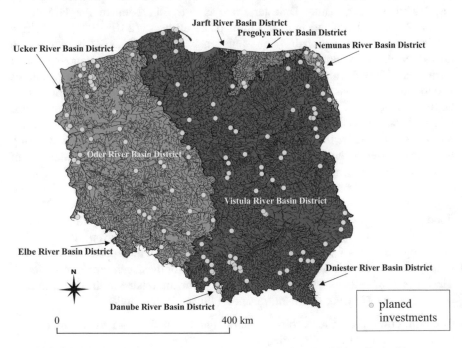

Fig. 12.9 Location of the planned investment in the second cycle of River Basin Management Plans on the background of SWB (map prepared by D. Szatten)

Table 12.3 Selected investments affecting the continuity of sediment transport listed in the updated River Basin Management Plans

Name of the investment	Description of the investment and discussion of the impact on the dynamics of the sediment
Protection against flood waters of the lower section of the Vistula River from Włocławek to its estuary to the Gdańsk Bay—reservoir and dam below Włocławek	In the area of the lower Vistula water region, the construction of a water stage (damming structure, flow reservoir, hydroelectric power plant) is planned in order to protect the Vistula River from its flood from Włocławek to its estuary to the Gdansk Bay. The planned completion date of the investment is 2021 (with a prospect until 2027). The cost of the investment is estimated at 3.02 billion PLN
	The reservoir will serve to improve water relations and counteract climate change, first and foremost to limit the effects of drought
	The main activities resulting in the achievement of environmental goals include, among others, stopping erosion below the dam Włocławek (through the construction of the weir with the use of the Jambora low-head dam and tributary basin) and elimination of unnatural daily fluctuations of the water level on the Vistula. Additional benefits include the creation of a Va class waterway along the length of a new reservoir, increased retention and production of electricity from a renewable source
	The following options were considered for this task:
	(a) a variant of worsening the existing state—intensifying the risk of construction disaster of the Włocławek water stage, riverbed erosion, the emergence of the ice jams, limitation of the possibility of exploiting the dam facilities, increasing problems with maintenance of river bank infrastructure, drainage of the river valley
	(b) variants of the flooding threshold—excluded, due to the increased risk of ice, ice, and frazil floods
	(c) variant of demolition of water stage in Włocławek—excluded due to social interest and environmental consequences (including reduction of groundwater level, utilization/risk of sediments flow deposited in the reservoir)
	(d) other technical and non-technical variants—construction of the polder is not possible due to the utilization of the surrounding areas
	(e) variant of construction of the water stage below Włocławek—the only one to meet environmental objectives

(continued)

Table 12.3 (continued)

Name of the investment	Description of the investment and discussion of the impact on the dynamics of the sediment
Construction of a dig channel connecting the Vistula Lagoon with the Gdansk Bay	In the area of the lower Vistula water region, the construction of a waterway connecting the Vistula Lagoon with the Gdansk Bay is planned. The planned completion date is 2023. The cost of the investment is estimated at 878 million PLN
	The investment is planned due to the overriding public interest, aimed at creating infrastructural guarantees of the Community freedom of free movement of goods and creating a new intermodal node between E60 and the TEN-T IA corridor
	On December 5, 2018, the Regional Director of Environmental Protection in Olsztyn determined the environmental conditions for the project entitled. "A waterway connecting the Vistula Lagoon with the Gdansk Bay - the location of Nowy Swiat" [100]. The investment is planned to include construction of a protective port from the side of Gdansk Bay, construction of a shipping channel with a sluice, construction of an artificial island on the Vistula Lagoon of the area 181 ha, construction of a fairway on the Vistula Lagoon and reconstruction of the fairway on the Elblag River. The analysis of the impact of the planned investment on the environment shows that with proper fulfillment of the conditions listed in the decision on environmental conditions, the implementation of the project should not have a significant negative impact on the environment

(continued)

Table 12.3 (continued)

Name of the investment	Description of the investment and discussion of the impact on the dynamics of the sediment
Construction of the Malczyce dam on the Oder River	In the area of the central Oder water region, the construction of the Malczyce water stage is planned to protect against flooding, prevent the effects of drought, protect the water dam in Brzeg Dolny from under-watering and loss of stability and to increase the use of the waterway by inland navigation on the section of Oder below the stage. The planned date of completion of the investment is 2021 (design work began in 1994, and construction works were commenced in 1997). The cost of the investment is estimated at 440 million PLN
	The investment includes three-span weir with hydraulically operated steel flap closures, fixed weir with a length of 256 m, a 190 m long shipping lock and a 9 MW hydroelectric power plant. Currently, the construction of the Malczyce water stage is at the final stage of implementation. The main activities resulting in the achievement of environmental objectives include among others: flood protection, protection of the Brzeg Dolny water dam against under-watering and loss of stability (erosion in the period 1925–1992 in the Brzeg Dolny cross-section is 314 cm and in Malczyce 216 cm), restoration of the level of groundwater lowered due to natural riverbed erosion below Brzeg Dolny. Additional benefits include development and improvement of the shipping conditions of the Oder River, strengthening its position in the Polish transport system

12.6 Recommendations

This publication aim is to help realize how important it is to have knowledge about the formation of sediments in the drainage basin and its transport across rivers. This hydrotechnical infrastructure has the strongest impact on the lack of continuity in sediment transport. Therefore, special attention should be paid to constant monitoring of the sediment circulation within the drainage basin area or along the riverbed. There is no guidelines for sediment management in the strategic documents of the institutions responsible for updating the RBMPs in Poland. A methodology to eliminate problems with excessive sedimentation in rivers and reservoirs has not yet been developed. The lack of a government sediment management policy generates huge costs each year related to the maintenance of hydrotechnical infrastructure. There are no guidelines as to which devices should be installed on reservoirs and the course of rivers in order to maintain the continuity of sediment transport for the water law permits under development. There is no national monitoring of reservoirs silting. Most frequently we learn about problems related to excessive sedimentation from open market ads for dredging works on reservoirs and river sections. Most often job advertisements involve reconstruction of the capacity of hydro-energy and recreational reservoirs. Available sources show that on average every 20 years dredging works were carried out on reservoirs, and every 8 years were carried out on navigable canals and rivers.

No data on sediment transport is currently being collected. Only until 1990, constant monitoring of suspended sediment transport was carried out by the Research Institute of Meteorology and Water Management (IMGW). The monitoring network was connected with the stations of the basic hydrological observation network.

New investments are being undertaken and planned for the coming years for which no specific requirements are indicated as to the continuity of sediment movement. For example for the Malczyce dam on the Oder river sediment replenishing below the dam was initially planned. Ultimately, this task was not started and development work on it ceased.

In our opinion, understanding the role of sediment and its proper management is necessary to achieve the environmental objectives of the Water Framework Directive, especially for water reservoirs to achieve a minimal good condition.

References

1. Weilbeer H (2014) Sediment transport and sediment management in the Elbe estuary. Die Kuste 81:409–426
2. Ramalhosa E, Pereira E, Vale C, Válega M, Monterroso P, Duarte AC (2005) Mercury distribution in Douro estuary (Portugal) Marine Pollut Bull 50:1218–1222
3. Mucha AP, Bordalo AA, Vasconcelos MTSD (2004) Sediment quality in the Douro river estuary based on trace metal contents, macrobenthic community and elutriate sediment toxicity test (ESTT). J Environ Monit 6:585–592

4. Syvitski JPM, Kettner AJ (2011) Sediment flux and the Anthropocene. Phil Trans R Soc 369:957–975
5. Aleksejevskij NI, Chalov RS (1997) Dviženije nanosov i rysłovyje processy, Moskovskij Gosudarstviennyj Universitet im MV Łomonosova, Moskwa
6. Knighton D (1998) Fluvial forms and processes: a new perspective. Co-published Wiley, New York
7. Summerfield MA (1991) Global geomorphology an introduction to the study of landforms, Longman Scientific and Technical. Wiley, New York
8. Klimaszewski M (1978) Geomorfologia. PWN, Warszawa
9. Bajkiewicz-Grabowska E, Mikulski Z (1996) Hydrologia i Oceanografia. PWN, Warszawa
10. Froehlich W (1982) Mechanizm transportu fluwialnego i dostawy zwietrzelin do koryta w górskiej zlewni fliszowej, Prace Geograficzne IGiPZ PAN, 143, s 144
11. Global Reservoir and Dam (GRanD) database—technical documentation, Global Water System Project, University of Bonn, Germany (2011), www.gwsp.org
12. Vörösmarty CJ, Meybeck M, Fekete B, Sharma K, Green P, Syvitski JPM (2003) Anthropogenic sediment retention: major global impact from registered river impoundments. Global Planet Change 39(1–2):169–190
13. Walling DE, Fang D (2003) Recent trends in the suspended sediment loads of the world's rivers Global Planet. Change 39:111–126
14. Meade RH, Moody JA (2010) Causes for the decline of suspended-sediment discharge in the Mississippi River system, 1940–2007 Hydrol Process 24:35–49
15. Giosan L, Syvitski JPM, Constantinescu S, Day J (2014) Protect the world's deltas. Nature 516:31–33
16. Syvitski JPM, Kettner AJ, Overeem I, Hutton EW, Hannon MT, Brakenridge GR, Day J, Vörösmarty CJ, Saito Y, Giosan L, Nicholls RJ (2009) Sinking deltas due to human activities. Nat Geosci 2(10):681–686
17. Babiński Z, Habel M (2017) Impact of a single dam on sediment transport continuity in large lowland rivers. In: Wieprecht S, Haun S, Weber K, Noack, M, Terheiden K (eds) River sedimentation. Taylor & Francis CRP Press
18. Szatten D, Habel M, Pellegrini L, Maerker M (2018) Assessment of siltation processes of the Koronowski Reservoir in the Northern Polish Lowland based on Bathymetry and Empirical Formulas. Water 10(11):1681–1694
19. Sundborg A (1992) Lake and reservoir sedimentation prediction and interpretation. Geogr Ann 74A(2–3):93–100
20. Toniolo H, Schultz J (2005) Experiments on sediment trap efficiency in reservoirs. Lakes Reservoirs: Res Manage 10(1):13–24
21. Vörosmarty CJ, Sharma KP, Fekete BM, Copeland AH, Holden J, Marble J, Lough JA (1997) The storage and aging of continental runoff in large reservoir systems of the world. Ambio 26:269–278
22. Gierszewski P, Habel M, Szmańda J, Luc M (2020) Evaluating effects of dam operation on flow regimes and riverbed adaptation to those changes. Sci Tot Environ, art. no 136202, ISSN: 0048-9697
23. Vannote RL, Minshall GW, Cummins KW, Sedell JR, Cushing CE (1980) The river continuum concept. Can J Fish Aquatic Sci 37(1):130–137
24. GRanD (Global Reservoir and Dams database)—Technical Documentation—Version 1.3 (2019) [on-line]. http://globaldamwatch.org/grand/. Accessed 10 September 2020
25. Faleńczyk-Koziróg K, Skubała P, Habel M, Waldon-Rudzionek B, Szatten D (2019) River islands as habitats for soil mites (Acari). River Res Appl 35:1–13
26. Cheng X (1992) Reservoir sedimentation at Chinese hydro systems. Water Power Dam Constr 44:44–47
27. Babiński Z (1994) Transport rumowiska unoszonego i wleczonego dolnej Wisły. Przegląd Geograficzny 67(3–4):82–95
28. Morris GL (1995) Reservoir sedimentation and sustainable development in India. In: Proceedings of the 6th international symposium on river sedimentation, New Delhi, India

29. Łajczak A (1995) Studium nad zamulaniem wybranych zbiorników zaporowych w dorzeczu Wisły. Monografie Komitetu Gospodarki Wodnej PAN 8:106
30. Gierszewski P (2007) Warunki transportu zawiesiny w Zbiorniku Włocławskim w świetle analizy jej składu i tekstury (in Polish). Nauka Przyroda Technologie 2:1–18
31. Gierszewski P, Zakonnov V, Kaszubski M, Kordowski J (2017) Transformacja właściwości wody i osadów w profilu podłużnym zbiorników zaporowych Kaskady Górnej Wołgi. Przegląd Geograficzny 89(3):391–412
32. Matysik M, Absalon D, Habel M, Maerker M (2020) Surface water quality analysis using CORINE data: an application to assess reservoirs in Poland. Remote Sens 12:979
33. Bergkamp G, McCartney M, Dugan P, McNeely J, Acreman M (2000) Dams, ecosystem functions and environmental restoration. WCD Thematic Rev Environ II(1):1–187, Cape Town, South Africa, World Commission on Dams
34. McCartney MP, Sullivan C, Acreman MC (2001) Ecosystem impacts of large dams, center for ecology and hydrology. UK IUCN—The World Conservation Union
35. Chen JG, Zhou WH, Chen Q (2012) Reservoir sedimentation and transformation of morphology in the lower yellow river during 10 year's initial operation of the Xiaolangdi reservoir. J Hydrodyn Ser B 24(6):914–924
36. Babiński Z (2002) Wpływ zapór na procesy korytowe rzek aluwialnych ze szczególnym uwzględnieniem stopnia wodnego Włocławek Wyd Akademii Bydgoskiej im Kazimierza Wielkiego Bydgoszcz
37. Kondolf GM, Gao Y, Annandale G, Morris G, Jiang E, Zhang J, Cao Y, Carling P, Fu K, Guo Q (2014) Sustainable sediment management in reservoirs and regulated rivers: experiences from five continents Earth's. Future 2:1–25
38. Rinaldi M, Wyżga B, Surian N (2005) Sediment mining in alluvial channels: physical effects and management perspectives. River Res Appl 21:805–828
39. Kodolf G (1997) Hungry water: effects of dams and gravel mining on river channels. Environ Manage 21(4):533–551
40. Babiński Z (1992) Hydromorphological consequences of regulating the lower Vistula. Poland Regulated Rivers: Res Manage 7(4):337–348
41. Habel M (2013) Dynamics of the Vistula River channel deformations downstream of Włocławek Reservoir. Kazimierz Wielki University Press, Bydgoszcz
42. Ciupa T (2009) Wpływ zagospodarowania terenu na odpływ i transport fluwialny w małych zlewniach na przykładzie Sufragańca i Silnicy (in Polish). Wyd Uniwersytetu Jana Kochanowskiego, Kielce
43. Szatten D, Habel M, Babiński Z, Obodovskyi O (2019) The impact of bridges on the process of water turbidity on the example of large lowland rivers. J Ecol Eng 20(10):155–164
44. Williams GP, Wolman MG (1984) Downstream effects of dams on alluvial rivers. Geological Survey Professional Paper, 1286, Washington
45. Obodovskyi O, Habel M, Szatten D, Rozlach Z, Babiński Z, Maerker M (2020) Assessment of the Dnieper Alluvial Riverbed stability affected by intervention discharge downstream of Kaniv Dam. Water 12:1104
46. Kondolf GM, Gao Y, Annandale GW, Morris GL, Jiang E, Zhang J, Cao Y, Carling P, Fu K, Guo Q, Hotchkiss R, Peteuil Ch, Sumi T, Wang HW, Wang Z, Wei Z, Wu B, Wu C, Yang CT (2014) Sustainable sediment management in reservoirs and regulated rivers: experiences from five continents. Earth's Future 2(5):256–280
47. Wang Z, Hu C (2004) Interactions between fluvial systems and large scale hydro-projects. In: Ninth international symposium on river sedimentation, October 18–21, 2004, Yichang, China
48. Habel M, Obodovskyi O, Szatten D, Babiński Z, Rozlach Z, Pochaievets O (2019) Using the raster calculator tool to appraise riverbed elevation changes nearby hydrotechnical objects on alluvial rivers. In: Monitoring 2019 conference—monitoring of geological processes and ecological condition of the environment, vol. 2019, pp. 1–5
49. Kantoush AS, Sumi T, Kubota A, Suzuki T (2010) Impacts of sediment replenishment below dams on flow and bed morphology of river. In: First international conference on coastal zone management of river deltas and low land coastlines, Alexandria, Egypt

50. Simons HEJ, Bakker C, Schropp MHI, Jans LH, Kok FR, Grift RE (2001) Man-made secondary channels along the River Rhine (The Netherlands); results of post-project monitoring. Regulated Rivers: Res Manage 17:473–491

51. Tockner K, Baumgartner C, Schiemer F, Ward JV (2000) Biodiversity of a Danubian flood plain: structural, functional and compositional aspects In: Gopal B, Junk W, Davies JA (eds) Biodiversity in wetlands: assessment, function and conservation, vol 1. Backhuys Publishers, Leiden, pp 141–159

52. Ward JV, Tockner K, Uehlinger U, Malard F (2001) Understanding natural patterns and processes in river corridors as the basis for effective river restoration. Regulated Rivers: Res Manage 17:311–323

53. Rollet AJ, MacVicar B, Piégay H, Roy A (2008) L'utilisation detranspondeurs passifs pour l'estimation du transport sédimentaire: premiers retours d'expérienceLa. Houille Blanche 4:110–116

54. Kondolf GM, Matthews WVG (1993) Management of coarse sediment regulated river of California. Report no 80 University of California, California

55. Goelz E (2008) Improved sediment-management strategies for the sustainable development of German waterways. In: Sediment dynamics in changing environments (proceedings of a symposium held in Christchurch, New Zealand, December 2008), IAHS Publ 325

56. Wedam G, Kellner R, Braunshofer R (2004) Innovative hydropower development in an urban environment. In: Proceedings of 19th World Energy Congress—Sydney, Australia, paper no 232

57. Ciupa T, Suligowski A, Łajczak A (2017) Rumowisko rozpuszczone. In: Jokiel P, Marszelewski W, Pociask-Karteczka J (ed) Hydrologia Polski. PWN, Warszawa, pp 146–152

58. Brański J, Banasik K (1996) Sediment yields and denudation rates in Poland, IAHS Publ, 236, s 133–138

59. Fal B (2004) Maksymalne przepływy rzek polskich na tle wartości zaobserwowanych w różnych rzekach świata. Gosp Wodna, nr 5, Wyd SIGMA-NOT Warszawa, s 188–196

60. Stachý J, Fal B, Dobrzyńska I, Hołdakowska J (1996) Wezbrania rzek polskich w latach 1951–1990, Mater Bad, seria Hydrologia i Oceanologia. IMGW, Warszawa, p 80

61. Ciupa T, Suligowski R, Łajczak A, Babiński Z (2017) Rumowisko klastyczne, In: Jokiel P, Marszelewski W, Pociask-Karteczka J (ed) Hydrologia Polski. PWN, Warszawa, pp 146–152

62. Łajczak A (1999) Współczesny transport i sedymentacja materiału unoszonego w Wiśle i jej dopływach (in Polish). Oficyna Wydawnicza Politechniki Warszawskiej, Warszawa

63. Brański J (1972) Zmącenie wody i transport rumowiska unoszonego w rzekach polskich w okresie 1956–1965. PIHM, Warszawa, p 108

64. Habel M (2018) Effects of flow regulation and river channelization on sandbar bird nesting availability at the Lower Vistula River. Ecol Questions 29(4):1–18

65. Świeca A (1998) Wpływ czynników antropogenicznych na rzeczny odpływ roztworów i zawiesin na międzyrzeczu Wisły i Bugu. Rozprawy habilitacyjne lXI, UMCS, lub-lin

66. Ciupa T (2009) The impact of land use on runoff and fluvial transport in small river catchments based on the Sufraganiec and Silnica Rivers. Jan Kochanowski University Press, Kielce, Poland, p 251

67. Łomniewski (1960) Ujście Wisły, Rocznik Polskiego Towarzystwa Geologicznego, t XXIX, Kraków, pp 391–416

68. Graniczny M, Janicki T, Kowalski Z, Koszka-Maroń D, Jegliński W, Uścinowicz S, Zachowicz J (2004) Recent development of the Vistula river outlet. Pol Geol Inst Spec Pap 11:103–107

69. Skibiński J (1976) Próba ilościowej oceny intensywności transportu rumowiska wleczonego w rzekach środkowej Polski Rozprawy Naukowe, 74, Zeszyty Naukowe SGGW – AR w Warszawie

70. Pasławski Z (1973) River hydrometric methods. Communication and Communications Press, Warsaw, Poland, p 338

71. Babiński Z, Habel M (2015) Międzinarodowa droga wodna E-40 Stan i możliwości zagospodarowania, w: Kwieciński Jerzy, Żegluga Śródladowa - Wisła, Global Gompact Network Poland, pp 212–217

72. Polski Związek Producentów Kruszyw [Polish Association of Aggregate Producers], Raport https://kruszpolpl/indexphp/publikacje
73. Czaja S, Pociask-Karteczka J (2017) Obieg wody na terenach zurbanizowanych i uprzemysłowionych. In: Jokiel P, Marszelewski W, Pociask-Karteczka, J (eds) Hydrologia Polski. Wydawnictwo Naukowe PWN, Warszawa
74. Babiński Z, Habel M, Szatten D & Dąbrowski J (2014) Influence of hydrological and sedimentological processes on the functioning of inland waterway transport on an example of Brdyujście Regatta Track. Sci J Marit University of Szczecin—Zeszyty NaukoweWyższa Szkoła Morska w Szczecinie—2014 37(109):16–21
75. Szatten D, Habel M, Dąbrowski J (2013) Oddziaływanie miast na zamulanie dróg wodnych - na przykładzie ujściowego odcinka Brdy w Bydgoszczy, Gospodarka Wodna, nr 6, Wyd Sigma-Not
76. Rozporządzenie Ministra Środowiska z dnia 21 lipca 2016 r w sprawie sposobu klasyfikacji stanu jednolitych części wód powierzchniowych oraz środowiskowych norm jakości dla substancji priorytetowych (Journal of Laws of 2016, item 1187)
77. Program monitoring środowiska (2009) Główny Inspektor Ochrony Środowiska, Warszawa (Environmental Monitoring Program, 2009)
78. Wren D, Barkdoll B, Kuhnle R, Derrow R (2000) Field techniques for suspended-sediment measurement. J Hydraul Eng 126(2):97–104
79. Polska Norma PN-EN 872 (2007) Jakość wody Oznaczanie zawiesin Metoda z zastosowaniem filtracji przez sączki z włókna szklanego. Polski Komitet Normalizacyjny, Warszawa
80. Born A (1958) Wleczenie materiału dennego w korytach rzek i potoków, Wiadomości Służby Hydrol-Meteorol, PIHM, Warszawa, 3, 3–29
81. Młynarczyk Z (1996) Transport materiału piaszczystego w korycie rzeki meandrującej I krętej, Acta Quaternaria, 1, UAM, Poznań
82. Ergenzinger P (1982) Uber den Einsatz von Magnettracern zur Messung des Grossgeschiebetransportes, Beitrage zur Geologie der Schweitz-Hydrologie, Bern, 28, 483–491
83. Babiński Z (2005) The relationship between suspended and bed load transport in river channels. In: Proceedings of the international symposium held at the 7th scientific assembly of the international association of hydrological sciences, pp 182–188
84. Aleksejevskij NI (1998) Formirovanije i dviženije recznych nanosov, Moskwa, p 202 s
85. Hart DA (1998) The status of planimetric and topographic mapping in Wisconsin's Lake Michigan Coastal Counties University of Wisconsin Sea Grant Institute, Wisconsin US
86. Lane SN, Westaway RM, Hicks DM (2003) Estimation of erosion and deposition volumes in a large, gravel-bed, braided river using synoptic remote sensing. Earth Surf Proc Land 28:249–271
87. Schwendel AC, Fuller IC, Death RG (2012) Assessing DEM interpolation methods for effective representation of upland stream morphology for rapid appraisal of bed stability. River Res Appl 28:567–584
88. Dawson JL, Smithers SG (2010) Shoreline and beach volume change between 1967 and 2007 at Raine Island, Great Barrier Reef. Aust Global Planet Change 72:141–154
89. Wheaton JM, Brasington J, Darby SE, Sear DA (2010) Accounting for uncertainty in DEMs from repeat topographic surveys: improved sediment budgets. Earth Surf Proc Land 35:136–156
90. Popek Z (2008) Zmienność natężenia ruchu rumowiska wleczonego w czasie wezbrania na małej rzece nizinnej. Przegląd Naukowy Inżynieria i Kształtowanie Środowiska, z 2(40)
91. Uchwała Rady Ministrów z dnia 14 czerwca 2016 r w sprawie przyjęcia „Założeń do planów rozwoju śródlądowych dróg wodnych w Polsce na lata 2016–2020 z perspektywą do roku 2030 r (Resolution of the Council of Ministers of June 14, 2016)
92. Szatten D, Babiński Z, Habel M (2018) Reducing of Water turbidity by hydrotechnical structures on the example of the Włocławek reservoir. J Ecol Eng 19(3):197–205
93. Felix D, Albayrak I, Abgottspon A, Boes RM (2016) Real-time measurements of suspended sediment concentration and particle size using five techniques. In: IOP conference series: earth and environmental science, vol 49, p 122006

94. Agrawal YC, Pottsmith HC (2000) Instruments for particle size and settling velocity observations in sediment transport. Mar Geol 168:89–111
95. Babiński Z, Habel M (2017) Wykorzystanie i zagrożenia rzek i wód rzecznych. In: Jokiel P, Marszelewski W, Pociask-Karteczka, J (eds) Hydrologia Polski. Wydawnictwo Naukowe PWN, Warszawa, pp 146–152
96. Ustawa z dnia 20 lipca 2017 – Prawo wodne (Dz U 2017, poz 1566) (Journal of Laws of 2017, item 1566)
97. Głuchowska B, Kasiorek-Godyń I (2010) Zarządzanie wodami w Polsce na przykładzie RZGW we Wrocławiu. In: Zarządzanie zasobami wodnymi w dorzeczu Odry – 2010 [red B Mońka] Materiały, wyd PZITS nr 894, pp 321–331
98. Rozporządzenie Rady Ministrów z dnia 18 października 2016 w sprawie Planu gospodarowania wodami na obszarze dorzecza Wisły (Dz U 2016, poz 1911) [Journal of Laws 2016, item 1911]
99. Rozporządzenie Rady Ministrów z dnia 18 października 2016 r w sprawie Planu gospodarowania wodami na obszarze dorzecza Odry (Journal of Laws of 2016, item 1967)
100. Obwieszczenie Regionalnego Dyrektora Ochrony Środowiska w Olsztynie z dnia 5 grudnia 2018 r, znak WOOŚ421112017AZ68 (Announcement, 2018)

Chapter 13
Environmental and Anthropogenic Determinants of Water Chemistry in the Carpathians

Joanna P. Siwek

Abstract The most important natural and anthropogenic factors affecting stream water and groundwater chemistry in the Carpathians were identified mainly on the basis of hydrochemical case studies from numerous small headwater Carpathian catchments. The most important natural factor determining water chemistry in the Carpathians is geology. In areas formed of poorly soluble granite and gneiss rocks (crystalline portion of the Tatra Mountains), the total dissolved solids (TDS) and concentration of main cations and anions are many times lower than that in areas formed of highly soluble carbonate rocks (sedimentary portion of the Tatra Mountains, Pieniny Mountains) and also clastic rocks forming so-called Carpathian flysch (Beskidy Mountains, Carpathian Foothills). Stream water and spring water chemistry in the Carpathians change primarily due to hydrologic factors—changes in discharge. As discharge increases, TDS declines and the concentrations of most main ions also decline, while the concentration of K^+, NO_3^-, and PO_4^{3-} increases. Elevation and geographic location of springs and streams in given climate zone and vegetation zone are additional natural factors affecting water chemistry, and this is particularly true of SO_4^{2-} and Cl^- concentration as well as pH. Anthropogenic factors affecting water chemistry in the Carpathians include acid rain, deforestation, agriculture, and tourist-generated wastewater. The effects of acid rain are felt in the form of low concentrations of main cations in stream and spring water in the western part of the Carpathians. This region was affected by very high acidic sulfur and nitrogen deposition in the second half of the twentieth century. Deforestation in the Carpathians impacts mainly spruce monocultures declining due to acid rain, strong winds, and bark beetle infestation, and is the main cause of increasing NO_3^- concentrations in stream water and spring water. Agricultural land use does not threaten stream water and groundwater due to low usage of mineral fertilizer. Threats that do exist are associated with unregulated releases of wastewater in rural areas. Wastewater generated by tourists is a major threat to stream water quality in areas with a high environmental value that are highly popular with tourists. Tourist lodges release wastewater

J. P. Siwek (✉)
Department of Hydrology, Institute of Geography and Spatial Management, Jagiellonian University in Kraków, Gronostajowa 7, 30-387 Krakow, Poland
e-mail: joanna.siwek@uj.edu.pl

© Springer Nature Switzerland AG 2021
M. Zeleňáková et al. (eds.), *Quality of Water Resources in Poland*, Springer Water,
https://doi.org/10.1007/978-3-030-64892-3_13

into Carpathian streams leading to excess nitrogen and phosphorus concentration in stream water.

Keywords The Carpathians · Water chemistry · Environmental and anthropogenic factors · Geology · Acid rain · Deforestation · Tourism activity · Poland

13.1 Introduction

Carpathian catchments are characterized by high specific runoff [L/s/km^2] relative to catchments in other parts of Poland. This is due to high atmospheric precipitation in the Carpathian region. Mean runoff depth in Poland equals a mere 164 mm, while mean runoff depth in the Tatra Mountains and the Podhale region (Carpathians) stands at 661 mm. The corresponding value for the western part of the Beskidy Mountains is 444 mm and for the Bieszczady Mountains 503 mm [1, 2]. Hence, the Carpathians determine the size of the water supply for the larger part of Poland [3, 4]. The discharge of streams in the Carpathians is characterized by very large changes over time due to the low retention capacity of soils and parent material in the region. Streams in the Carpathian region are characterized by the frequent occurrence of abrupt flooding on the one hand and severe droughts on the other hand [2, 5–7]. This large variability in stream discharge makes it very difficult to utilize the area's water resources [2]. Water quality is another potentially limiting factor in terms of how the area's waters can be utilized [8]. The quality of stream water and groundwater is affected by many different naturals and anthropogenic factors that may only be identified in small headwater catchments in which the number of factors is limited. Small headwater catchments play an immensely important role in the water chemistry of Carpathian river waters. It is these catchments that "generate" the water chemistry of river water further downstream. The purpose of this work is to use numerous case studies of small headwater Carpathian catchments to identify natural and anthropogenic factors that determine the stream water and groundwater chemistry in the Carpathians.

13.2 Environmental Determinants

13.2.1 Geology

Geologic structure is one of the most important elements of the natural environment determining stream water and groundwater chemistry in the Carpathians. Geology is noted to be the most important determinant of water chemistry in a variety of areas in the region: the Tatra Mountains [9], Śląski Beskid Mountains [10], Gorce Mountains [11], Carpathian Foothills [12], and the Bieszczady Mountains [13].

13.2.1.1   The Tatra Mountains

Large differences in geologic structure across the Polish Tatra Mountains leads to large differences in stream water and groundwater chemistry in the region. The southern part of the Tatra Mountains is formed of poorly soluble crystalline rocks known as the crystalline core [14, 15], where total dissolved solids (TDS) and conductivity of spring and stream water are very low (extremely fresh water) [16]. According to Żelazny [9], mean TDS and water conductivity in 489 springs located across the crystalline core region of the Tatra Mountains are 33.7 mg/L and 40.2 μS/cm, respectively. In the crystalline part of the Western Tatra Mountains, TDS and conductivity of spring waters are about twice as high as those in the crystalline part of the High Tatra Mountains (Fig. 13.1). This is due to the slightly better solubility of crystalline rocks in the Western Tatra Mountains versus those found in the High Tatra Mountains. The crystalline part of the Western Tatra Mountains is formed mainly of metamorphic rocks (gneiss and crystalline shale) and alaskite including fragments of granitoid rocks and mylonite [9, 14, 15, 17]. The crystalline rocks of the High Tatra Mountains consist mostly of highly resistant granitoids (granodiorites) accompanied by many veins of pegmatite and aplite [9, 18]. Stream waters in the crystalline core of the Polish Tatra Mountains are characterized by somewhat less TDS and lower conductivity—on average—than spring waters. Their TDS and conductivity do not exceed, respectively: 25 mg/L and 30 μS/cm [9]. TDS and stream conductivity are about twice that in the crystalline Western Tatra Mountains relative to the High Tatra Mountains.

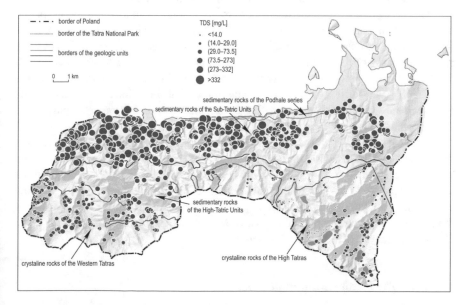

Fig. 13.1 Total dissolved solids (TDS) of spring water in the Tatra Mountains (after Żelazny [9], changed)

Groundwater and stream water in the northern sedimentary part of the Polish Tatra Mountains are characterized by a much higher TDS and conductivity than water in the southern crystalline part of the Tatra Mountains. Mean TDS and water conductivity in a total of 529 springs located in this part of the Tatra Mountains were found to be, respectively: 245.1 mg/L and 254.1 μS/cm [9]. Higher TDS and conductivity were noted for springs located among formations poorly resistant to weathering and leaching—i.e. so-called Sub-Tatric Units (dolomitic limestone, limestone, dolomite, sandstone, shale, conglomerates)—than across areas with more resistant formations—i.e. so-called High-Tatric Units (limestone, dolomite, shale, quartzitic sandstone) (Fig. 13.1). Streams with the highest mean water conductivity ranging from 260 to 270 μS/cm were those whose catchments are almost fully situated atop sedimentary formation of the High-Tatric Units and Sub-Tatric Units. Larger streams that flow through the sedimentary part of the Tatra Range are characterized by lower water conductivity—it is because of the fact that their springs and upstream sections are located across the crystalline parts of the Tatra Mountains [9].

The variable geologic structure of the Tatra massif produces a strong effect on the spatial distribution of main ion concentrations in groundwater and stream water. The largest content disproportions between streams draining Tatra areas with different geology occur in the case of Mg^{2+}, Ca^{2+} and HCO_3^-, while the smallest in the case of Na^+ and K^+ [9]. The first trend is associated with a very large difference in calcite ($CaCO_3$) and dolomite ($CaMg[CO_3]_2$) content in the formation of different tectonic units in the Tatra Mountains. These minerals are the main source of Mg^{2+}, Ca^{2+} and HCO_3^- in water. Rocks forming the crystalline Tatra core contain only trace amounts of calcite and dolomite, as opposed to rocks forming the sedimentary part of the Tatra massif. The second trend is the result of a relatively high proportion of minerals that contain sodium (plagioclase, albite $Na[AlSi_3O_8]$) as well as potassium (orthoclase $K[AlSi_3O_8]$, muscovite $KAl_2[AlSi_3O_{10}(OH)_2]$, biotite $K[AlSi_3O_{10}(OH)_2]$) in the crystalline rocks of the High Tatra Mountains and Western Tatra Mountains [19]. It results in relatively high concentrations of Na^+ and K^+ in the waters of the crystalline core. This is why differences in the concentrations of these ions in the crystalline core's stream water and streams flowing across the northern sedimentary part of the Tatra Mountains are not large.

The pH of groundwater and stream water is strongly correlated with differences in the geology of the Tatra Mountains (Fig. 13.2). The lowest pH of spring water occurs across the crystalline core and equals an average of 6.22 [9]. The pH of water in the crystalline part of the Western Tatra Mountains is higher than that in the crystalline part of the High Tatra Mountains [9, 18]. The highest pH is noted for spring water in the northern sedimentary part of the Tatra Mountains, especially for the Sub-Tatric Units, with a mean pH value of 7.95 [9].

13.2.1.2   The Pieniny Mountains

The Pieniny Mountains are a massif formed mostly of carbonate rock—marl limestone, limestone, and marl—whose chemical composition is dominated by calcite

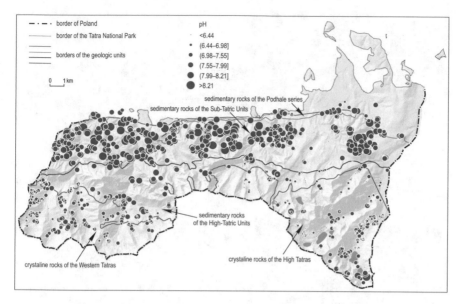

Fig. 13.2 pH of spring water in the Tatra Mountains (after Żelazny [9], changed)

and dolomite [20, 21]. Groundwater (springs) in the Pieniny Mountains is freshwater with mean TDS ranging from 343 to 496 mg/L [20–22]. These waters are characterized by a weakly basic pH. However, spring waters flowing through limestone have a somewhat higher pH, mean: 7.9, than spring waters flowing through marl, mean: 7.6 [20, 22]. The predominant ions in spring waters are Ca^{2+}, Mg^{2+} and HCO_3^-. Water chemistry is determined first and foremost by the dissolution of carbonate occurring in the water-bearing layer under shallow circulation conditions (fissure-karst circulation) in the presence of CO_2 [20, 21].

13.2.1.3 Beskidy Mountains and Carpathian Foothills

Spatial differences in stream water and groundwater chemistry in the Beskidy Mountains and the Carpathian Foothills are determined largely by the grain size of clastic rocks in so-called Carpathian flysch (sandstone, conglomerates, shale), binding agent type, and degree of cementation of the rock material [23]. Substantial spatial differentiation of flysch formations leads to large changes in water chemistry—even across relatively small geographic areas [24–26]. Example of small Ryjak catchment in Beskid Niski Mountains is shown in Fig. 13.3. Nevertheless, according to Kostrakiewicz [23], an increase in TDS and concentration of Ca^{2+}, Na^+, K^+, HCO_3^- and Cl^- is observed from west to east—along with a decrease in the concentration of Mg^{2+} and SO_4^{2-}—in both groundwater (fissures and layers) and stream water in this part of the Carpathians. This trend is associated with the occurrence of increasing concentrations of highly soluble carbonate from west to east in the parent

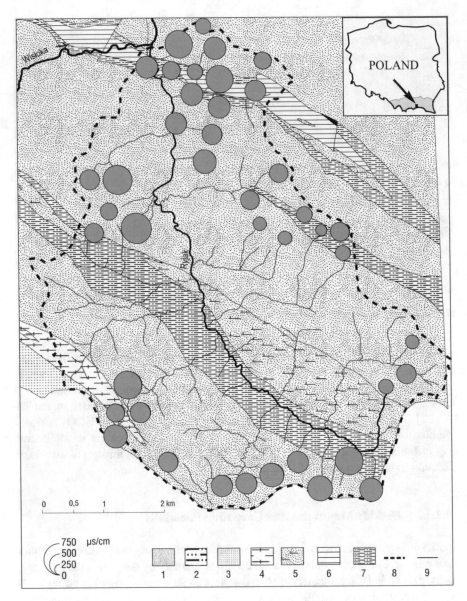

Fig. 13.3 Water conductivity of springs in the Ryjak stream catchment (Beskid Niski Mountains) according to geology (from Lasek [24], changed). 1—oligocen thick-layered sandstones and shales (magura layers), 2—eocenic shales with sanstone insert (magura layers), 3—eocenic thick-layered sandstones (magura layers), 4—eocenic shales and thin-layered sandstones (hieroglific and magura layers), 5—eocenic shales and thin-layered sandstones (hieroglific and beloweskie layers), 6—eocenic varicolored shales, 7—eocenic shales with sandstones-insert and shales with thin-layered sandstones (magura and hieroglific layers), 8—watershed, 9—streams

material [27]. Lowest TDS values are noted for waters flowing through Istebna-type and Godula-type sandstone—rocks that form the westernmost chains of the Beskidy Mountains (Beskid Śląski Mountains, Beskid Mały Mountains) as well as foothills (Pogórze Śląskie Foothills, Pogórze Wielickie Foothills) [23, 27]. The mean TDS of groundwater in this region oscillates from 180 to 490 mg/L [23]. A slightly higher TDS is noted for waters found across Magura Beds forming the following five mountain ranges: Beskid Żywiecki Mountains, Beskid Sądecki Mountains, Beskid Średni Mountains, Beskid Wyspowy Mountains, and Beskid Niski Mountains. The mean TDS of these waters ranges in between 151 and 523 mg/L [23]. According to Kostrakiewicz [23], the highest mean TDS ranges from 221 to 717 mg/L and occurs in Krosno Beds forming the Bieszczady Mountains as well as the Pogórze Ciężkowickie Foothills and Pogórze Dynowskie Foothills. However, according to case studies by Jasik et al. [11], Lasek [24], Rajda et al. [28], and Małek and Gawęda [29] mean TDS and conductivity of groundwater in selected parts of the Beskidy Mountains usually correspond to the lower extent of TDS and conductivity values provided by Kostrakiewicz [23]. Furthermore, Siwek et al. [13] and Kisiel et al. [30] found distinctly lower average TDS and conductivity of spring water in the Bieszczady Mountains than Kostrakiewicz [23] provided. Numerous case studies of groundwater chemistry in the Carpathians indicate distinctly higher conductivity of groundwater in the Carpathian Foothills than in the Beskidy Mountains (Fig. 13.4).

The geologic structure of the Beskidy Mountains and the Carpathian Foothills determines the occurrence of mineral waters: brine water, oxalic water, hydrogen sulfide water. Brine water of the Cl^--Na^+ type and $Cl^--HCO_3^--Na^+$ type are commonly found across the flysch Carpathians [2]. Brine water is considered a relict sedimentation type of water [31]. This is evidenced by the absence of rocks in the area that would contain easily soluble compounds of Cl^- and Na^+. The mineral content of brine water in the flysch Carpathians remains in the dozens of g/L and is highest in Skolska unit formations (Pogórze Dynowskie Foothills, Pogórze

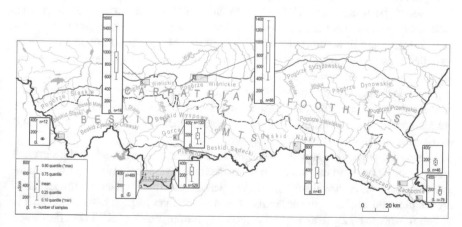

Fig. 13.4 Statistical characteristics of groundwater conductivity in the Carpathians on the basis of hydrochemical case studies from small Carpathian catchments [9, 11, 13, 24, 29, 30, 70, 81]

Przemyskie Foothills). For example, the Drohobyczka 4 drill site yields a value of 270,000 mg/L [32]. Oxalic waters or waters saturated with free carbon dioxide (>1000 mg/L) are considered to be young waters and occur in the southern part of the Magura Beds (Beskid Sądecki Mountains) in an area that is characterized by young Tertiary volcanic activity [2, 23]. Hydrogen sulfide waters (H_2S) are associated with the occurrence of tectonic dislocations that enable the migration and enrichment of groundwater in sulfate ions derived from the dissolution of gypsum, pyrite, and bitumen [2, 23].

13.2.2 Hydrology and Seasonality

13.2.2.1 Hydrologic Year

One very important factor driving water chemistry changes in the Carpathians, and especially in stream water, is changes in discharge (hydrologic factor). It is the most important factor shaping the seasonal water chemistry of stream water in the Carpathian Foothills [33] and in the Tatra Mountains [9]. Changes in stream discharge throughout the hydrologic year are triggered by the influx of meltwaters in the spring and rainwater in the summer. The influx of these waters characterized by low TDS, low conductivity, and low concentration of most main ions triggers a decline in these parameters for stream water with increasing discharge. This is due to a dilution of groundwaters with water characterized by a short time of contact with parent material. This was documented in the Homerka catchment in the Beskid Sądecki Mountains [34], Stara Rzeka and Wierzbanówka catchments in the Pogórze Wiśnickie Foothills [33, 35, 36], and in 23 Tatra area catchments [9, 37]. Nevertheless, the concentrations of some ions such as K^+, NO_3^-, and PO_4^{3-} increase with increasing stream discharge. This pattern is characteristic of ions whose main source is the soil [38–40].

Research performed in small foothill catchments shows that, at a given discharge, the conductivity of water and the concentrations of most main ions are lower in spring and summer than in autumn and winter [12, 33]. This is due to the occurrence of so-called season hysteresis (Fig. 13.5). Lower conductivity and lower ion concentrations in stream water in the spring and summer result from the flushing of chemical compounds from soils by snowmelt waters on the onset of snowmelt season (late winter). Higher conductivity and higher ion concentrations in the autumn and winter are associated with the influx of these compounds thanks to high rates of chemical weathering in the warm season of the year [12].

Seasonal changes in Carpathian spring discharge are determined by the seasonal influx of snowmelt and rain waters in water-bearing layers that also alter their chemical composition. At low discharge, spring waters are characterized by higher TDS, higher pH, and higher concentrations of main ions than at high discharge. This is shown in studies by Żelazny [9] in the Tatra Mountains, Jasik et al. [11] in the Gorce Mountains, Siwek et al. [13] for the Bieszczady Mountains, and Malata and Motyka [41] for Babia Góra Massif (Beskid Żywiecki Mountains).

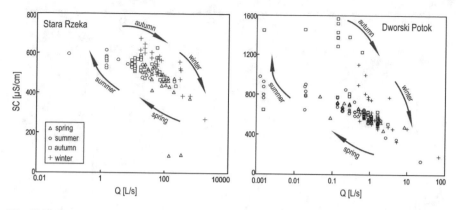

Fig. 13.5 Stream water conductivity (SC) vs. stream water discharge (Q) in two small catchments of the Carpathian Foothills—seasonal hysteretic effect (in part after Siwek [12], changed)

13.2.2.2 Low Flow Periods

At low discharge, streams are recharged primarily by groundwater, which explains their water chemistry and its dependence on groundwater chemistry predominant in each given area. It is usually assumed that stream water chemistry at low water stages is constant [27, 42]. However, studies by Siwek [12] and Raczak and Żelazny [43] in three rather small catchments in the Pogórze Wiśnickie Foothills show that diurnal changes in stream water chemistry and this is especially true in summer, may be quite large. They may be similar to changes during events (Fig. 13.6). The largest of these changes were observed in a stream draining an agricultural catchment where water conductivity ranged from 500 μS/cm in the evening to 826 μS/cm in the morning. These large changes are explained by a complex mechanism of stream recharge, which is shaped by the diurnal intensity of evapotranspiration. At increasing discharge at night and in the morning (low evapotranspiration), stream recharge includes a higher share of shallow alluvial water as well as water from springs and seepages in the middle and upper part of the catchment (Fig. 13.7a). These waters are characterized by high conductivity and high ion concentrations. The shallow alluvial water is highly polluted due to the extensive use of organic fertilizer in the agricultural catchment. Springs and seepages often transport household sewage discharged from farms located in the catchment. At decreasing stream discharge during the daytime or at high rates of evapotranspiration, the share of shallow alluvial waters and spring waters decreases (Fig. 13.7b). Streams are then recharged mostly with deeper alluvial waters characterized by lower water conductivity and lower ion concentrations [12]. Despite this pattern being explored in only three small foothill catchments, it is reasonable to infer that it occurs in other Carpathian catchments as well.

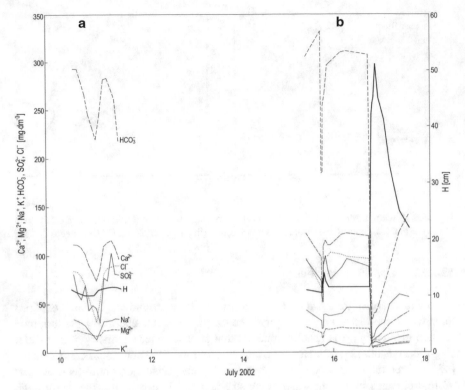

Fig. 13.6 Changes in the stream water chemical composition and water level [H] of stream draining agricultural catchment of the Carpathian Foothills: (**a**) diurnal changes, (**b**) changes during the rainstorm events (after Siwek [12])

13.2.2.3 Rainfall and Snowmelt Events

Stream water chemistry changes very rapidly at high water stages caused by rainfall or snowmelt [12, 34, 40, 44–46]. Stream channels become filled with not only groundwater but also throughflow and overland flow whose water chemistry is different from that of groundwater due to shorter contact with the parent material. Conductivity, pH and the concentration of most main ions declines at higher discharge—examples of streams draining the Pogórze Wiśnickie Foothills catchments are shown in Fig. 13.8. On the other hand, the concentration of K^+, NO_3^-, and PO_4^{3-} increases at the same time, as shown by studies conducted in small catchments scattered across the the Beskid Sądecki Mountains [34], Pogórze Wiśnickie Foothills [40, 44], and Beskid Śląski Mountains [46]. The decrease in TDS and accompanying declines in conductivity and concentration of most main ions with increases in discharge are associated with the dilution of groundwater with the overland flow and throughflow characterized by lower TDS, conductivity and concentration of most main ions. According to Froehlich [34], the greatest dilution occurs during snowmelt events. Research by Siwek [12] has shown that the degree to which catchment soils are frozen is highly

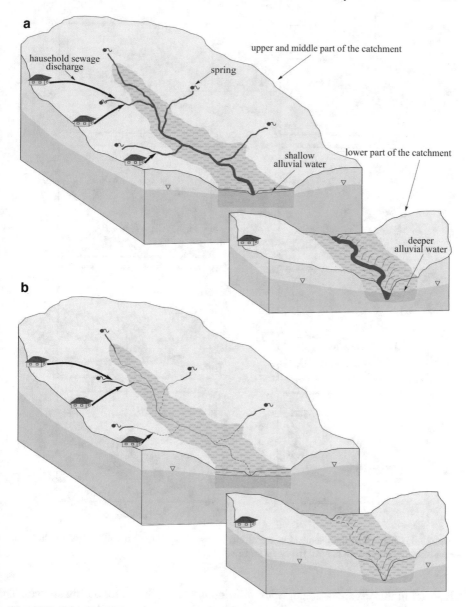

Fig. 13.7 Scheme of diurnal stream recharge mechanism during the nighttime (**a**) and daytime (**b**)—example of stream draining small agricultural catchment of the Carpathian Foothills

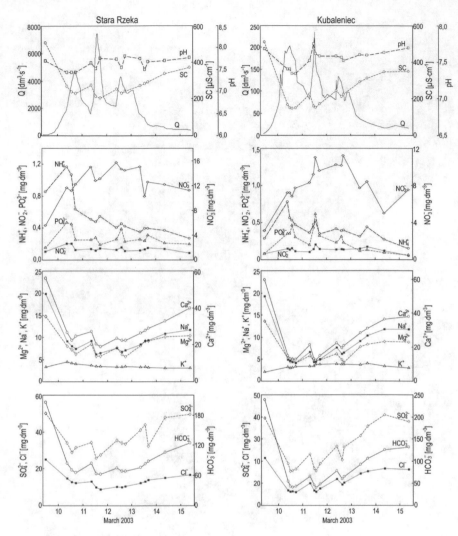

Fig. 13.8 Changes in the chemical composition of two Carpathian Foothill's streams during a rain-on-snow event with the soil frozen (after Siwek [12])

relevant in this regard—stream water dilution is markedly greater during snowmelt events with the soil frozen relative to snowmelt events with the soil not frozen.

Changes in water chemistry during rainfall and snowmelt events are often accompanied by the presence of a so-called hysteresis or ion concentration that is different for the rising limb versus falling limb of hydrograph [34, 44, 45]. The cause of a clockwise hysteresis is usually rapid flushing of chemical compounds in soils during the rising limb of hydrograph and their shortage during the falling limb. It results in higher ion concentrations during the rising limb than the falling limb of the hydrograph [44, 45]. Counterclockwise hysteresis tends to be associated with snowmelt events with the soil frozen—as documented by Siwek et al. [45] for changes in the K^+ concentration in foothill streams. In the course of snowmelt events with the soil frozen, streams were recharged first by snowmelt waters flowing across the frozen ground with a low K^+ concentration. The soil would unfreeze over time, and increasingly deep soil horizons would yield higher amounts of K^+ due to flushing processes carrying the ions to stream channels. Hence, K^+ concentrations in streams were higher during these types of events during the falling limb versus the rising limb of the hydrograph [45].

13.3 Climate and Vegetation Elevation Zones

Climate and vegetation zones based on elevation are yet another factor that affects the water chemistry of stream water and groundwater across the Carpathians. According to Siwek et al. [13], elevation effects follow the geologic structure in terms of their impact on spring water chemistry in the Bieszczady Mountains. Elevation effects first and foremost determine SO_4^{2-} and Cl^- concentrations—the higher elevation of the spring, the higher the SO_4^{2-} concentration and the lower the Cl^- concentration. This results from an increasing effect of precipitation rich in SO_4^{2-} on spring water chemistry with increasing elevations. Similar conclusions were drawn by Małecka [16] and Małecka et al. [47] for the Tatra Mountains: groundwater chemistry and stream water chemistry at higher elevations in the Tatra Mountains are 90% determined by the content of precipitation. At lower elevations, the impact of precipitation on Tatra water chemistry did not exceed 30%.

Research by Jasik et al. [11] in the Gorce Mountains indicates that the species composition of the forest changes with elevation and also strongly affects spring water chemistry. Springs whose recharge areas are located in the upper subalpine zone covered with spruce stands have a lower pH and lower concentration of Ca^{2+}, Mg^{2+} and HCO_3^- than springs whose recharge areas are found at lower elevations covered with beech-fir stands.

The location of a catchment in a given climate zone also determines water chemistry throughout the year via its hydrochemical regime. Żelazny [9] identified two types of stream hydrochemical regimes in the Tatra Mountains: (1) high mountain regime, (2) middle mountain regime. In the case of both regimes, the lowest TDS, conductivity, and concentration of most main ions occur during the spring snowmelt. However, in streams with a high mountain regime, the lowest TDS, lowest conductivity, and lowest ion concentrations occur later than in streams with a middle mountain regime. This is due to the snowmelt season occurring later and lasting longer at higher elevations versus lower elevations.

13.4   Anthropogenic Determinants

13.4.1   Acid Rain

The effect of acid rain on the chemistry of stream water and groundwater is observed first and foremost in the western and central parts of the Carpathians. This region experienced a significant influx of acidic emissions (SO_2 and NO_2) in the 1970s and 1980s from major industrial areas in Upper Silesia, Kraków, and the so-called Black Triangle, which was a vast geographic region consisting of southwestern Poland, northern Czechoslovakia, and southeastern East Germany [48]. The current level of gaseous pollution in the region is many times lower than in the 1970s and 1980s [49]. This is due to a large decline in SO_2 pollution and a smaller decline in NO_2 pollution in more recent years. In 1980 the mean rate of atmospheric SO_2 deposition in Poland was 132.1 kg/ha, while in 2016 it was 18.6 kg/ha. In 1985 the mean rate of atmospheric NO_2 deposition in Poland was 43.3 kg/ha, while in 2016 it was 23.2 kg/ha [50, 51]. The decrease in emissions of acidic gases led to a gradual increase in the pH of atmospheric precipitation in the Carpathians, although it remains acidic. For example, the mean weighted pH of precipitation in the Pogórze Wiśnickie Foothills in the years 1991–1994 was 4.42 while in the years 2002–2004 the pH was somewhat higher at 4.68 [48, 52]. The increasing trend in the pH of atmospheric precipitation was also documented by Rzychoń and Worsztynowicz [53] in the Tatra Mountains in the years 1992–2005.

Research by Kosmowska et al. [49] in the Skrzyczne Massif in the Beskid Śląski Mountains shows that the effect of acid rain on stream water chemistry remains pronounced even today—several decades after the era of the greatest atmospheric deposition of sulfur and nitrogen. The acidity of the natural environment remains the most important factor shaping the chemistry of stream water in the Beskid Śląski Mountains [49]. Key solutes have been flushed from soils by acidic rainfalls for many years. This has led to the current situation where TDS, water conductivity, and the concentrations of most main ions in streams draining the Skrzyczne Massif are several times lower than those for other Carpathian streams draining catchments in similar geologic structure but experiencing much less acid rain [24, 49, 54, 55]. Furthermore, the concentrations of main cations in the streams flowing across the Skrzyczne Massif in the 1950s and 1960s were much higher than they are at present [49]. Rzychoń and Worsztynowicz [53] documented the similar decline in the concentration of main cations in two ponds in the High Tatra Mountains associated with long-term acidification of precipitation in the area. The predominant anions in streams draining the Skrzyczne Massif today are SO_4^{2-} and NO_3^-, which is not encountered in other parts of the Carpathians characterized by much less acidic deposition [46]. According to Małek and Krakowian [56], the effect of acidic atmospheric deposition on stream water chemistry in the Beskid Śląski Mountains is most readily observed in spring snowmelt events when a large quantity of acidic precipitation becomes released over a short period in the form of melting snow. This leads to a large decline in water pH in streams.

13.4.2 Deforestation

The decline of Carpathian forests artificially dominated by spruce introduced in the nineteenth century began in the 1960s and proceeded from west to east, first affecting lower elevation forests in the Beskid Śląski Mountains, Beskid Mały Mountains, and Beskid Żywiecki Mountains [57–59]. The main reason for the decline of forests in the western part of the Carpathians was a high level of acidic gas emissions in the mid-twentieth century. Acid rain led to increased soil acidity (pH < 3.0) and loss of nutrients for trees (calcium and magnesium compounds) [60]. The decline of weakened spruce monocultures accelerated after 2006 due to long-lasting summer droughts, abrupt heavy winds, bark beetle infestation, and fungus expansion [59, 61–63].

Studies conducted in the Beskid Śląski Mountains by Małek and Gawęda [64] as well as Małek and Krakowian [65] indicate that deforestation caused by acidic sulfur and nitrogen deposition leads to lower pH, conductivity, and concentration of main ions in spring water as well as to increasing concentration of ammonia (NH_4^+). In streams draining deforested catchments in the Skrzyczne Massif (Beskid Śląski Mountains), the share of SO_4^{2-} in overall water chemistry is higher in than in streams draining forested catchments. In streams draining deforested catchments, SO_4^{2-} is very aggressively flushed out of soils during rainfall and snowmelt events, which yields a positive correlation between SO_4^{2-} concentration and discharge. In forested catchments, this flushing effect is less pronounced, which yields a negative relationship between SO_4^{2-} concentration and discharge. This suggests that the forest serves as a buffer that limits environmental acidification associated with acidic precipitation [49]. In streams draining deforested catchments, changes in water chemistry during events are larger than those in streams in forested catchments [46].

Excess atmospheric deposition of sulfur and nitrogen is not the only reason for the decline of tree stands in the Carpathians. The main cause of the decline of spruce monocultures in the Tatra Mountains over the last 100 years has been heavy winds and bark beetle infestations [66]. Research by Żelazny et al. [62, 63] conducted in the Kościeliski Potok catchment in the Tatra Mountains has shown that hillslope deforestation triggered by hurricane-force winds in 2013 and tree stand decline due to bark beetle infestation has led to significant changes in the water chemistry of springs and streams. One and a half year after deforestation, the concentration of NO_3^- (mean: 15.44 mg/L) was many times higher in spring and stream water in the deforested area than in waters in forested areas (mean: 3.26 mg/L). The increase in NO_3^- concentration in stream water and spring water caused by deforestation was more than 500% [62]. The mean concentration of NO_3^- in waters on deforested slopes continued to increase over time—the mean concentration of NO_3^- in the period 2015–2016 was 16.53 mg/L. At the same time, the mean concentration of NO_3^- in waters on forested slopes declined to a mean of 3.06 mg/L [63]. Such a large increase in the concentration of NO_3^- in deforested areas triggered a change in the overall water chemistry of these waters from one that is natural to forested carbonate-type catchments ($HCO_3^- > SO_4^{2-} > NO_3^- > Cl^-$) to the currently predominant $HCO_3^- > NO_3^- > SO_4^{2-} > Cl^-$ [62, 63].

13.4.3   Agriculture

Agriculture in the Carpathian region tends to be low volume at the present time; the use of fertilizer is very low relative to the rest of Poland. In the agricultural year 2016–2017, it did not exceed 100 kg NPK/ha [67]. The share of farmland has been declining steadily since the 1990s towards more pastoral uses including meadows. For example, in the upper Dunajec catchment (Tatra Mountains and Podhale region) the share of farmland in the early 1990s exceeded 60%, while in 2004 it was only 20% [68]. According to Smoroń et al. [69], the Carpathian region is unlikely to face the threat of stream water and groundwater pollution from nitrogen and phosphorus compounds used in agriculture. This is confirmed by Smoroń et al. [69] via low concentrations of NO_3^- and PO_4^{3-} in streams draining the agricultural areas of small headwater sub-catchments in the upper catchments of Dunajec (Tatra Mountains and Podhale region) and San (Bieszczady Mountains).

Research conducted in the Pogórze Wiśnickie Foothills by Siwek [12], Żelazny [70], Siwek et al. [71], and Jelonkiewicz et al. [72] has shown that unregulated wastewater releases represent the main threat to stream water and groundwater quality in rural areas in the Carpathians. Most rural households in the Stara Rzeka catchment, for example, were connected to water mains but did not have access to sewer lines. Sewage was collected in septic tanks, which were sometimes evacuated and the sewage taken to processing facilities. This is illustrated by the fact that only 2% of the water sold to customers in rural areas by the water company returned in the form of sewage to treatment plants [73]. The vast majority of the sewage remained in each given catchment. According to work by Żelazny [70], the migration of sewage from leaky septic tanks into the groundwater may be observed in the form of mosaic-type variability of nitrogen and phosphorus concentrations in well water across the Stara Rzeka catchment. Wastewater from the examined catchment was also found to end up in area ditches and local stream channels. Some wastewater was detected on fields in stream valleys. Streams burdened by wastewater possessed lower NO_3^- and PO_4^{3-} concentration at higher discharge. The concentrations of these ions in pre-event periods were high enough that the incoming event water—themselves rich in NO_3^- and PO_4^{3-}—would cause a dilution [12].

13.4.4   Tourism-Generated Wastewater

The development of the tourism sector in the Carpathians is associated with threats to the quality of stream water and groundwater. This problem is especially significant in environmentally valuable areas characterized by substantial environmental diversity, which are also attractive to tourists [74]. These areas include national parks, nature reserves, landscape parks, and protected landscape areas, which cover more than 65% of the area of the Polish Carpathians [75]. Research conducted in 2010 in the summer tourist season in Bieszczady National Park has shown an increase in the load of Cl^- and K^+ in streams downstream of popular tourist towns such as Ustrzyki Górne, Wołosate and Wetlina [55]. The impact of tourism on stream water quality was also documented by Smoroń and Twardy [76] as well as Smoroń et al. [77] in the region of Podhale. Concentrations of NO_3^-, PO_4^{3-} and Na^+ in the waters of the Biały Dunajec River draining a catchment with a very large number of tourists are much smaller than those in the waters of the Czarny Dunajec and Białka rivers draining catchments with much smaller tourist volumes.

Tourist lodges represent a major threat to water quality in the Carpathians via their frequent release of wastewater into local streams. This is shown by studies conducted in Tatra National Park by Siwek and Biernacki [78, 79]. Wastewater releases by three lodges in Tatra National Park in the years 2008–2009 were found to cause major changes in the concentration of nitrogen and phosphorus compounds in streams affected by these releases. Wastewater is released into Sucha Woda Stream from the Murowaniec Lodge. The same is true of wastewater from the Polana Chochołowska Lodge, which releases into the local Chochołowski Stream. Both streams experience large increases in the concentration of NH_4^+ and PO_4^{3-} (Fig. 13.9). The greatest effect of wastewater release into these streams on the concentration of these ions occurs at low water stages. The concentration of NH_4^+ in stream water downstream of the wastewater release point was about 200 times larger than the concentration upstream of the release point, while PO_4^{3-} concentrations were roughly 30 times higher. The concentration of NO_3^- in Roztoka Stream increased markedly following the release of untreated wastewater from the Lodge in the Valley of Five Polish Ponds.

Stream water pollution caused by wastewater releases from tourist lodges has been a key problem in Tatra National Park for a number of years. In 2009 the wastewater

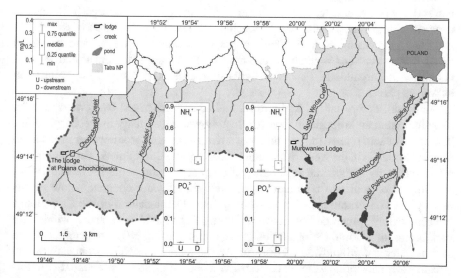

Fig. 13.9 Statistical characteristics of NH_4^+ and PO_4^{3-} concentrations of stream water upstream (U) and downstream (D) from the discharge point of wastewater produced by tourist lodges in the Tatra National Park (after Siwek and Biernacki [79], changed)

management situation of Tatra National Park improved markedly when wastewater treatment facilities at two lodges were fully modernized thanks to co-financing from the European Union as part of the project: "Branding of Mountain Tourism in the Polish Tatra Mountains – Green Lodges". The two tourist lodges were the Murowaniec Lodge and the Polana Chochołowska Lodge. A third lodge—the Lodge in the Valley of Five Polish Ponds—launched in 2009 a new mechanical-biological wastewater treatment facility [78]. In other areas of the Carpathians, tourist lodges were also equipped with new wastewater treatment facilities including Leskowiec Lodge in the Beskid Mały Mountains and the Stare Wierchy Lodge in the Gorce Mountains. Finally, some Carpathian lodges fully modernized their wastewater treatment facilities—i.e. the Markowe Szczawiny Lodge in the Babia Góra Massif in the Beskid Żywiecki Mountains [80].

13.5   Summary and Conclusions

Stream water and groundwater chemistry in the Carpathians is characterized by significant variances primarily due to complex geologic structure. For example, in the Tatra crystalline core formed of poorly soluble granite and gneiss rocks, the mean TDS of groundwater is many times smaller than that in the sedimentary part of the Tatra Mountains formed of an array of highly soluble carbonate rocks (33.7 mg/L and 245.1 mg/L, respectively). The water chemistry of the Carpathians is determined largely by local parent material, but also by a number of other natural factors. Changes in discharge (hydrologic factor) is the most important factor driving stream water and spring water chemistry changes in time. With increasing discharge, TDS and the concentrations of most main ions decreasing, while the concentration of K^+, NO_3^-, and PO_4^{3-} increasing. Elevation and geographic location of streams and springs in given climate zone and vegetation zone are additional natural factors affecting water chemistry in Carpathians. This is particularly true of SO_4^{2-} and Cl^- concentration as well as pH.

Acid rain, deforestation, agriculture, and tourist-generated wastewater are the most important anthropogenic factors affecting water chemistry in the Carpathians. The effect of acid rain on the chemistry of stream water and groundwater is observed first and foremost in the western and central parts of the Carpathians. This region was affected by very high acidic sulfur and nitrogen deposition in the 1970s and 1980s. The impact of acid rain on stream water and groundwater chemistry now results in very low concentrations of most of main ions. Deforestation in the Carpathians impacts mainly spruce monocultures declining due to acid rain, strong winds, and bark beetle infestation. Deforestation results in increasing NO_3^- concentrations in stream water and groundwater. For example, the increase in NO_3^- concentration in stream water and spring water in the Tatra Mountains one and a half year after deforestation was more than 500%. Agricultural land use does not threaten stream water and groundwater due to low usage of mineral fertilizer and low share of farmland areas in the Carpathians. However, threats that do exist are associated with unregulated releases of wastewater in rural areas. Wastewater generated by tourists is a major threat to stream water quality in areas with high environmental value such as Carpathian national parks that are highly popular with tourists. Tourist lodges release wastewater into local streams leading to excess nitrogen and phosphorus concentration in stream water. For example, the concentration of NH_4^+ in two streams in the Tatra National Park downstream of the wastewater release point was about 200 times larger than the concentration upstream of the release point. Since 2009, thanks to European Union support, the water and wastewater management situation in Carpathian protected areas improved significantly due to building or modernization numerous wastewater treatment plants at mountain lodges.

13.6 Recommendations

I believe that the most important fields of inquiry for future hydrochemical studies in the Polish Carpathians are as follows (i) identification of factors determining spring water chemistry on the basis of systematic methods employed in small catchments representing all major Carpathian regions, (ii) determination of ion migration pathways leading to Carpathian stream channels during rainfall and snowmelt events, (iii) determination of the impacts of tourist traffic and associated tourist infrastructure on stream water and groundwater quality throughout the Carpathians in order to more effectively protect them from degradation.

References

1. Stachy J (1985) Reżim hydrologiczny rzek polskich. PWN, Warszawa
2. Dynowska I (1995) Wody. In: Warszyńska J (ed) Karpaty polskie. Uniwersytet Jagielloński, Kraków, pp 49–67
3. Ziemońska Z (1973) Stosunki wodne w Polskich Karpatach Zachodnich. Prace Geogr PAN 103:1–124
4. Chowaniec J, Freiwald P, Patorski R, Witek K (2006) Zwykłe wody podziemne w zachodniej części polskich Karpat fliszowych. Gaz, Woda i Technika Sanitarna 11:14–17
5. Chełmicki W, Bieńkowski T (2005) Przepływy niżówkowe w dorzeczu górnego Dunajca w 2003 r. na tle wielolecia 1951–2003. Folia Geogr Ser Geogr Phys 35–36:65–75
6. Gutry-Korycka M, Woronko D, Suchożebrski J (2009) Uwarunkowanie regionalne maksymalnych prawdopodobnych przepływów rzek polskich. Pr i Studia Geogr 43:25–48
7. Bryndal T (2014) Powodzie błyskawiczne w małych zlewniach karpackich – wybrane aspekty zarządzania ryzykiem powodziowym. Annales Universitatis Paedagogicae Cracoviensis, Studia Geographica 7(170):69–80
8. Misztal A, Smoroń S (2003) Zasilanie środowiska makropicrwiastkami zawartymi w mokrym i suchym opadzie w rejonie bukowiny tatrzańskiej. Woda-Środowisko-Obszary Wiejskie 3(2):79–90
9. Żelazny M (2012) Czasowo-przestrzenna zmienność cech fizykochemicznych wód Tatrzańskiego Parku Narodowego. IGiGP UJ, Kraków

10. Astel A, Małek S, Makowska S (2008) Effect of environmental conditions on chemical profile of stream water in sanctuary forest area. Water Air Soil Pollut 195:137–149

11. Jasik M, Małek S, Żelazny M (2017) Effect of water stage and tree stand composition on spatiotemporal differentiation of spring water chemistiy draining Carpathian flysch slopes (Gorce Mts). Sci Total Environ 599–600:1630–1637

12. Siwek JP (2012) Naturalne i antropogeniczne uwarunkowania zmienności chemizmu wód powierzchniowych w małych zlewniach na progu Pogórza Wiśnickiego. Wyd IGiGP UJ, Kraków

13. Siwek J, Rzonca B, Jaśkowiec B, Plenzler J, Płaczkowska E (2013) Natural factors affecting the chemical composition of water in the catchment of Wołosatka stream (High Bieszczady Mts.). In Kozak J, Ostapowicz K, Bytnerowicz A, Wyżga B (eds) The Carpathians: integrating nature and society towards sustainability, environmental science and engineering. Springer, Verlag Berlin Heidelberg, pp 151–159

14. Michalik A (1985) Geologia (litologia i stratygrafia), skala 1:50000. In: Trafas K (ed) Atlas Tatrzańskiego Parku Narodowego, TPN, Zakopane-Kraków

15. Passendorfer E (1996) Geologia. In: Mirek Z (ed) Przyroda Tatrzańskiego Parku Narodowego. TPN, Kraków-Zakopane, pp 67–96

16. Małecka D (1989) Wpływ opadów atmosferycznych na kształtowanie chemizmu wód w obrębie masywu tatrzańskiego. Przegląd Geologiczny 37(10):504–510

17. Gawęda A (2009) Enklawy w granicie Tatr Wysokich. Prace Naukowe Uniwersytetu Śląskiego w Katowicach 2637:1–180

18. Oleksynowa K, Komornicki T (1996) Chemizm wód. In: Kotarba A (ed) Przyroda Tatrzańskiego Parku Narodowego a człowiek (t 1). Polskie Towarzystwo Przyjaciół Nauk o Ziemi. Oddział w Krakowie, Tatrzański Park Narodowy, Komitet Badań Naukowych, Zakopane–Kraków, pp 197–214

19. Gaweł A (1959) Zagadnienia petrograficzne trzonu krystalicznego Tatr Zachodnich. Biul Inst Geol 149:107–116

20. Kostrakiewicz L (1995) Stężenia jonowe i tło hydrochemiczne szczelinowych wód podziemnych Pienińskiego Pasa Skałkowego i jego przyległej części jednostki Magurskiej. Wszechświat 96(4):88–94

21. Szostakiewicz-Hołownia M (2012) Chemizm wód źródlanych zlewni Potoku Macelowego w Pieninach. Pieniny – Przyroda i Człowiek 12: 33–41

22. Kostrakiewicz L (1991) Charakterystyka fizyko-chemiczna oraz bakteriologiczna wybranych źródeł Pienińskiego Parku Narodowego i jego okolicy. Ochr Przyr 49(1):129–139

23. Kostrakiewicz L (1998) Regionalizacja hydrochemiczna źródeł polskich Karpat Zewnętrznych. Wszechświat 99(9):171–176

24. Lasek J (2008) Chemizm wypływów wód podziemnych w zlewni potoku Ryjak (Magurski Park Narodowy). In: Partyka J, Pociask-Karteczka J (eds) Wody na obszarach chronionych. Wyd IGiGP UJ, Kraków, pp 201–211

25. Satora S, Bugajski P, Satora P (2010) Zmienność reżimu wybranych źródeł występujących w obrębie Beskidu Wyspowego i Gorców. Infrastruktura i Ekologia Terenów Wiejskich 14:195–206

26. Malata M (2015) Sezonowa zmienność fizyko-chemiczna źródeł Babiogórskiego Parku Narodowego. Prz Nauk Inż Kszt Środ 67:26–39

27. Welc A (1986) Zróżnicowanie denudacji chemicznej w Karpatach fliszowych. Przegląd Geograficzny 55(1–2):99–118

28. Rajda W, Ostrowski K, Bogdał A (2001) Zawartość wybranych składników fizykochemicznych w wodzie opadowej i odpływającej z mikrozlewni leśnej. Zesz Nauk AR Krak 382 Inż Środ 21:21–31

29. Małek S, Gawęda T (2002) Charakterystyka chemiczna źródeł Potoku Dupniańskiego w Beskidzie Śląskim. Sylwan 3:39–46

30. Kisiel M, Bochnak D, Jastrzębska B, Mostowik K, Pufelska M, Rzonca B, Siwek J (2018) Skład chemiczny wód źródlanych w masywie Połoniny Wetlińskiej. Roczniki Bieszczadzkie 26:205–222

31. Chowaniec J (2009) Studium hydrogeologii zachodniej części Karpat polskich. Biul Państw Inst Geol 434:1–98
32. Rajchel L (2016) Surowce balneologiczne prowincji karpackiej. Acta Balneol 58(4/146):275–278
33. Żelazny M, Siwek JP (2012) Determinants of seasonal changes in streamwater chemistry in small catchments with different land use: case study from Poland's Carpathian Foothills. Pol J Environ Stud 21(3):791–804
34. Froehlich W (1982) Mechanizm transportu fluwialnego i dostawy zwietrzelin w górskiej zlewni fliszowej. Prace Geogr IGiPZ PAN 143:1–161
35. Drużkowski M, Szczepanowicz B (1988) Migracja pierwiastków w wodach powierzchniowych i opadach atmosferycznych na obszarze małej zlewni Pogórza Karpackiego. Folia Geogr Ser Geogr Physica 20:101–120
36. Drużkowski M (1998) Współczesna dynamika, funkcjonowanie i przemiany krajobrazu Pogórza Karpackiego. Inst Botaniki UJ, Kraków
37. Wolanin A (2013) Właściwości fizykochemiczne wody potoków tatrzańskich w okresie kwiecień – listopad 2011 roku. Prace Geogr 133:49–60
38. Gacek T (2001) Czynniki dostawy fosforu do wód powierzchniowych na Pogórzu Karpackim. In: Chełmicki W (ed) Przemiany środowiska na Pogórzu Karpackim (t. 1), Procesy, gospodarka, monitoring. IGiGP UJ, Kraków, pp 65–82
39. Siwek J (2001) Hydrochemiczna interpretacja wezbrań z lipca 1997 r. w zlewni Dworskiego Potoku (Pogórze Wiśnickie). In: Chełmicki W (ed) Przemiany środowiska na Pogórzu Karpackim (t. 1), Procesy, gospodarka, monitoring. IGiGP UJ, Kraków, pp 51–63
40. Siwek JP, Żelazny M, Chełmicki W (2011) Influence of catchment characteristics and flood type on relationship between streamwater chemistry and streamflow: case study from Carpathian Foothills in Poland. Water Air Soil Pollut 214:547–563
41. Malata M, Motyka J (2015) Chemizm wód źródeł w zlewni Markowego Potoku, Babiogórski Park Narodowy. Przegląd Geologiczny 63(10/2):912–918
42. Welc A (1985) Zmienność denudacji chemicznej w Karpatach fliszowych (na przykładzie zlewni potoku Bystrzanka. Dokumentacja Geograficzna 5:1–102
43. Raczak J, Żelazny M (2005) Diurnal fluctuations in stream-water chemical composition in small catchments of the Carpathian Foothills (Southern Poland). Tech Doc Hydrol 77:101–108
44. Siwek J, Siwek JP, Żelazny M (2013) Environmental and land use factors affecting phosphate hysteresis patterns of stream water during flood events (Carpathian Foothills, Poland). Hydrol Process 27(25):3674–3684
45. Siwek JP, Żelazny M, Siwek J, Szymański W (2017) Effect of land use, seasonality, and hydrometeorological conditions on the K^+ concentration-discharge relationship during different types of floods in Carpathian Foothills catchments (Poland). Water Air Soil Pollut 228:445
46. Kosmowska A, Żelazny M, Małek S, Siwek JP (2015) Wpływ wylesień na krótkoterminowe zmiany składu chemicznego wody w zlewni Potoku Malinowskiego (Beskid Śląski). Sylwan 159(9):778–790
47. Małecka D, Chowaniec J, Małecki J (2007) Region górnej Wisły. In: Paczyński B, Sadurski A (ed) Hydrogeologia regionalna Polski(t. 1), Wody słodkie. PIG, Warszawa, pp 108–158
48. Chełmicki W, Klimek M (1995) Odczyn wód opadowych w Łazach na Pogórzu Karpackim. In: Kaszowski L (ed) Dynamika i antropogeniczne przeobrażenia środowiska przyrodniczego Progu Karpat między Rabą a Uszwicą. IG UJ, Kraków, pp 249–254
49. Kosmowska A, Żelazny M, Małek S, Siwek JP, Jelonkiewicz Ł (2016) Effect of deforestation on stream water chemistry in the Skrzyczne massif (the Beskid Śląski Mountains in southern Poland). Sci Total Environ 568:1044–1053
50. Kaczor A, Brodowska MS (2008) Oddziaływanie emisji ditlenku siarki i tlenków azotu na zakwaszenie gleb Polski. Proceedings of ECOpole 2(1):191–195
51. GUS (2018) Statistical analysis. Environment. Chapter 4: Pollution and protection of air. www.stat.gov.pl/files/gfx/portalinformacyjny/pl/defaultaktualnosci/5484/1/18/1/ochrona_srodowiska_2017.pdf

52. Żelazny M (2005) Cechy fizykochemiczne opadów atmosferycznych. In: Żelazny M (ed) Dynamika obiegu związków biogennych w wodach opadowych, powierzchniowych i podziemnych w zlewniach o różnym użytkowaniu na Pogórzu Wiśnickim. IGiGP UJ, Kraków, pp 63–100
53. Rzychoń D, Worsztynowicz A (2008) What affects the nitrogen retention in Tatra Mountains lakes' catchments in Poland? Hydrol Earth Syst Sci 12(2):415–424
54. Satora S (2002) Zawartość niektórych składników chemicznych w wodach opadowych i powierzchniowych źródłowej części zlewni Skawy. Zesz Naukowe AR im. H. Kołłątaja w Krakowie 390(22): 67–74
55. Siwek J, Rzonca B, Płaczkowska E, Plenzler J, Jaśkowiec B (2011) Aktualne kierunki badań hydrologicznych w Bieszczadach Wysokich. Roczniki Bieszczadzkie 19:285–298
56. Małek S, Krakowian K (2009) The effect of environmental conditions on surface water quality in the Zimnik and Czyrna catchments of the Beskid Śląski. J Water Land Dev 13a: 205–223
57. Małek S (2010) Nutrient fluxes in planted Norway spruce stands of different age in Southern Poland. Water Air Soil Pollut 209:45–59
58. Małek S, Martinson L, Sverdrup H (2005) Modeling future soil chemistry at a highly polluted forest site at Istebna in Southern Poland using the "SAFE" model. Environ Pollut 137(3):568–573
59. Małek S, Barszcz J, Majsterkiewicz K (2012) Changes in the threat of spruce stand disintegration in the Beskid Śląski and Żywiecki Mts in the years 2007–2010. J For Sci 58(12):519–529
60. Materiały informacyjne na terenowe posiedzenie Sejmowej Komisji Ochrony Środowiska, Zasobów Naturalnych i Leśnictwa, Aktualna sytuacja lasów Beskidu Sądeckiego i Żywieckiego na przykładzie nadleśnictw Wisła i Ujsoły (2009) Warszawa
61. Małek S, Krakowian K, Jasik M, Dudek K, Bátor M (2014) Effect of deforestation on stream and spring water chemistry in Malinowski and Czyrna catchments in Beskid Śląski Mts. Folia For Pol, Ser A, Leśn 56(3):141–148
62. Żelazny M, Siwek JP, Fidelus J, Stańczyk T, Siwek J, Rutkowska A, Kruk P, Wolanin A, Jelonkiewicz Ł (2017) Wpływ wiatrołomu i degradacji drzewostanu na zróżnicowanie chemizmu wód w zlewni Potoku Kościeliskiego w obszarze Tatrzańskiego Parku Narodowego. Sylwan 161(1):27–33
63. Żelazny M, Kosmowska A, Stańczyk T, Mickiewicz M (2017) Effect of deforestation on water chemistry in the Kościeliska Valley in the Western Tatras in southern Poland. Ann Warsaw Univ Life Sci—SGGW, Land Reclam 49(3):223–235
64. Małek S, Gawęda T (2006) Charakterystyka chemiczna źródeł potoku Dupniańskiego w Beskidzie Śląskim. Sylwan 150(3):39–46
65. Małek S, Krakowian K (2012) The effect of deforestation on spring water chemistry on Skrzyczne (Silesian Beskid Mountains, Poland). J For Sci 58(7):308–313
66. Grodzki W, Guzik M (2009) Wiatro- i śniegołomy oraz gradacje kornika drukarza w Tatrzańskim Parku Narodowym na przestrzeni ostatnich 100 lat. Próba charakterystyki przestrzennej. In: Guzik M (ed) Długookresowe zmiany w przyrodzie i użytkowaniu TPN. Wydawnictwa Tatrzańskiego Parku Narodowego, Zakopane, pp 33–46
67. GUS (2018) Środki produkcji w rolnictwie w roku gospodarczym 2016/2017. www.stat.gov. pl/obszary-tematyczne/rolnictwo-lesnictwo/rolnictwo
68. Kopacz M, Twardy S (2006) Zmiany użytkowania ziemi w zlewni górnego Dunajca w aspekcie wybranych parametrów jakościowych wód powierzchniowych. Woda-Środowisko-Obszary Wiejskie 6(2):191–202
69. Smoroń S, Twardy S, Kuźniar A (2010) Bilans azotu i fosforu w rolniczych obszarach karpackich o niekorzystnych warunkach gospodarowania. Woda-Środowisko-Obszary Wiejskie 10(4):225–236
70. Żelazny M (2005) Cechy fizykochemiczne wód podziemnych. In: Żelazny M (ed) Dynamika obiegu związków biogennych w wodach opadowych, powierzchniowych i podziemnych w zlewniach o różnym użytkowaniu na Pogórzu Wiśnickim. IGiGP UJ, Kraków, pp 167–176

71. Siwek JP, Żelazny M, Chełmicki W (2008) Annual changes in the chemical composition of stream water in small catchments with different land-use (Carpathian Foothills, Poland). Soil Water Res 3:129–137

72. Jelonkiewicz E, Jelonkiewicz Ł, Wolanin A (2015) Czynniki kształtujące jakość płytkich wód podziemnych na Pogórzu Wiśnickim. Przegląd Geologiczny 63:786–790

73. Pietrzak M (2005) Gospodarka wodno-ściekowa. In: Żelazny M (ed) Dynamika obiegu związków biogennych w wodach opadowych, powierzchniowych i podziemnych w zlewniach o różnym użytkowaniu na Pogórzu Wiśnickim. IGiGP UJ, Kraków, pp 55–63

74. Soja R (2008) Wody w parkach narodowych południowej Polski. In: Partyka J, Pociask-Karteczka J (eds) Wody na obszarach chronionych. IGiGP UJ, Kraków, pp 43–52

75. Zawilińska B (2013) Landscape parks and the development of tourism in the protected areas of the Polish Carpathians. In: Kozak J, Ostapowicz K, Bytnerowicz A, Wyżga B (eds) The Carpathians: integrating nature and society towards sustainability, environmental science and engineering. Springer, Verlag Berlin Heidelberg, pp 461–476

76. Smoroń S, Twardy S (2003) Wpływ zmiennego nasilenia ruchu wczasowo-turystycznego na jakość wód Białego i Czarnego Dunajca. Woda-Środowisko-Obszary Wiejskie 3(2/8):91–102

77. Smoroń S, Twardy S, Janota D (2007) Jakość wód powierzchniowych w turystycznych obszarach Karpat Zachodnich. Cz. 2. Koncentracja działalności gospodarczej związanej z obsługą ruchu turystycznego. Woda-Środowisko-Obszary Wiejskie 7(2b): 167–176

78. Siwek JP, Biernacki W (2015) Wpływ ścieków odprowadzanych ze schronisk turystycznych na stężenia związków biogennych w potokach - odbiornikach tych ścieków na terenie Tatrzańskiego Parku Narodowego (lata 2008–2009). Gospodarka Wodna 7:202–209

79. Siwek JP, Biernacki W (2016) Effect of tourism-generated wastewater on biogenic ions concentrations in stream water in Tatra National Park (Poland). eco.mont 8(2): 43–52

80. www.schroniska-pttk.com.pl

81. Górnik M, Szczerbińska A (2015) Właściwości fizyczno-chemiczne i jakość wód podziemnych w rejonie Skawiny. Prace Geogr 140:79–89

Chapter 14
Surface Water Eutrophication in Poland: Assessment and Prevention

Elena Neverova-Dziopak

Abstract Presently the process of aquatic ecosystems transformation is much faster than before, because it is determined not so much by natural causes but mostly by anthropogenic factors. Anthropogenic eutrophication belongs to global processes which development has intensified significantly in recent decades and now affects all types of surface waters in many countries, leading to the disturbance of ecological balance and deterioration of water quality. In order to ensure the ecological safety of surface waters it is necessary to elaborate the reliable comprehensive abatement strategy based on appropriate system of eutrophication monitoring and assessment and the reduction of biogenic loads from all the sources. The paper presents the essence of eutrophication process, analysis of methodological problems of trophic state assessment and the assessment methodology based on simple and low-cost indicators which ensures the continuous eutrophication monitoring and control. The development of eutrophication of surface water in Poland and the principles of its prevention and management are also presented.

Keywords Eutrophication · Assessment methods · Trophic state · Nutrient loads · Poland

14.1 Definition, Essence and Consequences of Eutrophication

In modern times, the processes of aquatic ecosystems transformation run much faster than before, because they are conditioned not only by natural causes but also by anthropogenic factors. Anthropogenic eutrophication also belongs to global processes which intensity has increased significantly in recent decades, affecting

E. Neverova-Dziopak (✉)
Department of Environmental Management and Protection, Faculty of Mining
Surveying and Environmental Engineering, AGH University of Science and Technology, Al.
Mickiewicza 30, 30-059 Kraków, Poland
e-mail: elenad@agh.edu.pl

© Springer Nature Switzerland AG 2021
M. Zeleňáková et al. (eds.), *Quality of Water Resources in Poland*, Springer Water,
https://doi.org/10.1007/978-3-030-64892-3_14

Table 14.1 Negative consequences of eutrophication in water ecosystems

Ecological	Economic and social
Water blooms and the development of toxic algae species	Deterioration of water properties
Deterioration of aquatic organisms habitat (lack of oxygen, sunlight, algae toxins)	Increased costs of water and wastewater treatment
Reduction of species diversity	Reduced costs of land and real estate as a result of the loss of their aesthetic and recreational values
General ecological imbalance	Economic losses in fisheries and tourism sector
Loss of biospheric functions of aquatic ecosystems	Public health hazards

the surface waters in many countries and leading to deterioration of their ecological status and functional properties. The negative impact of this process concerns ecological, health and economic aspects (Table 14.1).

Relatively recently, 70–80 years ago, only a narrow group of specialists, mainly limnologists and hydrobiologists, was involved in the studies of biological productivity of water ecosystems and the methods of trophic state assessment. In recent decades, the significant deterioration of surface water quality as a result of eutrophication caused an increased interest. That's why now the studying of eutrophication development and the methods of assessment of water trophic status has ceased to be the object of interest only for hydrobiologists and limnologists. Now anthropogenic eutrophication has become the research focus for the specialists in the field of water protection, ecological engineering and water resources management.

Most of contemporary literature sources define the process of eutrophication as an enrichment of water with nutrients (biogenic substances) leading to the intensive development of aquatic vegetation and related negative consequences [1–4]. Only in a few fundamental studies it can be found the definition, which in more substantive way reflects the essence of this process and defines it as the disruption of the balance of processes of production and decomposition of organic matter produced by algae as a result of excessive nutrient supply of surface waters, leading to increased water productivity and secondary pollution [5–7]. Already at the beginning of research on lake ecosystems transformation, which was later defined as "anthropogenic eutrophication", it was established that the cause of these changes is the excessive enrichment of waters with nitrogen and phosphorus from anthropogenic sources. This statement doesn't not require any proving, because the dependence of aquatic autotrophic organisms development on nutrients content in water is well known and has been confirmed by numerous experiences and practice of using nitrogen and phosphorus fertilizers to increase the productivity of fish ponds.

Numerous long-term serious studies on surface water eutrophication have enabled the understanding and description of the essence of this phenomenon as the disturbance of the aquatic ecosystems functioning caused by the violation of their biotic balance, as a result of anthropogenic enrichment of waters with nutrients [6–9].

The supply of water vegetation with nutrients, the forms of their occurrence in water and their availability for primary produces are the most important aspects in investigation of eutrophication development. For autotrophic photosynthesis hydrobionts the main available forms of nutrients are inorganic nitrogen and phosphorus compounds. Many authors notice [10–12] that the increased nutrient content in water does not always lead to equivalent increase of aquatic vegetation development, that indicates the existence of other factors limiting the biomass production. Such factors include: temperature, degree of insolation, hydrodynamics, composition of dissolved substances, trophic chains and competition between organisms. The interaction of these factors sometimes leads to unpredictable effects, which complicates the assessment of individual factors impact.

With the increase of the rate of water plants biomass production, the rate of its decomposition increases accordingly. The balance between the process of production and decomposition of organic matter determines the trophic status of waters [3, 5, 6, 9]. The disturbance of this balance, which is called the biotic balance, its scale and consequences are determined mainly by climatic, morphometric, hydrological, hydrobiological and other characteristics of water ecosystems.

In the conditions of maintaining the homeostasis of aquatic ecosystems, these processes are self-regulating and provide the appropriate level of biological productivity and water quality in particular ecosystem. The introduction of additional anthropogenic nutrient loads to natural water environment, under the violation of self-regulation mechanisms threatens the stability of aquatic ecosystem. Natural eutrophication, as a rule, does not exceeds the homeostasis limits. But if eutrophication exceeds the limits of natural self-regulation, the changes may occur in biocenosis structures, which lead to the intensive development of blue-green algae, fish kill and other negative results. In the case of natural eutrophication, these changes are usually temporary and ecosystem renews itself [13–15].

One of the serious problems of anthropogenic eutrophication control is that at the initial stages of development the negative changes in aquatic ecosystems are hardly noticeable due to homeostatic mechanisms. Such changes are difficult to distinguish from natural changes, such as seasonal or annual fluctuations of different abiotic and biotic factors. It is worth noting, that "water blooms" arising as a result of anthropogenic impact on the ecosystem represent its adaptation reaction, which means that eutrophication can be considered a new stage of ecosystem development in new conditions. From the point of view of the impact on water ecosystems, the pollution and eutrophication are closely related and often mutually conditioned. In many cases, eutrophication occurs as a result of pollution, and pollution may results from eutrophication. In natural conditions, both processes usually occur together. However, in terms of the direction and degree of impact on aquatic ecosystem functioning, these processes are substantially different. It should be particularly emphasized, that the most important difference lies in the fact, that contamination is caused primarily by the discharge of harmful substances, that often reduces the biological productivity of waters, while eutrophication, conversely, leads to the increased productivity of water [7].

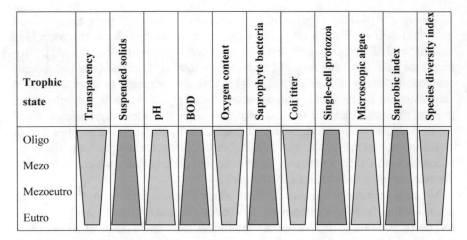

Fig. 14.1 Negative changes in water properties with trophic status increasing (on the base of [7])

The development of aquatic vegetation is a natural phenomenon and up to a certain level plays a positive role in water self-purification processes. However, the positive role of algae depends, to a large extent, on their concentration. When algae species multiplies to high concentrations, their positive role in water self-purification changes into self-polluting factor, leading to the disturbance of ecological equilibrium of water ecosystems, to the impoverishment of their species structure, reduction of resistance to negative impacts and deterioration of water quality [13, 16, 17]. The negative impact of eutrophication on physical, chemical and biological properties of water is presented in Fig. 14.1 [7].

When analysing the protection strategy against pollution and eutrophication and their impact on water ecosystems, it should be assumed that the pollution, as a kind of impact on ecosystems, should be excluded. However, the trophic state of surface water should be rationally regulated and maintained within the limits of its positive impact on ecosystems. It is very difficult but extremely important task.

The changes in ecosystem occur regardless of whether the increase of trophic level was achieved due to the natural development or under the influence of anthropogenic factors. But no matter, diagnostics of eutrophication causes and factors, the control and prognosis of its development are of great theoretical and practical importance. So, the elaboration of appropriate eutrophication assessment approaches, methodologies and indicators becomes a very important task.

14.2 Methodological Problems of Eutrophication Assessment

14.2.1 Assessment Approach According to Legislative Documents

To assess the dynamics of eutrophication development, it is necessary to collect extended reliable long-term monitoring data bank concerning the specific indicators of eutrophication. As shows the state of knowledge analysis, an appropriate assessment of water trophic status, the scale of negative changes in aquatic ecosystem and water quality is difficult due to the lack of sufficient long-term results of hydrobiological parameters observations. In addition, the system of eutrophication monitoring is unsatisfactory: there is no uniform methodological approach to trophic status assessment, the lack of reliable and low-cost eutrophication indicators.

Regarding the European legislation on the status of aquatic environment, it should be noted that many EU directives impose an obligation on member states to realize the monitoring of the parameters important for eutrophication assessment and appropriate recommendations on their limit values. Although only the Directive concerning urban waste-water treatment (91/271/EEC) [18] and Nitrates Directive (91/676/EEC) [19] contains explicitly the eutrophication assessment requirements: in 91/271/EEC Directive through the need to identify the "sensitive areas", i.e. "sensitive water bodies", and in the 91/676/EEC Directive by identifying "contaminated waters'" and consequently, the sensitive zones. The Water Framework Directive WFD (2000/60/EC) [20] maintains the recommendations of 91/676/EEC and 91/271/EEC Directives and additionally includes the recommendation to realize the eutrophication assessment as part of ecological status assessment. Recommendations for the assessment of eutrophication indicators are also included in some other EU Directives (Table 14.2) [18–23].

None of these directives gives a unified definition of the process, nor a methodology for its evaluation, except this, various indicators with different limit values are applied in different countries to assess the course of this phenomenon. In the interpretation of EU Directives, eutrophication is an undesirable phenomenon and does not require the precise definition of trophic state, but only the statement of the occurrence of adverse changes in water status. The comparison of the methods of surface water trophic status assessment according to the recommendations of WFD, Wastewater Directive and Nitrate Directive in terms of "ecological status", "sensitive area" and" contaminated waters" is presented in Table 14.3.

The assessment of trophic state according to the recommendations of different EU Directives often gives contradictory results. This is mainly the effect of the difference in adopted assumptions and approaches to assessing this phenomenon and practical objectives of the assessment. The different set of suggested trophic indicators and different interpretation of their limit values also contributes to the discrepancy of assessment results [23]. The disadvantage of these legislative recommendations lies

Table 14.2 Requirements of EU directives regarding surface waters eutrophication assessment and the scope of monitoring (on the base of [18–27])

Directive	Requirements for the assessment of eutrophication	Minimum requirements for monitoring of eutrophication
Water Framework Directive 2000/60/EU	Considered in the assessment of ecological status in case when the enrichment in biogenic substances effects the biological, physical and chemical quality elements	Phytoplankton (every 6 months), aquatic vegetation, macro invertebrates, fish (every 3 years) Hydromorphological elements (hydrology—continuous or monthly) other (every 6 years) Elements of physical and chemical quality (every 3 years)
Directive 91/271/EEC on Urban Wastewater Treatment	Conducted in order to identify the areas sensitive to eutrophication (i.e. eutrophic waters or which may become eutrophic in the near future)	Control of the sensitive areas or identification of new ones every 4 years
Directive 91/676/EEC concerning the protection of waters against pollution caused by nitrates from agricultural sources	Conducted in order to identify areas particularly exposed to pollution by nitrogen compounds from agricultural sources and eutrophication	Monitoring of nitrate content in surface waters and assessment of trophic status (every 4 years)
Directive 2006/44/EC Fish Directive	Does not contain the specific requirements for the assessment of eutrophication, but gives the recommended values of phosphorus concentrations	Ammonia, nitrites, pH, total phosphorus, dissolved oxygen (once per month)
Directive 79/923/EEC on the quality required of shellfish waters	Does not specify the requirements for the assessment of eutrophication	pH, dissolved oxygen (once per month)
Directive 2016/11/EU on pollution caused by some dangerous substances discharged to the aquatic environment	No special requirements for eutrophication assessment but qualitative requirements for phosphorus and substances affecting oxygen conditions, especially ammonia and nitrite	No special requirements
Directive 92/43/EEC on the conservation of natural habitats and of wild fauna and flora	Assessment of eutrophication only if there is a threat to the habitat or species	No special requirements

in the fact, that they enable only the descriptive assessment and can show only the tendency to eutrophication. For example, according to the Nitrate Directive, the trophic status of water body can be defined only as "sensitive or insensitive". According to the Water Framework Directive the eutrophication can be assessed only in the categories of "eutrophied or non-eutrophied waters". Also controversial

Table 14.3 Trophic state assessment according to EU directives (on the base of [18–23])

Ecological status	Ecological/trophic status definition		
	WFD	Wastewater directive	Nitrates directive
Very good	Almost undisturbed natural conditions	Eutrophication does not occur, identification of sensitive areas is not required	Not eutrophic, not contaminated. Identification of sensitive to nitrates areas is not required
Good	Conditions disturbed to a small extent as a result of anthropogenic impact	Eutrophication does not occur, identification of sensitive areas is not required	Not eutrophic, not contaminated. Identification of sensitive to nitrates areas is not required
Moderate	Moderate degree of disturbance of natural conditions and change in the structure of biocenosis	Eutrophic or may become eutrophic in the near future unless protective actions are taken. Identification of sensitive areas is required	Eutrophic or may become eutrophic in the near future unless protective actions are taken. Identification of sensitive to nitrates areas is required (unless the Member State considered the whole area of the country to be sensitive)
Bad	Significantly changed natural conditions, the structure of biocenosis differs significantly from the one usually found in a given type of water	Eutrophic. Identification of sensitive areas is required	Eutrophic. Identification of sensitive to nitrates areas is required (unless the Member State considered the whole area of the country to be sensitive)
Very bad	Seriously changed natural conditions in which there are no species typical for a given type of water	Eutrophic. Identification of sensitive areas is required	Eutrophic. Identification of sensitive to nitrates areas is required (unless the Member State considered the whole area of the country to be sensitive)

is the approach of the above Directives, specifying the need to take action in the case of waters classified as eutrophic. If high trophic state is conditioned by natural causes, the water body does not require protective measures. The measures are undertaken only if eutrophication is caused by anthropogenic reasons. The method of distinguishing between natural and anthropogenic eutrophication is also not specified in EU Directives [7, 21].

As a member of the European Union, Poland has implemented EU requirements to its environmental legislation: the Water Law Act [28] and appropriate implementing regulations. The main of them are: Regulation of the Minister of Environment of 23

Table 14.4 The principle of trophic status assessment according to HELCOM eutrophication assessment tool [23, 31]

Deviation from reference values	Acceptability	Trophic status
Lack	Acceptable	Very good
Slight		Good
Moderate	Unacceptable	Moderate
Significant		Bad
Strong		Very bad

December, 2002 on the criteria for determining waters sensitive to pollution with nitrogen compounds from agricultural sources [29] and Regulation of the Minister of the Environment of July 21, 2016 on the method of classification of the state of surface water bodies and environmental quality standards for priority substances [30]. After implementation of EU legislation requirements into Polish legislative documents the assessment of trophic status is made as the part of the assessment of general status of waters, and only indirectly in terms of the absence or occurrence of eutrophication.

Being a member of the European Union, Poland is also a signatory country of the Convention on the Protection of the Marine Environment of the Baltic Sea Area and has to meet its recommendations. For the purpose of monitoring the eutrophication of the Baltic Sea it was developed the HEAT index (HELCOM eutrophication assessment tool) [31, 32]. The HEAT index assumes a division into primary and secondary signs of eutrophication. The primary ones include: summer concentration of total nitrogen and phosphorus, winter concentration of dissolved mineral nitrogen and phosphorus, water transparency, chlorophyll content. Secondary signs include the depth of occurrence of sea grass and species diversity of macroalgae and invertebrates. For each of the mentioned parameters the reference conditions and acceptable deviations are established. The greater the deviation from the reference conditions the higher is trophic status, which is determined by the following descriptive definitions: very good, good, moderate, weak and bad. The final decision is adopted according to the rule "one out - all out", which is always imposed by the worst factor. The evaluation scheme based on the HEAT index is shown in Table 14.4.

14.2.2   Traditional Indicators of Trophic State

Traditionally, the process of eutrophication used to be assessed on the basis of the limit values of a set of indicators developed by different authors. Limit values of indicators for various trophic states are extremely difficult to determine. To a large extent it depends on the type of water bodies, their properties, geographical location and climatic conditions. The subjective preferences of the authors are often significant, which makes it impossible to develop uniform and unambiguous assessment system. The scale of variance of the numerical values of the basic widespread trophic indices is presented in Table 14.5.

Table 14.5 Traditional indicators of eutrophication and their reference values according to different authors (on the base of [12, 23, 33–36])

Criterion	Trophic state				Author
	Oligotrophic	Mesotrophic	Eutrophic	Hypertrophic	
1	2	3	4	5	6
Chlorophyll-*a* concentration, Chl *a*, μg/dm³	0.1–1.0	1–10	10–100	>100	Vinberg
	<4.3	4.3–8.8	>8.8	–	Dobsonetal
	<2.5	2–6	>6	–	Rast
	<1.5	1.5–10	10–50	>50	Trifinova
	<3.0	3–8	8–20	>20	Milius, Kyvasik
	0.1–1.0	1.10	>10	–	Bulion
	<4.0	4–10	10–100	>100	Hendrson-Selers
	<3	3.0–7.0	7.0–40.0	>400	Foresberg, Ryding
	<3	2.0–15	10–100	–	Likens
	0.3–3.0	4.0–10	10–100	>100	Welch
	0–4.0	5.0–20	20–100	>100	Sladecek
	<2.5	2.5–8.0	8.0–25	>25	OECD
	<2.9	2.9–5.6	>5.6	–	Chapry and Dobson
	<3.5	3.5–9.0	9.0–25.0	>25	Nurnberg
	<4.0	4.0–10.0	10–25	>25	ForsbergaiRydinga
	0.82–2.0	2.0–5.0	5.0–12.0	>31	Burns
	<2	2–5	>5	–	Dillon i Rigler
	<2	2–5	5–30	>30	Vant
	<0.95	2.6–7.3	20–56	56–155	Carlson
	<2.0	6.0	19.0	>24	Moss
Secchi disc transparency, H_{sc}, m	>4	2–3	<	–	Gantrbland
	11–6	6–2	<1	–	Thunmark
	>5	5–3	<2	–	Aberg, Roch
	>4	4–1	<3	–	Kitajev
	>6	6–3	<1	–	Hendersen-Selers
	9.9	4.2–2.4	<3	–	Vollenweider
	6.4	8–2	<1	–	Carlson
	>5	5–3	2.0–0.5	–	Chapry and Dobson
	>4	2–4	<3	<1	Nurnberg
	15–7.0	7.0–2.8	1–2	<0.4	Burnsa
	>5	5–2	2.8–1.1	–	Dillona i Riglera

(continued)

Table 14.5 (continued)

Criterion	Trophic state				Author
	Oligotrophic	Mesotrophic	Eutrophic	Hypertrophic	
1	2	3	4	5	6
	>10	5–10	<2	<1.5	Vant
	>8	4–2	1.5–5.0	0.25–0.5	Carlsona
	>6	6–3	0.5–1	<1.5	OECD
Total phosphorus concentration, TP.μg/dm^3	5–20	5–50	100	–	Romanienko
	5–10	10–30	30–100	>100	Vollenweider
	<15	15–25	25–100	>100	Foresberg, Ryding
	<10	10.0–35.0	35–100	>100	OECD
	<11.0	1.00–21.1	>21.7	>100	Chapry and Dobbson
	<10	10–30	30–100	>100	Nurnberg
	<25	15–25	25–100	>96	Forsberga i Rydinga
	4.1–9.0	9.0–20	20–43	–	Burnsa
	<9.9	9.9–18.5	>18.5	–	Dillona i Riglera
	<10	10–20	20–50	>50	Vant
	<6	12–24	48–96	96–192	Carlsona
	8.0	25.0	80.0	>100	Moss
Total nitrogen concentration TN, μg/dm^3	5–80	80–500	500–1500	–	Romanienko
	<400	400–600	600–1500	>1500	Foresberg, Ryding
	<350	350–650	650–1200	>1200	Nurnbrg
	<400	400–600	600–1500	>1500	Forsberga i Rydinga
	73–157	157–337	337–725	>1558	Burns
	<200	200–300	300–500	>500	Vant

However, due to the large number of factors determining the development of eutrophication, the criteria presented in Table 14.5 are not always recommended for practical use and may give insufficiently reliable or even contradictory results of assessment.

For example, in the conditions of high rate of organic matter decomposition which depends inter alia on hydrodynamic conditions, the high concentrations of nutrients not always lead to dangerous increase of water trophic level.

The concentration of chlorophyll "a" is a representative indicator of algae biomass. Therefore, this indicator is widely used to determine the trophic status of waters. The basic problem is that the concentration of chlorophyll at its value over 100 mg/dm^3, regardless of the increase in nutrient content, increases to a very small extent, because the lack of sunlight under conditions of high algae concentration can inhibit their further development [37].

The transparency according to Secchi disc is the oldest and the simplest method of trophic status assessment. However, this method does not always give an adequate assessment result, because the turbidity of water may also be caused by other factors and is not always associated with the concentration of algae. With a large turbidity of water, the penetrating possibility of sunlight decreases and turbidity may become a factor inhibiting the development of aquatic vegetation. So, the assessment based on a set of traditional eutrophication indicators does not guarantee an accurate result.

14.2.3   Integrated Numerical Indicators

An alternative to the approach based on the reference values of traditional eutrophication indices are the integrated numerical indicators, so-called trophic state indexes. The elaboration of such indexes is based on the established correlation between the basic factors of eutrophication, such as chlorophyll-a, nitrogen and phosphorus content, water transparency and other. For each trophic state the reference values of the indexes are developed. The chosen numerical indexes elaborated by different authors, that have found wide application in the assessment of water trophic status in different countries are presented in Table 14.6.

The aggregated indexes were developed by different authors on the basis of various methodological approaches and assumptions. The data base was the results of long-term monitoring carried out in various water bodies in different geographical regions. A common feature of the aggregated indexes is that they are based on the established correlation relationships and, to a greater or lesser extent, consider the key factors of eutrophication and the water ecosystem response to these factors.

14.3   Eutrophication of Surface Waters in Poland

It is generally recognized, that the main cause of eutrophication is the increased anthropogenic load of biogenic substances (nitrogen and phosphorus compounds) introduced into aquatic environment. At the same time, wastewaters, especially municipal, is considered one of the most dangerous source of nutrients. Therefore, in EU countries and the countries of the Baltic Sea catchment area, particular emphasis was placed on the elimination of these substances from wastewater. Joining the European Union Poland committed to fulfil the requirements of Council Directive 91/271/EEC concerning municipal wastewater treatment, therefore in 2003 the National Program of Municipal Wastewater Treatment (NPMWW) was created [42]. NPMWW contains a list of agglomerations which must be equipped with collective sewage systems and wastewater treatment plants within the time limits specified in the Program. Until now, five Program updates have been carried out in 2005, 2009,

Table 14.6 Numerical indexes for trophic state assessment with reference values for different trophic state [on the base of 7, 34, 38, 39, 40, 41]

Index	Numerical value					Author
	Ultraoligotrophic	Oligotrophic	Mesotrophic	Eutrophic	Hypertrophic	
TSI—Trophic State Index	<30	30–40	40–50	50–60	70–80	R. Carlson (American Lakes)
TRIX—Trophic Index	2.0	2–4	4–5	6–8	–	R. Vollenweider (Mediterranean, Black and Caspian Seas)
EI—Eutrophication Index	–	0.04–0.38	0.37	0.87	0.87–1.51	L. Primpas, Tsirtsis G. (Saronic Bay)
TLI—Trophic Level Index	2	2–3	3–4	4–5	5	N. Burns (lakes of New Zealand)
NIM—Nutrient Index Method	–	0–30	31–60	61–100	>100	National Chinese Center for Environmental Monitoring
ITS—Index of Trophic State	6.0–6.6	6.7–7.3	7.4–8.0	>8.0	–	E. Neverova-Dziopak (Gulf of Finland and low-mineralized waters)

2010, 2015 and 2017. Within NPMWW implementation in 2003 it was built 2016 84.8 thousand km of sewage network and 403 new municipal wastewater treatment plants; 1575 investments were carried out in the scope of modernization and/or extension of municipal wastewater treatment plants. The total investment cost was PLN 63.8 billion [43] (Figs. 14.2, 14.3 and 14.4).

As a result of the implemented investments in Poland, the improvement of wastewater management was observed: the amount of untreated wastewater discharged to water environment decreased significantly, and the load of nutrients was reduced accordingly (Fig. 14.5).

Among the countries signatory to the Helsinki Convention, Poland discharges the most significant loads of nutrients to the Baltic Sea and considered to be the greatest polluter of the sea. It is conditioned by a large catchment area, the share of agricultural lands (almost 50% of total agricultural areas of Baltic sea catchment) and high population density (almost 45% of the population of Baltic sea catchment live in Poland) [45].

Fig. 14.2 Number of population served by wastewater treatment plants in Poland, 2016 in % of total population (on the base of [4])

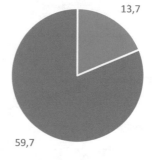

■ Biological   ■ With increased nutrients removal

Fig. 14.3 Number of population served by wastewater treatment plants in Poland in cities and villages, in % of total population (on the base of [4])

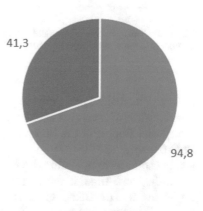

■ Cities   ■ Villages

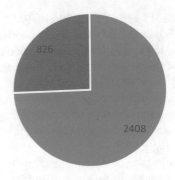

Fig. 14.4 Total number of wastewater treatment plants in Poland, 2016 (on the base of [4])

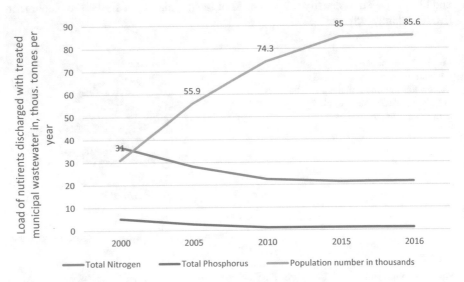

Fig. 14.5 Nutrient loads discharged to water receivers with treated municipal wastewaters and the number of population served by WWTP with EBNR (Enhanced Biological Nutrient Removal) (on the base of [4])

The percentage share in the total nutrient load discharged into the Baltic Sea by nine Helsinki Convention signatory countries is shown in Fig. 14.6. According to the special Report of the European Court of Auditors (2016) [46] and HELCOM PLC-6 project (2018) [47] Poland continued to be the largest exporter of nitrogen and phosphorus to the Baltic Sea.

At the same time, it can be noted that the load of nutrients discharged by the Polish rivers to the Baltic Sea has decreased as a result of the implementation of the National Program of Municipal Wastewater Treatment (Fig. 14.7).

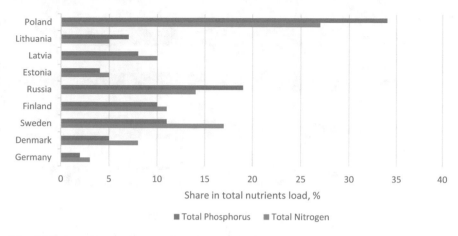

Fig. 14.6 Average annual proportion of total nutrients inputs into the Baltic Sea in the period 2001–2006 (on the base of [48])

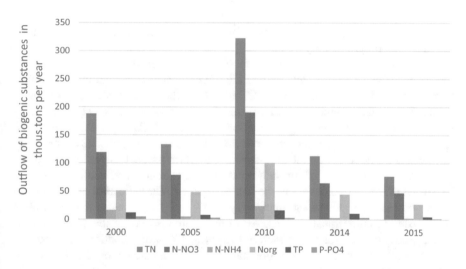

Fig. 14.7 Loads of biogenic substances carried by the rivers to the Baltic Sea (on the base of [44])

Summing up the results of the XIX International Environmental Forum «Baltic Sea Day» held in St. Petersburg in March 2018 it can be stated that positive trend in nutrient loads reduction is encouraging, but the state of the Baltic Sea is still alarming. With the development of agricultural sector more nitrogen and phosphorus is supplied to the Baltic Sea. The process of Baltic Sea eutrophication was diagnosed by experts half a century ago, and since then the scale of the process has been increasing. For this reason, the total annual losses amount to 3.8 - 4 billion euros, and 214 species of the Baltic flora and fauna now is affected by eutrophication. According to HELCOM

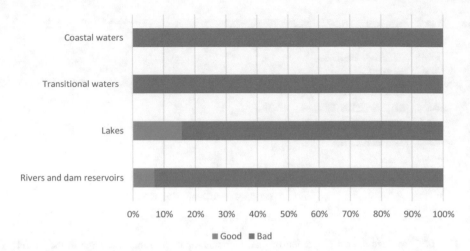

Fig. 14.8 The status of uniform surface water bodies monitored in 2011–2016 (on the base of [44])

the Baltic Sea Action Plan was fulfilled by 70%. But some countries implemented their regional programs only by 20% [49].

Unfortunately, the situation in Poland looks similar. Even after the implementation of capital-intensive investments in the field of wastewater treatment and the implementation of EU directives into Polish legislation, the state of water environment does not show the proportional improvement, especially in the aspect of eutrophication. Currently, after the implementation of WFD and EU legislation principles, the direct assessment of surface water trophic state is not required. The trophic state of waters is assessed within the assessment of uniform parts of surface water status, which in turn is assessed taking into account their chemical and ecological status. The worst indicator decides about the final assessment. Classification of the uniform water body to the status worse than "very good" or "good" means that the waters are exposed to eutrophication [50]. According to the results of the assessment of surface water current state presented in Fig. 14.8, 100% of transitional and coastal waters, about 75% of lakes and more than 90% of rivers are exposed to the risk of eutrophication.

14.4 Problem Analysis

Regardless of the broad-scale projects implemented in Poland and other Baltic Sea basin countries, aimed at reduction of biogenic loads discharged to the sea and related huge capital expenditures, the expected effects in eutrophication abatement were not achieved and costs-benefit ratio is quite unfavorable.

It seems that there are two basic reasons for failure: (1) the lack of an appropriate normative-legal base and assessment approaches, (2) the lack of an effective strategy for eutrophication control and managing.

14.4.1 Assessment Tool for Eutrophication Management

Prevention of eutrophication and its negative effects should be based on reliable information on actual water trophic state, obtained by means of low-cost, simple and easily interpreted eutrophication indicators which also suitable for application purposes. Such indicators form the basis for setting the permissible concentrations of nitrogen and phosphorus in surface water considering the individual properties of different ecosystems, for calculation of permissible loads of biogenic matter within the accumulative capacity of water body. The indicators can be also used for formulating the predictive mathematical models, forecasting and assessing eutrophication development and the effectiveness of protective measures, for the evaluation of wastewater impact and other sources of biogenic matter on water ecosystems and establishing of the necessary level of nutrients reduction. They could be the basis of a properly organized monitoring system and provide reliable information about the state of the environment, i.e. they will be an effective tool of eutrophication management.

The methodology of eutrophication assessment legally binding currently in Poland cannot be the basis for effective management of this process, because it does not allow to obtain an exact result of actual trophic state assessment and not useful for application purposes.

The analysis of the existing methodological problems in the assessment of eutrophication process presented in Sect. 14.2 allows to conclude, that presently the urgent task is to develop the universal methodological approach to eutrophication assessment based on such indicators, that would meet the objectives and tasks of water protection and management in accordance with sustainable development principles. The indicators of eutrophication should match the following modern requirements [51–53]:

- reflect changes in water ecosystems caused by anthropogenic factors;
- reflect the basic functions of aquatic ecosystems;
- based on a limited number of available data, and the cost of their assessment should be moderate;
- be suitable for forecasting of ecosystem state changes;
- benchmarks should be developed for indicators, that allow for the interpretation of possible changes or threats by different users of these indicators;
- they should be aggregated, integrating information in time and space;
- they should be well-founded theoretically.

Among the methods of eutrophication assessing, the aggregated indicators presented in Table 14.6 are the most suitable to achieve the above mentioned goals. A good example of such indicator may be the Index of Trophic State (ITS) elaborated by the Author of the chapter, which fully meets the requirements set for modern environmental indicators [12, 54]. The elaboration of this index was based on the following theoretical assumption: eutrophication is the process leading to deterioration of water ecosystem biotic balance, that in turn brings to the violation of gas balance in water and changes in oxygen and carbon dioxide content. ITS is based on the following correlation dependence established by the Author and can be calculated with the Eq. (1)

$$ITS = \sum_{i=1}^{n} pH_i/n + a\left(100 - \sum_{i=1}^{n} O_2/n\right) \qquad (1)$$

where pH and O_2% — the values of pH and oxygen saturation measured simultaneously; 100—normal 100% water saturation with oxygen; a-empirical coefficient; n—number of measurements [12, 15].

The following reference values of ITS index for waters of different trophic state have been determined empirically: for dystrophic water the ITS value is lower than 5.7 ± 0.3; for ultraoligotrophic water 6.3 ± 0.3; for oligotrophic water 7.0 ± 0.3; for mesotrophic water 7.7 ± 0.3; for eutrophic water over 8.3 ± 0.3 [7, 12].

The application of this index made it possible to assess not only the trophic status of all 26 types of rivers in Poland but also the tendency to eutrophication [23]. The final result of this assessment are presented in Table 14.7.

ITS applications enabled also the estimation of long-term dynamics of eutrophication processes in dam reservoirs in Poland. An example of the assessment of eutrophication development is presented in Table 14.8

The status of transitional waters of the Szczecin lagoon was also assessed using the ITS index. The results of the assessment in different measurement points in lagoon (Fig. 14.9) are presented in Table 14.9.

Applicability of ITS as reliable assessment tool was also confirmed by its usage for eutrophication assessment in various types of surface waters in different geographical regions e.g.: all types of rivers and over a dozen dam reservoirs and lakes in Poland, small Estonian rivers, the Szczecin lagoon, the Neva Bay and eastern part of the Gulf of Finland, the Saaler Bodden, Taganrog Bay of the Azov Sea and other water bodies e.g. [59–64]. The experience with practical application of the proposed method of trophic state assessment based on ITS index showed, that this method has an universal character and allows to obtain a reliable assessment of trophic status of surface waters of various types. Compared to the currently used methods, the assessment on the base of ITS is fast and cheap: time needed for assessment is approx. 50 times shorter and the cost of monitoring is 20–30 times less [23].

Table 14.7 Percent of measurement points with a tendency to eutrophication [23]

The abiotic type of running waters	Measurement points (%)
Unspecified type	**74**
Mountain landscape	**33**
Sudecki stream	33
Upland landscape	**61**
Silicate upland stream with a coarse substrate—west	46
Silicate upland stream with a fine-grained substrate—west	86
Upland carbonate with fine grain substrate on loess and loess-like	89
Upland stream of coarse carbonate substrate	40
Small uplandsilicate river—west	39
Small highland carbonate river	68
Medium upland river—west	64
Flysch stream	75
Small flysch river	63
Medium highland river—east	50
Lowland landscape	**72**
Loess or clay lowland stream	64
Sandy lowland stream	83
Gravel lowland stream	65
Sandy-clay lowland river	59
Gravel lowland river	69
Great lowland river	96
River estuaries under the influence of salt water	89
Independent of ecoregions	**74**
Stream in the area under the influence of peat-forming processes	73
Small and medium river in an area influenced by peat-forming processes	66
Watercourse connecting lakes	91
Watercourse in the valley of great lowland river	50

14.4.2   Eutrophication Abatement Strategy

The development of eutrophication abatement strategy aimed at prevention of aquatic ecosystems degradation should be based primarily on identifying all sources of nutrients discharge to surface waters and their share in the total external load. This approach will give the opportunity to identify the key goals of particular attention and manage the financial streams in a rational way to achieve the real reduction of total nutrient load. Currently, significant reduction of nutrient loads discharged into

Table 14.8 Results of trophic state assessment of chosen dam reservoirs [on the base of 55–58]

Dam reservoir	Assessment period	ITS value	Result of assessment
Tresna	2010	8.12	Eutrophic
	2013	7.94	Mesotrophic
	2014	7.80	Mesotrophic
	2015	8.18	Eutrophic
Międzybrodzie	2010	7.89	Mesotrophic
	2012	7.75	Mesotrophic
	2013	7.89	Mesotrophic
Czaniec	2011	7.86	Mesotrophic
	2012	7.89	Mesotrophic
	2013	7.67	Mesotrophic
	2014	7.74	Mesotrophic
	2015	7.86	Mesotrophic
Sola cascade	2010	8.01	Meso-eutrophic
	2011	7.86	Mesotrophic
	2012	7.82	Mesotrophic
	2013	7.83	Mesotrophic
	2014	7.77	Mesotrophic
	2015	8.02	Meso-eutrophic
Dobczyce	2010	8.27	Eutrophic
	2011	8.43	Eutrophic
	2012	8.16	Eutrophic
	2014	8.34	Eutrophic
	2015	8.34	Eutrophic
Solina	2010	8.56	Eutrophic
	2011	8.34	Eutrophic
	2012	8.41	Eutrophic
	2013	8.50	Eutrophic
	2014	8.42	Eutrophic
	2015	8.45	Eutrophic

surface waters was achieved as a result of focusing on point sources, which was associated with huge cost of investments in wastewater treatment plants. As was shown in the previous chapter, this has not lead to the prevention of eutrophication.

According to the data presented in the HELCOM Report the total nutrients load introduced into the Baltic Sea from Poland decreased as follows: total nitrogen—from 262,159 t in 1995 to 169,941 t in 2014 and total phosphorus—from 14,846 t in 1995 to 12,776 t in 2014 [47]. Whereas the contribution of various sources to the total load of these substances is presented as follows (Fig. 14.10).

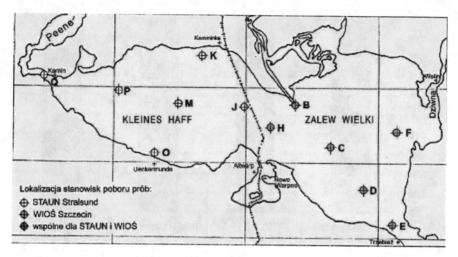

Fig. 14.9 Location of measurement points in Szczecin lagoon. *Source* https://www.wios.szczecin. pl/bip/

Table 14.9 The results of trophic state assessment in different monitoring points in Szczecin lagoon [58]

Measurement point	ITS value	Trophic state
B	8.54	Eutrophic
C	8.53	Eutrophic
D	8.50	Eutrophic
E	8.39	Eutrophic
F	8.48	Eutrophic
H	8.54	Eutrophic
J	8.51	Eutrophic
K	8.62	Eutrophic
M	8.76	Eutrophic
O	8.76	Eutrophic
P	8.68	Eutrophic

Despite the fact that in the period 1995–2014 the total nitrogen and phosphorus load decreased by 35 and 15% accordingly, it can be noticed that river runoff still plays a dominant role in the supply of nutrient loads to the Baltic Sea [47]. Riverine loads of nutrients in Poland are conditioned mainly by point-sources and diffused agricultural sources (Fig. 14.11).

The reduction of nutrient loads discharged into the Baltic Sea from Poland was achieved mainly as a result of the construction and modernization of wastewater treatment plants in the framework of implementation of the National Program of Wastewater Treatment [42, 65]. Diffuse agricultural sources while having the largest

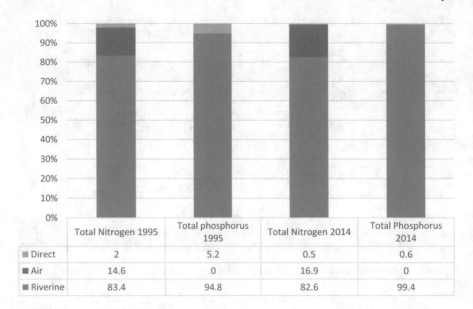

	Total Nitrogen 1995	Total phosphorus 1995	Total Nitrogen 2014	Total Phosphorus 2014
■ Direct	2	5.2	0.5	0.6
■ Air	14.6	0	16.9	0
■ Riverine	83.4	94.8	82.6	99.4

Fig. 14.10 The share of different sources in total loads of biogenic substances, % (on the base of [47])

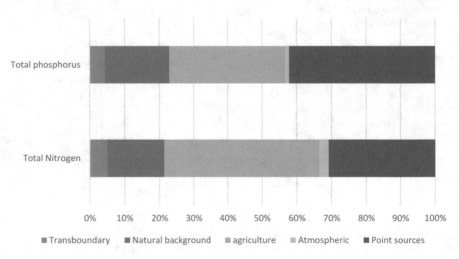

Fig. 14.11 The share of different sources in total load of nutrients discharged into the Baltic Sea (on the base [47])

share in these charges still remain without proper attention [66]. Such situation does not allow for effective implementation of protective strategies.

14.5 Conclusion

Currently, the significant successes have been achieved in the field of the abatement of anthropogenic eutrophication, but only in some directions. In general, the problem of biogenic loads reduction discharged with municipal wastewaters has been mainly solved. But the methods of elimination of negative impact of diffuse sources, especially, from agricultural land, are developing very slowly and are still at the margins of interest in existing protection strategies. It should be stated that the solution of the priority problem of eutrophication abatement requires, above all, a comprehensive approach. It becomes clear that one-sided approach, such as development and implementation of expensive wastewater treatment technologies, without taking into account other sources of nutrients does not give the desired result. The solution of this problem requires the systematic approach to the development and implementation of comprehensive technical, technological, organizational and legal measures covering the whole catchment and all sources of biogenic matter discharge.

However the basis for the development of protective measures directed to achievement and maintaining the good status of surface waters is the reliable information on their actual trophic state. This information can be obtained within properly organized system of monitoring based on appropriate methods of trophic status assessment.

References

1. Council Directive 91/676/EEC of 12 December 1991 concerning the protection of waters against pollution caused by nitrates from agricultural sources
2. Dziennik Ustaw Nr 115 poz. 1229 Ustawa z dn. 18 grudnia 1991 dotycz ochrony wód przed zanieczyszczeniami powodowanymi przed azotany pochodzenia rolniczego (91/676/EWG)
3. Dojlido J (1995) Chemia wód powierzchniowych. Wydawnictwo Ekonomia i Środowisko
4. Nixon S (1995) Coastal marine eutrophication:a definitione, social cuses and future concerns. OPHELIA 41, s. 199–219
5. Hutchinson GA (1967) Tretise on limnology. V.2. Introduction to lake biology and the limnoplankton, 2nd edn. New York
6. Rossolimo LL (1977) Izmienienie limniczeskich ekosystem pod wozdiejstwiem antropogennogo faktora. Izd. „Nauka", Moskwa
7. Neverova-Dziopak E (2010) Podstawy zarządzania procesem eutrofizacji antropogenicznej. Wydawnictwa AGH, Kraków
8. Ansari AA, Gill SS, Khan FA (2011) Eutrophication: threat to aquatic environment eutrophication: causes, consequences and control. Springer Science + Business Media B.V., ISBN 978-90-481-9624-1, Springer Dordrecht Heidelberg London New York. https://doi.org/10.1007/978-90-481-9625-8_7
9. Odum EP (1971) Fundamentals of ecology, 3rd edn. W.B. Saunders Company, Philadelphia, London and Toronto
10. Kajak Z (1979) Eutrofizacja jezior. PWN, Warszawa
11. Lund JW (1989) Phytoplankton. In: Eutrophication: causes, sequences, corrective. Proceedings of the symposium, Washington, pp 173–181
12. Neverova-Dziopak E (2007) Ekologiczne aspekty ochrony wód powierzchniowych. Monografia, Oficyna Wydawnicza Politechniki Rzeszowskiej, Rzeszow
13. Sirenko LA, Gawrilenko VJ (1978) Cwietienie wody i ewtrofirowanie. Naukowa dumka, Kijev
14. North American Lake management Society (2001) Managing lakes and reservoirs. Third edition

15. Tsvetkova LI, Neverova-Dziopak E, et al (2001) Ekologia. Uczebnoje posobie dla techniczes-kich wuzow. Moskwa, ACB, Chimizdat, St. Petersburg
16. Fiodorow WD (1970) Pierwicznaja produkcja kak funkcjia struktury fitoplanktonnogi soob-szestwa. Dokl. AN RSRSR, 192, Ser. Biol. N 4, s. 901–904
17. Chislock MF Doster E, Zitomir RA (2013) Eutrophication: causes, consequences, and controls in aquatic ecosystems. Nat Educ Knowl 4(4):10
18. COUNCIL DIRECTIVE of 21 May 1991 concerning urban waste water treatment (91/271/EEC)
19. COUNCIL DIRECTIVE of 12 December 1991 concerning the protection of waters against pollution caused by nitrates from agricultural sources (91/676/EEC)
20. Directive 2000/60/EC of the European Parliament and of the Council of 23 October 2000 establishing a framework for Community action in the field of water policy
21. Soszka H (2009) Problemy metodyczne związane z ocena stopnia eutrofizacji jezior na potrzeby wyznaczania stref wrażliwych na azotany. Water-Environment-Rural Areas, IMUZ, t.9, z.1 (25), s. 151–159
22. Loga M (2009) Raport D-I Analiza możliwości wykorzystania ocen środowiska wodnego wykonanej na podstawie Dyrektyw unijnych innych niż RDW do oceny stanu. WIOŚ, Warszawa
23. Kowalewski Z (2012) Weryfikacja możliwości zastosowania integralnego kryterium do oceny stanu troficznego wód płynących. PhD Thesis. AGH University of Science and Technology. Scientific supervisor: Elena Neverova-Dziopak, Krakow
24. Directive 2006/44/EC of the European Parliament and of the Council of 6 September 2006 on the quality of fresh waters needing protection or improvement in order to support fish life
25. Council Directive of 30 October 1979 on the quality required of shellfish waters (79/923/EEC)
26. Directive 2006/11/EC of the European Parliament and of the Council of 15 February 2006 on pollution caused by certain dangerous substances discharged into the aquatic environment of the Community
27. Council Directive 92/43/EEC of 21 May 1992 on the conservation of natural habitats and of wild fauna and flora
28. Ustawa Prawo wodne z dnia 20 lipca 2017 r. (Dz.U. 2017r. poz. 1566, ze zm.)
29. Rozporządzenie Ministra Środowiska z dnia 23 grudnia 2002 r. w sprawie kryteriów wyznaczania wód wrażliwych na zanieczyszczenie związkami azotu ze źródeł rolniczych (Dz. U. z dnia 31 grudnia 2002 r.)
30. Rozporządzenie Ministra Środowiska z dnia 21 lipca 2016 r. w sprawie sposobu klasy-fikacji stanu jednolitych części wód powierzchniowych oraz środowiskowych norm jakości dla substancji priorytetowych (Dz.U. 2016, poz. 1187)
31. HELCOM (2009) Eutrophication in the Baltic sea. Baltic Sea Environment proceedings No 115A
32. HELCOM (2007) Danish EPA: testing of the draft HELCOM eutrophication assessment (HEAT) in 45 basins and coastal water bodies of the Baltic Sea. HELCOM EUTRO-PRO 5/2007
33. Dmitriew VV, Frumin GT (2004) Ekologiczeskoje normirowanie i ustojciwosc prirodnych sistem. St Petersburg
34. Carlson RE (1977) A trophic state index for lakes. Limnol Oceanogr 22:361–369
35. Hooper FF (1969) Eutrophication indices and their relation to other indices of ecosystem charge. In: Eutrophication: causes, consequences, correctives. Proceedings of a symposium, Washington, DC
36. Vollenweider RA, Krekes JJ (1980) Background and summary results of the OECD cooperative program on eutrophication. In: Proceedings of the international symposium on inland waters and lake restoration
37. Dmitrijew WW (1995) Diagnostyka i modelirowanie wodnych ekosystem. SPbGASU, St. Petersburg
38. Giovanardi F, Vollenweider R (2004) Trophic conditions of marine coastal waters: experience in applying the Trophic Index TRIX to two areas of the Adriatic and Tyrrhenian seas. J. Limnol. 63(2):199–218

39. Burns NM, Rutherford JC, Clayton JS (2001) A monitoring and classification system for New Zealand lakes and reservoirs. J Lakes Res Manage 15(4):255–271

40. Primpas I, Tsirtsis G, Karydis M, Kokkoris G (2009) Principal component analysis: development of a multivariate index for assessing eutrophication according to the European Water Framework Directive. Ecol Ind. https://doi.org/10.1016/j.ecolind.2009.04007

41. Xiao Y, Ferreira JG (2007) Trophic assessment in Chinese coastal systems—review of methods and application to the Changjiang (Yangtze) Estuary and Jiaozhou Bay. Estuaries Coasts 30(6):901–918

42. KPOŚK (2003) Krajowy program oczyszczania ścieków komunalnych. Rada Ministrów, Warszawa

43. Gospodarka ściekowa w Polsce w latach 2015–2016. Państwowe Gospodarstwo Wodne „Wody Polskie". Krajowy Zarząd gospodarki wodnej, Warszawa. Available: www.wody.gov.pl

44. Environment (2017) Statistical information and elaborations. Central Statistical Office, Warsaw

45. Pastuszak M, Zalewski M, Wodzinowski T, Pawlikowski K (2016) Eutrophication in the Baltic Sea – necessity of a holistic approach to the problem. Stan środowiska południowego Bałtyku. Morski Instytut Rybacki, Gdynia

46. Europejski Trybunał Obrachunkowy (2016) Sprawozdanie specjalne. Przeciwdziałanie eutrofizacji w Morzu Bałtyckim – wymagane są dalsze, bardziej skuteczne działania. Urząd Publikacji Unii Europejskiej, Luksemburg

47. HELCOM PLC-6 Project (2018) Sources and pathways of nutrients to the Baltic Sea Environment proceedings No. 153, Helsinki

48. HELCOM (2009) Eutrophication in the Baltic Sea. An integrated thematic assessment of the effects of nutrient enrichment in the Baltic sea region. Baltic sea Environment proceedings no 115B, Helsinki

49. Юшковский В (2018) Балтика без прописки Санк-Петербургские ведомостию Available at: http://www.helcom.fi/news/Pages/Environmental-dialogue-at-international-Baltic-forum-in-Russia.aspx

50. Sygnały-2016. Stan środowiska w Polsce (2016) GIOŚ, Biblioteka Monitoringu Środowiska, Warszawa

51. Economic Commission for Europe, Committee on Environmental Policy (2007) Making monitoring and assessment an effective tool in environmental policy. ECE/CEP/2007/7, Geneva

52. Working Group on Environmental Monitoring and Assessment (2008) A practical guide. IX session of the Working Group, Geneva, 4–5 September

53. Belfiore S, Balgos M, McLean B, Galofre J et al (2003) Reference guide on the use of indicators for integrated coastal management. ICAM, UNESCO

54. Tsvetkova LI, Neverova E et al (1995) G.I.Patent N 2050128 FR, Rospatent. Sposob opriedielenia ekologiczeskogo sostojania priesnowodnych wodojemow. Moskwa

55. Neverova-Dziopak E, Droździk A (2016) Weryfikacja możliwości zastosowania wskaźnika ITS do oceny stanu troficznego zbiornika Dobczyckiego Budownictwo i Inżynieria Środowiska. 2016.166. https://doi.org/10.7862/rb.2016.166

56. Neverova-Dziopak E, Droździk A (2017) Analiza porównawcza stanów fizyko-chemicznego i troficznego wód Jeziora Chanieckiego i jeziora Goczałkowickiego w latach 2011–2015. Ochrona Środowiska, vol. 30, nr 2. Wrocław

57. Neverova-Dziopak E, Droździk A (2017) Analiza wieloletnich zmian stanu troficznego zbiornika Tresna w aspekcie jego lokalizacji i roli w kaskadzie zbiorników. Inżynieria Ekologiczna 18(1):s. 135–148, Politechnika Lubelska, Lublin

58. Droździk A (2016) Ocena stanu troficznego wód zbiorników zaporowych Jeziorsko i Sulejowskiego. Inżynieria i Ochrona Środowiska. T. 19, nr 1:127–140, Wrocław

59. Neverova-Dziopak E, Kowalewski Z (2014) Optymalizacja procedury oceny stanu troficznego wód przejściowych na przykładzie Zalewu Szczecińskiego. Budownictwo i Inżynieria Środowiska 54. https://doi.org/10.7862/rb.2014.54

60. Loigu EO (1982) Биогенные вещества и качество воды малых рек Эстонии. Автореферат; Гидрохимический институт, Ростов-на-Дону

61. Jagus A (2011) Assessment of trophic state of water in sola cascade dam reservoirs. Proc ECOpole 5(1):233–238, Opole
62. Gruca-Rokosz R, Koszelnik P, Tomaszek JA (2011) Ocena stanu troficznego trzech nizinnych zbiorników zaporowych Polski południowo-wschodniej. Inżynieria Ekologiczna, Nr 26:196–205, Lublin
63. Gruca-Rokosz R (2013) Stan troficzny zbiornika zaporowego Rzeszów. Journal of civil engineering, environment and architecture JCEEA, t. XXX, z. 60(3/13):279–291, Rzeszow
64. Zhidkova AJ (2017) Геоэкологическая оценка эвтрофирования вод Таганрогского залива. Диссертация на соискание ученой степени кандидата географических наук, Таганрог
65. Michalkiewicz M, Kruszelnicka I, Ginter-Kramarczyk D (2020) Ścieki i ich oczyszczanie - dane statystyczne. Technologia wody, No 3–4(71–72) (2020) Wydawnictwo Seidel-Przywecki Sp. z o.o
66. Pihlainen S, Zandersen M, Andersen HE et al (2020) Impacts of changing society and climate on nutrient loading to the Baltic Sea. Sci Tot Environ 731:138935, 20 August 2020. https://doi.org/10.1016/j.scitotenv.2020.138935

Chapter 15
Monitoring of Small Catchments in Poland Under the Integrated Environmental Monitoring Programme: The Functioning of the Struga Toruńska River Agricultural Catchment

Marek Kejna, Ireneusz Sobota, Joanna Uscka-Kowalkowska, and Henryka Wojtczak

Abstract This chapter analyses the functioning of the agricultural catchment area of the Struga Toruńska river (Central Poland). The observations were carried out as part of the Integrated Environmental Monitoring Programme in the years 1994–2017. A considerable increase in air temperature (0.48 °C/10 years) was found along with variability in precipitation totals (413.0–762.0 mm). Meteorological conditions contributed to substantial variability in water stages and flow rates in the river. The mean annual flow rate at Koniczynka ranged from 0.10 to 1.04 m^3s^{-1} (0.55 m^3s^{-1} on average). An average of 175 mm of water was drained from the catchment (32% of the precipitation). The specific runoff was 5.60 dm^3s^{-1} km^{-2}, which was similar to typical values observed in the Polish lowlands. The river regime is nival and moderately developed (the spring flow accounted for 166% of the mean monthly value). Extensive farming affects the quality of water in the Struga Toruńska. Increased concentrations of biogenic substances, mineral substances, and suspended matter were found in the agricultural catchment area, in particular after spring thaw when surface flushing increases. These substances are increasingly transported by a dense drainage network. In the summer, the biogenic content falls due to increased vegetation processes. The value of BOD5 also increases, and aerobic conditions decline. The catchment of the Struga Toruńska is considered an area at particular risk of pollution by nitrogen from agricultural sources.

M. Kejna (✉) · I. Sobota · J. Uscka-Kowalkowska
Faculty of Earth Sciences and Spatial Management, Nicolaus Copernicus University, Lwowska 1, 87-100, Toruń, Poland
e-mail: marek.kejna@umk.pl

I. Sobota
e-mail: irso@umk.pl

J. Uscka-Kowalkowska
e-mail: Joanna-uk@umk.pl

H. Wojtczak
Voivodship Inspectorate of Environmental Protection, Skargi 2, 85-018, Bydgoszcz, Poland
e-mail: henryka.wojtczak@wios.bydgoszcz.pl

© Springer Nature Switzerland AG 2021
M. Zeleňáková et al. (eds.), *Quality of Water Resources in Poland*, Springer Water,
https://doi.org/10.1007/978-3-030-64892-3_15

Keywords Integrated Environmental Monitoring Program · Climate change · Flow rate · Runoff · Water chemistry · Struga Toruńska. Poland

15.1   Introduction

Small river or lake catchments allow comprehensive research to be conducted into the state of an environment subject to continuous changes from natural and anthropogenic processes. Rapid climate changes in the world (IPCC 2018) and in Poland [1–3] and increasing environmental anthropopressure call for regular observations and monitoring of the environment.

Multi-year series of observations make it possible to conduct not only diagnostic but also predictive studies, and to develop environmental scenarios [4]. An understanding of the operating principles of analysed ecosystems makes it possible to identify risks and threats to the environment. Environmental protection is only practicable through sustainable management of its natural resources and respect for the rights of both animate and inanimate nature.

Particular risks occur in urban and agricultural areas. In Poland, agricultural land takes up as much as 60% of the area [5]. Since the end of World War II its extent has been decreasing due to afforestation [6], but it is paramount to look into the functioning of those ecosystems. Natural and artificial fertilisers are widely used in the cultivation of plants [7]. In the Kujawsko-Pomorskie province, the level of artificial fertiliser use is among the highest: in 2015 the nitrogen balance reached 73.3 kg/ha and the phosphate balance was 4.4 kg/ha [5]. Farming has a strong impact on the state of the environment, especially on the soil and on the surface- and groundwater [8].

The purpose of this chapter is to provide an analysis of the functioning of the catchment of the Struga Toruńska river (Chełmno Lakeland, Central Poland), which represents a typical agricultural catchment, in the years 1994–2017.

15.2   Integrated Environmental Monitoring in Poland

Integrated Environmental Monitoring (IEM) in Poland started in 1992, as a part of the State Environmental Monitoring Programme. The purpose of the IEM is to collect data on the condition of the natural environment that is representative for Polish geo-ecosystems and to interpret transformations undergoing in the environment [4]. The flow of energy and the circulation of matter in the environment cause changes in the landscape. Using identified regularities, it is possible to determine threshold values and elaborate development scenarios for the monitored geo-ecosystems where climate changes and various forms of anthropopressure are being considered. Comprehensive IEM studies have been carried out in the form of 18 observation projects, including biotic and abiotic elements of the environment [9].

Fig. 15.1 Integrated environmental monitoring stations in Poland (*Source* http://zmsp.gios.gov. pl/?page_id=20)

The IEM programme is in realise at 11 base stations in Poland (Fig. 15.1) and 1 station located on Spitsbergen in the Arctic. The observations are being carried out in representative small catchment of rivers or lakes.

In 2017, some of the Polish IEM stations were incorporated into the network of the European Integrated Monitoring programme [10].

15.3 Environmental Monitoring in the Struga Toruńska River Catchment

The Chełmno Lakeland IEM station at Koniczynka represents a geo-ecosystem of a glacial morainic upland, forestless, and used for agricultural purposes [11]. Intensive farming in that area is an extra burden on the environment, degrading soil, and polluting surface water. Due to the clayey soils and specific hydrological conditions

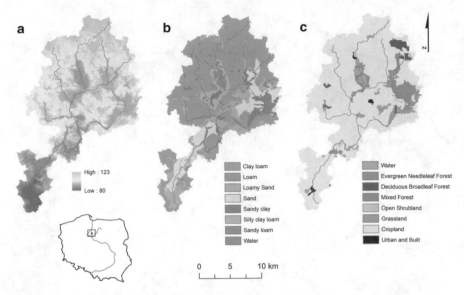

Fig. 15.2 The catchment area of the Struga Toruńska: A—hypsometry, B—surface sediments, C—land use (after Kejna and Strzyżewski [14])

of the morainic upland, most of the fields are drained. The vicinity of a city Toruń is also noticeable for the state of the environment [12].

The Struga Toruńska is a small river, at 51.3 km long, with its source in Lake Wieldządz and its outlet in the Vistula River in Toruń (Fig. 15.2). The area of its river basin is 371.0 km². In its upper part, the Struga Toruńska flows through a number of lakes, which considerably affects its water regime and the chemical properties of the water [13]. The river has been controlled along most of its length.

A 35.173 km² part of its river basin has been sectioned off as a representative catchment for the IEM programme. The area stretches from the river gauge at Lipowiec in the north to the river gauge at Koniczynka in the south (Fig. 15.3).

The area of the representative catchment is a morainic plateau, divided by a belt of thaw water plain of the Poznań phase of Weichselian glaciation. The surface of the area is flat and occasionally undulating. The highest point reaches 101.4 m in the north, whereas the lowest (77.8 m a.s.l.) is located at the Koniczynka river gauge (Fig. 15.3). The area is characterised by a varied geological structure with a prevalence of morainic clay, 3 to 27 m thick [15]. There are three lithological and soil areas within the representative catchment. In the morainic plateau, lessive soil prevails, and black-earth soils occur in land depressions [16]. The area was used for farming as early as in the Middle Ages, and today arable land accounts for 86.6% of the area, and meadows take up 9.1%. There are no forests in the area, parks or mid-field trees stands cover approx. 1.8% of the land [11] (Fig. 15.4).

The methodology of observations under the IEM programme is specifically prescribed [17]. There are 13 ongoing observation projects in the area of the

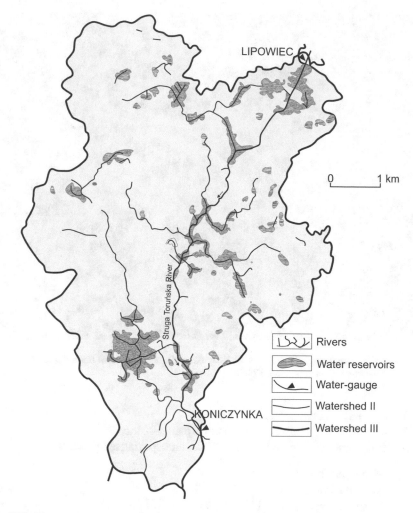

Fig. 15.3 The representative catchment of the Struga Toruńska

representative catchment of the Struga Toruńska. They cover abiotic and biotic elements of the environment enabling analysis of the flow of energy and matter:

A1. Meteorology—since 1993;
B1. Air chemistry—since 1996;
C1. Precipitation chemistry—since 1993;
D1. Heavy metals in lichens (transplantation)—since 2003;
E1. Soil—since 1993;
F1. Soil solution—since 2010;
F2. Groundwater—since 1995;
H1. Surface water—rivers (Struga Toruńska)—since 1993;
H2. Surface water—lakes (Lake Kamionkowskie)—since 2015;

Fig. 15.4 The meteorological station at Koniczynka and the agricultural landscape of the Struga Toruńska river catchment (Photo M. Kejna)

J1. Flora and vegetation—since 1993;
J3. Monitoring of invasive alien species—plants—since 2015;
M1. Tree epiphytes (lichens)—since 2001, and specialised projects:

> Drainage basins—(1993–2000);
> Radiation balance—since 2011;
> Soil fauna monitoring—(1993–2017);
> Monitoring of the breeding population of the barn swallows *Hirundo rustica* (*L.*) and the house martin *Delichon urbica* (*L.*)—since 1996.

In this study, data from the meteorological station in Koniczynka (Photo 13.1) was used from the period 1993–2017. Initially, observations were conducted three times a day and, from 2000 onwards, automatic registrations were introduced. Precipitation was measured daily using Hellmann rain gauge, and snowfall was measured using a snow gauge.

Hydrologic measurements were taken at the entrance to the representative catchment (Lipowiec) and the exit (Koniczynka) (Figs. 15.3 and 15.5).

In 1994–2017, hydrological researches of the Struga Toruńska were carried out and comprised such parameters as water stage, flow rate, runoff, specific runoff, runoff coefficient with regard to general hydrometeorological conditions of the analysed area. At first, water stages were measured daily using a staff gauge, but from

Fig. 15.5 River gauges on the Struga Toruńska: Lipowiec (left) and Koniczynka (right) (Photo M. Kejna)

2000 onwards electronic recorders were introduced. Flow rates are measured every month using a current meter.

The chemical analysis of the water in the Struga Toruńska was performed by the Provincial Inspectorate of Environmental Protection in Bydgoscz on monthly samples collected at Lipowiec (since 1996) and Koniczynka (since 1993). The following elements are analysed: water temperature; electrolytic conductivity; pH value; dissolved oxygen; BOD5; and the content of hydrocarbons, sodium, potassium, calcium, magnesium, ammonia nitrogen, chlorides, sulphates, total phosphorus, phosphates, total suspended matter, silica, dissolved organic carbon (DOC), total aluminium, cadmium, copper, lead, manganese, iron, zinc, nickel, arsenic and chromium.

15.4 The Variability of Climate Conditions at Koniczynka

Variability is an inherent characteristic of the global climate. It is caused by a number of factors related to astronomy, radiation, circulation, geography, ecology, and anthropogeny [18]. More recently, the influence of man has been increasingly stressed in research into climate change [19, 20–22].

In the analysed period (1994–2017), climate conditions were observed to vary considerably (Table 15.1). Sunshine duration at Koniczynka, registered from 1997, averaged at 1,635.7 hours in 1997–2017, which is a typical value for the whole of Poland [23]. However, the sunshine duration observed in the analysed period revealed a statistically significant positive trend (+97.7 h/10 years). In the annual course, the least sunshine was observed from November to January, with a minimum of 23.6 h in December. That was connected with a maximum cloud cover at the time, but also with the shortest duration of daytime. The period with the most sunshine (over 200 h per month) was from May to August, with a maximum of 239.2 h in May (Fig. 15.6). That was when the day was long, and there was relatively low cloudiness.

Table 15.1 Mean annual values of selected climate elements at Koniczynka in 1994–2017 (hydrological years)

Lata	U (h)	T (°C)	f (%)	P (mm)	SC (cm)	V (m•s^{-1})
1994	–	8.3	80	498.7	20.0	–
1995	–	9.1	81	577.8	12.0	2.7
1996	–	6.1	83	405.0	18.0	2.2
1997	1480.6	7.6	79	444.4	0.0	2.7
1998	1384.8	8.5	81	562.8	4.0	4.1
1999	1663.8	8.5	78	579.6	8.1	3.8
2000	1638.3	9.2	76	558.7	8.9	3.9
2001	1423.6	8.9	81	762.0	38.5	3.3
2002	1566.1	9.3	76	593.8	29.8	4.1
2003	1688.4	7.5	76	468.6	9.8	3.2
2004	1554.6	8.3	77	496.1	7.5	3.3
2005	1784.4	8.7	77	413.0	13.3	3.4
2006	1653.0	8.2	79	497.8	35.5	2.9
2007	1665.1	10.0	79	670.7	9.1	3.8
2008	1591.1	9.2	78	592.6	13.6	3.5
2009	1696.9	8.7	78	500.2	2.4	3.2
2010	1635.4	7.9	79	674.9	40.0	3.2
2011	1781.6	8.4	74	647.2	30.0	3.5
2012	1818.5	8.9	74	484.2	19.0	3.5
2013	1595.4	8.1	77	618.6	18.0	3.1
2014	1792.2	9.9	81	463.9	10.0	3.2
2015	1711.1	9.4	80	437.3	5.0	3.3
2016	1593.8	9.8	83	542.3	7.0	3.3
2017	1631.8	8.5	85	727.2	4.0	3.2
1994–2017	1635.7	8.6	79	543.1	40.0	3.3
STD	115.0	0.9	2.8	97.6	11.7	0.4

Abbreviations: U—sunshine duration, T—mean air temperature, f—mean relative humidity, SC—maximum snow cover thickness, P—precipitation, V—mean wind speed, STD—standard deviation

Air temperature is one of the key elements of climate. Compared to other areas at the same latitude, Poland is thermally privileged, which is a result of the inflow of air masses from the Atlantic Ocean. A growing trend in air temperature is observed in Poland; it amounted to 0.24 °C/10 years in 1951–2008 [24]. In the years 1994–2017, the mean annual air temperature at Koniczynka was 8.6 °C and had a growing trend (0.48 °C/10 years), but was statistically insignificant ($p = 0.056$). The analysed period, as compared with the multi-annual period of 1981–2010 (Toruń data), appeared much warmer. Only in four out of all the years considered was the air temperature lower than average, and in eight years it was higher; the years 2007

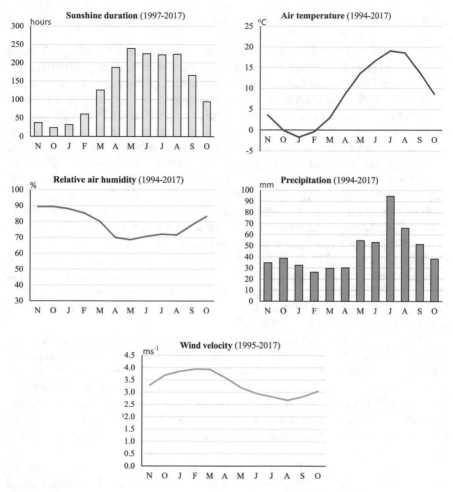

Fig. 15.6   Mean annual course of selected climate elements at Koniczynka, 1994–2017

and 2014 turned out to be very warm. In the analysed period, hot and very hot days showed a growing trend, whereas frost days, cold and freezing cold days occurred less frequently. The identified trends are not statistically significant.

In the annual course in 1994–2017, the lowest mean air temperature occurred in January (−1.7 °C), but the other winter months (December and February) also had negative mean values. The highest mean temperature in the annual course was observed in July (19.1 °C) (Fig. 15.6).

The relative humidity of air is an important parameter indicating the saturation of air with water vapour. In Poland, its standard values range from approx. 78% in the west part of the country (Poznań and Zielona Góra) to 87% at the top of Mount Śnieżka (1951–1980) [25]. In comparison, the area of the Koniczynka station is dry, as the mean relative humidity in 1994–2017 was 79% there. In the annual course, the driest month in terms of relative humidity was May (69%), and the most humid—December (89%) (Fig. 15.6).

Atmospheric precipitation plays an important role in the environment, being as it is part of the cycle of water and various chemical substances and supplies surface and groundwater. Precipitation in the majority of the area of Poland is rather small. In the central part of the country—the lowlands—annual precipitation does not exceed 550 mm [25]. At Koniczynka the mean annual precipitation total was 550.7 mm in the years 1994–2017. Precipitation is a weather element that reveals a high variability over time: the mean annual values may be different in different years. In 2001 and 2017, the precipitation totals were greater than 700 mm (762.0 mm and 727.2 mm, respectively), whereas in 1997, 2005 and 2015 they were less than 450 mm, with respective values amounting to just 444.4 mm, 413.0 mm and 437.3 mm. Such fluctuations may result in water excess or shortage in the ecosystem. At Koniczynka, meteorological droughts [25] and hydrological droughts were frequent. In the annual course, the precipitation total was smallest in February (26.3 mm) and biggest (94.8 mm) in July (Fig. 15.6). The summer maximum was due to storm rainfalls. In the multi-annual period, there is a slightly growing trend in precipitation total. It is similar in NW Poland, where most stations showed a slight upward trend (1951–2000) [26]. In the analysed period, the number of precipitation days revealed a statistically significant growing trend (+17.04 days/10 years).

Snow cover is formed through snowfalls, but its development and duration also depend on air temperature. The mean depth of snow cover in the lowland part of Poland in the years 1970/1971–1999/2000 ranged from less than 6 cm in the west to more than 12 cm in the north-east [25]. In the multi-year period (1960/1961–2009/2010) in Poland, there was a slight falling trend in the snow cover depth, but the changes were generally statistically insignificant [27, 28]. Similarly, the snow cover depth at Koniczynka was also a little reduced (-0.43 cm/10 years) and was statistically insignificant ($p < 0.05$).

The mean wind speed in Poland is low and amounts to 3–4 m·s^{-1} in most parts of the country [23, 25]. Over the multi-annual period of 1951–2005 in Central Europe wind speed had a growing trend [29]. On the one hand, a higher wind speed ensures better airing and improves air quality conditions. On the other hand, more frequent high winds may cause damage to the economy and environment. In the years 1995–2017 at Koniczynka, the wind speed trend was slightly upward, but statistically insignificant. In the annual course, the lowest atmospheric dynamics were found in the summer (July 2.7 m·s^{-1}), and the highest mean wind speeds were observed from January to March (3.9 m·s^{-1})—Fig. 15.6.

15.5 Water Flow Variability in the Struga Toruńska

The hydrological regime of a river depends on a number of factors. The runoff is determined by the amount and distribution of precipitation in a year and the duration and thickness of the snow cover. Air temperature is also relevant, as it affects evapotranspiration. The hydrological characteristics of the Struga Toruńska within the representative catchment area are largely affected by its geological structure and extensive agricultural activities. The catchment area is mostly drained, which, given the local hydrogeological conditions, substantially affects the development of the river's hydrological regime. The river relatively quickly responds to the springtime snow melting and any major precipitation.

The flow rate of water in the Struga Toruńska demonstrates a considerable year-on-year variability, which depends on the pattern of meteorological conditions (Table 15.2, Figs. 15.7 and 15.8).

Table 15.2 Precipitation at Koniczynka and hydrological characteristic of the Struga Toruńska River, 1994–2017

Year	Precipitation (mm)	Flow ($m^3 s^{-1}$)		Outflow (mm)	Flow rate (%)	Specific runoff ($dm^3 s^{-1} km^{-2}$)
		Lipowiec	Koniczynka	Representative catchment		
1994	**498.7**		**0.62**			
1995	577.8		0.45			
1996	405.0		0.66			
1997	444.4		0.31			
1998	562.8		0.51			
1999	579.6		0.84			
2000	558.7		0.28			
2001	762.0		0.54			
2002	593.8		0.73			
2003	468.6		0.39			
2004	496.1	0.51	0.61	84.9	0.17	2.69
2005	413.0	0.63	0.77	129.3	0.31	4.10
2006	497.8	0.55	0.65	92.2	0.19	2.92
2007	670.7	0.82	1.04	196.4	0.29	6.23
2008	592.6	0.36	0.52	142.2	0.24	4.55
2009	500.2	0.29	0.37	73.7	0.15	2.34
2010	674.9	0.33	0.57	200.1	0.30	6.60
2011	647.2	0.42	1.00	503.5	0.78	15.97
2012	484.2	0.26	0.47	176.2	0.36	16.20
2013	618.6	0.25	0.35	88.8	0.14	2.90
2014	463.9	0.14	0.19	43.5	0.09	1.40
2015	437.3	0.08	0.10	23.7	0.05	0.80
2016	542.3	0.09	0.23	125.4	0.23	4.00
2017	727.2	0.25	0.89	570.4	0.78	18.30
1994–2017	543.1	0.35	0.55	175.0	0.32	5.60

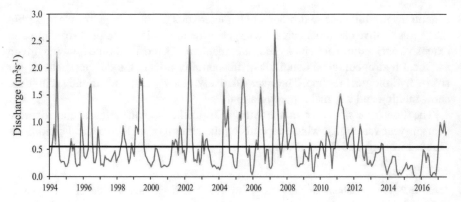

Fig. 15.7 The course of mean monthly flow values in the Struga Toruńska at Koniczynka in 1994–2017. The horizontal line marks the mean flow in the multi-annual period

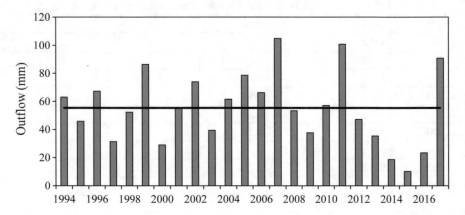

Fig. 15.8 The course of mean annual outflow values in the Struga Toruńska at Koniczynka in 1994–2017. The horizontal line marks the mean flow in the multi-annual period

In 1994–2017 the average water flow was 0.35 m^3s^{-1} at Lipowiec and 0.55 m^3s^{-1} at Koniczynka. Variable flows are typical of Polish rivers [30]. The maximum mean annual flow was recorded at Koniczynka in 2007 and amounted to 1.04 m^3s^{-1}. That year the precipitation total was also high (670.7 mm). In 2011, high flow in the spring was caused by snow cover melting (the snow cover thickness in December 2010 reached 30 cm). Conversely, the minimum annual flow (0.10 m^3s^{-1}) occurred in 2015, a year of low total precipitation (437.3 mm) and high mean air temperature (9.4 °C)—the scorching hot summer of that year favoured intensive evaporation. In that summer, after a long meteorological drought, the river dried up for the first time since 1994.

The mean specific runoff at Koniczynka in the analysed period was 1.76 dm^3s^{-1} km^{-2}. This is a low value for Poland [31]. The maximum specific runoff was 3.33 dm^3s^{-1} km^{-2}. On average, the river took 55.4 mm of water in a year (Fig. 15.9),

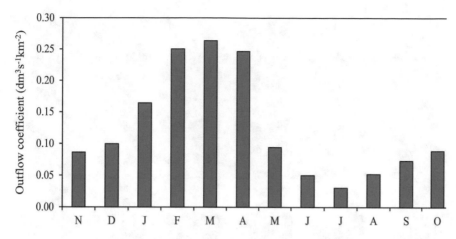

Fig. 15.9 The annual course of outflow values in the Struga Toruńska at Koniczynka in 1994–2017

which is just 10% of the precipitation total. This is due to the retentive activity of lakes in the upper part of the catchment.

The representative agricultural catchment between Lipowiec and Koniczynka has different hydrological characteristics. In the years 2004–2017 the specific runoff there averaged $5.60 \, dm^3 s^{-1} \, km^{-2}$. On average, 175 mm of water were driven off the catchment area, which accounted for 32% of the precipitation total.

In the annual course, high water stages in the Struga Toruńska, along with the highest flow rates, were usually recorded in the spring. This is connected with increased water supply, mainly from melting snow. The lowest flow rate was characteristic of the Polish summer season, when, despite more rainfall, evapotranspiration increases, and the climatic water balance is negative [32].

According to the river regime classification described by Dynowska [33], Dynowska and Pociask-Karteczka [34] and Wrzesiński [35], the regime of the Struga Toruńska can be classified as nival and moderately developed, in which the mean monthly flow in a spring month equals 130–180% of the mean annual runoff. In 1994–2017, it was 166%, but in individual years the annual runoff patterns varied depending on the prevailing meteorological factors (Fig. 15.10).

In recent years, the poorly developed nival regime has been increasingly common. A decreased share of spring runoff has also been noticed in other rivers of Poland [36]. This is an effect of a substantial increase in air temperature, especially in the winter season, and a shorter duration of the snow cover that is an important source of water supply for rivers and groundwater courses. This was the case in, for example, the Struga Toruńska in 2011 (Fig. 15.11).

Short-term fluctuations in the water stage are particularly significant for the river's hydrological regime. They are connected with sudden stormy rainfalls, which are often observed in the summer. In 2009, for example, after heavy rainfall on 18 July (38.7 mm), the water stage suddenly and markedly increased. Such a quick transformation of precipitable water into river runoff is enabled by the drainage

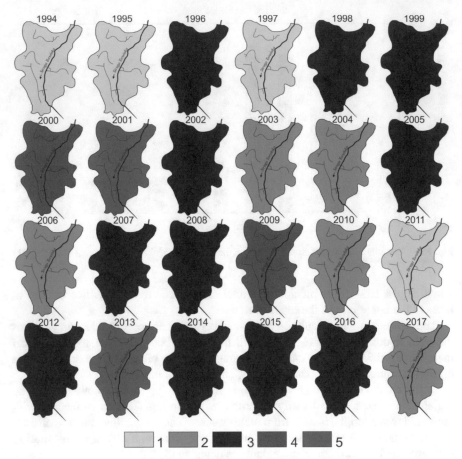

Fig. 15.10 The runoff regime of the Struga Toruńska at Koniczynka in 1994–2017, determined using the criteria proposed by Dynowska [33], Dynowska and Pociask-Karteczka [34] (Key 1—nival, poorly developed; 2—nival, moderately developed; 3—nival, well-developed; 4—nival-pluvial; 5—pluvio-nival)

system in the representative catchment area. Long-term meteorological droughts may cause the water flow to stop, as happened in the summer of 2015 (Fig. 15.11).

15.6 The Water Chemistry of the Struga Toruńska

The quality of water in the Struga Toruńska is largely affected by agricultural activities. The water flowing into the catchment area has specific chemical properties influenced by Lake Mlewieckie situated 3.5 km upstream of the river gauge at Lipowiec. The Struga Toruńska flows through the lake, and the lake serves as a buffer absorbing ample quantities of biogenic compounds [37]. Long-term agricultural pressure on

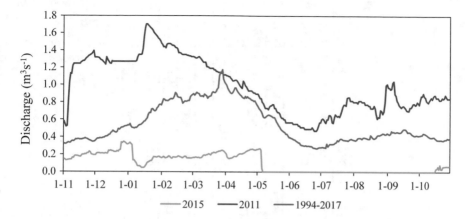

Fig. 15.11 The course of mean diurnal flow values in the Struga Toruńska at Koniczynka for a wet year (2011) and a dry year (2015), and the mean flow in the years 1994–2017

surface water quality has caused the lake to overgrow, become excessively eutrophicated and subject to intense and prolonged phytoplankton bloom and hypoxias in the summer.

The water flowing from the lake has a significant influence on the quality of water in the Struga Toruńska at the upper part of the representative catchment and has much lower concentrations of biogenic substances and increased organic matter content.

This was confirmed by analyses of the water at the Lipowiec site in 1993–2017. Aerobic conditions in the summer were unfavourable, and the organic content expressed using the five-day biochemical oxygen demand (BOD5) was increased. The highest BOD5 occurred in the first years of observations when it often exceeded 8 mg $O_2 \cdot dm^{-3}$. On the other hand, the content of biogenic elements was reduced as they were assimilated in the lake water (Table 15.3). The influence of the lake also contributed to a reduced concentration of dissolved substances.

The measuring site at Koniczynka is located 10 km downstream of the Lipowiec site. Along that stretch, the Struga Toruńska is subject to considerable anthropopressure, both from agricultural activities and from the effects of the nearby city of Toruń Wojtczak. The results of observations indicated increased concentrations of biogenic and mineral substances and suspended matter. Aerobic conditions in the water and the organic matter content were more favourable.

Tests for the presence of heavy metals—aluminium, cadmium, copper, lead, manganese, zinc, nickel, arsenic, iron and chromium—were carried out starting in 2007 at Lipowiec and Koniczynka. The heavy metal content was generally maintained below the detection limit or in quantities that did not pose a threat to the aquatic environment.

Analysing changes in the concentrations of elementary parameters of the water in the experimental catchment area of the Struga Toruńska in 1993–2017, it was found that aerobic conditions improved along the course of the river. On average, at Lipowiec, it was 7.5 mg $O_2 \cdot dm^{-3}$, and at Koniczynka 8.6 mg $O_2 \cdot dm^{-3}$. In the

Table 15.3 Mean concentrations of elementary parameters monitored in the water of the Struga Toruńska in 1994–2017

Parameter	HCO_3 [mg dm^{-3}]	$S\text{-}SO_4$	$N\text{-}NO_3$	$N\text{-}NH_4$	Cl	Na	K	Mg	Ca	SUS	pH	$P_{ogól.}$ [µg dm^{-3}]	$P\text{-}PO_4$	SEC [mS m^{-1}]	O_2 [mg dm^{-3}]	BOD_5	Temp [°C]
Lipowiec																	
Mean	281	31,1	1,34	0,31	41,8	17,2	8,96	17,3	120	9,3	7,9	148	51,1	67,9	7,5	3,9	10,0
STD	18	14,3	0,77	0,16	14,1	2,5	0,95	3,5	18	2,8	0,1	50	19,0	8,0	1,3	1,1	1,6
Min	234	15,7	0,27	0,09	4,8	12,9	6,62	13,6	96	5	7,6	90	29,7	54,3	5,7	2,2	6,5
Max	301	90,6	3,72	0,70	87,5	21,5	10,50	31,3	175	16	8,1	918	92,4	80,5	10,2	6,2	13,0
Koniczynka																	
Mean	325	36,0	4,8	0,17	46,3	17,3	9,51	18,5	137	11,2	7,9	188	73,8	77,6	8,6	2,8	10,3
STD	21	10,8	3,0	0,10	9,5	2,3	0,57	2,8	17	4,4	0,1	92	18,2	8,0	1,0	0,7	0,9
Min	297	20,9	1,1	0,04	35,7	13,5	8,60	12,6	114	2	7,7	110	39,9	64,8	6,8	1,4	8,9
Max	365	70,1	16,6	0,43	73,0	24,2	10,40	27,2	183	25	8,1	560	102,3	97,3	11,0	4,6	12,3

STD—Standard deviation

summer, the oxygen content would often drop to 1 mg $O_2 \cdot dm^{-3}$ at Lipowiec, whereas at Koniczynka it remained at a good level. The BOD5 indicated that organic content was higher at the Lipowiec site than at Koniczynka (Fig. 15.12). That was due to a substantial content of organic matter in Lake Mlewieckie, using the oxygen dissolved

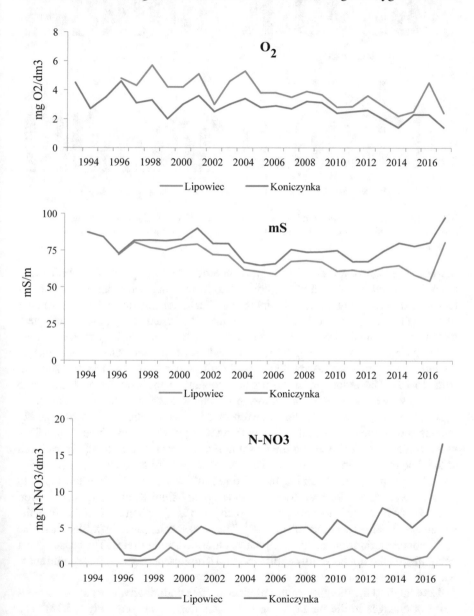

Fig. 15.12 Mean annual concentrations of O_2, electrolytic conductivity (mS) and nitrate nitrogen (N-NO$_3$) in the water of the Struga Toruńska in 1994–2017

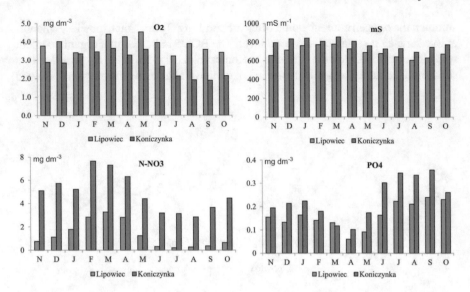

Fig. 15.13 Mean annual course of O_2, electrolytic conductivity (mS), nitrate nitrogen (N-NO$_3$) and phosphate concentration (PO$_4$) in the water of the Struga Toruńska in 1993–2017

in water. The highest values of BOD5 were recorded in the early period of observation when it reached 9 mg $O_2 \cdot dm^{-3}$ at both sites. In the analysed years the parameter tended to decrease (Fig. 15.12), except the year 2016 at Lipowiec, when—as a result of an exceptionally low water stage—only half of the usual samples were collected, and the mean annual values from the other samples were not entirely representative.

Analysing BOD5 in individual months, values of this indicator were higher in all months at Lipowiec. A particularly big difference was noted in the summer (Fig. 15.13). The reduction in the organic content in the summer at Koniczynka indicated very good aerobic conditions, in which the oxygen demand is smaller.

The content of dissolved solids, as expressed by electrolytic conductivity, clearly increased with the course of the river within the experimental catchment area. This was a result of soil erosion and the washing out of mineral compounds, introduced at the time of agricultural activities. The mean electrolytic conductivity at Lipowiec was 67.9 mS·m^{-1}, and at Koniczynka 77.6 mS·m^{-1} (Fig. 15.12). Over the analysed period the parameter was not found to undergo any significant changes. However, there was an increase in electrolytic conductivity in 2017, when most of the time, the parameter remained at about 100 mS·m^{-1}. That was accompanied by a slight increase in the concentrations of calcium, sulphates, chlorides, and nitrates. The cause of that phenomenon has not been fully explained, and subsequent observations should show whether or not the change persisted.

Looking into the monthly values of electrolytic conductivity, one can see a regularity of changes in the parameter at both sites during the year (Fig. 15.12). The highest values of dissolved substances, as expressed by electrolytic conductivity, were noted in the winter, whereas in the growing season they tended to fall and reach a minimum in August (Fig. 15.13).

The agricultural nature of the experimental catchment of the Struga Toruńska was reflected in a rising content of biogenic substances (Fig. 15.12). In the stretch of the river between Lipowiec and Koniczynka, increased concentrations of the following compounds were observed every year: nitrate nitrogen (mean increase from 1.3 to 4.8 mg $N-NO_3·dm^{-3}$), phosphate phosphorus (from 51 to 74 μg $P-PO_4·dm^{-3}$) and total phosphorus (from 148 to 188 μg $P·dm^{-3}$).

Throughout 2017 high concentrations of nitrate nitrogen were observed, reaching 28.9 mg $N-NO_3.dm^{-3}$ at the Koniczynka site in the winter (Fig. 15.13). As mentioned, the reason for the increase of a number of mineral parameters is not known and may be connected with the nature of certain agricultural practices in the fields within the catchment carried out at the time.

A high concentration of nitrate nitrogen in winter and early spring months is a characteristic of the quality of water subject to strong agricultural pressure. Mean monthly values in the multi-year period show maximum nitrogen concentrations in February–April at both sites (Fig. 15.13). In the growing season, they decreased and reached a minimum in the summer. A comparative analysis of nitrate nitrogen concentrations between the observation sites indicated that the parameter increased more in the runoff from the experimental catchment in the winter than in the summer.

The phosphate index, which is regularly monitored at both sites, verifies the agricultural influence of the experimental catchment, as shown in an increase in phosphate concentration along the course of the river (Fig. 15.12). A major increase was observed in the summer. The concentration declined at the beginning of the growing season when the biogenic compound is in maximum demand. The highest phosphate values were found between July and September, when the demand for the most easily absorbed form of nitrogen (nitrate nitrogen) was increasing (Fig. 15.13).

Analysing the load of substances outflow from the experimental catchment in the years 2004–2017, an evident annual cyclicality of nitrate nitrogen was found (Fig. 15.12). An accelerated nitrogen runoff was observed at the end of winter and the beginning of spring when nitrogen concentrations and flows were the highest in a year. The smallest loads of nitrate nitrogen were observed in the summer.

Phosphate loads did not demonstrate the same regularity as nitrogen. The highest quantities of phosphate occurred in September and January (Fig. 15.13). The lowest loads were observed in the autumn (October–November), and their concentrations reached a minimum in April and May.

Regarding current standards of water quality prescribed in the Regulation of the Minister of Environment on 5.08.2016 (Dz.U. 2016.1187) *"on the classification of the status of surface water bodies and environmental quality standards for priority substances"*, the mean annual values for the water in the Struga Toruńska in 1993–2017 were found to have exceeded the good status (class II) due to excessive concentrations of chlorides, calcium, manganese and BOD5 at the Lipowiec site, and nitrate nitrogen at Koniczynka. The other parameters, including those related to heavy metals, met the requirements for a very good and good status (classes I and II).

15.7   Closing Remarks

The functioning of the representative catchment of the Struga Toruńska is determined by its location within early glacial landforms with a diverse geology. The
prevalence of morainic clay in surface deposits hinders infiltration, and the area is
therefore largely drained, which accelerates water runoff in the spring and contributes
to droughts in the growing season.

In the analysed period, the runoff of the Struga Toruńska varied considerably due
to weather conditions. The water chemistry, however, was subject to strong anthropopressure connected with agricultural activities. The meteorological conditions in
1994–2017 were very changeable. The mean annual air temperature for 1994–2017
was 8.6 °C and its mean annual values ranged from 6.1 °C in 1996 to 10.0 °C in
2007. Extreme values of air temperature at Koniczynka varied between—28.3 °C
in January 2006 and 36.9 °C in July 1994. No statistically significant ($p = 0.056$)
temperature increase of 0.48 °C/10 years was observed. The trend was ascertained in
nearby Toruń, where in 1947–2017 the temperature statistically increased by 1.6 °C
(0.25 °C/10 years)–Fig. 15.14.

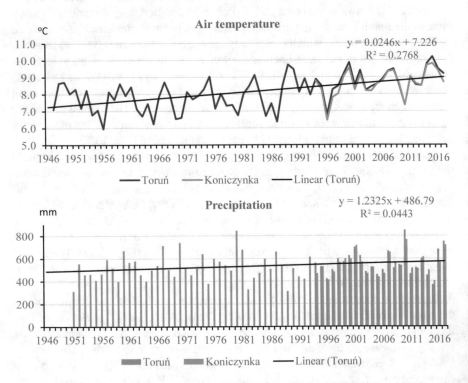

Fig. 15.14 The courses of air temperature and precipitation in Toruń (1947–2017) and Koniczynka
(1994–2017)

In 1994–2017 at Koniczynka the mean annual precipitation total was 543.1 mm. Annual totals ranged from 405.0 mm (1996) to 762.0 mm (2001). In the annual course, the highest total values of precipitation were recorded in the summer (July 93.0 mm), and the lowest in the winter (February 26.4 mm). In individual months, the totals were quite varied: from 2.1 mm in January 1997 to 190.3 mm in July 2003. There were 178.7 precipitation days ($p \geq 0.1$ mm) in a year, including 96.0 days with $p \geq 1.0$ mm and 11.9 days with $p \geq 10.0$ mm. No statistically significant trends in precipitation were observed, but there were some series of wet and dry years (Fig. 15.14). In the analysed years there were 15 meteorological droughts during which there was no precipitation for at least 15 days in a row.

The snow cover at Koniczynka formed from early October to the end of April, but its mean duration was only 29 days. Snow cover persisted the longest during the winter of 2012/2013 (80 days). The average thickness of the snow was a few to a dozen-or-so centimetres, but in 2010 it reached a record 40 cm. Winters without long-term snow cover have been increasingly observed, which has a significant influence on the environmental water resources.

The water stages and flow rates in the Struga Toruńska are very variable and dependent on weather conditions. There is a clear dependence of the outflow from the atmospheric circulation associated with the North Atlantic Oscillation (NAO) phases [38]. The highest flow rates are observed in the spring. However, the water regime includes numerous flow disturbances, surges, and lows connected with specific weather. The river quickly responds to torrential rain, snow melting and atmospheric droughts due primarily to anthropogenic transformations and man-made drainage of the catchment. This is particularly evident in periods of long-term high air temperatures, which favour low flows and contribute to an increased soil moisture deficit. Moreover, the complex geology of the land affects water infiltration and retention [14]. The local drainage system and lack of forests limit the retention capacity of the catchment. As a result, water stages are subject to rapid changes after precipitation events because of quick surface runoff and drainage.

The mean specific runoff from the entire catchment of the Struga Toruńska to Koniczynka in 1994–2017 was 1.76 dm^3s^{-1} km^{-2} and was among the lowest in Poland. However, in the separated representative catchment between Lipowiec and Koniczynka, the specific runoff was much greater (5.60 dm^3s^{-1} km^{-2}) and typical of Central Poland (5.4 dm^3s^{-1} km^{-2}). It was different again for different parts of the catchment (ranging from less than 2.0 dm^3s^{-1} km^{-2} to approx. 9.0 dm^3s^{-1} km^{-2}) and temporally varied [31, 35]. The smallest specific runoff in Poland is 2–4 dm^3s^{-1} km^{-2} and occurs in the belt of Middle Polish Plains [31, 38]. For example, the runoff from the agricultural catchment of the Zgłowiączka river in 1951–2012 was 2.51 dm^3s^{-1} km^{-2} [39]. Similar values (2.1 dm^3s^{-1} km^{-2}) were observed in the agricultural catchment of the Borucinka river in 2010 [40]. To compare, in the catchments of mountainous tributaries of the Vistula, the runoff exceeds 12 dm^3s^{-1} km^{-2} [39]. Jokiel and Stanisławczyk [41] stated that seasonal structures of runoff undergo nearly parallel changes in neighbouring rivers, and are similar in the case of rivers supplying the same regions. Water resources in Poland are small and do not exceed 1800 m^3 per capita per year [42–44], showing significant seasonal and year-to-year variability [41].

The agricultural development of the representative catchment of the Struga Toruńska affects the quality of its water. In the upper stretch of the river, the water is affected by the heavily eutrophicated Lake Mlewieckie, through which the Struga flows. The water coming out of the lake has an increased organic matter content and low concentrations of biogenic compounds. In the summer, hypoxia is observed. The water monitored at Koniczynka, unlike that at Lipowiec, demonstrated a good level of dissolved oxygen content and was low in organic compounds but the concentration of biogenic compounds, especially nitrate nitrogen, significantly increased along with mineral substances and suspended matter.

Seasonal variations in the analysed chemical parameters of the water are evident. The concentration of nitrate nitrogen was found to increase in winter and early spring months and to fall to a minimum in the summer. That was due to increased demand for the most easily absorbed form of nitrogen at the time of dynamic growing. On the other hand, the minimum concentration of phosphates took place in April, at the start of the growing season, and greatly increased as the season continued. Such a pattern was observed earlier by Sullivan and Drever [45] and Clark et al. [46]. High concentrations of phosphorus in the summer are related to reduced water supply and thus, the development of anaerobic conditions, which favour the release of phosphorus compounds accumulated in river bed deposits [47]. In the spring and summer, there are also enhanced processes of organic matter decomposition. In the water of the Struga Toruńska, the values of BOD5 are highest—and aerobic conditions most unfavourable—in the summer.

The increased biogenic content in the winter and spring may be related to intensive flushing during river surges caused by melting snow and winter and early spring rainfalls [48]. In agricultural catchment areas, surface runoff from nearby fields fertilised using mineral and organic substances increases, especially in the spring. As a result, ample quantities of dissolved substances are introduced [49, 50].

The drainage systems within the catchment of the Struga Toruńska are instrumental in conveying biogenic compounds, as they make so-called "preferred pathways" carrying large quantities of dissolved solids into the river [51]. This was confirmed in research by Krasowska and Banaszuk [52] concerning agricultural catchments with developed drainage systems.

Looking into the changes of primary concentrations of the water in the Struga Toruńska in 1993–2017, a considerable year-on-year variability was found, which was determined by hydro-meteorological factors. Aerobic conditions improved and BOD5 decreased, but the content of dissolved substances expressed by electrolytic conductivity clearly increased due to soil erosion and washing out of mineral compounds introduced during agricultural activities. Similar tendencies were observed in other rivers of the region [53].

The impact of agriculture on the quality of surface water is so substantial that the whole catchment of the Struga Toruńska was considered in 2012 to be an area of particular exposure to pollution with nitrates from agricultural sources [54].

15.8 Conclusions

The Struga Toruńska catchment represents the dominant form of land use in Poland—arable land. The catchment is under strong pressure from agricultural activity. The influence of the nearby city—Toruń is also marked. Here, the processes typical of the suburban area take place: land use changes, residential houses are being built, the industry develops, service facilities are created, the network of roads and highways is expanding.

The functioning of the agricultural catchment of Struga Toruńska is influenced by climate change, air temperature increases, there is often no thermal winter, and the growing season has extended. Atmospheric precipitation is characterized by considerable variability from year to year. Droughts or months with high rainfall are more and more frequent. Lack of permanent snow cover changes the water cycle and affects the functioning of the biosphere, including the cultivation of plants.

Despite the intensification of anthropopression, positive changes in the catchment of Struga Toruńska are observed. The air pollution has decreased (mainly SO_2), however, the NO_x level shows an increase related to the intensity of car traffic. The pH of atmospheric precipitation rises and their electrolytic conduction decreases. The quality of surface water is strongly affected by agricultural activity.

15.9 Recommendations

In the Struga Toruńska catchment, it is necessary to continue research and monitor the state of the environment. The problems are:

1. Changes in the functioning of the catchment ecosystem caused by the increase of air temperature and intensification of extreme weather phenomena.
2. Pollution of air by local emissions from industrial, municipal and communication facilities as well as agricultural areas.
3. Disorders of natural water outflow caused by melioration treatments.
4. Pollution of surface and underground water, mainly as a result of agricultural activity.
5. Soil degradation as a result of excessive agricultural use, including soil erosion.
6. Reduction of biodiversity in a forestless area, expansion of invasive species.
7. Threats caused by the development of construction in the suburban area of Toruń and the communication network and the increase of car traffic.

Acknowledgements The research and article were accomplished as part of the Integrated Environmental Monitoring programme financed by the National Fund for Environmental Protection and Water Management.

References

1. Skowera B, Kopcińska J, Kopeć B (2014) Changes in thermal and precipitation conditions in Poland in 1971–2010. Ann Warsaw Univ Life Sci SGGW, Land Reclam 46(2):153–162
2. Ilnicki P, Farat R, Górecki K, Lewandowski P (2015) Long-term air temperature and precipitation variability in the Warta River catchment area. J Water Land Dev 27:3–13
3. Pospieszyńska A, Przybylak R (2018) Air temperature changes in Toruń (central Poland) from 1871 to 2010. Theor. Appl. Climatol 135:1–18. https://doi.org/10.1007/s00704-018-2413-9
4. Kostrzewski A (2018) Założenia metodologiczne, merytoryczne i organizacyjne programu Zintegrowanego Monitoringu Środowiska Przyrodniczego. In: Kostrzewski A, Majewski AM (eds) Stan i przemiany środowiska przyrodniczego geoekosystemów Polski w latach 1994–2015 w oparciu o realizację programu Zintegrowanego Monitoringu Środowiska Przyrodniczego. Biblioteka Monitoringu Środowiska, Warszawa, pp 19–28
5. Rolnictwo (2017) In: Główny Urząd Statystyczny. http://stat.gov.pl/obszary-tematyczne/rolnic two-lesnictwo/rolnictwo/rolnictwo-w-2017-roku,3,14.html
6. Poławski ZF (2009) Zmiany użytkowania ziemi w Polsce w ostatnich dwóch stuleciach. Teledetekcja Środowiska 42:69–81
7. FAO (2015) World fertilizer trends and outlook to 2018. Rome. www.fao.org/3/a-i4324e.pdf
8. Głodowska M, Gałązka A (2018) Intensyfikacja rolnictwa a środowisko Naturalne. Zeszyty Problemowe Postępów Nauk Rolniczych 592:3–13. https://doi.org/10.22630/zppnr.2018.592.1
9. GIOŚ (2018) Zintegrowany Monitoring Środowiska Przyrodniczego. http://zmsp.gios.gov.pl/?page_id=20_
10. Integrated Monitoring (2018). http://www.syke.fi/nature/icpim
11. Wójcik G, Marciniak K (eds) (1996) Zintegrowany monitoring środowiska przyrodniczego— Stacja Bazowa w Koniczynce. Biblioteka Monitoringu Środowiska, Warszawa, pp 1–272
12. Kejna M, Uscka-Kowalkowska J (2018) Stan i przemiany środowiska przyrodniczego Geoekosystemu zlewni Strugi Toruńskiej. In: Kostrzewski A, Majewski M (eds) Stan i przemiany środowiska przyrodniczego geoekosystemów Polski w latach 1994–2015 w oparciu o realizację programu Zintegrowanego Monitoringu Środowiska Przyrodniczego. Biblioteka Monitoringu Środowiska, Warszawa, pp 193–235
13. Wojtczak H (2012) Jakość wód Strugi Toruńskiej w granicach zlewni reprezentatywnej ZMŚP w Koniczynce w latach 1993–2010. Wyd. Zintegrowany Monitoring Środowiska Przyrodniczego. Funkcjonowanie geoekosystemów w różnych strefach krajobrazowych Polski, Biblioteka Monitoringu Środowiska
14. Kejna M, Strzyżewski T (2014) Water resources of the Struga Toruńska river (Central Poland) in the context of climatic changes in 1994–2012. Int J Earth Sci Eng 7(1):74–79
15. Niewiarowski W (1996) Budowa geologiczna i rzeźba terenu. In: Wójcik G, Marciniak K (eds) Zintegrowany Monitoring Środowiska Przyrodniczego—Stacja Bazowa w Koniczynce Biblioteka Monitoringu, Warszawa, pp 41–57
16. Bednarek R, Szrejder B (2004) Struktura pokrywy glebowej zlewni reprezentatywnej Strugi Toruńskiej. In: Kejna M, Uscka J (eds) Zintegrowany monitoring środowiska przyrodniczego: funkcjonowanie i monitoring geoekosystemów w warunkach narastającej antropopresji. Biblioteka Monitoringu Środowiska, Toruń, Oficyna Wydawnicza Turpress, pp 242–250
17. Kostrzewski A, Kruszyk R, Kolander R (2006) Zintegrowany Monitoring Środowiska Przyrodniczego. Zasady organizacji, system pomiarowy, wybrane metody badań. http://zmsp.gios. gov.pl/?page_id=69
18. Niedźwiedź T (2003) Słownik meteorologiczny. IMGW, Warszawa
19. Crowley TJ (2000) Causes of climate change over the past 1000 years. Science 289:270–277
20. Barnett TP, Adam JC, Lettenmaier DP (2005) Potential impacts of a warming climate on water availability in snow-dominated regions. Nature 438(7066):303–309 https://doi.org/10.1038/nature04141
21. Kundzewicz ZW, Juda-Rezler K (2010) Zagrożenia związane ze zmianami klimatu. Nauka 4:69–76

22. IPCC (2018) Special Report. Global Warming of 1.5 °C. https://www.ipcc.ch/sr15
23. Lorenc H (ed) (2005) Atlas klimatu Polski. Instytut Meteorologii i Gospodarki Wodnej, Warszawa
24. Biernacik D, Filipiak J, Miętus M, Wójcik R (2010) Zmienność warunków termicznych w Polsce po roku 1951. Rezultaty projektu KLIMAT. In: Bednorz E, Kolendowicz L (eds) Klimat Polski na tle klimatu Europy. Zmiany i ich konsekwencje. Bogucki Wydawnictwo Naukowe, Poznań
25. Woś A (2010) Klimat Polski w II połowie XX wieku. Wydawnictwo Naukowe UAM, Poznań
26. Bąk B, Kejna M, Uscka-Kowalkowska J (2012) Susze meteorologiczne w rejonie stacji ZMŚP w Koniczynce (Pojezierze Chełmińskie) w latach 1951–2010. Woda-Środowisko-Obszary Wiejskie 12(2):19–28
27. Farat R (2010) Zmienność opadów atmosferycznych w Polsce Północno-Zachodniej. In: Bednorz E (ed) Klimat Polski na tle klimatu Europy. Warunki termiczne i opadowe. Bogucki Wydawnictwo Naukowe, Poznań
28. Czarnecka M (2012) Częstość występowania i grubość pokrywy śnieżnej w Polsce. Acta Agrophysica 19(3):501–514
29. Araźny A, Przybylak R, Vizi Z, Kejna M, Maszewski R, Uscka-Kowalkowska J (2007) Mean and extreme wind velocities in Central Europe 1951–2005 (on the basis of data from NCEP/NCAR reanalysis project). Geogr Pol 80(2):69–78
30. Jokiel P (2016) Przepływy i odpływy maksymalne w środkowej Polsce. Geogr Tourism 4(2):7–21
31. Michalczyk Z (2009) Średnie i skrajne odpływy z obszaru Polski. Zasoby i ochrona wód. In: Bogdanowicz R, Fac-Beneda J (eds) Obieg wody i materii w zlewniach rzecznych. Fundacja Rozwoju Uniwersytetu Gdańskiego, Gdańsk, pp 37–46
32. Ziernicka-Wojtaszek A (2015) Klimatyczny bilans wodny na obszarze Polski w świetle współczesnych zmian klimatu. Woda-Środowisko-Obszary Wiejskie 15(4) (52):93–100
33. Dynowska I (1994) Reżim odpływu rzecznego, plansza 32.3. Odpływ rzeczny. In: Atlas Rzeczpospolitej Polskiej, IGiPZ PAN, PPWK im. E. Romera S.A., Warszawa
34. Dynowska I, Pociask-Karteczka J (1999) Obieg wody. In: Starkel L (ed) Wydawnictwo Naukowe PWN, Warszawa, pp 343–373
35. Wrzesiński D (2017) Typologia reżimu odpływu rzek w Polsce w podejściu nadzorowanym i nienadzorowanym. Badania Fizjograficzne. Geografia Fizyczna A68:253–264
36. Piętka I (2009) Wieloletnia zmienność wiosennego odpływu rzek polskich. Prace i Studia Geograficzne 43:81–95
37. Wojtczak H (2004) Chemizm wód Strugi Toruńskiej w rejonie zlewni eksperymentalnej Koniczynka. In Kejna M, Uscka J (eds) Zintegrowany Monitoring Środowiska Przyrodniczego. Funkcjonowanie i monitoring geoekosystemów w warunkach narastającej antropopresji. Biblioteka Monitoringu Środowiska, pp 227–236
38. Stanisławczyk B (2017) Wieloletnia dynamika odpływów charakterystycznych z wybranych zlewni Polski w świetle zmian indeksu NAO. Przegl Geogr 89(3):413–428
39. Bartczak A, Glazik R, Tyszkowski S (2014) Czasowe i przestrzenne zróżnicowanie odpływu jednostkowego w zlewni rzeki Zgłowiączki (wschodnia część Kujaw). Nauka Przyr. Technol 8(3):28
40. Cieśliński R, Komkowska E (2016) Reakcja zlewni rolniczej na wielkość opadu atmosferycznego na przykładzie zlewni rzeki Borucinki (Pojezierze Kaszubskie). Prace i Studia Geograficzne 61(3):7–26
41. Jokiel P, Stanisławczyk B (2016) Zmiany i wieloletnia zmienność sezonowości przepływu wybranych rzek Polski. Prace Geograficzne 144:9–33
42. Jokiel P (2004) Zasoby wodne środkowej Polski na progu XXI wieku. Wydawnictwo Uniwersytetu Łódzkiego, Łódź
43. Fal B, Bogdanowicz E (2002) Zasoby wód powierzchniowych Polski. Wiadomości Instytutu Meteorologii i Gospodarki Wodnej 25(2):3–38
44. Gutry-Korycka M, Sadurski A, Kundzewicz Z, Pociask-Karteczka J, Skrzypczyk L (2014) Zasoby wodne i ich wykorzystanie. Nauka 1:77–98

45. Sullivan AB, Drever JI (2001) Spatiotemporal variability in stream chemistry in a high-elevation catchment affected by mine drainage. J Hydrol 252:240–253
46. Clark MJ, Cresser MS, Smart R, Chapman PJ, Edwards AC (2004) The influence of catchment characteristics on the seasonality of carbon and ni-trogen species concentrations in upland rivers of Northern Scotland. Biogeochemistry 68:1–19
47. Rauba M (2016) Wpływ obszarów użytkowanych rolniczo na właściwości fizykochemiczne wód płynących na przykładzie rzeki Rokietnica. Zeszyty Problemowe Postępów Nauk Rolniczych 584:71–80
48. Krasowska M (2017) Sezonowe zmiany składu chemicznego wód rzecznych w zlewni rolniczej. Inż Ekol 18(3):175–183 https://doi.org/10.12912/23920629/69378
49. Dąbkowski SL, Pawłat-Zawrzykraj A (2003) Jakość wód Raszynki i jej dopływów. Woda-Środowisko- Obszary Wiejskie 3(6):111–123
50. Skorbiłowicz M, Ofman P (2014) Seasonal chang-es of nitrogen and phosphorus concentration in Supraśl river. J Ecol Eng 15(1):26–31
51. Lin HS (2010) Linking principles of soil forma-tion and flow regimes. J Hydrol 393:3–19
52. Krasowska M, Banaszuk P (2015) Drogi migracji biogenów w zlewni rolniczej. Inż Ekol 43:35–41
53. Informacja o stanie środowiska województwa kujawsko-pomorskiego w 2017 roku (2018) Biblioteka Monitoringu Środowiska, Bydgoszcz. http://www.wios.bydgoszcz.pl/publikacje/raporty
54. Dziennik Urzędowy województwa kujawsko-pomorskiego z dnia 6 March 2017, poz. 1005

Part IV
Water and Wastewater in Urban Areas

Chapter 16
The State of Water and Sewage Management in Poland

Katarzyna Kubiak-Wójcicka and Monika Kielik

Abstract This chapter describes the state of water and wastewater management in Poland in the years 1999–2015. The study presents the state of Poland's water resources, the structure of their use and the amount of financial outlays allocated to water and sewage management. The analysis was based on statistical data obtained from the Central Statistical Office (GUS). The basic infrastructure related to water and sewage management is the water supply and sewerage network together with sewage treatment plants. The results of the analysis show that availability of water and sewerage infrastructure has significantly improved in recent years, however, in some provinces it is still not sufficient. Major changes in the financing of water and wastewater management resulted from Poland's accession to the European Union which increased the possibility of financing activities that led to the improvement of wastewater treatment.

Keywords Water intake · Water supply network · Sewerage network · Financial outlays · Poland

16.1 Introduction

Water resources in many regions of the world are insufficient and water demand constantly grows. In many countries, the issue of water resources shortage becomes a matter of national safety, in which water is an important and indispensable element [1, 2]. An issue of water supply for the population is especially noticeable in cities, where high population density generates high water demand [3].

Water demand surge and against limited water resources requires rational water resource management at all levels: country, regional and local [4]. Water supply depends not only on its geologically available amount but also on legal, economic and social frames related to water distribution system [5]. Reduction of water intake and consumption is possible by obtaining effective technologies and appropriate

K. Kubiak-Wójcicka (✉) · M. Kielik
Department of Hydrology and Water Management, Faculty of Earth Sciences and Spatial Management, Nicolaus Copernicus University, Lwowska 1, 87-100 Toruń, Poland
e-mail: kubiak@umk.pl

© Springer Nature Switzerland AG 2021
M. Zeleňáková et al. (eds.), *Quality of Water Resources in Poland*, Springer Water,
https://doi.org/10.1007/978-3-030-64892-3_16

375

technical infrastructure related to water intake and sewage disposal [6]. In some countries, due to the existing water shortage, the possibility of buying water is being considered, and water is becoming a product for which there is demand. On the other hand, countries with large water resources are interested in selling water. There are already international water markets around the world that deal with the transport and supply of water for areas with water shortages [7, 8]. In the near future, water shortage will be noticeable in a larger area due to population growth, economic growth and climate change. That is why many water managers and decision-makers are actively implementing supply and demand management strategies [9]. Both developed and developing economies are facing an imbalance of water supply and demand, which will potentially affect the price of water, especially during a water shortage [10–13]. Only optimal water prices can ensure sustainable consumption, production and protection of water [14, 15]. Water is therefore a product for which there is a high demand, but it cannot be treated solely as a product for sale.

In most European countries, there applies the Water Framework Directive, which treats water in a special way. As the Water Framework Directive says, "Water is not a commercial product like any other, but rather an inherited good that must be protected, defended and treated as such ..." (from the preamble of the Water Framework Directive). Poland became a member of the European Union on May 1, 2004. That required adaptation of national regulations in the area of water resources management to the principles applied in the Water Framework.

Poland has small renewable surface water resources compared to other European countries [16, 17]. In addition, water resources in Poland are characterized by high variability in time and space, which requires rational management of water resources [18]. The variability of water resources over time results from climate change that has been observed in recent years. Droughts appear more and more often [19]. Spatial variability results from the uneven distribution of both, the hydrographic network and the amount of precipitation [20].

Water and sewage management is an important element of the national economy in Poland. It is implemented both at the local and regional level. Water management in Poland is carried out by specific entities. Their organization, skills and equipment as well as external conditions (environment) affect the rationality of this management [21]. In terms of water and wastewater management, extremely important changes took place as a result of Poland's accession to the European Union on May 1, 2004. Poland was preparing for this process a few years earlier, among others by amending the Water Law in 2001. Upon accession to the EU, there began the implementation of national provisions, taking into account the provisions of the Water Framework Directive. Accession also meant access to EU funds to finance both government and local government investments in the field of water and wastewater management.

The purpose of this study is to present the state of water and wastewater management in Poland in the years 1999–2015. A detailed analysis of the state of water and wastewater management was presented for 16 voivodships in 3 selected years: 1999, 2007 and 2015. These years reflect the state of water and wastewater management before Poland's accession to the European Union (1999), in the initial period of

Poland's functioning in the Union (2007) and 11 years after Poland's accession to the European Union (2015).

16.2   Materials and Methods

The study was based on statistical information, Poland's legal provisions and reports on the implementation of national programs available on ministerial websites.

The elaboration uses data collected in the Local Data Bank of the Central Statistical Office [22] and in the annuals "Environmental Protection", which cover the period 1999–2015 [23]. The study also used the information contained in the Report on the implementation of the National Program for Municipal Sewage Treatment (KPOSK) [24].

Data on water and wastewater management have been presented in general for Poland and in regional terms, taking into account the division of the country into administrative units (voivodships). To characterize Poland's water resources and the way they were used, there were utilized the data regarding the volume of water intake, the length of the water supply and sewage networks, the number of sewage treatment plants and the amount of financial outlays allocated for water management.

The selected data has been processed the way, which allowed the use of the indicator method. The volume of water resources and water intake was normalized to the population of Poland, which allows comparison of obtained results with other countries.

The network density index was calculated according to the formula:

$$W_g = L/P$$

W_g—network density index
L—the length of water or sewerage network (km)
P—voivodship area ($100\ km^2$)

The applied density indices of the water supply and sewage network express the length of the network per $100\ km^2$ of surface. The degree of equipment of a given area with municipal infrastructure was determined by the ratio of the length of water supply network referenced to a 1 km of sewerage network.

The index has been calculated on the base of the formula:

$$W_d = L_w/L_k$$

W_d—network equipment index
L_w—water network length (km)
L_k—sewerage network length (km)

The high value of the index indicates a low level of investment in sewerage network [25]. The results obtained for 3 indicators were presented in the form of cartograms using ArcGis software. This allowed for an analysis of the spatial diversity of selected elements of water and sewerage infrastructure in voivodships.

A detailed analysis of the data was carried out in three selected years, i.e. 1999, 2007 and 2015, which include the years before Poland's accession to the European Union and in the years of its functioning within the structures of the EU.

16.3 Characteristics of the Study Area

The total area of the country is 312,679 km^2. There is a three-level administrative division of the country in Poland. According to this division, which has been in force since January 1, 1999, Poland has been divided into 16 voivodships and divided into smaller administrative units such as poviats and communes. As of January 1, 2015, there are 314 poviats in Poland, 66 cities with poviat rights and 2,478 communes [22].

The population living in Poland at the end of December 2015 was 38.437 million. The largest population lives in the voivodships Mazowieckie and Śląskie, while the smallest—in Opolskie voivodship (Table 16.1). The average population density in Poland is 123 people per 1 km^2. The largest population density is there in Śląskie voivodship (372 people/1 km^2), the next one is Małopolskie voivodship (222

Table 16.1 Population and area of voivodships (own study based on the data from [22])

No	Voividship	Population			Area (km^2)
		1999	2007	2015	
1.	Dolnośląskie	2,917,139	2,878,410	2,904,207	19,947
2.	Kujawsko-pomorskie	2,068,864	2,066,136	2,086,210	17,972
3.	Lubelskie	2,209,083	2,166,213	2,139,726	25,122
4.	Lubuskie	1,007,967	1,008,481	1,018,075	13,988
5.	Łódzkie	2,637,438	2,555,898	2,493,603	18,219
6.	Małopolskie	3,217,865	3,279,036	3,372,618	15,183
7.	Mazowieckie	5,112,652	5,188,488	5,349,114	35,558
8.	Opolskie	1,074,205	1,037,088	996,011	9,412
9.	Podkarpackie	2,098,771	2,097,338	2,127,657	17,846
10.	Podlaskie	1,212,269	1,192,660	1,188,800	20,187
11.	Pomorskie	2,166,230	2,210,920	2,307,710	18,310
12.	Śląskie	4,776,856	4,654,115	4,570,849	12,333
13.	Świętokrzyskie	1,302,518	1,275,550	1,257,179	11,711
14.	Warmińsko-mazurskie	1,424,772	1,426,155	1,439,675	24,173
15.	Wielkopolskie	3,339,749	3,386,882	3,475,323	29,826
16.	Zachodniopomorskie	1,696,925	1,692,271	1,710,482	22,892

people/1 km^2). Voivodships with the lowest population density are: Podlaskie (59 people/1 km^2) and Warmińsko-Mazurskie (60 people/1 km^2). Less than 100 people per 1 km^2 also live in the following voivodships: Lubuskie, Zachodniopomorskie and Lubelskie. The largest population is recorded in cities. In nine most densely populated ones there are over 3000 people per 1 km^2. In Poland, on average, there are 52 people per 1 km^2 in rural areas [22]. The cities with the largest population include: Warsaw (1,744,351 people), Kraków (761,069), Łódź (700,982), Wrocław (635,759), Poznań, Gdańsk, Szczecin, Bydgoszcz, Lublin and Katowice—as of December 31, 2015.

Water resources accumulated in rivers, lakes and underground waters depend on rainfall, influencing their supply and their natural renewal. The average annual amount of precipitation in Poland in the years 1951–2015 was 625 mm. The total renewable resources of Poland's surface waters were described as the average annual surface water runoff. Depending on the assumed period, the value of the average annual outflow from the territory of Poland varies and fits in the range of 60–64 km^3 [26, 27]. The average annual total outflow from the territory of Poland in the years 1951–2015 was about 61 km^3, while it ranged from 37.5 km^3 in 1954 to 89.9 km^3 in 1981 [28]). There is an explicit irregularity in the total outflow, which manifests itself in the form of alternating cycles of years with high and low outflows. Clear periods of high outflows occurred in 1977–1982, 1997–2002 and 2010–2011, while low outflows occurred in 1988–1993 and 2003–2009. An increased frequency of high and low outflows alternately has been recorded since 2001 [20]. In 2010, the outflow was exceptionally high and amounted to 86.9 km^3, while in 2015 it was extremely low—40.8 km^3 [28].

In the analyzed period of 1999–2015, the average outflow for Poland was 61.1 km^3. The highest values were recorded in 2010 and they amounted to 86.9 km^3, when a rainfall flood occurred in Poland [25, 29]. The lowest outflow values were recorded in 2015 and it amounted to 40.8 km^3, which was the result of hydrological drought occurring almost all over Poland [19]. In comparison to the other European countries, the outflow from Poland is small (Michalczyk and Sposób in this volume, Kubiak-Wójcicka in this volume). The countries with lower water resources include Belgium, the Czech Republic, Denmark, Ireland, Luxembourg and Switzerland [23]. However, due to the population of particular countries, the size of water resources is not so favorable in Poland. On average, in the years 1999-2015, there were 1601 m^3 of water per capita. In Europe, this rate is three times higher, and in the world even four times higher than in Poland [18, 25, 26].

16.4   The State of Water and Sewage Management in Poland in the Years 1999–2015

Depending on the analyzed period, the volume of water intake in Poland was changing. The largest total amount of water intake in Poland was 14.03 km^3 (1990), while the smallest was 10.5 km^3 (2015) [17]. Total water consumption in Poland

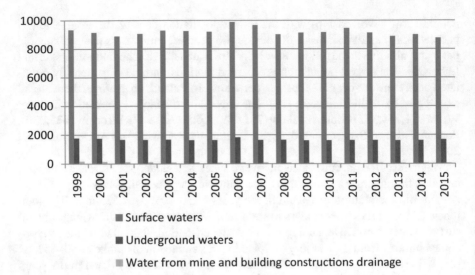

■ Surface waters

■ Underground waters

▩ Water from mine and building constructions drainage

Fig. 16.1 Water intake for the national economy, divided by intake sources, in hm^3 (own study based on the data from [22])

in the years 1999–2015 averaged 10,955 hm^3. The reason for the decrease of water intake and consumption by the economy and the population was the closure of many industrial and agricultural plants, or the limitation of their production, the introduction of water-saving technologies and the introduction of water charges. Surface waters taken from rivers and lakes, which covers 84% of the country's needs (9198.6 hm^3) and are used mainly for industrial purposes, play a key role in the water supply. Groundwater accounts for an average of 15.2% (1665.1 hm^3) and is intended mainly for supplying the public with good quality drinking water. A small share—on average 0.8% (91.8 hm^3) in the analyzed period of 1999–2015 comprised water from drainage of mining areas and buildings (see Fig. 16.1).

The impact on the volume of water intake depends mainly on the population living in the area and the economic situation. In recent years, there has been a trend towards a decrease of water resources per capita, as seen in the attached diagram (Fig. 16.2). In 2015, there were only 1,100 m^3 per 1 inhabitant of Poland. Hence, many studies state that Poland's water resources are among the lowest in Europe, and Poland is called "Egypt of Europe" [30].

The average water consumption in the years 1999–2015 was 287 m^3 per capita. The highest consumption was 310 m^3 and the lowest 276 m^3 per capita. The volume of water intake per capita has remained constant over the last 17 years, the variation in the volume of water intake varies from 4 to 8% of the average value for the years 1999–2015.

The water deficit indicator which determined the ratio of water intake to the amount of available water in the analyzed period varied in Poland from 12.5% (2010) to 25.7% (2015). This indicator was defined by UNESCO (2000) as an indicator of

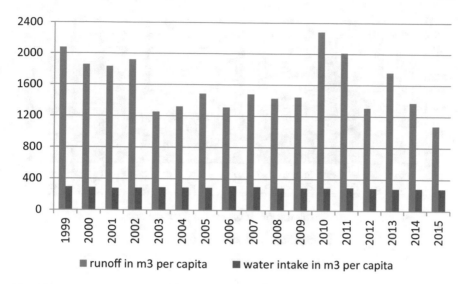

Fig. 16.2 Volumes of runoff and of water intake in m³ per capita in the years 1999–2015 (own study based on the data from [22])

water deficit. In Poland, it is on average at the level of 10–20%, which locates our country in a moderate water stress range. In the event of an extremely dry year, the amount of surface water that is available is significantly reduced. In this situation, the water deficit indicator is in the range of 20–40%, which places Poland among countries with medium to high water stress. In such a case there is competition between water users, which can cause potential conflicts [31].

In the GUS reports, the structure of water intake takes into account the division into production purposes, irrigation in agriculture and forestry, filling and supplementing fish ponds, as well as the operation of the water supply network. Usually, production purposes include water intake for industrial purposes. Irrigation in agriculture and forestry data pertain to areas over 20 ha and filling ponds pertain to ponds with an area of at least 10 ha. The data is obtained by the GUS from water law consent documents issued by respective self-government units and does not include irrigation of smaller areas. The data regarding the operation of the water supply network includes all units that supervise the water supply network, including housing associations, water companies, water service companies and other production plants.

The largest water abstractions were recorded for production purposes and they constituted on average 74.2% of total water abstractions (Fig. 16.3). Irrigation purposes are on average 10.4%, while the operation of the water supply system on average is 15.4% of total water abstraction. During the analyzed years 1999–2015, total water consumption showed a significant downward trend. Water consumption for production purposes showed a slight downward trend, however it was not statistically significant. On the other hand, water intake for the needs of water supply network operation showed a clear decrease. There was a noticeable increase in the

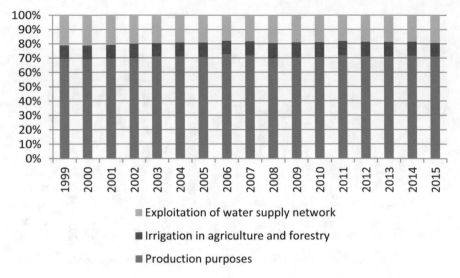

Fig. 16.3 Water intake division in Poland in the years 1999–2015 in % (own study on the basis of the data from [22])

amount of water taken for irrigation purposes. Despite the noticeable slight downward and upward trends, the share of water intake for particular sectors of the national economy was relatively stable, and the changes noted were small.

According to [32], infrastructure is the basic factor in the development of a given area. It affects the location of investments and the related economic situation of the population. The systematic increase in the length of the water supply network was spread over time. Construction of the sewerage network did not occur at the same pace as the construction of the water supply network. The growth in the length of the sewerage network accelerated at the beginning of the twenty-first century and in the years 1999–2015 amounted to as much as 320%, while in the water supply network it was 146%. Construction of the sewerage network together with sewage treatment plants, accelerated more after 2007, when the expenditures increased significantly.

Expenditure on water management in the years 1999–2003 ranged between EUR 600 and 800 million. In later years (2004–2010) a significant increase in financial outlays was recorded. In 2011–2015, financial outlays slightly decreased, however, they remained at a high level, from EUR 1.3 billion to 1.6 billion (Fig. 16.4). The volume of financial outlays allocated to water and wastewater management was associated with the possibility of obtaining EU funds.

The most important source of financing projects related to environmental protection in Poland in the years 2007–2013 was the Operational Program Infrastructure and Environment (POIiŚ). In terms of the amount of available funds, it was the largest operational program not only in Poland, but also in the whole European Union. For purposes of its implementation in the years 2007–2013, Poland received around EUR

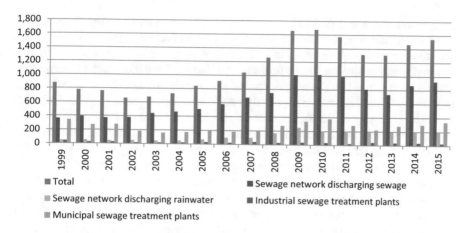

Fig. 16.4 Financial outlays on water management in EUR million in the years 1999–2015 (own study on the basis of the data from [22])

27.9 billion from the EU budget, of which approximately EUR 5.5 billion was allocated for investments in environmental protection. Funds for the implementation of the Infrastructure and Environment Program came from two sources of financing: from the Cohesion Fund (EUR 22.2 billion) and the European Regional Development Fund (EUR 5.7 billion) [24]. Water and wastewater investments not only allow for faster socio-economic development, but also give the opportunity to meet the requirements of the EU directives that are important for human life and health. The development of water and sewerage infrastructure directly and indirectly contributed to fulfilling the requirements arising from the following EU environmental directives:

- Directive 1976/464/EEC of May 4, 1976 regarding pollution caused by certain hazardous substances discharged to the aquatic environment,
- Directive 91/271/EEC of May 21, 1991 regarding the treatment of household sewage,
- Directive 2000/60/EU of October 23, 2000 which established a framework for Community's actions in the scope of common water policy.

The provisions of the Accession Treaty obliged Poland to reduce biodegradable pollution loads by 2015. The program instrument aimed at coordinating activities and monitoring in the field of construction, extension and modernization of sewage systems and municipal sewage treatment plants is the National Program for Municipal Sewage Treatment [24]. In addition to the national ecological funds and own resources, the financial instrument that allows the implementation of the KPOŚK are European Union aid funds [33].

At the end of 2015, there expired the eligibility period for expenditure in projects co-financed by the European Union, among others, under the Operational Program Infrastructure and Environment (POIS) in the financing perspective 2007–2013. Despite enormous financial outlays allocated for this purpose in the 2007–2013 perspective, the needs of local governments in the area of investing in water and wastewater infrastructure and improvement of water and wastewater management are still significant. Investments in this area not only lead to faster socio-economic development, but also lead to the fulfillment of the requirements of the EU directives. Therefore, it seems that the new financing perspective 2014–2020, under which the most funds (EUR 27.4 billion), were allocated to the Infrastructure and Environment Program, will play an important role in this area. Investments affecting environmental protection and low-carbon economy, development of technical infrastructure and energetic safety will continue to be priorities of this program. As part of this action, there still can be funded the investments in the construction or modernization of sewage treatment plants and sludge management, construction or modernization of the sanitary sewerage network, including the division of combined sewerage system into rainwater and sanitary sewage, construction and modernization of water supply systems and construction of storm water drainage. The EU funds from this program will also be allocated, to a limited extent, to investments in the area of health protection and cultural heritage.

16.5 Spatial Diversity of Water Intake in Individual Voivodships

The state of development of water and wastewater management in Poland is varied. Water intake and infrastructure investments show spatial diversity and largely depend on the economic development of the region, which is already visible in the division voivodship (Figs. 16.5, 16.6, and 16.7).

Water abstraction in Poland varies between regions. The largest water intake was recorded in 1999 in the Mazowieckie voivodship (2,275,087 dam^3), Wielkopolskie (2,018,592 dam^3), Zachodniopomorskie (1,684,515 dam^3) and Świętokrzyskie (1,200,162 dam^3). In total, these 4 voivodships took 67.2% of water abstracted in Poland in 1999. In subsequent years, a further increase of that share was recorded— in 2007 it was 69.2% and in 2015 it was 70.7%. A large amount of abstracted water was used for production purposes.

The lowest water consumption was recorded in Podlaskie and Lubuskie voivodships (less than 90,000 dam^3) (Figs. 16.4, 16.5, and 16.6).

The largest amount of water was abstracted in the Mazowieckie voivodship, where the food, fuel and chemical industries are developing. 84 to 88% of the abstracted water was allocated for production purposes in this province. In Świętokrzyskie voivodship, there dominate the mining industry and the production of building materials (88–92% of the water taken for production purposes). In Zachodniopomorskie voivodship, the chemical industry dominates (93–94% of abstracted water), while in the Wielkopolskie voivodship, lignite mining plays a big role (83–86% of abstracted water).

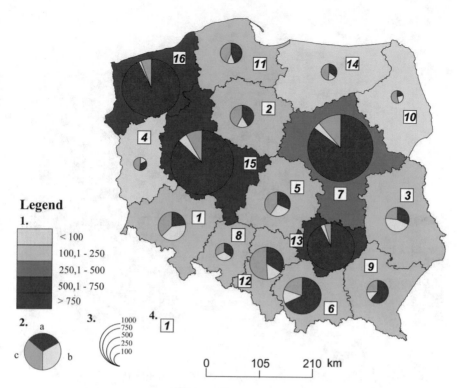

Fig. 16.5 Structure of water intake in Poland by voivodship in 1999 (own study on the basis of the data from [22]). Legend: 1—water consumption in m^3 per 1 person, 2—% share of water intake for: a—production, b—water supply network operation, c—irrigation, 3—total water consumption in hm^3, 4—voivodship names according to numbering shown in Table 16.1

Taking into account the population of particular voivodeships, the largest water abstractions per capita was recorded in 2015 in the Świętokrzyskie (1063 m^3 per capita) and Zachodniopomorskie (865.7 m^3 per capita) voivodships, with an average intake of 261.6 m^3 for Poland. In these voivodships, water was mainly abstracted for production purposes, respectively 91.9% and 92.8%.

In turn, the largest share of water intake for irrigation took place in the Dolnośląskie, Lubelskie and Kujawsko-Pomorskie voivodships. This was related to the predominance of agricultural activity in these areas over industrial activity [25].

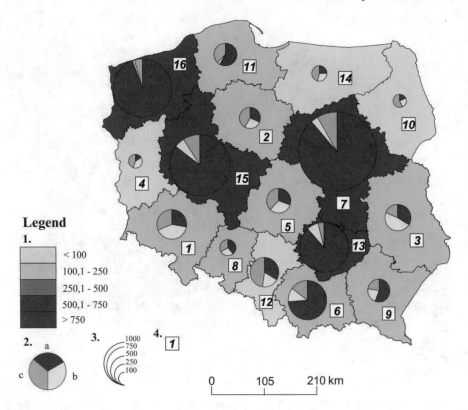

Fig. 16.6 Structure of water intake in Poland by voivodship in 2007 (own study on the basis of the data from [22]). Legend: 1—water consumption in m^3 per 1 person, 2—% share of water intake for: a—production, b—water supply network operation, c—irrigation, 3—total water consumption in hm^3, 4—voivodship names according to numbering shown in Table 16.1

16.6   The Degree of Water and Wastewater Infrastructure Development

The development of water supply infrastructure and the universality of water supply services is an important distinguishing feature and determinant of general economic development and the standard of living of the population. In addition, the structure of water supply and unit consumption is not only a measure of the achieved civilization standard, but also a determinant of rational management, determined by appropriate technical, investment, organizational and economical solutions [34].

The degree of development of water supply and sewerage infrastructure was spatially diverse. The largest increase in the length of the water supply network in the analyzed period 1999–2015 was recorded in Mazowieckie voivodship (193%) and Warmińsko-Mazurskie voivodship (183%), while the largest increase in the length of the sewerage network was recorded in Podkarpackie voivodship (483.8%) and Świętokrzyskie voivodship (453%). The situation is similar to the number of water

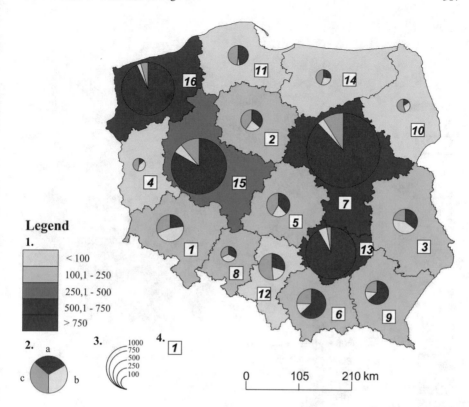

Fig. 16.7 Structure of water intake in Poland by voivodship in 2015 (own study on the basis of the data from [22]). Legend: 1—water consumption in m^3 per 1 person, 2—% share of water intake for: a—production, b—water supply network operation, c—irrigation, 3—total water consumption in hm^3, 4—voivodship names according to numbering shown in Table 16.1

pipe connections to residential buildings, which recorded an average increase of 147%. The number of connections to the sewerage network has grown twice as quick as the water supply network—by 296.7%. Despite this increase, there is still a significant disproportion between water supply and sewerage networks saturation in individual voivodships (Figs. 16.8, 16.9, and 16.10).

The indicator of the length of the water supply network per 1 km of the sewerage network during the analyzed period of 1999–2015 decreased by 54%. This means that the disparities between the length of the water supply and sewerage network have decreased. It resulted from the increase of financial outlays for the development of the sewerage network. A significant advantage in the length of the water supply network over the length of the sewage network in 2015 was recorded in the Podlaskie, Łódzkie and Lubelskie voivodships. The province that shows the best condition in this scope is Podkarpackie voivodeship (0.91).

Clear disproportions between the equipment in the water supply and sewerage network result mainly from the amount of financial outlays on the equipment and

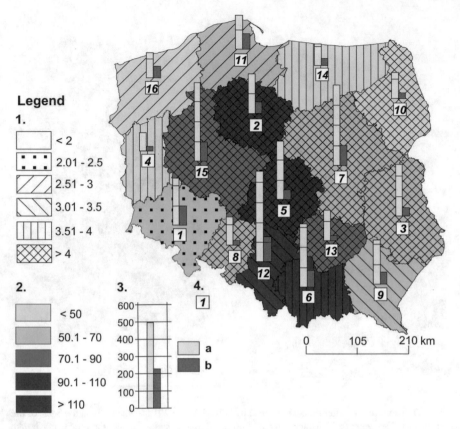

Fig. 16.8 Technical infrastructure indicators in 1999 (own study on the basis of the data from [22]). Legend: 1—the number of km of the water supply network per 1 km of the sewerage network, 2—the density of the water supply network per 100 km^2, 3—the number of connections to residential buildings in thousands: a—water supply, b—sewerage, 4—names of voivodships in accordance with the numbering presented in Table 16.1

maintenance of this infrastructure. Sewerage systems, like other systems constituting technical infrastructure, are characterized by a complex network structure, high capital intensity, long service life and related to its significant differences in the age of the devices used. Particularly large neglects in this area are visible in rural areas with too dispersed buildings. Connecting network devices there generates high costs and is not always possible due to technical reasons. In this situation, the best solution is to build individual treatment plants, but this requires even more support from local authorities.

The degree of development of water and wastewater infrastructure in individual voivodships depended on the amount of financial outlays allocated to it (Figs. 16.11, 16.12, and 16.13). The largest financial outlays on water and wastewater management were allocated in the following voivodships: Wielkopolskie (1999), Śląskie (2007) and in Mazowieckie, Śląskie, Wielkopolskie and Małopolskie (2015).

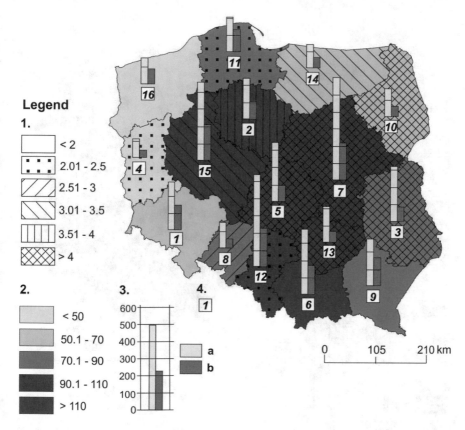

Fig. 16.9 Technical infrastructure indicators in 2007 (own study on the basis of the data from [22]). Legend: 1—the number of km of the water supply network per 1 km of the sewerage network, 2—the density of the water supply network per 100 km², 3—the number of connections to residential buildings in thousands: a—water supply, b—sewerage, 4—names of voivodships in accordance with the numbering presented in Table 16.1

In the scope of removing biodegradable pollutants contained in municipal sewage, the National Program for Municipal Sewage Treatment has been developed. This program defined agglomeration lists that should be equipped with collective sewerage systems and sewage treatment plants as well as the size of biodegradable loads from these agglomerations, municipal sewage treatment plants and deadlines for their implementation [24].

The National Program for Municipal Wastewater Treatment [24] was approved by the Cabinet on December 16, 2003. The Program includes a list of agglomerations together with the necessary list of projects to be carried out in these agglomerations in the field of construction and modernization of collective sewage systems, by the end of 2015. The implementation of the entire KPOŚK was divided into four-time horizons, i.e. the years 2003–2005, 2006–2010, 2011–2012 and 2014–2015. Program

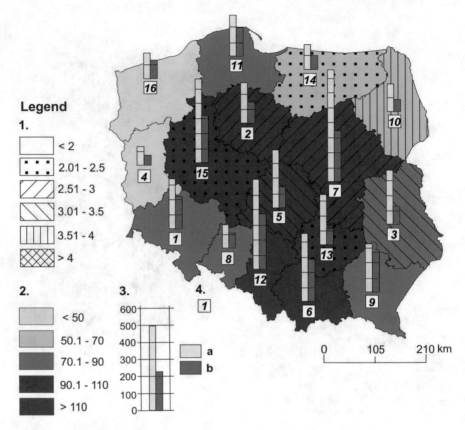

Fig. 16.10 Technical infrastructure indicators in 2015 (own study on the basis of the data from [22]). Legend: 1—the number of km of the water supply network per 1 km of the sewerage network, 2— the density of the water supply network per 100 km^2, 3—the number of connections to residential buildings in thousands: a—water supply, b—sewerage, 4—names of voivodships in accordance with the numbering presented in Table 16.1

updates were carried out on an ongoing basis and implemented in years 2005, 2009, 2010 and 2015 [24].

As part of the Rural Areas Development Program (PROW), water and wastewater management projects were implemented for rural communes, urban-rural communes (excluding cities with more than 5,000 inhabitants) and urban communes (excluding cities with more than 5,000 inhabitants).

16.7　Discussion

The state of water and wastewater management in Poland significantly improved in the years 1999–2015. In the 1980s, the volume of water abstraction in Poland was definitely higher and the decline in water consumption in households was started

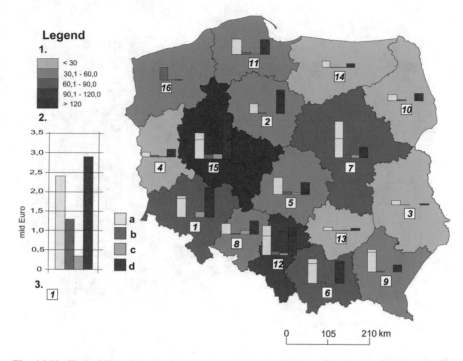

Fig. 16.11 Financial outlays on water management in 1999 (own study on the basis of the data from [22]). Legend: 1—total expenditure on water management in EUR million, 2- expenditure on water management depending on funding directions in EUR million: a—sewerage network draining sewage, b—sewerage network draining rainwater, c—treatment of industrial sewage, d—municipal wastewater treatment, 3—voivodship names in accordance with the numbering presented in Table 16.1

after 1990 along with the transition from a centrally planned economy to a free-market economy [35]. The new socio-economic situation in Poland resulted in a decrease in water consumption rates by all groups of its recipients. The decrease in water consumption for municipal purposes was mainly caused by the increase in water and sewage charges, mass installation of water meters, and consequently the reduction of wastage, water-saving and care for the quality and condition of equipment and installations by water users [18]. After a period of dynamic decline in water consumption, consistently with forecasts and analyses, small changes and some stabilization have been observed since 1999.

The main factor that contributed to a significant improvement in the state of water management was Poland's accession to the European Union on May 1, 2004 and the obligation to adopt national regulations to the Water Framework Directive. The main goal of the WFD is to prevent the deterioration of surface and groundwater quality by limiting the load of substances introduced into the environment.

The objectives of the WFD were implemented through the National Program for Municipal Waste Water Treatment [24] for urban areas and the related construction

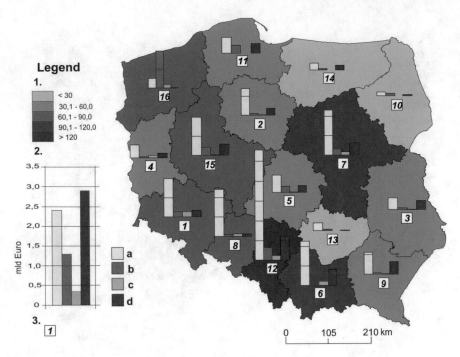

Fig. 16.12 Financial outlays on water management in 2007 (own study on the basis of the data from [22]). Legend: 1—total expenditure on water management in EUR million, 2—expenditure on water management depending on funding directions in EUR million: a—sewerage network draining sewage, b—sewerage network draining rainwater, c—treatment of industrial sewage, d—municipal wastewater treatment, 3—voivodship names in accordance with the numbering presented in Table 16.1

of sewage treatment plants for which significant financial outlays were allocated. In turn, investments in the field of water and wastewater management in rural areas with significantly lower outlays were implemented through the Rural Areas Development Program (PROW).

In recent years, there has been a leveling of differences in technical infrastructure in Poland in urban and rural areas. The administrative unit responsible for collective water supply is the commune, which is regulated by the Act of 7 June 2001 on collective water supply and collective sewage disposal (Journal of Laws of 2018, item 650). Since the development of technical infrastructure is a particularly costly and time-spreading process, local authorities need external assistance in this area, because neither the rural community nor local governments are able to allocate adequate financial resources for this purpose [32]. Low-income communes do not have adequate financial resources that they could spend on the construction of new water intakes and modernization of water and sewerage infrastructure in the near future. Water and sewerage infrastructure was implemented with partial support from the EU funds, but also loans taken out by local governments. These loans are an important element of the municipality's budget in the coming years. Hence, the

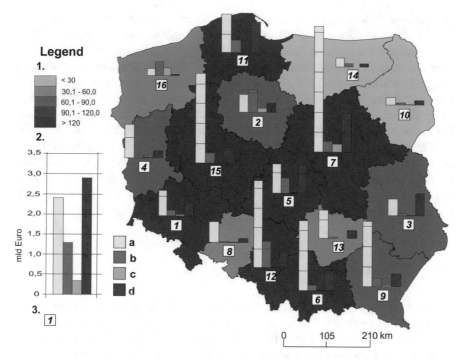

Fig. 16.13 Financial outlays on water management in 2015 (own study on the basis of the data from [22]). Legend: 1—total expenditure on water management in EUR million, 2—expenditure on water management depending on funding directions in EUR million: a—sewerage network draining sewage, b—sewerage network draining rainwater, c—treatment of industrial sewage, d—municipal wastewater treatment, 3—voivodship names in accordance with the numbering presented in Table 16.1

limited financial capabilities of communes largely constitute the main brake on the future development of rural areas in Poland. Further expansion of the water supply and sewerage network in the future must take into account the factors of its maintenance and support, which in effect translates into the cost for end-users. With less demand and fewer users, maintaining the infrastructure can be cumbersome. This problem is significant because there is an increase in the migration of people from cities to the countryside. In addition, urban population migration to rural areas causes increased demand for water, most often using existing water intakes, which are no longer sufficient. This problem has been noticed not only in Poland but also in other European countries [36]. In turn, in cities there is a problem with maintaining technical infrastructure in the light of the decreasing population [37, 38]. Therefore, it is advisable to conduct interdisciplinary studies that take into account demographic issues. Demographic trends indicate complex interactions between population changes and transformations in infrastructure and supply systems [36]. According to many authors, greater emphasis should be put on infrastructure development and the involvement and education of decision-makers, employees, society and other groups that can help local governments meet the challenges related to drinking water supply [39–41].

Reducing water consumption requires not only individual restrictions but systemic solutions of water consumption adapted to living conditions and using the latest technology in households. An effective tool is a proper management of water consumption, and water consumption at home should be seen in the context of social practices rather than individual behavior [12, 42, 43]. Unfortunately, the provisions in force in Poland do not have preferential conditions for plants or residential buildings that have installations for the recovery of technical water. In addition, the relatively low cost of water does not encourage its users to save water. The lack of a subsidy system for plants that want to invest in modern installations in the field of technological water recovery and its re-use causes the lack of interest in it. This issue should be resolved in a systemic way.

16.8   Conclusions

In recent years, progress in improving the condition of water and wastewater management has become visible. The main reason for this is the implementation of the WFD regulations, which resulted from Poland's accession to the European Union on May 1, 2004. This gave the opportunity to benefit from the EU funds, which in the first place constituted support for the construction of water and sewerage infrastructure. The analysis shows that the availability of water supply and sewerage infrastructure in Poland is not sufficient. The disproportion between the length of the water supply system and the sewerage system is particularly evident. Despite a significant improvement in the condition of the infrastructure, which consisted in the increase in the length of the sewerage network and the population served by sewage treatment plants, the process is slow and does not completely solve the problems associated with sewage disposal. This is particularly evident in the countryside. It should be kept in mind that the problem will increase as the population of people who live in rural areas grows, which is partly a result of the natural growth, but also of the systematically increasing migration from urban to rural areas. In the next financial perspective, emphasis should be put on building sewerage infrastructure in rural areas, but also on modernizing the existing infrastructure in urban areas. That is why systemic solutions should play an important role, which will allow to introduce rational water management at the commune level, without limiting expenses from the budget for other activities.

The second major challenge is to adapt water intake to climate changes. Despite the moderate climate in Poland, water stress is a growing challenge, especially in the years with low precipitation. Individual and collective consumers should be included in a system of incentives to save water and invest in the installation of modern technologies for its recovery. Such actions require the intervention of the state and efficient demand management in line with the expectations of society and directions of economic development.

16.9   Recommendation

This chapter describes the state of water and wastewater management in Poland in the years 1999–2015. Based on the analysis, areas in Poland have been identified in which the water and sewerage infrastructure requires significant improvement. Lack of financing or small expenditure on investments related to water management indicate areas that require state aid in financing water and sewage management. It also shows which elements of the policy have contributed to the development of water and wastewater management.

References

1. Kim MS, Park SW, Jang NJ, Kwak DH (2016) Statistical Analysis of Water Infrastructure Characteristics: case Study of Saemangeum Watershed. Water Resour 43(1):58–65. https://doi.org/10.1134/S0097807816120071
2. Lee K-H (2016) Korea has no water scarcity! Water Resour 43(3):579–582. https://doi.org/10.1134/S0097807816030088
3. McDonald RI, Weber K, Padowski J, Martina Flörke M, Schneider Ch, Green PA, Gleeson T, Eckman S, Lehner B, Balk D, Boucher T, Grill G, Mark Montgomery M (2014) Water on an urban planet: urbanization and the reach of urban water infrastructure. Glob Environ Change 27:96–105. https://doi.org/10.1016/j.gloenvcha.2014.04.022
4. Hu P, Chen N, Li Y, Qiwei Xie Q (2018) Efficiency evaluation of water consumption in a Chinese Province-level region based on data envelopment analysis. Water 10(6):793. https://doi.org/10.3390/w10060793
5. Bower KM (2014) Water supply and sanitation of Costa Rica. Environ Earth Sci 71:107–123. https://doi.org/10.1007/s12665-013-2416-x
6. Radonić L (2019) Re-conceptualising water conservation: rainwater harvesting in the desert of the Southwestern United States. Water Altern 12(2):699–714
7. Danilov-Danilyan VI, Demin AP, Pryazhinskaya VG, Pokidysheva IV (2015) Markets of water and water management services in the world and the Russian Federation: Part II. Water Resour 42(3):378–388. https://doi.org/10.1134/S0097807815030033
8. Venkatachalam L (2015) Informal water markets and willingness to pay for water: a case study of the urban poor in Chennai City, India. Int J Water Resour Dev 31(1):134–145. https://doi.org/10.1080/07900627.2014.920680
9. Katz D (2016) Undermining demand management with supply management: moral Hazard in Israeli water policies. Water 8:159. https://doi.org/10.3390/w8040159
10. Hoekstra AY, Chapagain AK (2007) Water footprints of nations: Water use by people as a function of their consumption pattern. Water Resour Manage 21:35–48. https://doi.org/10.1007/s11269-006-9039-x
11. Mansur ET, Olmstead SM (2012) The value of scarce water: measuring the inefficiency of municipal regulations. J Urban Econ 71:332–346
12. Soto Rios PC, Deen TA, Nagabathla N, Ayala G (2018) Explaining water pricing through a water security lens. Water 10:1173. https://doi.org/10.3390/w10091173
13. Heino O, Takala A (2015) Social norms in water services: exploring the fair price of water. Water Altern 8(1):844–858
14. Clere J (2016) True Cost of Water: monetization of water risks, shared value creation, and local acceptality of extractive projects. Field Actions Science Reports. 14. http://journals.openedition.org/factsreports/4069

15. Arfanuzzaman Md A, Atiq Rahman A (2017) Sustainable water demand management in the face of rapid urbanization and ground water depletion for social–ecological resilience building. Glob Ecol Conserv 10:9–22. https://doi.org/10.1016/j.gecco.2017.01.005

16. Gutry-Korycka M, Sadurski A, Kundzewicz ZW, Pociask-Karteczka J, Skrzypczyk L (2014) Zasoby wodne a ich wykorzystanie. Nauka 1:77–98. http://www.pan.poznan.pl/nauki/N_114_07_Gutry.pdf

17. Kubiak-Wójcicka K, Piątkowski K (2015) Analiza zmian poboru wody w woj. kujawsko-pomorskim na tle kraju. Ekologia i Technika XXIII(5):397–404

18. Hotloś H (2010) Badania zmian poboru wody w wybranych miastach Opolski w latach 1990–2008. Ochrona środowiska 32(3):39–42

19. Kubiak-Wójcicka K, Bąk B (2018) Monitoring of meteorological and hydrological droughts in the Vistula basin (Poland). Environ Monit Assess 190:691. https://doi.org/10.1007/s10661-018-7058-8

20. Kubiak-Wójcicka K (2019) Long-term variability of runoff of Vistula River in 1951–2015. "Air and Water—Components of the Environment" Conference Proceedings, Cluj-Napoca, Romania, pp 109–120. https://doi.org/10.24193/awc2019_11

21. Kaca E, Drabiński A, Ostrowski K, Pierzgalski E, Cz Szafrański (2011) Gospodarowanie wodą w sektorze rolno-żywnościowym i obszarach wiejskich w warunkach nowych wyzwań i ograniczeń. Pol J Agron 7:14–21

22. Główny Urząd Statystyczny. Central Statistical Office. http://stat.gov.pl/

23. Ochrona Środowiska (Environment) (2009) Główny Urząd Statystyczny, Warszawa

24. Krajowy Program Oczyszczania Ścieków Komunalnych (KPOŚK) (2003) http://www.kzgw.gov.pl/index.php/pl/materialy-informacyjne/programy/krajowy-program-oczyszczania-sci ekow-komunalnych

25. Kubiak-Wójcicka K, Piątkowski K (2017) Evaluation of water and wastewater infrastructure in communes od kujawsko-pomorskie voivodeship. Infrastruktura i ekologia terenów wiejskich, Nr III/1/2017, Polish Academy of Sciences, Cracow Branch, pp 907–922. http://dx.medra.org/10.14597/infraeco.2017.3.1.070

26. Wałęga A, Chmielowski K, Satora S (2009) Stan gospodarki wodno-ściekowej w aspekcie wdrażania Ramowej Dyrektywy Wodnej. Infrastruktura i Ekologia Terenów Wiejskich 4:57–72

27. Michalczyk Z, Sposób J (2011) Water resources of Poland and their state in lublin district. Teka Komisji Ochrony Środowiska Przyrodniczego – OL PAN 8:104–111

28. Kubiak-Wójcicka K, Machula S (2020) Influence of climate changes on the state of water Resources in Poland and their usage. Geosciences 10(8):312. https://doi.org/10.3390/geosci ences10080312

29. Maciejewski M, Ostojski MS, Walczykiewicz T (ed) (2011) Dorzecze Wisły. Monografia powodzi maj-czerwiec 2010. IMGW-PIB Warszawa

30. Syposz-Łuczak B (2010) Gospodarowanie wodami powierzchniowymi i podziemnymi w Polsce. Technika Poszukiwań Geologicznych. Geotermia. Zrównoważony Rozwój 1–2:151–160

31. Kowalczak P (2008) Zagrożenia związane z deficytem wody. Wydawnictwo Kurpisz SA, Poznań. ISBN 978-83-75249-79-8

32. Krakowiak-Bal A, Salamon J (2011) Public-private partnership as a method of implementing technical infrastructure investments. Infrastruct Ecol Rural Areas 12:179–188

33. Gospodarka wodno-ściekowa (2011) Narodowy Fundusz Ochrony Środowiska i Gospodarki Wodnej. ISBN: 978-83-915678-1-4

34. Gorczyca M (1998) Stan i rozwój infrastruktury wodociągowej w Polsce na tle wybranych krajów. Studia i prace. Zakład Badań Statystyczno-Ekonomicznych. Główny Urząd Statystyczny, Warszawa, p 257

35. Gorączko M, Pasela R (2015) Causes and effects of the water consumption drop by the population of cities in Poland—selected aspects. Bull Geogr. Socio-economic Ser 27:67–79. https://doi.org/10.1515/bog-2015-0005

36. Hummel D, Lux A (2007) Population decline and infrastructure: the case of the German water supply system. Vienna Yearbook of Population Research, pp 167–191

37. Faust KM, Mannering FL, Abracham DM (2015) Statistical analysis of public perceptions of water infrastructure sustainability in shrinking cities. Urban Water J. https://doi.org/10.1080/1573062X.2015.1011671
38. Goemans Ch, Costanigro M, Stone J (2012) The interaction of water restriction and pricing policies: Econometric, managerial and distributional implications. J Nat Resour Policy Res, 61–77. https://doi.org/10.1080/19390459.2012.665668
39. Eledi SB, Minnes S, Vodden K (2017) Source Water Protection in Rural Newfoundland and Labrador: Limitations and Promising Actions. Water 9:560. https://doi.org/10.3390/w9080560
40. Minnes S, Vodden K (2017) The capacity gap: understanding impediments to sustainable drinking water systems in rural Newfoundland and Labrador. Canadian Water Resou J / Revue canadienne des ressources hydriques 42(2):163–178. https://doi.org/10.1080/07011784.2016.1256232
41. Lennartsson M, McConville J, Kvarnström E, Hagman M, Kjerstadius H (2019) Investments in innovative, urban sanitation—decision-making processes in Sweden. Water Altern 12(2):588–608
42. Rajala RP, Katko T (2004) Household water consumption and demand management in Finland. Urban Water J 1:17–26. https://doi.org/10.1080/15730620410001732080
43. Corbella HC, Pujol DS (2009) What lies behind domestic water use? A review essay on the drivers of domestic water consumption. Boletín de la A.G.E.N., 50:297–314

Chapter 17
Water Problems in Urban Areas

Tadeusz Ciupa and Roman Suligowski

Abstract The chapter presents selected issues of the water environment of Polish cities. Attention is drawn to the causes of periodic excess of rainwater and surface runoff, with particular emphasis on sealed areas. The influence of these areas on the shaping of dynamics and the size of flood waves, and on flood risks as a consequence is demonstrated. The occurrence of floods of various genesis (precipitation, snow thawing, ice jams, frazil ice, sea storms, caused by failures of damming facilities) is discussed. The importance of small streams in the geoecosystem of a city with regard to their functioning and management is presented. The importance of surface water (including lakes and water reservoirs) in the land use structure and their use, mainly for recreational purposes (bathing areas) is presented. The issues of groundwater are also outlined along with the determination of selected threats to their quality. The structure of water usage in Polish cities in the years 1989–2016 is characterized in spatial terms. The importance of the human-water environment interaction with the maintenance of the sustainable development of the city in terms of ensuring the safety of its inhabitants is referred to. On the example of selected cities, a series of activities of an administrative, technical and pro-ecological nature are presented. These have a spatial and socio-economic dimension, anchored in the European and Polish legal system.

Keywords Human—Aquatic environment interactions · Urban floods · Water consumption · Water policy · Poland

T. Ciupa (✉) · R. Suligowski
Department of Environmental Research and Geo-Information, Institute of Geography and Environmental Sciences, Jan Kochanowski University in Kielce, Uniwersytecka 7, 25-406 Kielce, Poland
e-mail: tadeusz.ciupa@ujk.edu.pl

R. Suligowski
e-mail: roman.suligowski@ujk.edu.pl

© Springer Nature Switzerland AG 2021
M. Zeleňáková et al. (eds.), *Quality of Water Resources in Poland*, Springer Water,
https://doi.org/10.1007/978-3-030-64892-3_17

17.1   Introduction

Most cities in Poland, like all over the world, are located by rivers [1]. This is because they have always been a city-forming function. The proximity of a river has brought many socio-economic, political, military and other benefits, depending on its size and the topographic location of a city [2]. The importance of rivers in the formation of cultures, civilizations and history—including many Polish cities, is discussed interestingly in a multi-volume monographic work edited by J. Kułtuniak [3]. The issues of water circulation in Polish cities in historical terms have also been taken up by Czaja [4, 5], Drwal et al. [6], Pociask-Karteczka [7], Kaniecki [8], Magnuszewski and Gutry-Korycka [9], Cieśliński and Bastian [10].

The problem of the impact of urbanization on water conditions, due to the rapid development of urban areas, was noticed in Poland in the 1970s [11–16]. However, the intensive development of hydrological research took place in the next decade. It mainly covered strongly urbanized areas of existing and newly developing industrial regions: Upper Silesian, Konin and Bełchatów, as well as the agglomerations of Łódź, Warsaw and Cracow. Initially, the scientific work concerned observed changes in the surface hydrographic network and groundwater levels [17–24]. Furthermore, it has been numerously pointed out that runoff is influenced by the land use and management in industrialized and urbanized areas [4, 25–43], as well as in suburban catchments [44–48]. Changes in water environment in urban areas are often very diverse and have both quantitative and qualitative character [37, 49–56].

The increase in interest in the hydrology of urban areas in Poland also results from practical aspects [57–62]. Knowledge of the impact of transformations in urbanized catchments on runoff processes, sediment supply and fluvial transport is indispensable when preparing hydrotechnical projects and forecasting their impact on the environment. This knowledge is also used in expertises and analyzes related to maintaining the capacity of riverbeds [63], limiting surface runoff [64–66], protecting the land from excessive erosion [35], the silting up of water reservoirs and the maintenance of drainage systems [67, 68], flood protection and population risk [69, 70], the use of river waters in industrial processes [71], etc.

Studies show that the greatest hydrological consequences of anthropogenic transformations are observed in small catchments with an area up to several hundred square kilometers. It has been proven that in urbanized areas disturbances of various water balance elements occur, which directly shape the size of surface runoff, its relation to underground runoff and they significantly affect the height and course of flood waves as well as mechanisms and intensity of fluvial processes [72, 73].

The resources of surface and underground water are the fundamental basis for the functioning and economic development of cities and the quality of living of their inhabitants. There are 930 cities in Poland (as of January 1, 2018) [74], including 1 exceeding a million inhabitants (Warsaw—1.758 million), 15 with 200–1000 people, 206 inhabited by 20–200 thousand people. The most numerous group (708) are, however, cities in which the number of inhabitants does not exceed 20,000. In Poland, as many as 60.1% of the population (23.1 million people) live in cities on an area of 6.9% of the entire country (21.6 thousand km^2). For comparison, in Europe, the urban population is 74.2% of the total population [75]. The quoted data indicate an advanced state of urbanization and industrialization processes in Poland, and this contributes significantly to changes in the water cycle and the degradation of the aquatic environment and related ecosystems. Therefore, there are conflicts between human activities, water protection as well as ensuring the safety of people and property. Therefore, it is necessary to conduct sustainable development in these areas, taking into account the multidirectional needs of water and socio-economic environments [71, 76, 77].

The chapter presents selected issues of the water environment of Polish cities regarding:

- periodic excess of rainwater and surface runoff,
- flood risk,
- the importance of small streams and reservoirs,
- water usage and wastewater treatment.

17.2 Periodic Excess of Water

The specificity of urban areas is their large sealing. It results mainly from the coverage of a significant amount of area by residential buildings, commercial buildings, warehouse halls, industrial facilities, etc. The buildings are accompanied by impermeable road surfaces or poorly permeable surfaces of sidewalks, squares and car parks. This means that the sealed area with not-rough and impervious surfaces without vegetation increases. There are then clear hydrological and geomorphological consequences in small urban catchments. The former result mainly from the reduction of infiltration and increase of the surface runoff component, and the second from significant spatial differentiation and intensification of erosion processes, including surface wash, especially along artificially oriented zones, most often of a linear nature and accumulation. The extensive drainage and hydrotechnical network enable quick drainage of excess water along with the carried matter to a river or a collecting channel. Here, due to the large and almost simultaneous supply of water and matter from different areas of a catchment, there is a rapid increase in flows and fluvial transport. Rapid reaction of a drainage system in an urbanized area is conducive to an increase in the energy of flowing waters, which initiate and intensify the processes of erosion, transport and accumulation in river channels [25–28, 30, 35, 78, 79].

The share of built-up and urbanized lands in the total area of Poland in 2016 was 5.37%, and it shows an upward trend [80]. This indicator in all Polish cities in 2014 reached 27.8%, and in individual cities, it showed a significant variation, especially in small cities: from 1.5% in Krynica Morska (Pomorskie Voivodeship) to 87.7% in Piastów (Mazowieckie Voivodeship). In capitals of voivodeships, the share of sealed areas in 2016 ranged from 31.3% in Szczecin to 55.4% in Warsaw, with its average value of 42.1% (Fig. 17.1).

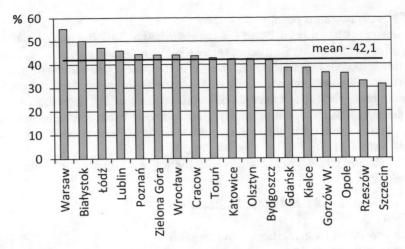

Fig. 17.1 Build-up and urbanized lands in the capitals of voivodeships in 2016. Based on: Central Statistical Office [134]

In the situation of a changing climate, the frequency and amount of heavy rainfall increase especially in cities constituting islands of heat [81]. Cities generate intense convection processes, which combined with an increase in the concentration of dust pollutants in the air (condensation nuclei) leads to the formation of clouds with vertical structure (Cumulonimbus) and then to high-intensity rainfalls. Thus, in urban catchments, in conditions of excessive sealing of the ground, periodical water excess appears, and the unit runoff exceeds 500 L/s/km^2, e.g. from the Silnica catchment in Kielce [35], or Sokołówka catchment in Łódź [38]. This leads to the acceleration of the water cycle, and this is reflected in the formation of the average flow velocity in the regulated compact channel (about 2 m/s). The consequence of these processes is more and more frequent overload of the surface and underground drainage network, which necessitates its expansion. During periods of intensive surface runoff, small rivers flowing through cities fulfill the function of a collector collecting water from numerous mouths of covered rainwater channels, surface drainage gutters and roads [35, 40, 48, 82]. Stormwater sewerage along with roadways creates an efficient two-level drainage system for surface runoff from the city area. In Kielce, the average density of storm water drainage is 8,2 km/km^2. In small sewered catchments its density even exceeds 40 km/km^2, and roads—22 km/km^2 [83].

The shape of a flood wave, especially in an urbanized section, is also significantly affected by the compact cross-section of a riverbed and its reduced roughness. The beds of these watercourses are adapted to receive large volumes of water of significant dynamics flowing in a short time. For example, in Kielce, it was found that in the urbanized catchment of the Silnica river, the maximum flow of the flood wave was almost 5 times higher than the maximum in an adjacent suburban (forestal and agricultural) catchment, characterized by similar surface and physiogeographical

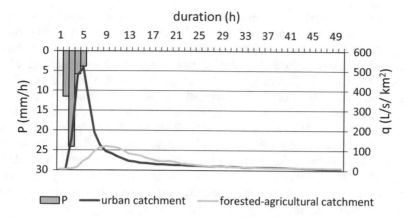

Fig. 17.2 Differences in specific runoff hydrographs (q – L/s/km^2) at cross sections closing off particular urban catchment (Silnica) and forested-agricultural catchment (Sufraganiec) during flood event (July 15, 1999). P (mm)—areal amount of precipitation. *Source* [35]

features [35]. Hydrograms of typical flood waves during the summer surge in 1999 caused by heavy rainfall differs significantly in these catchments (Fig. 17.2).

Another significant hydrological effect of urbanization is a significant reduction in the time of wave formation (from the beginning of the effective precipitation to the wave maximum). In the urbanized segment of the Silnica catchment, during the analyzed flood, it was 7 h shorter than in the section closing the suburban catchment (Fig. 17.2). An unnaturally short time of the concentration of flood waves of small streams (about 1 h) flowing through a city center causes flood threat and limits the possibility of reaction of emergency services in the protection of life and property of city residents [84]. Another results is bank defenses and hydrotechnical infrastructure being damaged. Such local flash floods in watercourses and in depressions occurring in large and small towns and bringing economic losses are called urban floods. Due to their short duration and small spatial range, it is often difficult to document them, hence there are very few items on the subject in Polish hydrological literature, although in recent years there has been an increase in interest in this topic [85–87]. The highest frequency of flash floods with disastrous consequences, qualified as natural disasters in cities in the years 1971–2010, occurred in southern Poland, i.e. in Outer Western Carpathia (Bielsko-Biała, Tarnów, Rzeszów, Andrychów, Krosno), Sudety Mountains and Sudety Foreland (Strzelin, Strzegom, Kamieniec Ząbkowicki, Boguszów, Chojnów), in the Polish Uplands (Katowice, Kielce, Skarżysko-Kamienna), and in some large cities: Warsaw, Łódź, Poznań, Szczecin, Białystok, Olsztyn [85]. Detailed case studies taking into account hydrological conditions and analyzes as well as environmental effects, relate to floods, among others, in Gdańsk: Strzyża [88], Radunia and Motława streams [89, 90], in Wojcieszów: Bełczek [91], in Bogatynia: Miedzianka [92], in Łódź: Sokołówka [38], in Kielce: Silnica [34], in Warsaw: Służewiecki stream [64], in Lublin: Bystrzyca [93], as well as in several cities of the Carpathians and the Sandomierz Basin [94]. The official reports of state and local government institutions, as well as numerous reports in the mass media, are an important source of information reaching a wide range of society. According to media reports, in Polish cities only in the summer half of 2018 there were several dozen flash floods, for example: 5/10—Bolesławiec, 5/11—Gdańsk, Łódź, 5/16—Brzeziny, 6/2—Szczecin, 7/11—Poznań, Luboń, 7/15—Bydgoszcz, 7/16—Białystok, 7/17—Radomsko, 7/18—Łódź, Siedlce, 7/20—Szczawnica; 7/27—Pleszew, 7/30—Pabianice, Bytom Jaworzno, Ruda Śląska, Tarnowskie Góry, 8/1—Pleszew, 8/14—Limanowa, 9/1—Ruda Sląska, Opole, 9/4—Białystok, Słupsk, 9/9—Ostrowiec Świętokrzyski, 9/17—Zakopane, Nowy Targ, Krościenko. Some of them concerned descriptions of the effects of events related to flooding of roads, parking lots, metro stations, residential and commercial buildings. In general, they were the result of insufficient capacity and patency of rainwater drainage, capacity of hydraulic engineering structures and valleys, etc.

Much greater scale of threats in some Polish cities is caused by transit rivers, especially during regional floods with rainy genesis, characteristic for the summer season [95]. They have a large spatial coverage of the inundation, a long duration of the flood, and severe economic losses are the consequence. A typical mechanism of their formation is the inflow of warm and humid sea air from the northern part of the Mediterranean (mainly the Adriatic Sea) or the Black Sea, controlled by a low-pressure center with a center over mid- or southern Poland. There is long-lasting high-intensity precipitation covering a large area of Poland then. Typical examples of this large-scale mechanism in the last few decades were the floods in July 1970 [96] and 1997—millennial flood [97, 98], August 1972 [99], May and June 2010 [100, 101]. The most severe effects were related to the two largest rivers: the Vistula—1020 km long and the Odra—840 km and their upper tributaries [102, 103]. The most frequently mentioned cities located by the Vistula River that have been affected by catastrophic floods in history are: Kraków, Sandomierz, Warszawa, Tczew, Gdańsk, and Malbork [104, 105], and by Odra: Racibórz, Kędzierzyn-Koźle, Opole, Wrocław, Słubice, Nowa Sól, Krosno Odrzańskie [106, 107]. Flooded rivers inundated urban areas located behind the river shafts numerously, and the resulting floodplains persisted for many days. The flood in July 1997 caused 55 people to die in Poland. About 100,000 people were evacuated at that time, and 47,000 of residential buildings, 1,700 bridges and culverts, 71 hospitals, almost 1,000 schools and kindergartens, 100 intakes of drinking water were flooded and damaged [108].

Cities in Poland are less and less threatened by mid-winter and spring thaw floods. A large event of this type was in March-April 1979, due to the very intense long-lasting snowfall at the turn of 1978/1979. The cities located in the Bug and Narew basins (including Ostrołęka, Pułtusk) and Warta basin (including Poznań, Kalisz) were at risk. Recently, the thaw flood occurred at the turn of February and March 2011 in the Noteć and lower Warta catchments (Poznań).

Small rivers draining cities during the mid-winter and spring thaw are not a flood hazard, because due to the accelerated melting of the snow cover and increased evaporation within roads, parking lots, built-up areas, etc. there is a significant loss of water resources accumulated in the snow cover. For example, in the urbanized catchments of Kielce, the average maximum specific runoff in March (1998–2003) was significantly lower than at the same time in the neighboring suburban catchments [35].

In cities located by large rivers, there is a possibility of winter floods: ice jam and frazil ones. A significantly greater threat occurs in the lower reaches of the Vistula River than the Odra River. On the Vistula during this type of flood, the absolute historical maximum water levels were recorded [109]. After building the dam of the water reservoir in Włocławek and putting it into use in 1970, the risk of flooding associated with the river icing and the occurrence of congestion distinctly increased above the reservoir, and decreased below it [110]. The largest disastrous ice-jam flood in Poland occurred in January 1982 in the upper part of the Włocławek reservoir, covering, among others, left-bank part of the city of Płock [111, 112]. In turn, a huge frazil jam formed on Wisła cross-cut in Gdańsk (Świbno) in March 1956 and caused a unique high water accumulation in the river and the breaking of the flood embankment. On the regulated section of the Odra, congestions threatening Szczecin are eliminated by icebreaking actions (e.g. in March 2018). Floods of this type also occur on the Warta and Noteć rivers (including Wieleń at the turn of 2011/2012).

Storm floods occur in cities located at river estuaries on the Baltic Coast, on Żuławy and the Vistula Lagoon and the Szczecin Lagoon [113, 114]. They are a hydrological effect of the movement of low pressure from the Norwegian Sea to the east or south-east, mainly in the period from November to February. There are inflows of sea waters into riverbeds and river valleys, and water levels rise up to 2 m in them [115, 116]. An example of such an extreme event was a storm flood in January 1983, which occurred in cities along the entire Baltic coast (Świnoujście, Kołobrzeg, Ustka, Władysławowo, Gdańsk), and was caused by a wind of 8–10° on the Beaufort scale [117]. The storm flood was also lately reported (October 2009) in Elbląg on the Vistula Lagoon. A backwater in the Elbląg River bed was formed then.

Massive material losses in cities over the last dozen or so years have been caused by floods resulted from failures of water accumulation structures on rivers. The reasons were extremely high sums of atmospheric precipitation in their catchments, and in some cases design, implementation and exploitation errors of a dam and reservoir. Examples of such events come from all over Poland. On the night of February 2, 2000, the reservoir dam on the Młynówka River (Warmian-Masurian Voivodeship) was destroyed and a wave of a height of approx. 7 m flooded Górowo Iławeckie [67]. Unfortunately, three fatalities were the result of this flood. Another incident was a failure of the structural coffer of the Wióry reservoir on July 24, 2001, on the Świślina River in the Holy Cross Mountains. After an extremely high rainfall (daily sum in Kielce—155.2 mm), with high humidity of the ground in the catchment, the water accumulated and the embankment was destroyed. The surge wave exceeding the height of 8 m caused significant damage in two cities far away from the event, i.e. in Kunów and Ostrowiec Świętokrzyski (8.5 km and 17.0 km away, respectively). Another spectacular example of the collapse of an accumulating construction, the effects of which were also felt in two cities (in Bogatynia and Zgorzelec), was the destruction of the dam in Niedów on the Witka river (Dolnośląskie Voivodeship) in

August 2010. In the first city, it was deprived of the supply of drinking water from the reservoir, and in the second there was flooding.

17.3 Small Streams—Status, Functioning, Management

The majority of small permanent and periodic streams commonly found in urbanized areas are not covered by continuous quantitative and qualitative monitoring. They are a very sensitive system reflecting natural and anthropogenic changes in a catchment. The latter influence, e.g. current conditions of the channel and the bed as well as disturbances of water circulation, which in turn affects the surface runoff, its relation to underground runoff and flood wave features, as well as the quality of outflowing waters [26, 35, 38]. Small urban streams are often the only receiver of surface runoff waters, and sometimes even municipal and industrial sewage of a legal nature (on the basis of water-law permits) and illegal ("wild"). These streams, in terms of nature and ecology, constitute an important living space of animals and plants. In a situation of strong anthropogenic pressure, they quickly lose their ecological balance, because nature itself is not able to counteract increasing burdens. Therefore, in some cities assessments of the condition of selected riverbeds and valley bottoms are carried out. Resistance to anthropogenic pressure is taken into account in relation to technical and investment activities, e.g. regulation and change of the course of a river channel, hydraulic engineering [63]. This problem is so important that it has found its place in the water policy program in the European Union countries [118] and has been implemented into Polish law [119]. These documents impose requirements aimed at improving the quality of water discharged through these watercourses. This state can be recognized by field mapping specifying the current hydromorphological situation of the valley bottom. Particular attention is paid to the identification of the structure and stability of biotopes: channel, bank and terrestrial and their ecological capacity. Values of the riverbed morphometry, hydrodynamic conditions, the method of bank strengthening and hydrotechnical development are also taken into consideration when valorising the channel and bank biotope. Within selected homogeneous sections of the river, a visual assessment of the purity of water, sediments, living space of organisms is made, taking into account the outbreaks of pollution. In turn, in the valorization of a terrestrial biotope, features of a valley landscape are taken into account, including land use, diversity of living spaces and the nature of the relationship between a valley and a watercourse. For the separation and evaluation of homogeneous sections, several methods are used (e.g. RHS—English, LAWA—German, HEM—Czech) with varying degrees of details depending on the needs and scale of the study [e.g. 120, 121]. In urban conditions with strongly differentiated anthropopressure, it is often necessary to modify and adapt them to local conditions [54, 63]. So far, in Poland, this type of mapping has been used mainly for the needs of medium-sized rivers [122], and sporadically small rivers—e.g. in Kielce [63, 78], Dąbrowa Górnicza [123]. Common features that determine their low level of conservation and ecological efficiency include, among others: regulation and technical structures of the riverbeds and banks, covering and transferring the trough (change of course), mechanical pollution and numerous places of sewage discharge.

In small urban catchments, a linear delivery to riverbeds of water, dissolved substances and sediment is dominant, which determines the high dynamics of runoff and fluvial transport. Extensive supply areas can reach a watershed and locally even exceed it. There are numerous sources of supply of clastic and dissolved matter of an aerial, point and linear nature [35]. In the latter case, the most important role is played by roads where a significant mass of matter is accumulated, mainly traffic pollution of anthropogenic and natural origin. This material in the periods of occurrence of surface runoff with great ease is activated and transported on the surface of roads and a system of rain drains. In a short time, it is introduced to the riverbed [37, 38, 73, 93]. Below the city centers, as a result of increased accumulation of discharged clastic matter, river beds are often silted and shallow, and, in the situation of excessive concentration of nutrients, they are intensively overgrown. As a result of this process, the cross-sectional areas and the bottoms of the valleys are significantly reduced. This, in turn, contributes to more frequent overflows [35, 124].

Storm sewers provide water with a high concentration of suspended and dissolved matter to urban watercourses [55, 125]. Their high concentrations have been documented in the winter season in the Silnica basin (Kielce). During the mid-winter thawings, the concentration of suspended matter found in water flowing out of the storm sewer ($1377 \, mg/dm^3$) was almost 100 times higher than it was at the same time in the river bed ($14 \, mg/dm^3$) [35]. The concentration of dissolved material reached then even higher values, i.e. at the outlet of the storm sewer—a dozen thousand or so mg/dm^3, and in the river—over 3 thousand mg/dm^3. It is a hydrochemical effect of the delivery of mainly communication pollutants, which are associated with the use of different salts (NH_4Cl, $CaCl_2$, $MgCl_2$) for winter road maintenance [126]. Urban areas against a background of neighboring areas are the "salt islands" [35]. Excessive salinisation of urban areas is reflected not only in fluvial transport but also affects environmental pollution. It has a negative impact on fauna and flora [127] and hydrotechnical constructions. For example, in Sokołówka (Łódź) during a sudden thaw in January 2013, the concentration of chlorides reached $3750 \, mg/dm^3$ [38]. A measurable and troublesome effect of excessive concentration of chlorides in water in urbanized areas is the decay and corrosion of hydrotechnical structures. In many small urban streams, the ecological capacity, which is also affected by the nitrogen, phosphorus, phenolics, detergents and oil derivatives, is in danger [35, 38, 128, 129]. Within administrative borders of cities, sections of watercourses being relatively similar or far from natural conditions extend up to 70% of the total length of streams [63].

17.4 The Use of Water

There are 166 cities in Poland, within which there are 314 lakes and 31 artificial water reservoirs (in whole or in part) with an area larger than 1 ha. Most lakes are in the northern part of the country (e.g. in the cities of the West Pomeranian Voivodeship—77, Warmian-Masurian Voivodeship—67), and they have mainly postglacial genesis.

In particular Polish cities, there is a very large diversity of the share of land being under waters (flowing and lentic, without taking into account internal marine waters) in their total areas. The largest values of this ratio are in small cities, e.g. 71.7%—Otmuchów (Opole Voivodeship), which is determined by a large dam reservoir at Nysa, 66.6%—Moryń, 62.3%—Ińsko (both in West Pomeranian Voivodeship), 61.5%—Zalewo (Warmian-Masurian Voivodeship). In the last case, Ewingi Lake covers more than half of the entire city solely [130]. It is worth noting that as many as 73 cities do not have any land under waters, with the largest number of them in the Wielkopolskie Voivodeship (20), and the smallest in the Lubelskie Voivodeship (1—Świdnik) [80]. Among voivodship capitals, this ratio reaches a maximum of 21.5% in Szczecin (a significant share—Dąbie Lake). The next three places in this ranking are Olsztyn—9.6% (14 lakes), Toruń—6.4% (Vistula River) and Bydgoszcz—4.2% (Bydgoszcz Water Node). The smallest share is found in Kielce—0.45% and in Łódź—0.46%. For cities over 100,000 inhabitants (39 cities), the average value of this ratio is 3.6%. In turn, the largest share of land under waters is in the cities of the Warmian-Masurian Voivodeship—9.3% and West Pomeranian Voivodeship—9.0%, and the smallest in the cities of the Łódź Voivodeship (1.1%) and Świętokrzyskie Voivodeship (1.3%).

Within the administrative borders of 19 Polish cities, there is land under internal marine waters with its total area of 231.3 km^2. In this classification, two cities stand out, i.e. Świnoujście—98.8 km^2 (Pomeranian Bay and Szczecin Lagoon) and Krynica Morska—92.8 km^2 (Gdańsk Bay and Vistula Lagoon). The largest values of the percentage ratio of these waters in the total area of a city are as follows: Krynica Morska—80.0%, Świnoujście—49.7%, Nowe Warpno—39.2%, Wolin—25.3%.

Surface waters within cities have mainly recreational functions. In 2018 there were 266 bathing areas around them, including 170 new ones [131]. Such a significant increase indicates a huge demand for services related to leisure by water. Most of them were organized in cities by lakes (111), by the Baltic Sea and its bays (67) and artificial water reservoirs (62). In addition, 26 bathing places functioned on rivers within cities. Cities that stay out in this respect are located by the sea: Kołobrzeg (11), Sopot (10), Gdańsk (8), Władysławowo (5), Gdynia (4) and by lakes: Olsztyn and Poznań (5), Szczecinek and Grudziądz (3). In the summer season of 2018, the authorities of the State Sanitary Inspection showed that water suitable for bathing was in 254 bathing areas, and only 12 of them did not meet the water quality requirements, mainly due to microbiological contamination and excessive bloom of cyanobacteria. Excellent quality was found in 68 facilities.

Water reservoirs within cities are also used for shipping purposes (e.g. in Ostróda, Elbląg, Kruszwica), as flood defenses (e.g. Otmuchów, Sulejów), industrial purposes (e.g. Konin, Olsztyn, Janikowo), municipal purposes (e.g. Dobczyce), etc. It is worth noting that many cities use surface water resources located outside their administrative borders, e.g. Szczecin (about 80% of water is taken from Miedwie Lake), Cracow (about 55%—Dobczyce Water Reservoir on Raba River), Upper Silesian conurbation (about 35%—Goczałkowicki Reservoir on the Vistula).

17.5   Groundwater

Most cities in Poland use groundwater resources, whose conditions of occurrence and the degree of recognition are very diverse. Because of their great economic and social importance, two extensive monographs were devoted to them, in which the characteristics of resources, their use and threats were presented. The first study concerns voivodeship capitals [132], and the later one—Polish cities above 50,000 inhabitants [133]. The analysis, which takes into account only the capitals of voivodeships (18) shows that the share of water extracted from groundwater intakes is about 56.5%. Within all these cities there are the Main Underground Water Reservoirs (33), with the fact that two are located under the entire city (Warsaw, Opole). Water intake in three cities takes place only from groundwater intakes (Gorzów Wielkopolski, Kielce, Opole), and a very high level of exploitation of these resources is in cities located outside the valleys of large rivers (Białystok—99.9%, Lublin—94.0%, Łódź—90.0%). Only two big cities (Katowice, Rzeszów) are supplied entirely from surface waters. In areas having a high degree of urbanization and industrialization, there are many potential outbreaks of groundwater pollution. They are connected with the multidirectional economic activity, housing development, the impact of means of transport and various forms of land use. The most endangered groundwater is within the cities of Upper Silesia, where mining and industrial activities are carried out. Exploitation of hard coal contributes to the lowering of the land and, consequently, to the apparent rise in the groundwater table and the formation of floods, which affect the quality of the Quaternary aquifer. A significantthreat to groundwater also occursin cities in the strip of Polish lowlands, where Pleistocene aquifers dominate. There is a very high and high degree of danger here, mainly due to the lack of or minimal-natural insulation layer (e.g. Gorzów Wielkopolski, Olsztyn, Bydgoszcz). High risk of groundwater exists in the center of cities located directly above the GZWP (e.g. Łódź, Cracow, Kielce, Gdańsk, Toruń). Despite the large threat to usable aquifers, groundwater captured in most of these cities have relatively good physicochemical and bacteriological quality.

17.6   Water Usage and Wastewater Treatment

Water usage in Polish cities [134] shows a downward trend from the beginning of political changes, i.e. after 1989 (in 1990—8.8 bcm of water, in 2000—6.7, in 2012—4.7, 2016—4.3), while in entire Poland in this period it decreased from 15.1 bcm (1989) to 10.1 bcm (2016). This decline has been caused by the processes of socio-economic transformation, with the biggest consequences in the industrial sector (liquidation of unprofitable and water-intensive plants, introduction of new industrial technologies and modernization of transmission installations), as well as metering of water intake and a significant increase in its price [135]. In 2016 in Polish cities about 75% of water was used for industrial purposes (the most in: Konin—1.32 bcm, Ostrołęka—0.42, Skawina—0.26, Police—0.17, Warsaw—0.16).

The average water consumption per capita in Poland in 2016 was 272 m^3/person and in all Polish cities—168 m^3/person [134]. This indicator showed significant variation among individual cities. It was the lowest in small towns in the Małopolskie Voivodeship (Ryglice—2.2 m^3/person, Zakliczyn—7.0 m^3/person), and the highest in cities with large power stations (Konin—17.7 thousand m^3/person, Skawina—10.8 thousand m^3/person), pulp and paper factories (Ostrołęka—8.1 thousand m^3/person) and chemical industry plants (Police—5.2 thousand m^3/person, Puławy—2.1 thousand m^3/person). According to Piasecki [136], the consumption of water in cities over 500,000 inhabitants is higher than in cities below 10,000 inhabitants. In the latter, it is determined by low water consumption for industrial purposes resulting from the small number and sizes of industrial plants. Water consumption for irrigation (lawns, gardens, and vegetable crops) has also been increasing frequently.

Based on to the data collected from the Central Statistical Office [134], in cities of individual voivodeships, the average water consumption per capita in 2016 was also strongly diversified, i.e. from 52 m^3/person in the Łódź Voivodeship, up to 1042 m^3/person in the Wielkopolskie Voivodeship. In capitals of voivodeships, this indicator varies within this broadrange as well. The lowest values are found in voivodeship capitals in eastern Poland (Białystok—52 m^3/person, Lublin—57 m^3/person), and the highest in cities on the Baltic Sea coast, i.e. in Szczecin (421 m^3/person) and Gdańsk (174 m^3/person), as well as in Warsaw (174 m^3/person) and in Wrocław (141 m^3/person) (Fig. 17.3).

The supply of water to municipal recipients by the water supply system creates a number of problems related to, among others, with its failures, and thus the emergence of leaks. Water losses usually result from the improper technical condition and

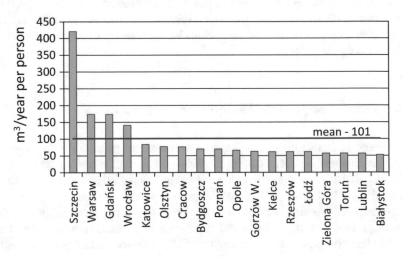

Fig. 17.3 Average annual water consumption per capita in voivodeship capitals in the period 2003–2016. Based on: Central Statistical Office [134]

excessive pressure in the network. According to Dohnalik [137], the losses in the transmission of tap water in 195 cities in Poland amounted to 18.6%. In comparison with other European countries, this was a satisfactory level, because in Spain it was at the level of 20% and in France—30%. In extreme cases, these losses were recorded at 60–75%, as it was the cases of Croatia and Albania. There is a sizeable spatial diversity of water losses in Poland, e.g. 80% of cities in the West Pomeranian Voivodeship province showed losses above 15%, whereas in 75% of cities in Silesian Voivodeship they exceeded 25% [138]. After 2000, the reduction in the percentage of water loss (PWS) is observed [139]. It is caused by the modernization of existing water supply systems in many cities, including the use of plastic pipes (HDPE, PVC), displacing traditional materials such as grey iron and steel. Failures of the water supply network in cities located on loess areas cause particularly dangerous effects in the municipal infrastructure. Then, the suffosion processes are initiated and, as a consequence, the soil is washed out, voids and channels are created, the foundations are washed, and construction damage is done, e.g. during a failure in the old town of Sandomierz in 2007.

The reduction in water demand after 1989 resulted in a decrease in the volume of municipal wastewaters generated and discharged into rivers [41]. However, according to data from the Central Statistical Office [134], in 2003, treated sewage accounted for 89% of the volume of all sewage generated in cities in Poland, and in 2016 it was as high as 96%. In as many as 794 cities, this rate reached 100%, and only in 9 did not exceed 50% (the smallest: Libiąż—7.1%, Brzeszcze—10.5%). Bearing in mind, however, more stringent purification criteria taking into account increased biogen removal, this rate for all cities in Poland in 2016 was on average considerably lower, i.e. 65% (Fig. 17.4).

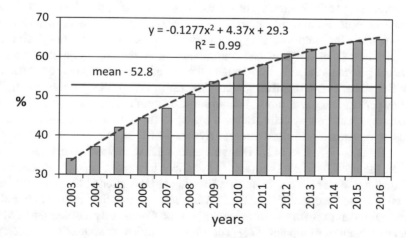

Fig. 17.4 Average annual share of treated wastewater with increased biogen removal (%) in the total volume of wastewater generated in Polish cities in years 2003–2016. Based on: Central Statistical Office [134]

The conducted analysis for the years 2003-2016 indicates a constant increase in the ratio of the average share of treated wastewater with increased biogen removal in the total volume of generated wastewater. It is described by a polynomial equation of the 2nd degree with a high coefficient of determination $R^2 = 0.99$ (Fig. 17.4). In 308 cities, 100% biogens along with other pollutants were removed, while in 378 cities this process was not carried out. In 2016, the voivodeship capitals showed a large variation in the characteristics described, i.e. from 49.8% (Katowice) to 100% (Warsaw, Łódź). It is worth noting that in 2003, in Szczecin this rate had a value of 16%, in Cracow—24.5% and it reached 100% in none of the voivodeship capitals (maximum in Lublin—99.6%). At that time, Szczecin was on the infamous "black list" of the biggest pollsters of the Baltic Sea—announced by the Commission for the Protection of the Marine Environment of the Baltic Sea (HELCOM). Visible improvement is a result of putting many modern, highly efficient sewage treatment plants into operation (e.g. in 2010 "Pomorzany" in Szczecin) or modernization of previously existing ones (among others, 2010—"Sitkówka" in Kielce, 2012—"Czajka" in Warsaw).

17.7 Water Policy

While carrying out the tasks of sustainable development of a city concerning surface and underground waters, shaping the interaction between the human and the water environment is of particular importance, taking into account the safety of its inhabitants. In many Polish cities, a number of administrative, technical and pro-ecological activities are undertaken that have a spatial and socio-economic dimension, anchored in the European and Polish legal system.

Additionally, at the beginning of the twenty-first century, the Council of Europe adopted the Water Framework Directive [118] establishing the directions of Community actions in the field of water policy and the Flood Directive [140] aimed at reducing the risk of flooding and minimizing its effects. It is complemented by numerous programs that take into account the sustainable development of cities (e.g. New Urbanism, Eco-Urbanism, Green Urbanism). As well as programs aimed at limiting periodic excess of rainwater and surface runoff associated with city sealing, protection of water ecosystems, rational management and use of existing resources including *Sustainable Urban Drainage System*—SUDS [141] and *Water Sensitive Urban Design*—WSUD [142]. Design trends related to architecture taking into account the need for direct contact between city inhabitants and water have also appeared, e.g. Water Urbanism or even Floating Urbanization [143]. With Poland's accession to the European Union, the Polish legal system in the field of water policy is gradually adapting to the existing regulations and Community requirements, which prefer sustainable development [77]. The effective implementation of water policy in the city in the field of water management and defining the terms of water resources consumption is currently provided by the Water Law [119]. The rules contained therein determine the possibilities of full use of the law by the municipal authorities of a city in order to undertake comprehensive pro-ecological activities, including limiting sources of pollutant emissions and water resources management or restrictive enforcement of the conditions for obtaining water permits for facilities harmful to the water environment. Detailed rules and conditions for the use of water resources in the city in terms of planning, design, construction and technological solutions, including ownership rights, controls and appropriate sanctions are regulated by implementing regulations. Local governments of Polish cities currently pay the greatest attention to ways of reducing the risk of flooding. The documents that underlie the spatial policy are flood hazard maps (flood risk areas coverage, depths, ordinates of water table), flood risk maps (flood losses, population and objects at risk of flooding) and flood risk management plans. These maps were handed to administrative units by the National Water Management Authority in 2015. Governments of many cities, knowing the extents of the above areas, include them in local spatial development plans and in decisions on determining the location of public-purpose investments or development conditions. These maps have become an essentialsource of information as well as an effective tool for setting priorities and making further decisions of a technical and financial nature [66, 144, 145].

According to Romanowicz et al. [146] technical investment projects limiting the effects of flooding in cities mainly include: multi-purpose retention reservoirs (e.g. in Włocławek, in Dobczyce, in Nysa), polders (e.g. Racibórz, Wrocław), flood embankments on major rivers (e.g. in Warsaw with a length of 69 km, including Vistula River—52 km) and their tributaries in backflow zone in cities around the country (e.g. Narew River—Nowy Dwór Mazowiecki, Czarna Staszowska—Połaniec). In addition, these are relief canals (e.g. Opole, Wrocław, Konin, Śrem, Gorzów Wielkopolski), reconstruction of large river beds (e.g. the Flood Canal and the Odra Miejska in Wrocław) as well as the introduction of natural reinforcements of the banks of small watercourses. Other technical activities include, above all, repairs and modernization of existing hydraulic engineering (including bridges, culverts—eliminating their insufficient capacity causing accumulation of flood waters) and damming devices (often exploited for several decades and is in poor condition).

A critical issue to be solved in Polish cities is to undertake systematic actions aimed at reducing surface runoff and increasing retention, especially in a city center. This can be achieved by replacing the traditional "gray" infrastructure with "blue-green" elements [147–149]. Such solutions include: construction and reconstruction of small water reservoirs (e.g. the Arturówek reservoir in Łódź) and the creation of absorptive (infiltrative) traps, green terraces and roofs (e.g. the Warsaw University Library in Warsaw, Opolska Business Park and Kärcher in Cracow), rain gardens (e.g. in Katowice, Gdańsk, Łódź) and open drains (rain gutters) with a cascade system delaying the runoff. A model example of the use of most of the solutions mentioned above is the modern Marina Mokotów housing estate in Warsaw, where the public space has been combined attractively with the blue-green rainwater management system. In the central part of the estate, there is a reservoir fed with rainwater from the roofs of 87 residential buildings, surrounded by greenery and multi-level walking paths with small architecture—bridges, terraces, piers, cascades, water walls and fountains [150]. Similar interesting solutions for limiting surface runoff appear in other cities at the design stage (e.g. a park in Wola Duchacka in Kraków) or implementation stage (e.g. in Radom).

Environmental activities in the field of protection and renewal of water ecosystems in Polish cities include, first of all, restoration of degraded areas, including improvement of water quality and maintenance of hydrogenic habitats in a state similar to natural ones, supported mainly by local governments [151, 152]. From the beginning of the twenty-first century, the trend of "returning to the river" has been observed in many Polish cities, expressed by the development of coastal areas, e.g. natural marinas (e.g. Bydgoszcz, Wrocław, Konin, Sandomierz), walking promenades (e.g. Płock, Chełm), bicycle routes (e.g. Kazimierz Dolny, Kielce), beaches, boulevards and promenades (e.g. Warsaw, Toruń, Poznań). These investments stimulate not only the recreation of residents but also have practical significance because they stimulate the development of various services, including catering. An important element of local government activities is to inform local communities, including school youths about the state, threats and protection of the aquatic environment through publications on this topic in bulletins and mass media, organizing seminars and lectures, city games, happenings, whiteboard films broadcasted in public transportation (e.g. Warsaw). Moreover, World Water Day has been actively celebrated for over a dozen years in Polish cities. Concerts, trips, workshops and competitions for children and youths are organized, and the topic is different each year, e.g. in 2011 the slogan was "Water for cities", and in 2017—"Sewage as an unlimited source of water". In 2018, whose main theme was "Nature for water", events took place, among others, in Poznań, Warsaw, Wrocław, Opole, Katowice, Cracow, Kielce, Białystok, Olsztyn, Gdańsk and Dąbrowa Górnicza. Organizing pro-ecological events for several years as part of the international PARK(ing) Day is another interesting initiative, where much attention is devoted to topics related to water in the city (including Łódź, Poznań, Olsztyn, Warsaw). Information and educational campaigns (leaflets, brochures, meetings, workshops) are also systematically carried out among residents, and information boards and didactic paths are placed in places that are valuable and exposed to degradation. These initiatives are increasingly supported financially by ecological foundations and the European Union.

17.8 Conclusion

The assessment of water resources in Poland would be incomplete if it does not take into account the importance of urban areas, which occupy 6.9% of the territory of the country (930 cities) inhabited by 60.1% of the total population (about 23 million people). The advanced state of urbanization processes affects significantly changes in water cycle and leads to the degradation of water environment. In cities, what gains special significance is learning the man–water environment interactions treated as pressure $\rightarrow$ state $\rightarrow$ reaction. It is necessary to conduct sustainable development in these areas, taking into account the multidirectional needs of population, economy as well as the water environment.

The above discussion can be summarized as follows:

- High and constantly increasing share of sealed surfaces in cities (average 27.8%) results in reduced infiltration and increased surface runoff component, especially after downpours in the summer half-year. The consequence of these processes is the more and more frequent congestion of the surface and underground drainage network as well as the occurrence of flash floods. This forms an unexpected threat to the life and property of residents.
- Much larger scale of threats in large Polish cities is caused by transit rivers, especially during regional floods with a rainy genesis. The cities affected by the catastrophic floods in the past include: on the Vistula River—Cracow, Sandomierz, Warsaw, Tczew, Gdańsk and Malbork, and on the Oder River— Racibórz, Kędzierzyn-Koźle, Opole, Wrocław, Słubice, Nowa Sól and Krosno Odrzańskie.
- The threat to cities with floods of a different origin (thaw, jam, frazil, sea storm and caused by failures of damming facilities on rivers) has a rather local range and it happens relatively rarely.
- Small permanent and periodic watercourses have significant importance, often underestimated in urbanized areas. They are sometimes the only receivers of surface runoff waters, and even municipal and industrial sewage. Meanwhile, from the point of view of nature and ecology, these streams constitute an important living space of animal and plant organisms. In recent years, a number of activities have been carried out in many Polish cities for the revitalization of riverbeds and valleys of small watercourses.
- Interesting is the fact that within 166 Polish cities there are about 350 lakes and artificial water reservoirs with an area larger than 1 ha, having mainly recreational functions, on which 173 bathing areas are located. Objects of this type also exist in cities on the Baltic coast (67) and on rivers (26).
- Water consumption in Polish cities shows a downward trend from the beginning of political changes, i.e. after 1989, and currently it is 168 m^3/person. Another positive phenomenon is the decrease in the volume of municipal sewage discharged into rivers, and at the same time, a very high degree of water treatment (96% in 2016). In as many as 85% of cities this ratio reached 100%, and only in a few cities it did not exceed 50%.
- Local authorities of Polish cities show the necessity of modern water management by undertaking numerous initiatives and actions resulting in the development and implementation of programs aimed at rational, effective management and usage of existing resources, including those that appear periodically (e.g. New Urbanism, Eco-Urbanism, Green Urbanism, Sustainable Urban Drainage System, Water Sensitive Urban Design, Water Urbanism). A measurable element of these activities is the fact that environmental programs are dominated by tasks related to the water environment.

The presented chapter is a very on-point answer to the significant increase in interest, in recent years, in the water environment of cities in Poland—mainly in relation to the need to improve the quality of life of their residents and the need to conduct flood protection. In addition, the identification of such problems is necessary when preparing hydrotechnical projects and forecasts of investments' impact on the environment. This information can be used to develop expert opinions and analyzes related to the maintenance of river channels throughput, limiting surface runoff, protecting the land from excessive erosion, silting of water reservoirs, maintenance of drainage systems, etc.

References

1. Liszewski S (1995) Geografia miast nadrzecznych. In: Rzeki – kultura, cywilizacja, historia (vol. 4). Wyd. Śląsk, Katowice, pp 127–151
2. Pancewicz A (2004) Rzeka w krajobrazie miasta. PŚ, Gliwice
3. Kułtuniak J (ed) (1994–1999) Rzeki – kultura, cywilizacja, historia (vol. 1–8). Wyd. Śląsk, Katowice
4. Czaja S (1995) Zmiany użytkowania ziemi i powierzchniowej sieci hydrograficznej na obszarze miasta Katowice w latach 1801–1985. Geographia. Studia et Dissertationes 19:7–23
5. Czaja S (2011) Powodzie w dorzeczu górnej Odry. UŚ, Katowice
6. Drwal J, Fac J, Borowiak M, Głogowska M (1996) Zmiany stosunków wodnych w granicach fortyfikacji miasta Gdańska w okresie od początku XVII wieku do końca XIX wieku. In: Dziejowe przemiany stosunków wodnych na obszarach zurbanizowanych. UAM, UŚ, Poznań-Sosnowiec, pp 31–41
7. Pociask-Karteczka J (1996) Woda w dziejach miasta (zarys hydrografii Krakowa). In: Dziejowe przemiany stosunków wodnych na obszarach zurbanizowanych. UAM, UŚ, Poznań-Sosnowiec, pp 131–151
8. Kaniecki A (2004) Poznań – dzieje miasta wodą pisane. PTPN, Poznań
9. Magnuszewski A, Gutry-Korycka M (2009) Rekonstrukcja przepływu wielkich wód Wisły w Warszawie w warunkach naturalnych. Prace i Studia Geograficzne 43:141–151
10. Cieśliński R, Bastian J (2014) Powodzie w Gdańsku w czasach historycznych. In: Problemy badań wody w XX i XXI wieku. UG, Gdańsk, pp 109–122
11. Dynowska I, Zbadyńska E (1974) Wpływ działalności człowieka na zmianę pierwszego horyzontu wód gruntowych. Prace Geograficzne 37:9–20
12. Hojda K (1974) Zmiany stosunków wodnych w obszarze silnie uprzemysłowionym na przykładzie zlewni Chechła. Prace Geograficzne 37:21–34
13. Soczyńska U (1974) Hydrologiczne skutki urbanizacji. Wiadomości Meteorologii i Gospodarki Wodnej 22(4):11–22
14. Hydrological effects of urbanization: Studies and Reports in Hydrology, vol. 18. UNESCO, Paris (1974)
15. Dobija A (1975) Wpływ urbanizacji na stosunki wodne. Czasopismo Geograficzne 46(1):73–78
16. Uchnast-Nowicka B (1981) Wpływ urbanizacji na stosunki wodne na przykładzie zlewni Brzeźnicy. Notatki Płockie 26(3):46–48
17. Rayzacher Z (1980) Niektóre problemy hydrologiczne obszarów zurbanizowanych. Czasopismo Geograficzne 51(2):215–217
18. Mikulski Z, Nowicka B (1982) Wpływ urbanizacji na stosunki wodne w świetle badań polskich. Gospodarka Wodna 7:100–105
19. Lenart W (1984) Wpływ urbanizacji na ustrój hydrologiczny i jakość wody. Przegląd Geofizyczny 29(2)
20. Jankowski AT (1986) Antropogeniczne zmiany stosunków wodnych na obszarze uprzemysłowianym i urbanizowanym (na przykładzie Rybnickiego Okręgu Węglowego). Prace Naukowe UŚ 868, Katowice

21. Jankowski AT (1987) Wpływ urbanizacji i uprzemysłowienia na zmianę stosunków wodnych w regionie śląskim w świetle dotychczasowych badań. Geographia. Studia et Dissertationes 10:62–97

22. Jankowski AT (1996) Przemiany stosunków wodnych na obszarze Górnego Śląska. In: Dziejowe przemiany stosunków wodnych na obszarach zurbanizowanych. UAM, UŚ, Poznań-Sosnowiec, pp 42–53

23. Choiński A (1978) Analiza zmian układu sieci wód powierzchniowych i podziemnych w południowej części Konińskiego Zagłębia Węglowego. Badania Fizjograficzne nad Polską Zachodnią. Seria A: Geografia Fizyczna 31:33–55

24. Choiński A, Ziętkowiak Z (1991) Przeobrażenia stosunków wodnych i aktualny stan czystości wód rejonu Turku. In: Przemiany środowiska geograficznego obszaru Konin-Turek. UAM, Poznań, pp 195–203

25. Czaja S (1988) Zmiany stosunków wodnych w zlewni Brynicy w wyniku gospodarczej działalności człowieka. Geographia. Studia et Dissertationes 11:60–93

26. Czaja S (1999) Zmiany stosunków wodnych w warunkach silnej antropopresji (na przykładzie konurbacji katowickiej). Prace Naukowe UŚ, Katowice 1782

27. Bielawski Z (1994) Określanie charakterystyk odpływu z terenów zurbanizowanych na podstawie cech fizycznych zlewni i opadu. In: Modelowanie procesów hydrologicznych. Monografie Komitetu Gospodarki Wodnej PAN (vol. 5). PW, Warszawa

28. Gutry-Korycka M (1993) Wpływ urbanizowania i uprzemysłowienia. In: Przemiany stosunków wodnych w Polsce w wyniku procesów naturalnych i antropogenicznych. UJ, Kraków, pp 344–354

29. Pociask-Karteczka J (1994) Przemiany stosunków wodnych na obszarze Krakowa. Prace Geograficzne 96:7–53

30. Absalon D (1998) Antropogeniczne zmiany odpływu rzecznego w zlewni Rudy. UŚ, Katowice

31. Jankowski AT (1998) Antropogeniczne uwarunkowania obiegu wody na terenie Górnośląskiego Zagłębia Węglowego. In: Hydrologia u progu XXI wieku. Komisja Hydrologiczna PTG, Warszawa, pp 139–156

32. Soja R (2002) Hydrologiczne aspekty antropopresji w polskich Karpatach. Prace Geogr. IGiPZ PAN 186

33. Biernat T, Ciupa T, Suligowski R (2008) Wody powierzchniowe i podziemne miasta Kielce w świetle Mapy Hydrograficznej Polski w skali 1:50,000. Problemy Ekologii Krajobrazu 32:297–304

34. Ciupa T (2008) Influence of urbanisation on the runoff exemplified on the Silnica and Sufraganiec catchments (Kielce, Poland). In: Hydrological extremes in small basins. Institute of Geography and Spatial Management. Jagiellonian University, Cracow, pp 85–89

35. Ciupa T (2009) Wpływ zagospodarowania terenu na odpływ i transport fluwialny w małych zlewniach na przykładzie Sufragańca i Silnicy (Kielce). UJK, Kielce

36. Bartnik A, Tomalski P (2012) The influence of small reservoirs of different functions on seasonal oscillations of the selected physico-chemical parameters of the urban river waters (the Sokolowka catchment, Lodz case). In: Anthropogenic and natural transformations of lakes. IMGW-PIB, Poznań, pp 19–33

37. Michalczyk Z, Chmiel S, Głowacki S, Kowal P (2014) Spływ wody kanalizacją burzową ze Świdnika w czerwcu i lipcu 2013 roku. In: Woda w mieście. Monografie Komisji Hydrologicznej PTG (vol. 2). IG UJK, Kielce, pp 165–172

38. Bartnik A (2017) Mała rzeka w dużym mieście. Wybrane aspekty obiegu wody na obszarze zurbanizowanym na przykładzie łódzkiej Sokołówki. UŁ, Łódź

39. Graf R (2014) Przestrzenne zróżnicowanie spływu powierzchniowego w zlewniach zurbanizowanych na przykładzie miasta Poznania. In: Woda w mieście. Monografie Komisji Hydrologicznej PTG (vol. 2). IG UJK, Kielce, pp 59–71

40. Barszcz M (2015) Influence of applying infiltration and retention objects to the rainwater runoff on a plot and catchment scale—case study of Służewiecki Stream subcatchment in Warsaw. Pol J Environ Stud 24(1):57–65

41. Absalon D, Matysik M (2016) Zmiany odpływu w silnie zurbanizowanych zlewniach województwa śląskiego. Monografie Komitetu Gospodarki Wodnej PAN 39:175–186
42. Machowski R, Rzętała MA, Rzętała M, Solarski M (2016) Geomorphological and hydrological effects of subsidence and land use change in industrial and urban areas. Land Degrad Dev 7:1740–1752. https://doi.org/10.1002/ldr.2475
43. Matysik M (2018) Wpływ zrzutów wód kopalnianych na odpływ rzek Górnośląskiego Zagłębia Węglowego. UŚ, Katowice
44. Michalczyk Z, Łoś MJ (1996) Przemiany stosunków wodnych w okolicy Lublina. In: Dziejowe przemiany stosunków wodnych na obszarach zurbanizowanych. UAM, Poznań, pp 96–107
45. Jokiel P, Moniewski P (2000) Warunki gromadzenia i drenażu oraz kierunki ochrony zasobów wód podziemnych w strefie podmiejskiej Łodzi, na przykładzie zlewni Dzierżąznej. Acta UL – Folia Geogr Phys 5:29–48
46. Jokiel P (2002) Woda na zapleczu wielkiego miasta. Możliwości wykorzystania i problemy ochrony zasobów i obiektów wodnych w małej zlewni strefy podmiejskiej Łodzi. UŁ, Łódź
47. Michalczyk Z (2003) Stan i zmiany stosunków wodnych w rejonie Chełma. In: Człowiek i woda. PTG o. Katowice, Sosnowiec, pp 91–100
48. Sikora M (2015) Cieśliński, R.: Kształtowanie się odpływu w zlewni zurbanizowanej na przykładzie zlewni Strzyży. Ecol Eng 41:69–78. https://doi.org/10.12912/23920629/1830
49. Mikulski Z (1990) The effect of urbanization on the hydrological system and water quality. Misc Geogr 4:107–112
50. Molenda T (2005) O niektórych specyficznych właściwościach fizyczno-chemicznych wód zbiorników antropogenicznych. In: Jeziora i sztuczne zbiorniki wodne – procesy przyrodnicze oraz znaczenie społeczno-gospodarcze. UŚ, Sosnowiec, pp 161–168
51. Jekatierynczuk-Rudczyk E (2008) Threats to a small river and its urban catchment: hydrological and hydrochemical assessment of Jaroszówka River in Białystok, Poland. Ecohydrology Hydrobiology 8(1):77–87. https://doi.org/10.2478/v10104-009-0006-z
52. Rzętała M (2008) Funkcjonowanie zbiorników wodnych oraz przebieg procesów limnicznych w warunkach zróżnicowanej antropopresji na przykładzie regionu górnośląskiego. UŚ, Katowice
53. Rzętała M, Jaguś A (2011) New lake district in Europe: origin and hydrochemical characteristics. Water Environ J 26:108–117. https://doi.org/10.1111/j.1747-6593.2011.002 69.x
54. Ciupa T, Suligowski R, Biernat T (2011) Ocena stanu dna doliny rzeki Chodczy (Kielce) i propozycje działań w celu jego poprawy. Problemy Ekologii Krajobrazu 31:33–42
55. Jaromin-Gleń K, Widomski M, Lagod G, Mazurek W (2012) Stężenia zanieczyszczeń w ściekach deszczowych dla wybranej zlewni miasta Lublin. Proceedings of ECOpole 6:725–730. https://doi.org/10.2429/proc.2012.6(2)099
56. Woźnica A, Absalon D, Libera M, Łozowski B, Siudy A (2018) Wyzwania związane z wodą na Śląsku – wspólnie zadbajmy o jakość wód. In: Aktualne Problemy Gospodarki Wodnej. Monografie Śląskiego Centrum Wody (vol. 1), Katowice, pp 11–30
57. Kaniecki A (2003) Problem ochrony wód rzecznych na obszarach miejskich na przykładzie Poznania. Człowiek i woda. PTG o. Katowice, Sosnowiec, pp 64–73
58. Banasik K (2009) Wyznaczanie wezbrań powodziowych w małych zlewniach zurbanizowanych. SGGW, Warszawa
59. Wiatkowski M, Czamara A, Kosierb R (eds) (2008) Zarządzanie kryzysowe – ochrona przed powodzią (rozwiązania praktyczne). UO, Opole
60. Twaróg B, Kęsek Z (2015) Floodpolis – powodziowe miasto przyszłości. J Civil Eng Environ Arch 62(2):553–564. https://doi.org/10.7862/rb.2015.84
61. Graf R, Pyszny K (2016) Zintegrowana gospodarka wodna na obszarze metropolitalnym. In: Integracja planowania przestrzennego w Metropolii Poznań – problemy, metody, osiągnięcia. Bogucki Wydawnictwo Naukowe, Poznań, pp 45–64
62. Januchta-Szostak A (2017) Podejście zlewniowe w urbanistyce jako narzędzie zapobiegania powodziom miejskim. Przegląd Budowlany 9:30–33

63. Ciupa T, Suligowski R, Biernat T (2012) Stan, zagrożenia i ochrona małych cieków na obszarze miasta Kielce. In: Gospodarowanie wodą w warunkach zmieniającego się środowiska. Monografie Komisji Hydrologicznej PTG (vol. 1). UMK, Toruń, pp 63–74
64. Barszcz M (2010) Ocena zagrożenia powodziowego i możliwości jego ograniczenia w zurbanizowanej zlewni Potoku Służewieckiego. Monografie Komitetu Inżynierii Środowiska PAN 68:219–230
65. Januchta-Szostak A (2011) Woda w miejskiej przestrzeni publicznej. Modelowe formy zagospodarowania wód opadowych i powierzchniowych. PP, Poznań
66. Bryndal T (2014) Znaczenie map zagrożenia oraz ryzyka powodziowego w ograniczeniu skutków powodzi błyskawicznych w miastach. In: Woda w mieście. Monografie Komisji Hydrologicznej PTG (vol. 2). IG UJK, Kielce, pp 29–38
67. Ambrożewski Z (2001) Stan techniczny małych obiektów piętrzących wodę na tle katastrofy zapory w Górowie Iławeckim. Gospodarka Wodna 11:472–476
68. Mrozik K, Przybyła C, Pyszny K (2015) Problems of the integrated urban water management. The case of the Poznań Metropolitan Area (Poland). Ann Set Environ Prot 17:230–245
69. Dąbkowski S, Górski J, Bodulski J (2009) Problemy ochrony przed powodzią opadową z deszczów nawalnych na przykładzie Kunowa. Gospodarka Wodna 10:389–396
70. Działek J, Biernacki J, Konieczny R, Fiedeń Ł, Franczak P, Grzeszna K, Listwan-Franczak K (2017) Zanim nadejdzie powódź. Wpływ wyobrażeń przestrzennych, wrażliwości społecznej na klęski żywiołowe oraz komunikowania ryzyka na przygotowanie społeczności lokalnych do powodzi. IGiGP UJ, Kraków
71. Bondaruk J, Kwapuliński J (2007) Zasady rozwoju zrównoważonego w działalności zakładów przemysłowych w zakresie obiektów wodno-ściekowych. Problemy Ekologii 5:263–270
72. Świeca A (1998) Wpływ czynników antropogenicznych na rzeczny odpływ roztworów i zawiesin na międzyrzeczu Wisły i Bugu. Rozprawy Habilitacyjne (vol. 61). UMCS, Lublin
73. Ciupa T (2014) Miasto jako wydajny system zasilania rzek i dostawy ładunków fluwialnych na przykładzie Kielc. In: Woda w mieścieMonografie Komisji Hydrologicznej PTG (vol. 2). IG UJK, Kielce, pp 39–48
74. Edukacja i wychowanie (2018) Ludność: Bank Danych Lokalnych. Warszawa: GUS. Retrieved from https://bdl.stat.gov.pl/BDL. Accessed on 30 Nov 2018
75. World Urbanization Prospects (2018) The 2018 Revision: United Nations, Department of Economic and Social Affairs, Population Division. Retrieved from https://population.un.org/wup/Publications/Files/WUP2018-KeyFacts.pdf. Accessed on 30 Nov 2018
76. Kowalczak P (2010) Wodne dylematy urbanizacji. Wyd. Poznańskie, Poznań
77. Rybka A, Kozłowska K (2016) Reintegracja zdegradowanych sieci wodnych w mieście. Rocz Ochr Sr 18:543–554
78. Kupczyk E, Biernat T, Ciupa T (1998) Przyrodnicze podstawy naturalnej regeneracji rzeki antropogenicznie przekształconej. In: Hydrologia u progu XXI wieku. UW, Warszawa, pp 167–180
79. Parzonka W, Głowski R, Kasperek R (2002) Wpływ zmian użytkowania doliny Widawy w obrębie miasta Wrocławia na warunki przepływu w tej rzece. Obieg wody w zmieniającym się środowisku 7:123–131
80. Powierzchnia geodezyjna kraju według kierunków wykorzystania: Bank Danych Lokalnych (2017) Warszawa: GUS. Retrieved from https://bdl.stat.gov.pl/BDL. Accessed on 30 Nov 2018
81. Lorenc H, Cebulak E, Głowicki B, Kowalewski M (2012) Struktura występowania intensywnych opadów deszczu powodujących zagrożenie dla społeczeństwa, środowiska i gospodarki Polski. In: Klęski żywiołowe a bezpieczeństwo kraju. Monografie IMGW–PIB, Warszawa, pp 7–32
82. Nowicka B, Cudna N (2002) Wpływ zabudowy miejskiej na warunki odpływu w zlewni Srebrnej. Prace i Studia Geograficzne 31:189–204
83. Wałek G (2019) Wpływ dróg na kształtowanie spływu powierzchniowego w obszarze urbanizowanym na przykładzie zlewni Silnicy w Kielcach. UJK, Kielce

84. Weinerowska-Bords K (2010) Czas koncentracji w uproszczonych obliczeniach odpływu ze zlewni zurbanizowanej. Monografie Komitetu Inżynierii Środowiska PAN 68(1):367–378
85. Ostrowski J, Czarnecka H, Głowacka B, Krupa-Marchlewska J, Zaniewska M, Sasim M, Moskwiński T, Dobrowolski A (2012) Nagłe powodzie lokalne (flash flood) w Polsce i skala ich zagrożeń. In: Klęski żywiołowe a bezpieczeństwo kraju. Monografie IMGW-PIB, Warszawa, pp 123–149
86. Pociask-Karteczka J, Żychowski J (2014) Powodzie błyskawiczne (flash floods) – przyczny i przebieg. In: Woda w mieście. Monografie Komisji Hydrologicznej PTG (vol. 2). IG UJK, Kielce, pp 213–226
87. Pociask-Karteczka J, Żychowski J, Bryndal T (2017) Zagrożenia związane z wodą – powodzie błyskawiczne. Gospodarka Wodna 2:37–42
88. Szpakowski W, Szydłowski M (2018) Evaluating the catastrophic rainfall of 14 july 2016 in the catchment basin of the urbanized Strzyża Stream in Gdańsk, Poland. Pol J Environ Stud 27(2):861–869. https://doi.org/10.15244/pjoes/75962
89. Majewski W (2001) Measures and solutions for flood management. A local case: flash flood 2001 in Gdańsk, Poland. Irrig Drain 55:S101–S111. https://doi.org/10.1002/ird.252
90. Majewski W (2001) Urban flash flood in Gdańsk—2001: case study. Meteor Hydrol Water Manage 4(2):41–49. https://doi.org/10.26491/mhwm/64636
91. Franczak P, Działek J, Fiedeń Ł, Biernacki W (2017) Powódź błyskawiczna jako zdarzenie przyrodnicze i społeczne na przykładzie powodzi w Wojcieszowie 5 lipca 2012 roku. Prace Geogr 151:27–51
92. Franczak P, Listwan-Franczak K (2016) Powódź w zlewni Miedzianki (zlewnia Nysy Łużyckiej) w sierpniu 2010 roku. Dobra praktyka w redukcji ryzyka powodziowego w małych zlewniach górskich, w których wystąpiła powódź błyskawiczna. In: Współczesne Problemy i Kierunki Badawcze w Geografii (vol. 4). IGiGP UJ, Kraków, pp 55–84
93. Michalczyk Z, Chmiel S, Głowacki S, Paszczyk J, Siwek K (2012) Hydrologiczna ocena spływu powierzchniowego z obszaru Lublina i ze zlewni kanałów burzowych. In: Ocena warunków występowania wody i tworzenia się spływu powierzchniowego w Lublinie. UMCS, Lublin, pp 161–180
94. Bryndal T (2015) Local flash floods in Central Europe: a case study of Poland. Norsk Geogr Tidssk 69(5):288–298
95. Dobrowolski A, Ostrowski J, Żelaziński J (2005) Powodzie opadowe w Polsce w okresie 1946-2001. In: Ekstremalne zjawiska hydrologiczne i meteorologiczne. Monografie IMGW, PTGeof., Warszawa, pp 221–230
96. Powódź w lipcu 1970 (1972) Monografia. WKiŁ, Warszawa
97. Kundzewicz ZW, Szamałek K, Kowalczak P (1997) The great flood of 1997 in Poland. Hydrol Sci J 44(6):855–870. https://doi.org/10.1080/02626669909492285
98. Kundzewicz ZW (2002) Flood of the floods—Poland, summer 1997. IAHS Publ 271:147–153
99. Powódź w sierpniu 1972 (1973) Monografia. IMGW, WKiŁ, Warszawa
100. Maciejewski M, Ostojski MS, Tokarczyk T (eds) (2011) Dorzecze Odry: monografia powodzi 2010. IMGW-PIB, Warszawa
101. Maciejewski M, Ostojski MS, Walczykiewicz T (eds) (2011) Dorzecze Wisły: monografia powodzi maj – czerwiec 2010. IMGW-PIB, Warszawa
102. Kundzewicz ZW, Pińskwar I, Brakenridge GR (2013) Large floods in Europe, 1985–2009. Hydrol Sci J 58(1):1–7
103. Kundzewicz ZW, Dobrowolski A, Lorenc H, Niedźwiedź T, Pińskwar I, Kowalczak P (2012) Floods in Poland. In: Changes in Flood Risk in Europe. IAHS Press, Wallingford, Spec. Publ. 10:319–334. https://doi.org/10.1201/b12348
104. Cyberski J, Grześ M, Gutry-Korycka M, Nachlik M, Kundzewicz ZW (2006) History of floods on the River Vistula. Hydrol Sci J 51(5):799–817
105. Nachlik E, Kundzewicz ZW (2016) History of floods on the Upper Vistula. In: Flood risk in the Upper Vistula basin. GeoPlanet: Earth and Planetary Sciences, Springer. https://doi.org/10.1007/978-3-319-41923-7_13

106. Dubicki A, Malinowska-Małek J, Strońska K (2005) Flood hazards in the upper and middle Odra River basin—a short review over the last century. Limnologica 35:123–131

107. Czaja S, Machowski R, Rzętała M (2014) Floods in the upper part of Vistula and Odra river basins in the 19th and 20th centuries. Chemistry-Didactics-Ecology-Metrology 19:127–134. https://doi.org/10.1515/cdem-2014-0012

108. Dubicki A, Słota H, Zieliński J (eds) (1999) Dorzecze Odry: monografia powodzi lipiec 1997. IMGW, Warszawa

109. Gorączko M (2010) Wezbrania w obrębie Bydgoskiego Węzła Wodnego i ich wpływ na zagrożenie powodziowe miasta. Monografie Komitetu Inżynierii Środowiska PAN 68, pp 241–250

110. Grześ M (1991) Zatory i powodzie zatorowe na dolnej Wiśle: mechanizmy i warunki. IGiPZ PAN, Warszawa

111. Grześ M, Banach M (1983) Powódź zatorowa na Wiśle w styczniu 1982. Przegląd Geograficzny 55(1):91–113

112. Pawłowski B, Gorączko M, Szczerbinska A (2017) Zjawiska lodowe w rzekach Polski. In: Hydrologia Polski. PWN, Warszawa, pp 195–200

113. Dziadziuszko Z, Majewski A, Wiśniewska A (1983) Monografia powodzi sztormowych 1951–1975. WKiŁ, Warszawa

114. Przygrodzki P, Letkiewicz B (2015) Charakterystyka wezbrań sztormowych wzdłuż polskiego wybrzeża Morza Bałtyckiego. Inżynieria Morska i Geotechnika 3:158–165

115. Wiśniewski B, Wolski T (2008) Katalog wezbrań i obniżeń sztormowych na polskim wybrzeżu Bałtyku. In: Morze – ląd wzajemne relacje. US, Szczecin, pp 101–126

116. Jokiel P, Bartnik A (2017) Wezbrania i powodzie. In: Hydrologia Polski. PWN, Warszawa, pp 167–175

117. Makowski E (1983) Powódź sztormowa na Żuławach w województwie elbląskim w styczniu 1983. Gospodarka Wodna 9

118. Directive 2000/60/EC of the European Parliament: establishing a framework for Community action in the field of water policy. Official Journal of the European Communities, L32 (2000)

119. Prawo wodne: Ustawa z dnia 20 lipca 2017 r. Dz. U. 2017, poz. 1556 (2017)

120. Lewandowski P (2012) Polish investigation of river hydromorphology. Pol J Environ Stud 21(4):957–965

121. Wiatkowski M, Tomczyk P (2018) Comparative assessment of the hydromorphological status of the rivers Odra, Bystrzyca, and Ślęza using the RHS, LAWA, QBR, and HEM methods above and below the hydropower plants. Water 10(7):855. https://doi.org/10.3390/w10070855

122. Osowska J, Kalisz J (2011) Wykorzystanie metody River Habitat Survey do waloryzacji hydromorfologicznej rzeki Kłodnicy. Górnictwo i Geologia 6(3):141–156

123. Trząski L, Mana V (2008) Ocena hydromorfologicznego stanu rzeki Bobrek metodą RHS (River Habitat Survey). Prace Naukowe GIG. Górnictwo i Środowisko 1:53–62

124. Szatten D, Habel M, Dąbrowski J (2013) Oddziaływanie miast na zamulanie dróg wodnych – na przykładzie ujściowego odcinka Brdy w Bydgoszczy. Gospodarka Wodna 6:224–227

125. Zawilski M, Sakson G (2013) Assessment of total suspended solid emission discharged via storm sewerage system from urban areas. Ochr Sr 35(2):33–40

126. Chełmicki W (2001) Woda, zasoby, degradacja, ochrona. PWN, Warszawa

127. Gałuszka A, Migaszewski ZM, Podlaski R, Dołęgowska S, Michalik A (2010) The influence of chloride deicers on mineral nutrition and the health status of roadside trees in the city of Kielce, Poland. Environ Monit Assess 176(1–4):451–464. https://doi.org/10.1007/s10661-010-1596-z

128. Rajda W, Kanonik W (2007) Some water quality indices in small watercourses in urbanized areas. Arch Environ Prot 33(4):31–38

129. Sikora M, Cieśliński R, Pietruszyński Ł (2014) Wpływ Gdańska na stosunki wodne zlewni rzeki Strzyża. In: Woda w mieście. Monografie Komisji Hydrologicznej PTG (vol. 2). IG UJK, Kielce, pp 253–262

130. Marszelewski W, Sokalska A (2014) Próba wyznaczenia kryteriów i klasyfikacji „jezior miejskich". In: Woda w mieście. Monografie Komisji Hydrologicznej PTG (vol. 2). IG UJK, Kielce, pp 145–154

131. Kąpieliska (2018) Warszawa: GIS. Retrieved from https://sk.gis.gov.pl. Accessed on 30 Nov 2018
132. Nowicki Z (ed) (2007) Wody podziemne miast wojewódzkich Polski. PIG, Warszawa
133. Nowicki Z (ed) (2009) Wody podziemne miast Polski – miasta powyżej 50,000 mieszkańców. PIG, Warszawa
134. Zużycie wody i oczyszczalnie ścieków: Bank Danych Lokalnych (2017) Warszawa: GUS. Retrieved from https://bdl.stat.gov.pl/BDL. Accessed on 30 Nov 2018
135. Roman M, Kłoss-Trębaczkiewicz H, Osuch-Paździńska E, Kałużna M, Mikulska E (2001) Zmiany zużycia wody w miastach polskich w latach 1987–1998. Ochr Sr 3:3–6
136. Piasecki A (2014) Analiza wielkości i struktury zużycia wody w miastach Polski. In: Woda w mieście. Monografie Komisji Hydrologicznej PTG (vol. 2). IG UJK, Kielce, pp 197–204
137. Dohnalik P (2000) Straty wody w miejskich sieciach wodociągowych. Wyd. Polskiej Fundacji Zasobów Ochrony Zasobów Wodnych, Bydgoszcz
138. Bergel T, Pawełek J (2006) Straty wody w systemach wodociągowych – charakterystyka, wielkość, wykrywanie i ograniczanie. In: III Konferencja Naukowo-Techniczna „Błękitny San". Dubiecko, pp 123–137
139. Kwietniewski M (2013) Zastosowanie wskaźników strat wody do oceny efektywności jej dystrybucji w systemach wodociągowych. Ochr Sr 35(4):9–16
140. Directive 2007/60/EC of the European Parliament and of the Council of 23 October 2007 on the assessment and management of flood risks: Official Journal of the European Communities, L 288/27 (2007)
141. Woods-Ballard B, Kellagher R, Martin P, Jefferies C, Bray R, Shaffer P (2007) The SuDS manual. CIRIA, London
142. Hoyer J, Dickhaut W, Kronawitter L, Weber B (2011) Water sensitive urban design—principles and inspiration for sustainable stormwater management in the city of the future. HafenCity Universität, Hamburg
143. Nyka L (2013) Architektura i woda. Przekraczanie granic. PG, Gdańsk
144. Drożdżal E, Grabowski M, Kondziołka K, Olbracht J, Piórecki M, Radoń R, Ryłko A (2009) Mapy ryzyka powodziowego – projekt pilotażowy w zlewni Silnicy. Gospodarka Wodna 1:19–29
145. Szewrański S, Kazak J, Szkaradkiewicz M, Sasik J (2015) Flood risk factors in suburban area in the context of climate change adaptation policies—case study of Wrocław, Poland. J Environ Eng 16(2):13–18. https://doi.org/10.12911/22998993/1854
146. Romanowicz RJ, Nachlik E, Januchta-Szostak A, Starkel L, Kundzewicz ZW, Byczkowski A, Kowalczak P, Żelaziński J, Radczuk L, Kowalik P, Szamałek K (2014) Zagrożenia związane z nadmiarem wody. Nauka 1:123–148
147. Green Infrastructure (2014) Guide for Water Management: ecosystem-based management approaches for water-related infrastructure projects: United Nations Environment Programme (UNEP). Retrieved from http://www.unepdhi.org/media/micrositeunepdhi. Accessed on 30 Nov 2014
148. Surma M (2015) Sustainable urban development through an application of green infrastructure in district scale—a case study of Wrocław (Poland). J Water Land Develop 25:3–12
149. Trząski L (2018) Błękitno-zielona infrastruktura (B-ZI): mało znany i niedoceniany kapitał naszych miast. In: Aktualne problemy gospodarki wodnej. Monografie Śląskiego Centrum Wody (vol. 1). Katowice, pp 47–58
150. Burszta-Adamiak E (2012) Wody opadowe w miastach. Rynek Instalacyjny 5:35–38
151. Kobojek, E (2015) Antropogeniczne przekształcenia i możliwości renaturyzacji małych rzek oraz dolin w miastach. In: Badania miejskie i regionalne: Potencjały rozwojowe oraz kierunki przemian w miastach i regionach. UE, Katowice, pp 145–161
152. Ciupa T, Suligowski R, Ciupa S (2017) Problematyka hydrologiczna w Programie Ochrony Środowiska miasta Kielce. Studia Miejskie 28:167–182

Chapter 18
The Multidimensional Aspect of Water Resources Management in Metropolitan Areas (a Case Study of the Poznań Metropolis)

Renata Graf and Krzysztof Pyszny

Abstract Metropolitan areas are characterised by a high density of population, which implies increasing problems related to the development and functioning of the urbanised area. These include problems concerning water supply, the use and protection of water resources, the collection and purification of wastewater and sewage, and also the protection of citizens from extraordinary hydrological and climatic phenomena. The implementation of water management goals in harmony with environmental protection objectives is known to evoke negative phenomena and strong conflicts. A characteristic feature of metropolitan areas is the continuous transformation instigated by a multitude of factors—political, economic and demographic, but also technological and environmental. The need of achieving a number of goals in the same area implies their necessary integration and the implementation of appropriate strategies, concepts and models or programs which make it possible to take correct decisions and realise planned tasks. It is in this specific regard that the model of integrated management of water resources, the concept of iterative and multi-level e-governance, and the so-called "catchment area" management concept have been presented and discussed. Moreover, consideration was given to the problem of limiting anthropogenic pressure with reference to the protection of disposable resources (mainly useful aquifers), of the quality of waters as regards reducing the risk of non-attainment of environmental objectives, and of the integrated minimisation of flood risk on the example of the Poznań metropolis. The analysis was performed in the context of the EU Water Framework Directive, river basin management plans, and flood risk management plans. The multidimensional nature of water resources management in a metropolitan area was considered in its spatial, functional, ecological and social aspects as an activity that includes the planning, development, distribution and optimisation of usage of water resources.

R. Graf (✉)
Department of Hydrology and Water Management, Institute of Physical Geography and Environmental Planning, Adam Mickiewicz University, Bogumiła Krygowskiego 10, 61-680 Poznań, Poland
e-mail: rengraf@amu.edu.pl

K. Pyszny
EnviMap, Piątkowska 118/31, 60-649 Poznań, Poland
e-mail: pyszny@envimap.pl

M. Zeleňáková et al. (eds.), *Quality of Water Resources in Poland*, Springer Water,
https://doi.org/10.1007/978-3-030-64892-3_18

423

Keywords Metropolitan areas · Water management · EU water framework directive · Sustainable development · Poznań metropolis · Poland

18.1 Introduction

More than half of the world's population (3.5 billion) currently lives in cities, and by 2030, according to forecasts, the number of city residents will make up almost 60% of the population [1]. The development of urban agglomerations contributes to the introduction of new water management models. This is related to the growth in the water supply and sewage services to the population, the increasing impact of the city on water resources and the increase of threats related to the effects of extreme weather and hydrological events [2].

Over the historical development of cities, water management paradigms have changed significantly, from the function of providing access to drinking water and sewage disposal, flood protection and finally water-cycle city [3]. The National Water and Environment Programme Guidance emphasize water resource efficiency based on "sustainable communities" and "healthy catchment areas" and indicate various implementation strategies to bring appropriate solutions and benefits [4].

According to the functional classification of urban area, based on population size (presented by the Organisation for Economic Co-operation and Development OECD) four types of urban areas were proposed [5]:

- Small urban areas, with a population below 200,000 people;
- Medium-sized urban areas (200,000–500,000);
- Metropolitan areas (500,000–1,500,000);
- Large metropolitan areas, with a population of 1,500,000 or more.

In Poland, 10 metropolitan centers have been distinguished [6]: the capital of Warsaw, the Upper Silesian conurbation (with the center in Katowice), Łódź, Cracow, Tricity ("Tricity" comprising Gdańsk, Sopot and Gdynia with the center in Gdańsk), Wrocław, Poznań, Szczecin, Bydgoszcz and Lublin (Fig. 18.1). As a metropolitan area[1] (legal definition until 2017), a spatially coherent area of influence of a city located in a voivodeship or a voivodeship council, with strong functional connections and advanced urbanization processes, with at least 500,000 inhabitants, was recognized.

One of the essential elements of the European Union's Europe 2020 Strategy is the principle of sustainable development, to which the development of metropolitan functions should refer. The development of metropolitan functions requires coordination of social and economic development, activities in the field of shaping and

---

[1]Pursuant to the provisions of Article 5 of the Act of 9 October 2015 on Metropolitan Unions (Journal of Laws Item 1890 with subsequent amendments)—this regulation has been repealed, and currently only one metropolitan area, that of "Upper Silesia and the Dąbrowa Basin", is legally operational; this was established by the Act of 9 March 2017 on the Metropolitan Union in the Province of Silesia (Journal of Laws of 2017, Item 730).

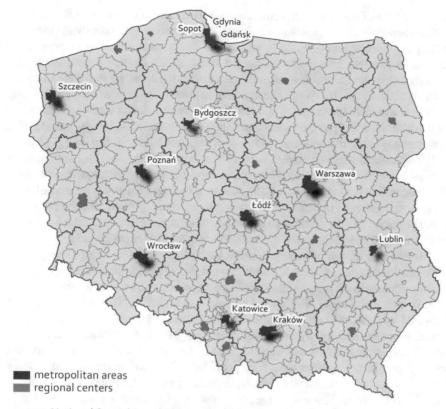

source: National Spatial Development Concept 2030 (NSDC 2030)

Fig. 18.1 Metropolitan areas and regional centers in Poland. *Source* Based on [6]

protecting the natural environment and water management. Adaptation to the requirements of the environment and prevention of its further degradation to the maximum possible extent is a necessary condition for achieving the expected level of social and economic development [7]. Creating spatial structures which are conducive to maintaining the high quality of the natural environment and landscape values and increasing the resistance to natural threats are indicated as the main aims of the spatial development of Poland [6]. The integration of the spatial development policy with the protection of environmental resources and appropriate quality of life is created by appropriate legal standards, which regulate the principles of water management and create protective areas, on the other hand, by instruments adapted to assess the level of environmental impact and mitigate conflicts [8].

A constituent element of the metropolitan area is the system of water resources management, which depending on the functional-spatial relations develops in high coherence with the spatial layout of the city [9]. The structure of the metropolitan area is usually very complex. It is determined by such units as communes (different status, degree of development, resources and water and economic needs), between which there is often a conflict of interest. Urbanization and suburbanization are a frequent phenomenon, which is manifested by the concentration of population in ever larger areas adjacent to the largest Polish cities. In such situations the city directly influences the adjacent areas and determines the development of the so-called urban functional area [6]. Water management in urbanized areas, where demographic growth is associated with spatial and economic development, is viewed as "the art of difficult choices" [10] or the "wicked problem" for the solution of which changes in water management methods and the urban system are usually necessary [11]. Carter [12] emphasized that very rarely is there any cooperation between communes or regions in the field of water resources management in terms of catchment area, often this cooperation is simply missing. The problems registered in highly urbanized areas include: the increase in demand for good quality water, growing threats related to water pollution and the occurrence of periods of drought or flooding, which were the basis for the implementation of the common area and water policy [13].

We find a reflection of these principles in the new model of water management, the essential features of which are multifaceted and interdisciplinary management of water resources in the metropolitan area. The importance of water in the metropolitan area is higher the more differentiated its spatial layout is. This differentiation is provided by areas with a high degree of urbanization and investments, as well as by valuable natural and legally protected areas. Integration of activities at the spatial, functional, environmental and social levels in the new model of water resources management in metropolitan areas [9] is provided by new assumptions of water policy of the European Union Member States, defined in the Water Framework Directive [14].[2] The EU Water Framework Directive has been implemented into Polish law in the Water Law Act [15].

18.2   Water Management Problems in Metropolitan Areas

Metropolises are characterised by the multidirectional and multilevels of spatial development and management of a system comprised of units of differing statuses. Their characteristic feature is continuous transformation, which occurs under the influence of a countless number of factors of a political, economic and demographic, and also technological and environmental nature. Metropolitan areas are distinguished by zones of permanent impact and areas of potential developmental opportunities. As a result of progressing urbanisation, some 75–80% of the entire population of Poland lives within the administrative borders of cities and in zones located around large municipal centres. According to forecasts, this concentration of people will occur both in large municipal centres throughout the country, and in functional areas, with the greatest increase taking place in metropolitan areas [6].

---

[2]Directive 2000/60/EC of the European Parliament and of the Council of 23 October 2000 establishing a framework for Community action in the field of water policy.

The rise in the number of residents and the concomitant large population density lead to an exacerbation of problems connected with the development and functioning of urbanised areas. This also applies to the aspect of water provisioning, utilisation of water resources, and the collection, treatment and channelling away of sewage, the management of rainwater, the protection of residents against extraordinary hydrological and climatic phenomena, and the protection of waters and ecosystems against dependent waters (Fig. 18.2). In quasi-natural areas, the primary objective of water

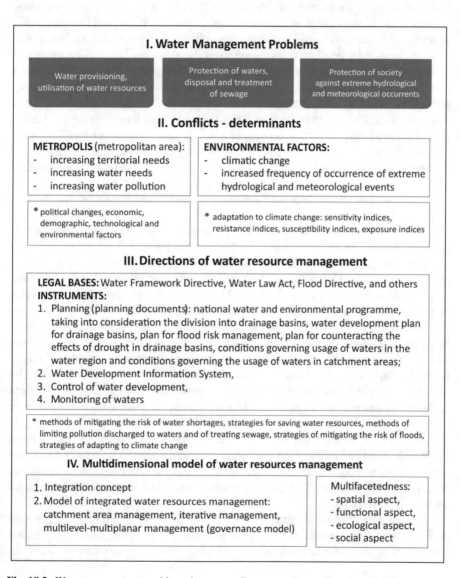

Fig. 18.2 Water management problems in metropolitan areas. *Source* Based on [9, 10]

development is the protection of waters and ecosystems dependent thereon, and this is achieved among others by pro-environmental methods of spatial development, compensatory actions and—to a lesser extent—corrective actions. In urbanised areas, in turn, there is observed an increase in the demand for potable water and water for manufacturing purposes, and the need to protect society and minimise hazards connected with the occurrence of extreme climatic and hydrological phenomena, which objectives are implemented in parallel with that of protection of waters. The basis for adapting socio-economic systems to climate change and preventing catastrophes is the appropriate management of water, soil and biological resources, their protection, and actions aimed at maintaining and restoring the efficient functioning of ecosystems. Metropolitan areas are amongst the main areas for actions connected with the protection of biological diversity and the adaptation of the water management, spatial management and construction sectors to climate change. In Poland, this purpose is served by the "Strategic Plan of Adaptation for Sectors and Areas Sensitive to Climate Change up to the Year 2020, with a Perspective up to the Year 2030" [16]. This plan forms a part of the adaptational actions planned out by EU strategy in the document entitled "White Paper—Adapting to Climate Change: Towards a European Framework for Action", elaborated in 2009 [17]. On the state level, an active climatic policy promotes a fundamental instrument, namely integrated spatial planning based on efficient spatial monitoring which utilises the Spatial Information System (SIS).

The following are considered as priority water development actions: ensuring the appropriate quantity and quality of water for the population and satisfying social and economically justified water-related needs of the economy when realising environmental objectives, which concern limiting the degradation of waters by attaining and maintaining the proper state and potential of waters and the ecosystems connected therewith (Fig. 18.2). As regards the quality and protection of waters, the Water Framework Directive [14] indicates the most important environmental objective as being the attainment by 2015 (and thereafter in 2021 and 2027) of the good ecological condition (naturally uniform parts of waters) and the environmental potential of waters (strongly altered and artificial uniform parts of waters). However, the majority of metropolises in Poland are representative of areas threatened by the non-attainment of the environmental objective (Fig. 18.3), which require a high degree of intervention and management, and are exposed to considerable anthropogenic pressure. In turn, the non-threatened group usually includes areas (catchment areas) which are in the main located on lands that are less urbanised and developed, with an insignificant degree of anthropogenic impact on the state of waters, or where this pressure has been minimised to a certain degree, or where corrective and compensatory actions have been performed.

The dynamic changes occurring under the influence of anthropogenic pressure in the natural environment of metropolitan areas bring about far-reaching transformations of their structure and spatial function. They concern among others the

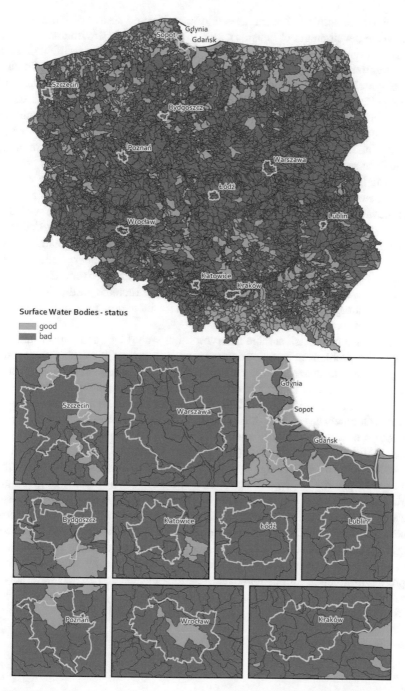

Fig. 18.3 The Surface Water Bodies status in Poland and in the largest cities of the metropolitan areas. *Source* Polish Water State Water Holding National Water Management Board, access: https://mapy.geoportal.gov.pl

reduction of the active biological area and the so-called sprawl of building development (the development of housing), also in the vicinity of environmentally valuable areas. Presently registered climate changes cause the ever more frequent occurrence of extreme weather phenomena, such as drought and torrential precipitation. Approximately 38% of the territory of Poland is affected by drought and the excessive exploitation of water resources [6]. The connection of these changes with an increase in urbanisation is associated among others with the problem of rainwater retention in cities, its appropriate management and utilisation in periods of resource shortages, and also with the necessity of collecting, treating and reusing municipal sewage. The surface run-off generated in cities in consequence of intense precipitation is usually channelled away through storm sewage systems to surface waters or storage (detention) reservoirs, which constitute an element of cities with a sustainable rainwater management system [18–21]. In the case of cities it is assumed that approximately 90–100% of precipitation is channelled away directly to rainwater receptacles, whereas on areas with single-family housing nearly one half of precipitation is discharged directly to a receptacle, with the remainder shaping ground run-off [22–26]. By speeding up the flow of precipitation to rivers, the municipal rainwater sewage system frequently brings about an increase in the flood hazard, however in areas with a low infiltration capacity its action may limit the risk of occurrence of local small-scale flooding.

Polish cities often lack storage reservoirs that would halt and collect rainwater, and thus the problem in question is analysed in terms of increasing the share of biologically active zones in the city structure [27, 28]. Through the intensification of rainwater retention and infiltration processes and the concomitant slowing down of water outflow, these areas contribute to the stabilisation of water circulation conditions in catchment areas with a high index of development. In municipal areas, the concept of rainwater management is implemented among others by the utilisation of Low Impact Development (LID), Water Sensitive Urban Design (WSUD), and Sustainable Urban Drainage Systems (SUDS) [29]. Depending on the nature of urban space, priority is given to actions contributing to an increase in green areas within fixed properties, e.g. the erection of green roofs, the setting up of local small ponds or the construction of car parks. These activities facilitates rainwater development and limits the need for channeling precipitation away through rainwater sewage systems. The sustainable utilisation of water resources in metropolises is made possible by comprehensive surface water management systems which enable the pro-environmental and economic reutilisation of water, and thereby water security management for both man and the natural environment.

18.3   The Multidimensional Model of Water Resources Management

Multidimensionality is the identifier of the structurally and functionally complex socio-economic systems known as metropolitan areas. Interconnected facts and their temporal and spatial measures are represented by the multidimensional model, which constitutes an integral part of the synthesis of data and the decision-making process. It is used among others in environmental management, which also has a complex dimension. One of its aspects is water resources management (development), which consists in their rational shaping and utilisation, taking into consideration their quantity and quality, and also the needs of users and consumers of water, and of water-dependent ecosystems. These actions are conducted within the framework of sustainable water management, which makes it possible to minimise the risk of occurrence of threats to the health and life of residents of metropolises, and to tangible property [30, 31].

The necessity of implementing many objectives on the territorially cohesive metropolitan area determines the need to introduce the appropriate strategies, models and programmes which facilitate the taking of correct decisions and the execution of planned tasks. One of the methods of resolving disputes—apart from compromises and domination—is integration, which consists in the unification of conflicted subjects and interests in order to solve a common problem. Integration is discussed in numerous dimensions, for example in the political, cultural, economic, or even European aspects. As regards the aspect of water management, integration constitutes a key element of the concept of integration as set out in the Water Framework Directive [14]. She is perceived as the basis for implementing water policy and coordinating activities in the field of management the quantity and quality of water resources in catchment areas and drainage basins (see Fig. 18.4).

The Integrated Water Resources Management (IWRM) refers to the coordination of political, economic and social actions connected with the implementation of the concept of sustainable development and the management of water, space and other resources. These actions have as their objective increasing the social and economic advantages in an equiponderant way, without it being necessary to disturb the natural equilibrium [6, 33]. The strategy of integrated water management is a complex process of reacting, which makes it possible to create and maintain the appropriate relations between the objectives of water management and water resources, and variable external conditions (e.g. environmental or climatic conditions)—Fig. 18.4. Water resources management should also be integrated with specific directional actions, which are indispensable for the attainment of goals connected with the spatial development of the metropolitan area. The essence of the IWRM concept is a systemic approach to the solution of water management problems, particularly on urbanised areas (IUWM—Integrated Urban Water Management, SUWM—Sustainable Urban Water Management, MWM—Metropolitan Water Management) with a large population density and a high degree of investment, which results in the creation of new and qualitatively different water resources development systems.

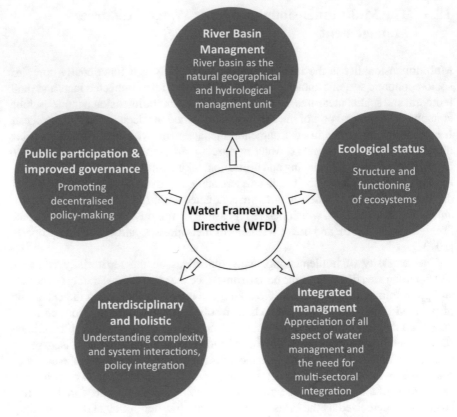

Fig. 18.4 The concept of integration of the Water Framework Directive (on the basis of [14, 32])

The following are taken into consideration in the systemic approach to the management process: (a) the perspective of objectives (national, regional, local, sectoral), (b) process management, which refers to permanent and systematic actions connected with the planning and monitoring of the process of water resources management, (c) the acceptance of a water resources management standard for all types of risk (e.g. occurring within the metropolitan area), and (d) the optimisation of the usage of resources [34].

18.3.1 Legal Bases and Environmental Determinants

In Poland, the development of integrated water resources management has been facilitated by a new system of legal norms, institutional solutions, and practical actions indicated in the Water Framework Directive [14]. Other documents pertinent to the issue under discussion include the Directive on the assessment and management of flood risks [35], the Directive on the protection of groundwater against pollution and deterioration [36], and the Directive on environmental quality standards in the field of water policy [37].

The concept of Integrated Water Resources Management, which constitutes the basis of water policy, has been introduced into national planning documents: the national water and environmental programme, water development plans for drainage basins, flood risk management plans, plans concerned with counteracting the effects of drought, and conditions governing usage of waters from water regions and catchment areas (Fig. 18.2). In water governance, planning is just one of the instruments of water resources management; others include water permits, fees and charges by way of the usage of waters and the water infrastructure, the water cadastre, and control of the development of waters.

The attainment of the strategic objectives set out in the EU and national documents is made possible by the execution of space and natural resources management goals at the local level. Unfortunately developmental priorities at the local self-government still too infrequently take into consideration the objectives of integrated water management. The lack of effective legal and planning tools continues also to be felt [38–40]. Strategic programmes elaborated at the level of communes and cities constitute the basis for efficient action on "one's own" area. However these should be supported through the use of legal and strategic tools—European, national, and regional. According to [39], strategic documents at the national and regional level currently take into consideration the principles of IWRM. However problems with the implementation of the principles of the concept appear at the level of local spatial planning. Effective action in this regards requires an efficient institutional and legal base, as well as representative and competent local authorities [10, 41].

18.3.2 Concepts and Strategies

The multifaceted (multidimensional) model of water resources management is based on the fundamental concepts set forward in the assumptions of the Water Framework Directive [14]—Fig. 18.4: the concept of integration, of "catchment area" management, of multiplanar and multilevel management (governance), and of "iterative" management. The adopted concepts (principles) serve to determine a set of guidelines for the taking of decisions and actions pertaining to individual fields in a coordinated manner, with reference to the volume and quality of resources, changes therein, and time-frames for implementation.

The integrated approach in water resources development constitutes a method for the comprehensive definition of problems connected with various aspects of the functioning of metropolitan areas. The concept of integration determines aspects relating to the efficient execution of environmental goals and points to the necessity of treating surface water and groundwater resources as a whole, of combining the applications of waters, their environmental functions and values, and also of integrating disciplines, analyses used, and geospatial data. The Integrated Water Resources Management concept promotes the coordinated development and management of waters, space and other resources in order to increase social and economic advantages in an equiponderant way, without it being necessary to disturb the equilibrium of ecosystems [33].

Pursuant to the assumptions of the Water Framework Directive [14], water resources development takes into consideration the so-called principle of "catchment area management" (Integrated River Basin Management IRBM), which means that it is implemented in areas with hydrographic borders (river catchment areas) and a fixed procedure for their protection (Fig. 18.4). The integration of the water development policy and the improvement of the planning system in water governance, together with the designation of priorities of planning actions, will ensure the division of a state into drainage basins, water regions and catchment areas.[3] This spatial approach should integrate spatial management and water management, and also formulate—among others—criteria governing access to water resources in the metropolitan area. The hydrological balances of specific catchment areas constitute a fundamental element of the plan for the development of waters in the catchment area [14]. They also form the basis for conducting discussions into the best solution, e.g. of a local water deficit problem, while respecting the principle of sustainable utilisation and non-impairment of the state of waters [25].

What is preferred in the metropolitan area is water management in the form of participative management (Fig. 18.4), which means the coordination of actions at the multisectoral level (multilevel governance). The competent authorities are involved in elaborating competences for multilevel cooperation: between self-governments and between sectors, as a tool of metropolitan development. Water resources management is also founded on base functions: planning, organising, administering and control, and thus possesses the features of an iterative process. The iterative management process, alternatively termed development management, includes the following: identification of the state, determination of a schedule and plan of action, implementation and monitoring, followed by a review of execution, assessment and a new planning cycle. The most effective model of integrated water resources management functions on the basis of the selection and adaptation of actions to specific situations and to the nature of the water management system.

Water development in the metropolitan area cannot be the subject solely of sectoral planning, for it requires integration with planning and spatial management, and also with urban planning and architecture, this due to the significant impact of the method of land development on the possibility of water resources management in the catchment area [28]. At the level of planning and design, a significant role is played by the discussion on the utilisation of the potential of water management as regards provisioning and regulating the state and environmental functions [42]. An important problem which concerns the integration of water management in the metropolitan area is the lack of coordination of strategic documents connected with water development at the national and regional levels with local spatial planning documents. The

---

[3] By the Order of the Council of Ministers of 27 June 2006 on the course of the boundaries of drainage basins and water regions; Journal of Laws No. 126, Item 878.

coordination of actions as regards the degree of integration of the water resources management system is further made difficult by the lack of conformity of the administrative division applicable to spatial planning with the division into drainage basins and water regions (the catchment area division) in water management. Polish legislation lacks provisions comprehensively regulating local water resource development practices which take into account the connection of metropolitan areas with the catchment area arrangement. As a result the problem is frequently marginalised by local self-government [27, 38, 39, 43, 44]. At the local level, there are no formal planning documents pertaining to water management, while spatial planning has not adopted the catchment area division as the basic area for planning and decision-making actions. This aspect should be considered at the stage of elaboration of the integrated water resources management model recommended for implementation in urbanised areas [28, 41].

18.4 The Metropolitan System of Water Resources Management—A Case Study of the Poznań Metropolis

18.4.1 Location and Administrative Structure of the Metropolitan Area

The Poznań metropolis, which is discussed as a case study, is located in the western part of Poland (Fig. 18.1) and covers an area of approximately 3,081 km^2. Using the OECD definition, the city of Poznań together with its 21 adjacent communes constitutes a metropolitan area (Fig. 18.5). The city of Poznań and suburban communities form a monocentric settlement (the Poznań metropolis) centred around the city of Poznań, which is surrounded by 21 communes. The Poznań metropolis occupies 10% of the Province of Wielkopolska (Greater Poland), inhabited by more than 1 million people representing 29% of the total number of inhabitants of the province. The population density is 320 residents per km, which is almost three times greater than the average for Poland.

18.4.2 Resource Potential

According to the regional approach, the Poznań metropolis is located in the central part of the Wielkopolska (Greater Poland) Lowland, within the area of two Lake Districts—those of Poznań and Gniezno. The area's low water abundance is attested to by the unfavourable structure of the hydrological balance, which is determined by the low sums of atmospheric precipitation (500–550 mm), high evaporation (in excess of 500 mm), and low run-off and the high water deficits associated therewith. The Poznań metropolis is located within reach of the Warta Water Region, primarily

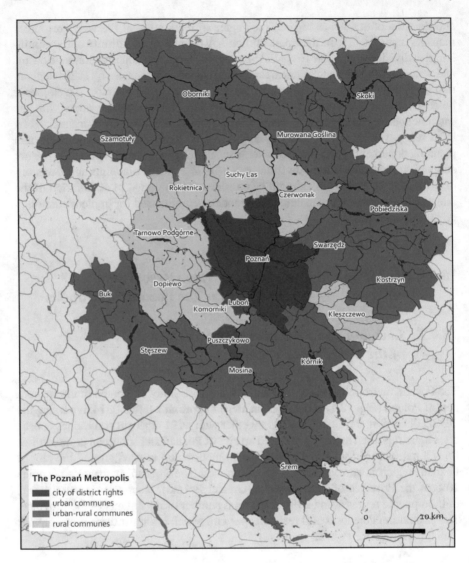

Fig. 18.5 The Poznań metropolis—administrative division into communes (on the basis of [45])

within the boundaries of the Poznań Catchment Area of the Warta P-X (source: Polish Water State Water Holding Regional Water Management Board in Poznań). The hydrographic axis of the area is formed by the River Warta and its tributaries along the Poznań Warta Gorge (among others the rivers Wirynka, Potok Junikowski, Bogdanka, Kopel, Głuszyna, Cybina and Główna), and in part along the Obornicka Valley of the Warta (the River Sama) and the Central Obra Valley (the Mosiński Canal)—Fig. 18.6. At the gorge section, which has a length of approximately 45 km

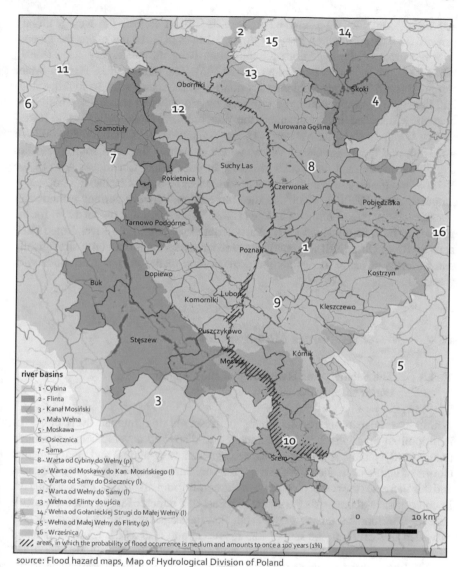

river basins

1 - Cybina
2 - Flinta
3 - Kanał Mosiński
4 - Mała Wełna
5 - Moskawa
6 - Osiecznica
7 - Sama
8 - Warta od Cybiny do Wełny (p)
10 - Warta od Moskawy do Kan. Mosińskiego (l)
11 - Warta od Samy do Osiecznicy (l)
12 - Warta od Wełny do Samy (l)
13 - Wełna od Flinty do ujścia
14 - Wełna od Gołanieckiej Strugi do Małej Wełny (l)
15 - Wełna od Małej Wełny do Flinty (p)
16 - Wrześnica

/// areas, in which the probability of flood occurrence is medium and amounts to once a 100 years (1%)

source: Flood hazard maps, Map of Hydrological Division of Poland

Fig. 18.6 The hydrographic network and the extent of the flood hazard zone in Poznań Metropolis (on the basis of [41])

(between Mosina and Oborniki), the River Warta forms the axis of a strip and node arrangement of lands with considerable natural and recreational values.

The resource potential is of considerable importance for the development of integrated water management functions within the area of the Poznań metropolis. Average unitary run-off in the Poznań Catchment Area of the River Warta is more or less $q = 4.0 \ dm^3 s^{-1} \ km^{-2}$ for the catchment area of the River Warta and $q = 2.5 -$

$3.0 \text{ dm}^3\text{s}^{-1} \text{ km}^{-2}$ for the catchment areas of its tributaries, with the average unitary run-off for Poland totalling $q = 5.5 \text{ dm}^3\text{s}^{-1} \text{ km}^{-2}$ [41]. In the main, the disposable resources of the region are groundwaters from the quaternary stage (69%), while 18% are resources from older stages [46]. The metropolitan Poznań Water Supply System functions within the boundaries of the metropolis. Its hydrogeological bases are determined first and foremost by groundwater resources, which have been duly identified and documented: the Wielkopolska Fossil Valley (Main Reservoir of Groundwaters, GZWP no. 144), the Warsaw-Berlin Glacial Valley (GZWP no. 150), and the Warta Gorge Valley, which constitute the main source of water supply for the Poznań metropolis [46, 47]. In the Poznań Catchment Area of the River Warta the degree of utilisation of groundwater resources available for development (source: PIG-PIB), which specifies the relation between current groundwater consumption and their resource, is low (15–30%) and very low (under 15%) for the remaining balance units: the River Warta from Obrzycko to the River Noteć (XII) and the River Obra (XIII). Groundwater resources are important both strategically and in terms of usage as regards the provisioning of the Poznań metropolis with water.

18.4.3 Water Resources Management

18.4.3.1 The Security and Reliability of the Poznań Water Supply System

Within the Poznań metropolis, there is observed the mutual overlapping of metropolitan functions: housing (compact and dispersed development), economic (water intake points, sewage systems, sewage treatment plants) and environmental (forms of nature protection). Water resources are coming under increasing pressure, and this increases the risk of disruption of the equilibrium of the hydrological balance. In consequence, there exists the potential threat of a deficit of water resources available for development arising in the event that the consumption of groundwaters is exceeded in relation to the size of this resource, and also as the result of planned investments in intermediate protection zones and water-bearing lands earmarked for the development of intake points. Priority in this regard has been given to the Mosina-Krajkowo water intake point, which is the leading supplier of the Poznań Water Supply System (Fig. 18.7), and also utilises intake points and water treatment plants located in Poznań (Dębina water intake point), Gruszczyn (commune of Swarzędz), Murowana Goślina, and in the communes of Kórnik and Suchy Las. Together, the Poznań Water Supply System and the Poznań Sewage System constitute fundamental elements of a water and sewage infrastructure that is metropolitan in nature, with its largest operator being the water supply and sewage company Aquanet S.A.; remaining networks are primarily local in nature [48].

To date, the quantitative and qualitative protection of water resources at the most important water intake point, i.e. Mosina-Krajkowo, was implemented through the

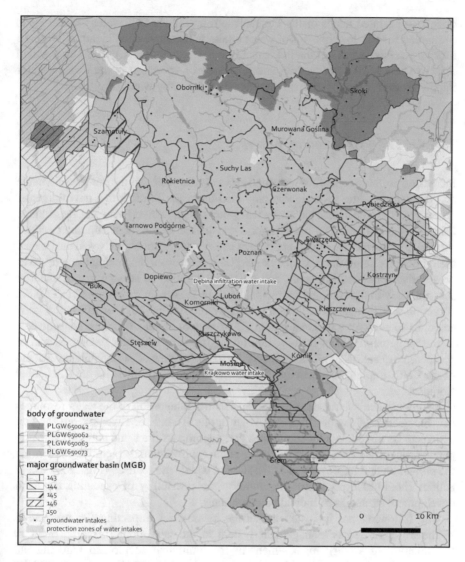

Fig. 18.7 Groundwater and water protection in Poznań Metropolis (on the basis of [41])

rationalisation and optimisation of water consumption, in parallel with the protection of nature. Currently, the Mosina-Krajkowo water intake point may be threatened by a change in the forms of spatial development of water-bearing areas, connected among others with the development of tourist functions, construction, and transport routes [49]. Investments and the conversion of this area for the achievement of objectives other than the protection of water-bearing lands may lead to the permanent loss of availability of areas constituting the natural developmental base of the water intake point, which from the beginning of its functioning was intended to satisfy the basic water-related needs of the Poznań metropolis [46, 50]. Water-bearing lands earmarked for the development of water intake points have been identified for both Mosina-Krajkowo and Dębina in Poznań (source: Aquanet S.A.), this with the twin objective of ensuring the security and reliability of operation of the Poznań Water Supply System and maintaining reserves of good-quality water (Fig. 18.7).

18.4.4 The Protection of Waters

Increasing pressure on surface water and groundwaters impacts the process of potable and production water provisioning, and also increases the risk that environmental objectives (the good state or ecological potential of waters) will not be attained (Figs. 18.4 and 18.8). The following are considered as the most important actions serving to lower the risk of non-attainment of environmental objectives in the Poznań metropolis: reducing the pressure exerted on water resources, maintaining reserves prospective water-bearing lands, and the administration of sewage management in accordance with the National Municipal Sewage Treatment Programme[4] (KPOŚK). Attaining and thereafter maintaining, this by the end of 2021 or 2027, the good ecological state (or ecological potential) of waters should occur in consequence of the implementation of actions minimising pressure on water resources, and of compensatory actions. This is confirmed by the principles of responsible investment of the UNO, implemented through the UNEP Finance Initiative and the UN Global Compact [9].

In order to achieve the optimisation and spatial integration of the water resources management system within the Poznań metropolis, it would be advantageous for joint actions to be undertaken by communes that encompass the same naturally uniform parts of waters (Surface Water Bodies) in which environmental objectives concerning the protection and improvement of the state of waters are realised (Fig. 18.9). These actions should also take into consideration the possibility of executing the local development strategies of individual communes.

The following are indicated as actions favourably impacting the integration of the water resources management system in the metropolis: bringing order to the situation of sewage management in the municipal sector and reducing the demand for water in the municipal sector and industry [41]. These should be served by the so-called demand management of water needs, which consists in the shaping of the water-related requirements of residents and the economy with the ultimate objective of their limitation (the limitation of water consumption). These assumptions have also been presented in the Development Strategy for the Poznań Agglomeration 2020, in the strategic axis "*Spatial Management and the Environment*" under programme 1.6: "The Quality and Disposability of Water Resources" [51]. Tasks specific to the communes comprising the Poznań metropolis

---

[4]National Municipal Sewage Treatment Programme together with five updates, from 2005, 2009, 2010, 2015 and 2017.

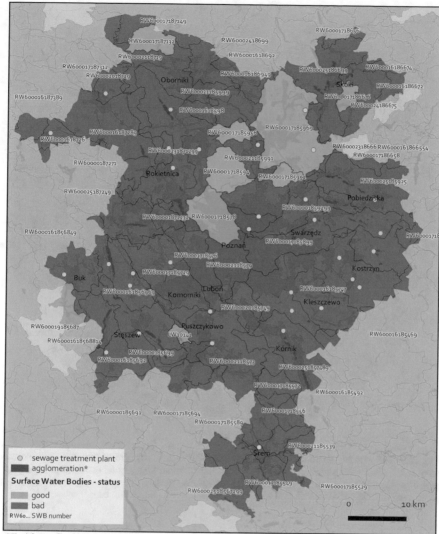

* The definition of 'agglomeration' is given in Article 2(4) of the Directive 91/271/EEC of 21 May 1991 concerning urban waste-water treatment

Fig. 18.8 Status of the Surface Water Bodies and the elements of sewage management in Poznań Metropolis (on the basis of [41])

include securing water resources and erecting an infrastructure that ensures their proper distribution, and also the safe channelling away and treatment of sewage, which is handled within the metropolis by sewage systems and 34 sewage treatment plants (Fig. 18.8). Under the National Municipal Sewage Treatment Programme (KPOŚK) 22 sewage system agglomerations, to which the planned development of the area of the metropolis should be limited, had been designated within its boundaries

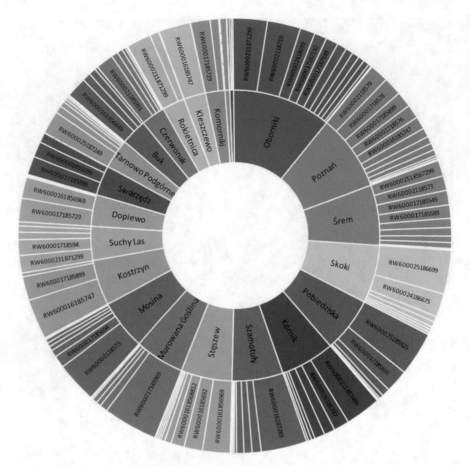

Fig. 18.9 Number and share the Surface Water Bodies in the structure of communes in Poznań Metropolis. *Source* Polish Water State Water Holding National Water Management Board, access: https://mapy.geoportal.gov.pl. *ID, the Surface Water Bodies, is according to the arrangement on the Fig. 18.8

by the end of 2014 (Fig. 18.8). This programme concerns the erection in agglomerations with in excess of 2,000 RLM (1 RLM = 1 resident) of collective sewage systems and sewage treatment plants. The total sum area of all designated sewage agglomerations accounts for approximately 22% of the area of the metropolis [48]. In order to improve the quality and quantity of water resources in the metropolitan area it is usually of considerable importance to introduce the principle of prevention (e.g. limiting the pollution of waters) with the objective of preventing degradation, and not only restoring the original state. In turn, the application of the principle of circumspection usually follows from the impossibility of obtaining and integrating information as to the causes and effects of anthropogenic pressure on water resources on the local and metropolitan scale (Fig. 18.8). Greater integration of metropolitan actions is assumed in the case of "internal" communes, that is ones located in the central part of the Poznań metropolis (Fig. 18.5), than in communes situated in the "boundary zone", which additionally neighbour on communes from outside the metropolitan area.

18.4.5   Flood Risk and Rainwater Management

Management of the space comprising the metropolitan area requires the establishment of safe living conditions, the provision of protection against floods and their effects, protection against local small-scale flooding and the inflow of pollution, and also the regulation of microclimatic conditions. Implementation within the Poznań metropolis of the objectives of the so-called Flood Directive, i.e. on the assessment and management of flood risks [35], should take into consideration the reach of flood risk zones and the scale of flood risk, together with a designation of the protective function.

Within the metropolis, areas threatened by flooding with a probability of occurrence of 1% (once every 100 hundred years) include those situated in the River Warta Valley along the following sections (Fig. 18.6): the Poznań Warta Gorge (between Śrem and Oborniki), the Obornicka Valley of the Warta (below Oborniki), and partially above Śrem (source: Hydroportal, Polish Water State Water Holding).

In the directions of spatial development of the metropolis, areas exposed to flood risk should be excluded from development or classified as land with special conditions of building development and utilisation. The nature of actions connected with flood risk management in the metropolis has been specified in the "Flood Risk Management Plan for the Drainage Basin of the River Odra", which was implemented in 2016. The catalogue of primary objectives of flood risk management in the drainage basin takes into consideration groups of both technical and non-technical actions connected with: halting the increase of flood risk, minimising the existing flood risk, and improving the system of flood risk management [52]. Within the area covered by the present study, these actions concern the entire River Warta Valley along the section between Śrem and Oborniki, and the catchment area of the River Sama (commune of Szamotuły), the River Mała Wełna (commune of Skoki), and the River Cybina (commune of Swarzędz)—Fig. 18.6. The minimisation of flood risk in the metropolitan area can be attained through the implementation of a cohesive programme targeted at changing the method of usage of the river valleys of the Warta and its tributaries. This should contribute to reducing susceptibility and exposure, and also the sensitivity of development and society to floods, as well as other extreme phenomena, for example drought. Limiting the area's exposure to floods by determining opportunities for the utilisation of floodlands in a manner insusceptible to the effects of flooding (e.g. parks, recreational areas) constitutes an alternative (ecological) method of providing a system of technical flood protection, which usually has a negative impact on the state of the natural environment. Spatial analyses of components of the hydrological balance in urbanised catchment areas of the Poznań metropolis have demonstrated, particularly in the city's centre, the low share of zones where the process of rainwater retention and infiltration is dominant; such zones shape the quantity of run-off and the volume of groundwater resources [53–55]. As regards maintaining the equilibrium of the hydrological balance in the municipal system, both impermeable surfaces, which in the main generate surface run-off and valley-like drainage zones, which may contribute to local increases in evapotranspiration, are equally unfavourable. The share of surface run-off in the structure of the hydrological balance of municipal catchment areas is approximately 10%, while in catchment areas with a high degree of urbanisation (e.g. Bogdanka, Potok Junikowski) it rises to as much as 18%. This value is higher than that measured for the catchment areas of tributaries of the River Warta located outside the city, where surface run-off accounts for approximately 5–6% of precipitation [53, 54]. The greatest surface run-off (>150–200 mm) is generated by catchment areas with a degree of urbanisation of 40–60%, where the considerable compactness of development may contribute to an increase in the risk of occurrence of small-scale flooding at points of accumulation of water run-off [55]. The proper management of water resources in the metropolitan area through the biotechnical method of development of precipitation water, afforestation, or limiting the sealing of surfaces of urbanised areas (e.g. green car parks) will exert a significant impact on increasing the biologically active surface, which regulates the state of surface waters and groundwaters in municipal catchment areas. As regards ensuring the security and protection against the effects of extreme climatic and hydrological phenomena in the context of the Poznań agglomeration, it would appear advantageous to bring about the integration of actions of many subjects on numerous planes, e.g. the spatial and functional.

18.5 Summary and Conclusion

The integration of the water resources management system in the metropolitan area is connected with the coordination of activities not only on the spatial and functional levels but also on the social and environmental. Integrated management of the metropolitan area is based on three pillars: equal access to a high-quality natural environment, the efficient utilisation of natural resources in order to achieve economic advantages, and maintaining the ecological equilibrium and the ability of the natural system to regenerate. Identified problems connected with the risk of non-attainment of environmental objectives, increasing pressure on existing water resources, and flood risk make it necessary to ensure high standards of spatial development in metropolises. This requires taking into account the taking into consideration of an interdisciplinary approach, multisectoral and long-term planning, and the broad cooperation of various groups of stakeholders, all the more so as the decided majority of metropolitan functions are located in Poznań [45]. An improvement of the quality of management in the field of water resources in the metropolitan system also requires the competent support of the authorities and necessitates the elaboration and adoption of common principles of water resources management for a common functional area. The competent support of the authorities refers to the creation of competences for multilevel cooperation between self-governments and sectors as a tool for metropolitan development, e.g. the establishment of cooperation with Polish Water State Water Holding, the Regional Water Management Board, the appropriate authorities and teams.

The objectives of actions concerning the integration of water management are precisely specified in the respective policies, both European (the Water Framework Directive, the Floods Directive) and national (the National Water-Environmental Programme, catchment areas water development plans, flood risk management plans). At the local level their implementation is made possible by strategic planning and water management tools, which are used by self-governmental authorities to create specific mechanisms for their integration with the metropolitan area development strategy. The attainment of multidimensional advantages, which is of significance from the point of view of long-term and integrated management, may be realised by the inclusion of water-related aspects in the development of all sectors of the metropolis (city) [28, 38]. In this regard use may be made of, among others, environmental protection programmes, studies of conditions and directions of spatial development, local spatial development plans, river development plans, small retention programmes and projects, sectoral policies concerning water resources and natural resources, and other development strategies for metropolitan areas.

18.6   Recommendations

(a) There is a pressing need to develop legal foundations that would support activities coordinating the process of integration of water management and spatial management in urbanised areas.

(b) The objectives of actions focusing on the integration of water management in the metropolitan area should be closely connected with the strategy of the area's development. The attainment of multifaceted advantages, which is of significance from the point of view of long-term and integrated water resources management, may be realised by the inclusion of water-related aspects in the development of all sectors of the metropolis.

(c) Catchment area management as the preferred direction of effective water resources management does not coincide with the administrative borders of municipalities, however it can be implemented through the furthering of inter-municipal cooperation and the coordination of objectives of various sector strategies.

(d) The integration of the water resources management system in the metropolitan area should take into account the coordination of activities at the spatial, functional, social and environmental levels.

(e) It is indispensable to implement a system for monitoring and evaluation of the executed tasks and planned goals that will contribute to the minimisation of the level of risk and degree of threat while at the same time helping to achieve maximum benefits for all stakeholders, with due recognition of the need to promote socio-economic development and maintain the environmental capital.

References

1. United Nations (2015) Sustainable development goals—17 goals to transform our world. United Nations. Available https://www.un.org/sustainabledevelopment/cities/. Accessed on 29 Aug 2018
2. Tortajada C (2008) Challenges and realities of water management of megacities: the case of Mexico city metropolitan area. J Int Affairs 61(2):147–166
3. Brown RR, Keath N, Wong THF (2009) Urban water management in cities: historical, current and future regimes. Water Sci Technol 59(5):847e855. https://doi.org/10.2166/wst.2009.029
4. Kim JH, Cho J, Keane TD, Bernard EA (2015) Fragmented local governance and water resource management outcomes. J Environ Manage 150:378–386
5. OECD, Organisation for Economic Co-operation and Development (2013) Definition of Functional Urban Areas (FUA) for the OECD metropolitan data base. https://www.oecd.org/cfe/regional-policy/Definition-of-Functional-Urban-Areas-for-the-OECD-metropolitan-database.pdf. Accessed on 30 Aug 2018
6. National Spatial Development Concept 2030 (2012) Ministry of Regional Development, Warsaw. http://www.espon-usespon.eu/dane/web_usespon_library_files/682/national_spatial_development_concept_2030.pdf. Accessed on 30 Aug 2018
7. Bosselmann K (2008) Principle of sustainability: transforming law and governance. Ashgate Publishing Group, Abingdon
8. Zalewski M (2011) Ecohydrology for implementation of the EU Water Framework Directive. Proc ICE—Water Manage 164(8):375–386
9. Graf R (2016) Spatial and functional integration of a water resources management system on the example of the Poznan Metropolis. In: Chaberek-Karwacka G, Malinowska M (eds) Geography in the Face of the Modern World Challenges. University of Gdańsk, Institute of Geography, LIBRON Publishing House, Kraków, pp 73–92
10. Kowalczak P (2011) Water dilemmas of urbanization [Wodne dylematy urbanizacji]. PTPN, Publishing House, Poznań
11. Krauze K, Wagner I (2014) Water in the city space and integrated city management [Woda w przestrzeni miejskiej a zintegrowane zarządzanie miastem]. In: Bergier T, Kronenberg J, Wagner I (eds) Water in the city. The sustainable development—application [Woda w mieście. Zrównoważony Rozwój – zastosowania], vol. 5. Sendzimir Fundation, Kraków, pp 95–114
12. Carter, J (2007) Spatial planning, water and the Water Framework Directive: insights from theory and practice. Geog J 173(4):330–342
13. Van den Brandeler FF, Gupta J, Hordijk M (2018) Megacities and rivers: scalar mismatches between urban water management and river basin management. J Hydrol. https://doi.org/10.1016/j.jhydrol.2018.01.001
14. Directive 2000/60/EC of the European Parliament and the Council of 23 October 2000 establishing a framework for Community action in the field of water policy (2000) (L 327/1 of 22.12.2000). https://eur-lex.europa.eu. Accessed on 28 Aug 2018
15. Water Law Act (2001) An Act on the amendment of the Water Law Act and certain other acts, Art. 9 item 1 pt. 10 of the Act of 18 July, 2001 Official Journal 2005 No. 130 item 1087
16. Strategic Plan of Adaptation for Sectors and Areas Sensitive to Climate Change up to the Year, with a Perspective up to the Year 2030 [Strategiczny Plan Adaptacji dla sektorów i obszarów wrażliwych na zmiany klimatu do roku 2020 z perspektywą do roku 2030] (2013) Ministry of Environment, Warsaw. https://www.mos.gov.pl/fileadmin/user_upload/SPA_2020.pdf. Accessed on 28 Aug 2018

17. White Paper - Adapting to Climate Change: Towards a European Framework for Action (2009) COM (2009) 147, Brussels. http://www.europarl.europa.eu/meetdocs/2009_2014/documents/com. Accessed on 30 Aug 2018

18. Russo B, Gómez M, Martínez P, Sánchez H (2005) Methodology to study the surface runoff in urban streets and the design of drainage inlets systems. Application in a real case study. 10th International Conference on Urban Drainage, Copenhagen/Denmark, 21–26 Aug 2005

19. Słyś D, Stec A (2008) Technical and aesthetic aspects of the open rain waters retention reservoirs design [Techniczne i estetyczne aspekty projektowania otwartych zbiorników retencyjnych wód deszczowych]. Gospodarka Wodna 10:411–414

20. Barszcz M (2009) Analysis of freshets caused by heavy rainfall on small urbanized drainage basin of Służew Stream. Studia Geotechnica et Mechanica 4:3–15

21. Hoyer J, Dickhaut W, Kronawitter L, Weber B (2011) Water sensitive urban design principles and inspiration for sustainable stormwater management in the city of the future. Jovis Verlag, Berlin

22. Marsh WM (2005) Landscape Planning: Environmental Applications. Wiley

23. Nowicka B (2002) Impact of urbanization on outflow conditions [Wpływ urbanizacji na warunki odpływu]. Prace Instytutu Geografii Akademii Świętokrzyskiej 7:77–86

24. Ciupa T (2009) The impact of landuse on runoff and fluvial transport in small river catchments. Based on the Sufraganiec and Silnica rivers (Kielce) [Wpływ zagospodarowania terenu na odpływ i transport fluwialny w małych zlewniach na przykładzie Sufragańca i Silnicy (Kielce)]. UJK Publishing House, Kielce

25. Haase D (2009) Effects of urbanisation on the water balance—a long-term trajectory. Environ Impact Assess Rev 29 I(4):211 − 219

26. Michalczyk Z (ed) (2012). Evaluation of water occurrence conditions and surface runoff in Lublin [Ocena warunków występowania wody i tworzenia się spływu powierzchniowego w Lublinie]. Wydawnictwo Uniwersytet Marii Curie-Skłodowskiej Publishing House, Lublin

27. Łomotowski J (ed) (2008) Problems of rainwater management [Problemy zagospodarowania wód opadowych], Seidel-Przywecki, Wrocław-Warszawa

28. Januchta-Szostak A (2014) The role of urban planning and architecture in water management [Rola urbanistyki i architektury w gospodarowaniu wodą]. In: Bergier T, Kronenberg J, Wagner I (eds) Water in the city. The sustainable development—application [Woda w mieście. Zrównoważony Rozwój – zastosowania], vol. 5. Sendzimir Fundation, Kraków, pp 31–47

29. EPA (2007) Reducing stormwater costs through Low Impact Development (LID) strategies and practices. U.S. Environmental Protection Agency, Washington, DC. https://nepis.epa.gov/Exe/ZyNET.exe. Accessed on 30 Aug 2018

30. Gabe J, Trowsdale S, Vale R (2009) Achieving integrated urban water management: planning top-down or bottom-up? Water Sci Technol 59(10):1999–2008

31. The Global Risks Report 2018, 13th Edition (2018) World Economic Forum, Committed to Improving the State of the World. Strategic Partner of the Report, Zurich

32. Giakoumis T, Voulvoulis N (2018) The Transition of EU Water Policy Towards the Water Framework Directive's Integrated River Basin Management Paradigm. Environ Manage 62:819. https://doi.org/10.1007/s00267-018-1080-z

33. Global Water Partnership (2000) Integrated Water Resources Management, Global Water Partnership Technical Advisory Committee. Background Paper 4

34. Romanowska M, Trocki M (2004) Process approach in management [Podejście procesowe w zarządzaniu]. Szkoła Główna Handlowa Publishing House, Warszawa

35. Directive, 2007/60/EC of the European Parliament and the Council of 23 October 2007 on the assessment and management of flood risks. (2007) (L 288/27 of 6.11.2007). https://eur-lex.europa.eu. Accessed on 28 Aug 2018

36. Directive, 2006/118/EC of the European Parliament and of the Council of 12 December 2006 on the protection of groundwater against pollution and deterioration (2006) (L 372/19 of 27.12.2006). https://eur-lex.europa.eu. Accessed on 28 Aug 2018

37. Directive, 2008/105/Ec of the European Parliament and of the Council of 16 December 2008 on environmental quality standards in the field of water policy (2008) (L 348/84, 24.12.2008). https://eur-lex.europa.eu. Accessed on 28 Aug 2018

38. Wagner I, Januchta-Szostak A, Waack-Zając SA (2014) Strategic planning and management tools for water in urban space [Narzędzia planowania i zarządzania strategicznego wodą w przestrzeni miejskiej]. In: Bergier T, Kronenberg J, Wagner I (eds) Water in the city. The sustainable development—application [Woda w mieście. Zrównoważony Rozwój –zastosowania], vol. 5. Sendzimir Fundation, Kraków, pp 17–29

39. Mrozik K, Przybyła Cz, Pyszny K (2015) Problems of the Integrated Urban Water Management. The case of the Poznań Metropolitan Area (Poland). Rocznik Ochrona Środowiska (Annual Set the Environment Protection) 17:230–245

40. Kowalczak P (2015) Integrated water management in urban areas. Part one: Hydrological and environmental foundations [Zintegrowana gospodarka wodna na obszarach zurbanizowanych. Cześć pierwsza: Podstawy hydrologiczno-środowiskowe]. ProDruk Publishing House Poznań

41. Graf R, Pyszny K (2016) Integrated water resources management [Zintegrowane zarządzanie zasobami wodnymi]. In: Kaczmarek T (ed) Spatial development concept for the Poznań Metropolis [Koncepcja Kierunków Rozwoju Przestrzennego Metropolii Poznań]. Metropolis Association, Metropolitan Research Center Adam Mickiewicz University, Poznań, pp 59–74

42. Kronenberg J (2012) Ecosystem services in cities [Usługi ekosystemów w miastach]. Zrównoważony Rozwój –Zastosowania 3:14–28

43. Kundzewicz Z, Gromiec M, Iwanicki J, Kindler J, Matczak P (2014) Report on water hazards—introduction [Raport o zagrożeniach związanych z wodą - wprowadzenie]. Nauka 1:59–195

44. Bernaciak A, Spychała M, Korytowski M, Powolna P (2015) Small water retention in local environmental protection programs of Warta river municipalities [Mała retencja wodna w programach ochrony środowiska gmin nadwarciańskich]. Inżynieria Ekologiczna 44:121–130

45. Kaczmarek T (ed) (2016) Spatial development concept for the Poznań Metropolis [Koncepcja Kierunków Rozwoju Przestrzennego Metropolii Poznań]. Metropolis Association, Metropolitan Research Center Adam Mickiewicz University, Poznań

46. Przybyłek J (1995) Hydrogeological principles of the water supply system for the Poznań agglomeration [Hydrogeologiczne podstawy systemu zaopatrzenia w wodę aglomeracji Poznania]. Współczesne problemy hydrogeologii 7(1):405–415

47. Dąbrowski S, Przybyłek J, Górski J (2007) Hydrogeological characteristics of the water regions, the Warta region, Warta lowland sub-region [Charakterystyka hydrogeologiczna regionów wodnych, region Warty, subregion Warty nizinnej]. In: Paczyński B, Sadurski A (eds) Regional hydrogeology of Poland [Hydrogeologia regionalna Polski]. T.I., PIG, Warszawa, pp 369–406

48. Pyszny K, Bednarek R, Binder M, Wróżyński R, Waldmann K, Wójcicki M (2016) The technical infrastructure [Infrastruktura techniczna]. In: Kaczmarek T (ed) Spatial development concept for the Poznań Metropolis [Koncepcja Kierunków Rozwoju Przestrzennego Metropolii Poznań]. Metropolis Association, Metropolitan Research Center Adam Mickiewicz University, Poznań, pp 262–279

49. Górski J, Przybyłek J, Kasztelan D (2011) Problems of land use changes and the protection of water-bearing areas accounting for the important source of drinking water supply based on the study of Mosina-Krajkowo waterworks [Problemy zagospodarowania i ochrony terenów wodonośnych o szczególnym znaczeniu dla zaopatrzenia w wodę na przykładzie ujęcia Mosina –Krajkowo]. PIG Bulletin 445 12(1):127–137

50. Górski J, Przybyłek J (1996) Geological building and groundwater [Budowa geologiczna i wody podziemne]. In: Poznan City Office. Environmental Protection Department (ed) The natural environment of the Poznań city. Poznań [Środowisko naturalne miasta Poznania], pp 23–44

51. Strategy for Development of the Poznań Agglomeration [Strategia Rozwoju Aglomeracji Poznańskiej] (2012) Poznań Metropolis 2020. Metropolitan Research Center, Adam Mickiewicz University, Poznań

52. Graf R, Jawgiel K, Łukaszewicz J (2017) Integration of a flood risk management and waters protection system on the example of the Warta Valley between Konin and Oborniki [Integracja systemu zarządzania ryzykiem powodziowym i ochroną wód na przykładzie doliny Warty między Koninem a Obornikami]. Prace Geograficzne 148:107–133

53. Graf R (2012) The structure and functioning of local groundwater circulation systems within the Poznań Upland [Struktura i funkcjonowanie lokalnych systemów krążenia wód podziemnych na obszarze Wysoczyzny Poznańskiej]. Studia i Prace z Geografii i Geologii 26, Bogucki Publisher, Poznań
54. Graf R (2014) Spatial differentiation of surface runoff in urbanised catchments on the example of Poznań [Przestrzenne zróżnicowanie spływu powierzchniowego w zlewniach zurbanizowanych na przykładzie miasta Poznania]. In: Ciupa T, Suligowski R (eds) Water in the city [Woda w mieście]. Monografia Komisji Hydrologicznej PTG 2, IG UJK, Kielce, pp 59–71
55. Graf R, Jawgiel K (2018) The impact of the parameterisation of physiographic features of urbanised catchment areas on the spatial distribution of components of the water balance using the WetSpass model. ISPRS Int J Geo-Inf 7:278

Part V
Conclusion

Chapter 19
Updates, Conclusions, and Recommendations for "Quality of Water Resources in Poland"

Katarzyna Kubiak-Wójcicka, Martina Zelenakova, and Abdelazim M. Negm

Abstract This volume discusses current problems and research topics carried out by scientists dealing with the state and quality of water resources in Poland. Based on many years of research conducted by authors from various scientific centers from all over Poland, the main conclusions and recommendations are presented, which are included in the chapters presented in the book. First, the general state of surface and underground water resources, their quality and possibilities of water resources protection were discussed. Then, the book focuses on water quality and its changes as a result of natural and anthropogenic factors. The next part of the study is devoted to water and sewage infrastructure and water management in urban areas. This volume highlights the problems of sustainable use of water resources for the needs of human activities.

Keywords Water quality · Pollution · Eutrophication · Assessment indicators · Poland

19.1 Introduction

Water resources comprise an integral element of the natural environment in which man plays an important role. As a result of the influence of natural and anthropogenic factors, the quality of water resources may change. Knowledge of the quality of waters is the basis for taking action to improve them and protect against pollution. Pollution

K. Kubiak-Wójcicka
Department of Hydrology and Water Management, Faculty of Earth Sciences and Spatial Management, Lwowska 1, 87-100 Toruń, Poland

M. Zelenakova
Department of Environmental Engineering, Faculty of Civil Engineering, Technical University of Kosice, Košice, Slovakia
e-mail: martina.zelenakova@tuke.sk

A. M. Negm (✉)
Water and Water Structures Engineering Department, Faculty of Engineering, Zagazig University, Zagazig 44519, Egypt
e-mail: amnegm@zu.edu.eg

© Springer Nature Switzerland AG 2021
M. Zeleňáková et al. (eds.), *Quality of Water Resources in Poland*, Springer Water,
https://doi.org/10.1007/978-3-030-64892-3_19

from the municipal sector and agriculture (biogenic pollution), as well as industrial pollution lead to changes in the quality of water. Various types of indicators, which determine the amount and type of impurities contained in water and the condition of aquatic biocoenoses, are used to assess the quality of water. There are biological, physical and chemical indicators. According to the Water Framework Directive, biological indicators are used to determine the ecological status of surface waters in detail, while other indicators (physicochemical, hydromorphological) have an auxiliary value. Much attention is paid to pollution that leads to eutrophication. Surface water status assessment is used for integrated water management. Having knowledge about the size and quality of water resources and their changeability over time, it is possible to properly manage and manage water resources.

This volume discusses the current problems and research topics implemented by scientists from various scientific centers from all over Poland dealing with the quality of water resources. Based on their many years of research, a summary of the most important findings and conclusions from the research on the quality and variability of surface and underground water resources has been presented. A set of recommendations extracted from all chapters was also presented to help scientists and decision-makers in the pursuit of sustainable use of water resources in Poland.

19.2 Update

In Poland, different systems for classifying water status were used at different times. From 1991 to the end of 2003, water purity classes were determined on the basis of monitoring results for the purity of rivers and lakes on a three-point scale. Class I included waters suitable for use as drinking water (also as a source of water for industry requiring clean water and for salmonid farming). Class II included waters suitable for use of farmed animals, including fish other than salmonids, as well as for sport and recreation. Class III is water suitable for use by industry and agriculture. Waters not meeting the norm for the 3rd class were called NON (an abbreviation of 'not conforming to the norms'). Detailed physicochemical and sanitary indicators together with admissible ranges for individual classes were specified in relevant ordinances. Some indicators were obligatory and they were: the amount of dissolved oxygen, BZT_5, $ChZT_{Mn}$, the concentration of phenols, chlorides, sulfates, solutes and suspensions. Guidelines on the principles of basic monitoring of lakes carried out by the State Inspectorate for Environmental Protection are presented in scientific publications [1, 2] and on the websites of institutions responsible for water monitoring. Poland's accession to the European Union in May 2004 required the adaptation of national regulations to EU regulations. In 2004, the ordinance of the Minister of the Environment "on the classification for presenting the status of surface waters and groundwater, how to monitor and how to interpret the results and presentation of these waters" was in force, in which five water quality classes were defined (classes IV and V have been added). This system operated in Poland until 2008 and was referred to as a temporary one. The introduction of the WFD required the development of an appropriate methodology for determining the biological monitoring of surface water in terms of macrophytes [3, 4].

Since 2008, national research on the quality of water resources in Poland and the possibilities of their economic use are conducted based on the guidelines contained in the Water Framework Directive 2000/60/EC (WFD) of the European Parliament and of the Council from October 23, 2000 [5]. For the classification of water bodies, completely new assessment criteria have been introduced, which differentiated not only by surface water categories (flowing and standing waters), but also taking into account the specificities of their various types. This directive, focusing mainly on elements of the biological quality of water, has fundamentally changed the methods of qualification.

The physicochemical indicators commonly used in the previous periods are of secondary importance. Therefore, comparing data that determines long-term water quality in accordance with the WFD requirements is impossible due to various parameters and indicators that were taken into account when assessing water quality. The monitoring and assessment system for surface water quality is carried out in Poland by the Inspectorates of Environmental Protection. They publish annual assessments of the status of surface water quality, which are made available in the form of reports on the websites of individual inspectorates. Many authors note that the water quality monitoring system in Poland is not satisfactory. This is due to the small number of lakes covered by the State Environmental Monitoring. In addition, the adopted system of water sampling frequency for monitoring water quality is not very accurate [6].

This volume focuses on assessing water quality, identifying the causes and counteracting the negative effects of water quality deterioration. This will allow the development of a course of action for water managers in the scope of water quality its protection in the near future. The current research problem is the eutrophication of lakes and artificial reservoirs [7, 8]. It is necessary to determine the trophy and natural susceptibility of lakes to degradation, which will allow determining the scope of activities in the catchment area, protecting individual lakes and determining the methods of their remediation.

The following are the major update for the book project based on the main book theme.

1. To ensure the quality of water resources in Poland is necessary to assess and manage them and their usability. The most practical solution of the reduction of water consumption is to invest in the most modern water recovery and reuse technology in industrial plants, agriculture, and newly built residential buildings.

2. Environmentally valuable areas protected by law are components of a broader system of environmental protection in Poland. To achieve sustainable development of water areas includes their protection also from a social and economic point of view.

3.   Artificial water bodies in Poland mostly serve numerous functions despite the many environmental protection problems related to the lentic water environment, e.g. eutrophication, salinisation.

4.   Irrigation and drainage systems should be rebuilt and modernized to inhibit and control the outflow of water from the systems and to create favourable conditions for water retention, for controlling water discharge and effective water management in agricultural land.

5.   Poland's main rivers varied significantly in pollutant contents along their courses. At the majority of study sites on the main rivers, the values of indicators of mineral, organic and biogenic pollutants did not exceed the water quality standards.

6.   Although continuous monitoring of water quality is not yet so perfect for excluding the necessity of measurements with traditional methods and expensive laboratory analyses, it shows the direction where modern water quality testing methodology is heading.

7.   Mathematical models for the simulation of water quality and ecosystems of lakes and reservoirs are continuously developed and used widely over the last decades proving their usefulness in studies of processes occurring in water bodies and in assessments of impacts.

8.   Wastewater management in cities and recreation centers around lakes should be regulated and controlled. Balanced vegetation should help as a barrier that limits nutrient migration to lakes.

9.   The Southern Baltic Sea coast is one of the regions in which the risk assessment of both fresh groundwater resources and intake of admissible volumes, defined as a safe maximum yield, is a complex problem. Salt and brackish waters that occur along the coastal Baltic lowlands in Poland originate from the seawater intrusion.

10. The groundwater of the Jurassic, Cretaceous as well as Palaeogene and Neogene aquifers in Poland has had a stable chemical composition. The greatest variability is revealed by the groundwater of the Quaternary aquifer.

11. Considering constant hydromorphological changes in river channels and water reservoirs caused by anthropogenic and natural influences, adaptation and optimization of sediment management strategies is an endless task. Sediment management in Poland is part of the tasks defined by River Basin Management Plans (RBMP).

12. The most important natural factor determining water chemistry in the Carpathians is geology. Stream water and spring water chemistry in the Carpathians change primarily due to hydrologic factors—changes in discharge. Anthropogenic factors affecting water chemistry in the Carpathians include acid rain, deforestation, agriculture, and tourist-generated wastewater.

13. The basis for the development of protective measures directed to achievement and maintaining the good status of surface waters is the reliable information on their actual trophic state. This information can be obtained within the properly organized system of monitoring based on appropriate methods of trophic status assessment.

14. Increased concentrations of biogenic substances, mineral substances and suspended matter were found in the agricultural catchment area, in particular after spring thaw when surface flushing increases. In the summer, the biogenic content falls due to increased vegetation processes.
15. Water supply and sewerage infrastructure in Poland is not sufficient, it requires significant improvement and also financial aid. The policy have contributed to the development of water and wastewater management.
16. Sustainable development of a city concerning surface and underground waters, shaping the interaction between the human and the water environment is of particular importance, taking into account the safety of its inhabitants.
17. Integrated water resources management of the metropolitan area is based on three pillars: equal access to a high-quality natural environment, the efficient utilisation of natural resources in order to achieve economic advantages, and maintaining the ecological equilibrium and the ability of the natural system to regenerate.

19.3 Conclusions

The **Water Resources in Poland: Quality** consists of 3 main parts in total plus an introduction (this chapter) and conclusions to close the book. Part II discusses "Key Issues of Water Resources Protection in Poland", part III discusses "Water Quality Evaluation". The part IV concerns "Water and Wastewater in Urban Areas", and the last part contains "Conclusions". Based on the chapters of this volume, the following findings could be reported:

1. Small water resources in Poland and their high variability in time require economic use of water and good management of water resources. The improvement of the situation may be caused by the construction of small retention reservoirs in areas with limited resources, but also by economic water management and reuse of previously used water. The introduced changes may encourage potential water consumers to use water more economically and to apply new technological solutions for proper rainwater management and recovery of used water.
2. Development and protection of water resources in protected areas in Poland is important for the sustainable development of the country. The principle of treating the duty of environmental protection as an element of the proper functioning of the economy should be introduced, and the breach of this obligation should be regarded as a betrayal of the rules of good management. It increases the role of local communities that decide on the development of water resources (subsidiarity principle).
3. Stagnant water resources are accumulated mainly in anthropogenic reservoirs. The water resources of the 50 largest reservoirs in Poland with a total area of 530 km^2 are just over 3.7 km^3. Limnic processes occurring in water reservoirs are an important condition for their use. Quantitative and qualitative changes in

the retention of genetically differentiated water reservoirs in Poland provide the basis for the development of scenarios for the use, protection and reclamation of standing inland waters in other areas of the world, especially in the temperate climate zone.

4. It is necessary to improve the organization of irrigation and drainage in Polish agriculture; water management and supervision over this management; improve the knowledge and skills of water management entities and increase the financial resources for maintenance of water melioration facilities and construction of new ones. Modernization of irrigation/drainage systems should take into account experience and forecasts of conditions in which they will operate. Modernization should take into account technical progress, economic conditions, management system and environmental protection requirements.

5. The quality of water in Polish rivers requires further improvement in the field of physicochemical elements, including mainly biogenic and biological compounds. An analysis of the spatial variability of the quality of river water resources has indicated regions where the water quality of water bodies is potentially threatened. The deepening water deficit associated with the observed climate change highlights the need to protect waters against point and areal pollution. This is important due to the retention of water that is planned in the governmental Retention Development Programme. In the case of rivers loaded with nutrients, retention may accelerate the eutrophication rate of newly established water reservoirs.

6. Regarding water quality in the main dam water reservoirs in Poland is possible to state that almost half of the dam reservoirs in Poland have good ecological status and good chemical status. Good water quality is mostly in dam reservoirs located in the Carpathians, in upper sections of rivers. In reservoirs that supply potable water, the aim should be to develop continuous monitoring of physical and chemical parameters of water, which would give new possibilities for proper interpretation of the values obtained and improve the speed of response to threats, as well as opens new perspectives for better assessment of water quality.

7. Mathematical models for the simulation of water quality and ecosystems of lakes and reservoirs were continuously developed and used widely over the last decades proving their usefulness in studies of processes occurring in water bodies and in assessments of impacts. Models are more and more often required by decision-makers in procedures related to water management. That is a general trend, evident also in Poland.

8. It is clear that wastewater management in cities and recreation centers around lakes should be regulated and controlled. A balanced vegetation band should be created as a barrier that limits nutrient migration to lakes. It is advisable to take reclamation measures to reduce the "internal" supply of nutrients by inactivating them in bottom sediments. Determining the activity in the catchment area, protecting the lakes and determining the method of their reclamation will allow to properly develop the concept of protecting the waters of the West Pomeranian Lake District. In the long term, conservation measures should lead to immediate measures eliminating or reducing the size of local pollution sources.

9. The next important topic in water management in Poland is groundwater quality on the Polish coast. The characteristic feature of the Polish coast is the presence of brines of the Cl–Na type within the Mesozoic strata, from where they ascend into the Cainozoic aquifers. "Changes in the chemical composition of waters in this area are strongly affected by human activity consisting of excessive exploitation of aquifers, mainly in the 1970s and 1980s. The process of saline water penetration into exploited aquifers was most extensive at that time. At present, it has been significantly inhibited as a result of reduced water extraction at the water intakes in this area. The efficient groundwater extraction in the coastal zone is an important issue. In holiday resorts, the demand for freshwater greatly increases in the summer season. Excessive exploitation of aquifers on the southern Baltic coast leads to the development of chloride anomalies and the closure of many groundwater intakes. Monitoring the volume and quality of freshwater extraction, as well as safely assessing the admissible volume of extracted water for all water management areas will prevent the future degradation of groundwater resources through salinity.

10. Poland's groundwater quality evaluation based on the points of network monitoring of the Polish Hydrogeological Survey has revealed a visible improvement in its quality between 2013 and 2017. Thus, it can be stated that groundwater chemical status in particular regions of Poland has been well recognized and that the groundwater conservation measures have been effective. Bearing in mind the economic development of the country, it is strongly recommended that further monitoring of mainly the areas of intensive agricultural economy or mining should be continued. The maintenance and further development of the monitoring network are also dictated by prospective geogenic threats resulting from the geological structure of the area of Poland, climatic conditions and water resources.

11. Each investment (already implemented or planned) must take into account subsequent changes of reference conditions. That is why learning the operation of the river system regarding all its components becomes such an important element, including debris transport. According to strategic documents in force in Poland, there is no coherent strategy for sediment management in the catchment. However, due to planned investments (recorded in RBMP), comprehensive sludge dynamics studies should be carried out so that decision-making body can make rational decisions regarding future water investments.

12. The impact of environmental and anthropogenic factors on the chemistry of groundwater and stream waters in the Carpathians is huge. Anthropogenic factors such as acid rain, deforestation, agriculture, and tourism affect water chemistry in the Carpathians in some extent. The most important issue for further hydrochemical studies in the Carpathians would be to identify ion migration pathways to springs and streams during rainfall and snowmelt events as well as to determine the impacts of tourist traffic and tourist infrastructure on water quality in order to more effective water protection.

13. The solution of eutrophication in surface waters of Poland requires the systematic approach to the development and implementation of comprehensive technical, technological, organizational and legal measures covering the whole catchment and all sources of biogenic matter discharge. The comprehensive action plan to combat eutrophication should first of all take into account the development of the scientifically justified and efficient system of measures to achieve water state improvement. This is of practical and economic importance and allows the proper identification of priority tasks and avoid unprofitable and expensive investments.

14. The problem of monitoring the quantity and quality of water in a small agricultural catchment on the example of Struga Toruńska could be illustrative for other regions in Poland. It is indicated that flow disorders are caused by natural factors (climate change) and human activity, mainly by intensive plant cultivation and the influence of the nearby city of Toruń. The water cycle is disturbed, environmental pollution is increasing, soil degradation is progressing, biodiversity decreases. Conducted since 1994 comprehensive environmental studies, including abiotic and biotic ones in the Struga Toruńska catchment, should be continued. The results of these studies allow to analyze the energy and matter flow in this area. They allow understanding of the threats caused by natural factors and human activity.

15. The analysis of Water And Wastewater Management Condition In Poland shows that the availability of water supply and sewerage infrastructure in Poland is not sufficient. The disproportion between the length of the water supply system and the sewerage system is particularly evident. In the next financial perspective European Union, emphasis should be put on building sewerage infrastructure in rural areas, but also on modernizing the existing infrastructure in urban areas. Based on the analysis, areas in Poland have been identified in which the water and sewerage infrastructure requires significant improvement. Lack of financing or small expenditure on investments related to water management indicate areas that require state aid in financing water and sewage management. It also shows which elements of the policy have contributed to the development of water and wastewater management.

16. The water cycle and degradation of the aquatic environment is the most important in urban areas. In Polish cities, what gains special significance is man—water environment interactions. Sustainable development is necessary for these areas, taking into account the multidirectional needs of the population, economy as well as the water environment. The identification of such problems is necessary when preparing hydrotechnical projects and forecasts of investments' impact on the environment. This information can be used to develop expert opinions and analyzes related to the maintenance of river channels throughput and drainage systems, together with limiting surface runoff, etc.

17. Multidimensional aspect of the water resources management in metropolitan areas is a significant issue that needs to discuss the problem related to the risk of non-attainment of environmental goals in metropolitan areas. The increasing pressure on existing water resources and the risk of flooding make it necessary

to ensure high standards of spatial development and to consider the interdisciplinary approach, multi-sectoral planning and broad cooperation of various stakeholder groups. The integration of water resources management in the metropolitan area should be closely connected with the area's strategy of development. It is necessary to implement a system for monitoring the results of the multidimensional model of management which will contribute to the minimisation of the degree of threat while at the same time improving the safety of waters and helping to achieve maximum benefits for all stakeholders.

19.4 Recommendations

Research confirms that water resources in Poland are diverse both in time and in space. Compared to other European countries, Poland has relatively small resources of renewable surface waters. Lack of adequate protection of surface and underground water resources may constitute a barrier to future economic development. Solving this problem requires a systematic approach to the development and implementation of a comprehensive action plan that will take into account technological, financial, organizational and legal issues in the field of water management throughout the entire catchment. Therefore, further research is recommended to develop detailed procedures for monitoring the quality of surface and groundwater. It is advisable to adapt the research to diverse physical and geographical conditions that require an individual approach, as well as to increase the number of monitored objects and the frequency of research, especially in different hydrological conditions. A separate problem is the issue of monitoring sources of pollution that get into the receivers from the catchment area. It is recommended to implement a pollution reduction program from various sectors of the national economy: industry, agriculture or municipal economy. Emphasis on further wastewater treatment and co-financing for modern technologies in the field of wastewater treatment is recommended. It is proposed to support the program for the development of new technologies and their transfer to relevant users through various incentive systems. Joint activities in the field of water resources protection should be carried out in cooperation with hydrologists, officials, planners and engineers. Practical conclusions, especially regarding regional development barriers, can significantly expand water management aimed at limiting operations and improving water quality on a regional and national scale.

This section contains also a set of recommendations, obtained from the chapters presented in this volume, for future research in water resources management in Poland.

1. The construction of small retention reservoirs in areas with limited water resources and economical water management—water reusing and proper rainwater management—will improve the situation in the variability of water resources in Poland.

2. The best way to maintain environmentally valuable areas is to grant them legal status based on environmental protection goals where it is possible. Each

national park needs to have a unique protection strategy based on differences in water availability.

3. Large spectrum of quantitative-qualitative changes in reservoirs of lentic waters in Poland at their large numerical force identified with the multiplicity of scenarios of their functioning and use, creates the possibility to predict the course of degradation processes of limnic waters quality in the objects, which contemporarily function under incomparably smaller environment transformation and to undertake the protection and recultivation activities.

4. Different actions should be taken to stimulate the development of irrigation and drainage systems in Poland, to adapt the existing systems to extreme meteorological events resulting from climate change, to enhance its operation, maintenance and management.

5. Due to Poland's water resources being very limited, and the possibility of them falling further as a result of climate change, further protection of waters against pollution is extremely important. Water quality can be improved by, among others, increasing the number of sewage treatment plants and by improving the performance of existing facilities.

6. Continuous monitoring of physical and chemical parameters of water shows the dynamics of the observed indexes, not only seasonal changes but also daily fluctuations. This improves the possibilities of the proper interpretation of the obtained values and the speed of reaction to threats.

7. It is necessary to prove a general trend that models are more and more often required by decision-makers in procedures related to water resources management of the country.

8. Determining the activity in the catchment area, protecting the lakes and determining the method of their reclamation will allow to properly develop the concept of protecting the waters of the West Pomeranian Lake District.

9. Monitoring the volume and quality of freshwater extraction, as well as safely assessing the admissible volume of extracted water for all water management areas will prevent the future degradation of groundwater resources through salinity.

10. The improvement in the quality of groundwater in Poland is visible in the comprehensive summary of quality classes concerning all water-bearing levels. This state is necessary to keep in sustainable conditions.

11. Sustainable economic use of rivers requires extensive knowledge about the dynamics of sediments and their quality within river basins. Measures are taken to manage the sediment on an ongoing basis.

12. Building or modernization numerous wastewater treatment plants at mountain lodges in Poland improve significantly the water and wastewater management situation in Carpathian protected areas.

13. To combat eutrophication should take into account the resolution of the proper organization of the monitoring of eutrophication, the developing of mathematical prognostic models of eutrophication for forecasting the eutrophication development trends and its ecological, economic and social consequences, the

development of an approach to assess the role of all the sources of nutrients within the catchment area.

14. An understanding of the operating principles of analysed ecosystems makes it possible to identify risks and threats to the environment, to water pollution. Small river or lake catchments allows comprehensive research to be conducted into the state of an environment subject to continuous changes from natural and anthropogenic processes.

15. In the future the emphasis should be put on building sewerage infrastructure in rural areas, but also on modernizing the existing infrastructure in urban areas.

16. The effective implementation of water policy in the city in the field of water management and defining the terms of water resources consumption is important. Also detailed rules and conditions for the use of water resources in the city in terms of planning, design, construction and technological solutions, including ownership rights, controls and appropriate sanctions are regulated by implementing regulations.

17. Environmental protection programs, studies of conditions and directions of spatial development, local spatial development plans, river development plans, small retention programs and projects, sectoral policies concerning water resources and natural resources, and other development strategies for metropolitan areas have to be used.

References

1. Kudelska D, Cydzik D, Soszka H (1994) Guidelines for basic monitoring of lakes (in Polish). State Inspectorate for Environmental Protection. Oficyna. „OIKOŚ" sp z o.o. Warszawa
2. Soszka H, Kolada A, Pasztaleniec A, Ochocka A, Kutyła S, Bielczyńska A (2016) Development of a collective assessment of the state of lakes (in Polish). Annex 4.8. to elaborate "Processing and verifying SEM data in the scope of lake monitoring from 2013–2015, together with preparation of condition assessment and substantive supervision', GIOŚ. http://www.gios.gov.pl/images/dokumenty/pms/monitoring_wod/Ocena_jezior_2010-2015.pdf
3. Ciecierska H, Kolada A, Soszka H, Gołub M (2006) Development of a methodology for field tests of macrophytes for the needs of routine water monitoring and a method for assessing and classifying the ecological status of waters based on macrophytes (in Polish). T. II—Jeziora. Ministry of the Environment, Warszawa
4. Szoszkiewicz K, Zbierska J, Jusik S, Zgoła T (2006) Development of methodological bases for biological monitoring of surface waters in the area of macrophytes and their pilot application for water bodies representing selected categories and types (in Polish). Stage I-II (in terms of rivers). Ministry of the Environment, Warszawa-Poznań-Olsztyn
5. Directive 2000/60 / EC of the European Parliament and of the Council of 23 October 2000 establishing a framework for Community action in the field of water policy
6. Loga M, Jeliński M, Kotamäki M (2018) Dependence of water quality assessment on water sampling frequency—an example of Greater Poland rivers. Archives of Environmental Protection 44(2):3–13. https://doi.org/10.24425/119688
7. Neverova-Dziopak E, Droździk A (2017) Analysis of long-term changes in the trophic state of the Tresna reservoir in the aspect of its location and role in the cascade of reservoirs (in Polish). Inżynieria Ekologiczna 18(1):135–148. https://doi.org/10.12912/23920629/67001

8. Siuda W, Grabowska K, Kaliński T, Kiersztyn B, Chróst RJ (2020) Trophic state, eutrophication, and the threats for water quality of the Great Mazurian Lake System. In: Korzeniewska E, Harnisz M (eds) Polish river basins and lakes—Part I. The handbook of environmental chemistry, vol 86. Springer, Cham

Printed in the United States
by Baker & Taylor Publisher Services